Nanochemistry

A Chemical Approach to Nanomaterials

Dedication

We (Geoff and André) dedicate this book to our wives, Linda and Charlene. They are truly the inspiration, energy, and dedication contained in this text, for without them we would have none of these things

Nanochemistry
A Chemical Approach to Nanomaterials

Geoffrey A. Ozin and André C. Arsenault
Chemistry Department, University of Toronto, Canada

RSCPublishing

Though this book has been diligently created and every effort made to ensure accuracy, the authors and publishers do not warrant the contents of this manuscript to be free from errors or omissions. The readers are reminded that statements, data, illustrations, references, experimental details, or other items may inadvertently be inaccurate.

ISBN 0-85404-664-X

A catalogue record for this book is available from the British Library

Published by The Royal Society of Chemistry,
Thomas Graham House, Science Park, Milton Road,
Cambridge CB4 0WF, UK

Registered Charity Number 207890

For further information see our web site at www.rsc.org

Typeset by Alden Bookset, Northampton, UK
Printed by Sun Fung Offset Binding Company Ltd, China

PREFACE

In the Beginning There was Nano...

"Imagination is more important than knowledge"
Albert Einstein (1879–1955)

1 Nano – The Interdisciplinary Science

In December of 1959, the eminent physicist Richard Feynman described the future in a groundbreaking talk about the physical possibilities for "making, manipulating, visualizing and controlling things on a small scale," and imagining that in decades to come, it might be possible to arrange atoms "the way we want." As an example of what would be possible, Feynman used physics to demonstrate theoretically that all of the accumulated information in all of the books in the world could be written "in a cube of material one two-hundredth of an inch wide – which is the barest piece of dust that can be made out by the human eye." Feynman did not suggest how this might be done, that task would have to be left to others as nanoscience evolved over the next 40 years, overcoming the early skepticism of much of the scientific community.

You might imagine nanoscience as a pretty specialized field, when in reality it is quite the opposite. In nano, the idea is to do something really big and important with objects that are really small and distinct in their properties and behavior from the large objects with which we are all familiar. It is a field full of enormous challenges that cannot be met by any one specialist. It requires collaboration on the part of many specialists, just as fashion designers rely on weavers, tailors, and cutters to accomplish their designs.

Nano is a word that has now left the science reservation and entered the public consciousness. The field of nanoscience with its promise of amazing nanotechnologies is one of today's most challenging, exciting, mutli-disciplinary and competitive fields, and one of the most highly funded. It is considered poised to revolutionize the world as we know it, and transform in into something better.

Some researchers still argue that nanotech, as a coherent field with clear goals, does not actually exist. The current leap of faith by so many ill-informed folk into a new field seemingly fraught with so much uncertainty can create fear of the unknown. This is clearly the case with nano, and it is what lends it a science fiction quality. Recently, in some circles, a dread of nanotechnology has emerged. This hysteria is not unlike the fears surrounding biotechnology when it first gained public attention. We believe that it is very important to counter this type of paranoia, especially in a field with such potential to do good for society. We must use effective means of communication to educate the public about the scientific and technological issues of nano, as well as the ramifications of nano for the social sciences and humanities, biology and medicine. It is up to us, scientists, to dissuade the public that the ultimate goal of nanotechnology is to fabricate self-replicating nanorobots which will overrun the earth.

In order to be globally competitive, emerging nanotechnology companies and universities need creative and competent personnel who "know it all." These are the people who in their education and training have been wise and brave enough to cross borders between fields. By doing so, they can really identify and understand what the problems are and how to solve them. Moreover, they can communicate them, not just to other scientists, but also to those in the social sciences, health sciences, humanities, the business world and the media, so that they too can get a perspective on the problem and how it affects their fields.

2 Plenty of Room at the Top and Bottom

To gain perspective of the nature and scope of the field of nanochemistry it is important to recognize the evolution of materials science with respect to the creation of new technologies. In the latter half of the 20th century, materials engineering science enabled innovations in electronics, communications, construction, transportation, energy, biomedicine and space research. Here for instance, the synthesis of solid-state materials led both to a new brand of physics and electronic devices. However, to satisfy the 21st century demand for new materials in fields like nanotechnology, information technology and biotechnology, solid-state synthesis approaches to preparing materials are gradually being supplanted by molecular methodologies, particularly the self-assembly of materials with structures that can approach the complexity of those observed in nature.

In *Nanochemistry* the authors describe how recent research in materials synthesis and self-assembly has helped fuse "top–down" solid-state physics ways of making structures and "bottom–up" molecular-chemistry methods of making materials. They show how the integration of materials chemistry, self-assembly, templating and chemical patterning strategies, from nm to μm length scales, is being actively pursued in university and industrial laboratories to make new materials having structures, properties and functions suited to a range of nascent technologies.

An attempt is made in this textbook to paint a picture of modern materials research founded upon a building block approach to materials synthesis through the paradigm of self-assembly, comprising shape controlled nanometer to micron scale objects, such as spheres and wires, tubes and sheets, made of organic and inorganic, organometallic and coordination, polymer and ceramic compounds. These are orchestrated through chemistry to auto-construct into functional architectures with designed utility. This approach to making materials fuses "top–down" solid-state engineering physics ways of making structures, and "bottom–up" molecular-chemistry methods.

One of the main challenges in this new approach to making materials is the actual synthesis of the building blocks with pre-determined bulk and surface composition, and shape, and in a size regime straddled between that of molecules and bulk matter. The building blocks, making and perfecting them, are the key to success. By having access to them through the amalgamation of organic and polymer, inorganic and organometallic chemistry one is able to develop a new way of doing solid-state chemistry that is distinctive to what has emerged over the past 50 years. It is no longer the structure and composition of the material that is new, but rather its form and scale, and guiding its arrangement and integration, that is new and which yields new materials with new function and new utility.

Nanochemistry presents a blueprint for the future development of an exciting teaching and research program in materials chemistry that is organized around the theme of tailored materials and the tenet that "all matter has a shape and all shapes matter." The central idea for developing a new materials chemistry with a bright future is that it is pivotal to know which size and shape a material must have for it to exhibit a particular property, function and utility, and that designing and synthesizing a shaped material with the right length scale can be just as important as selecting its structure and composition. However, synthesizing a material with particular forms and dimensions requires a different way of thinking than synthesizing matter with a specific structure and composition.

The latter approach that focuses just on structure and composition of a material represents the classical way of making materials, whose properties and function relate to structure and composition with size and shape as an afterthought. It has worked well for more than half a century during which time the development of solid-state materials synthesis, X-ray diffraction structure of solids, electronic properties of solids and function of defects in solids, have enabled utility of solid-state materials in a myriad of advanced technologies – almost everything around us these days! This is however, the old way of thinking about and acting upon the relations between the synthesis, structure, property, function and utility of materials. The challenge of this classical approach for making materials has always been how to choose a synthetic method that targets a product with structure and composition, to provide materials properties and functions that can be orchestrated to create utility when made into a suitable shape.

In this textbook, the authors make it clear that the approach of making materials where size and shape are a "fore thought" rather than a "postscript" is just as important in synthetic design as the structure and composition of the material. They present a "bottom–up" chemistry approach to synthetic size and form rather than a "top–down" engineering physics way of sculpting the shape. In many of today's advanced technologies, it is not just choosing the right process for synthesizing the material that counts, but also making sure that the synthesis can also form the material into the desired shape and at the appropriate length scale. It is always surprising to realize that a material of the proper composition can be totally useless if not fashioned into exactly the right form. To be successful in this new approach to making materials, one needs to have wide-ranging skills in chemical synthesis, traversing organic and polymer, inorganic and organometallic chemistry, together with a deeply analytical appreciation of materials diagnostics using the methods of modern diffraction, microscopy, spectroscopy and thermal, electrical and optical, magnetic and mechanical methods.

One should ask, is this approach to synthetic shape new, or is it just an obvious, incremental and natural extension of the classical way of making and forming materials into functional shapes? Traditional methods for synthesizing solids often involve direct reaction of solid precursors. Their chemical reactivity mainly depends on their form and physical size as well as their structure and defects. As-synthesized materials made in this way are often crystalline or glassy micron scale powders, which may need to be post-engineered into pellet, plate, tube, rod or wire shapes. If the material needs to be fashioned as a single crystal with specific dimensions and orientations, special growth processes have to be devised. These crystals may be polished or cut to expose particular faces and cut into wafers. If single crystal or polycrystalline films are required special deposition processes must be formulated. Films may be grown on lattice-matched substrates to create epitaxial films or artificial superlattice films. All of these materials' morphologies can be accessed by classical methods. However, when the physical size of shaped materials, like spheres and clusters, wires and rods, tubes and sheets, drops down to the nanoscale where their properties vary with size according to mathematically defined quantum mechanical scaling laws, then new approaches have to be devised. Both the synthesis of materials on this minute length scale, their size tunable properties and surface functionalization, as well as their organization and inter-connection into functional constructs, can pose unexpected difficulties.

The ramifications of reconstructing matter into shaped building blocks has profound scientific and technological ramifications in application areas that include batteries, fuel cells and photovoltaics, digital imaging and printing, microelectronic packaging and controlled chemical release, chemical sensing and molecule sepa-rations, catalysis and photocatalysis, combinatorial materials chemistry, micro-fluidics and lab-on-chip, nanoelectronics, nanophotonics and nanomagnetics. In fact, one would be quite hard pressed to find a field which nanotechnology will not influence.

The authors of *Nanochemistry* present a basic chemistry approach for making nanomaterials and describe some of the principles of materials self-assembly over "all" scales. It is demonstrated how nm to μm scale building blocks with a wide range of shapes, compositions and surface functionalities can be coerced through chemistry to organize spontaneously into unprecedented structures, which can serve as tailored functional materials. This approach to materials discovery utilizes modular synthesis of hierarchical materials, according to which molecular-scale building blocks self organize into complex structures that span the entire hierarchy of length scales. Through a series of purposeful synthesis strategies, the authors illustrate how truly revolutionary advances in nanoscience and nanotechnology can result from this approach to materials discovery.

Nanochemistry research is showing the way to a world of materials that had not previously existed. The work surveyed by the authors involves a "global" way of thinking about solid-state materials that introduces the notions of complexity and hierarchy into materials chemistry, which not so long ago were deemed appropriate only for the incredible mineral-based materials made by living organisms. In essence, the authors show that the self-assembly of inorganic, organic and polymeric building blocks may occur spontaneously or be directed by molecules, aggregations of molecules, microphase-separated block copolymers, colloidal crystals or a variety of forces. Self-assembly can provide an effective pathway to new materials whose structure at "all" levels of construction, from the nanoscale to the overall macroscopic form determine materials properties, desired function and practical utility. The broadly tunable length scales and dimensionalities, platonic and curved morphologies, close-packed and open-framework structures, elemental compositions and physicochemical properties of solid-state materials to emerge from this work are of interest in a myriad of areas as diverse as those mentioned above.

But truly the greatest contributions still lay dormant in the thoughts of the students training right now, those brave enough to cross disciplines, entice collaborators, be creative, and Go Nano!

3 Nano Safety

No chemical is safe unless proven otherwise; even water can be dangerous when experienced in the wrong quantities! There are national and global regulated standards in health and safety for chemicals by which all governments, industries and universities abide. Every laboratory working with chemicals must have a comprehensive and current laboratory safety manual that describes recommended protocols for safe handling, storing and disposing of chemicals as well as access to materials safety data sheets (MSDS) that document toxicology of chemicals and recommended treatment for exposure.

Do we need to treat nanomaterials differently than other chemicals? Some might argue that nanomaterials are only versions of already classified chemicals and can be treated with the same caution. Others recognize the added fact that nanomaterials, particularly nanochemicals, are not only characterized by their chemical composition but also by their special physical dimensions, which are below the micron scale of common particles (*i.e.*, sizes between a nanometer and a micrometer). This means that the toxicity of both the bulk forms and the activity and interactions of these nanomaterials, or their intermediates, need rigorous testing. Therefore, to ensure safety to our health and our environment, new protocols on how to work safely with nanomaterials need to be developed and we should evaluate and test nanomaterials under new safety criteria (Figure 1).[1–3] Any new nanomaterial, with known or unknown composition, should be treated with as much caution as any new and untested "non-nano" chemical. Paradoxically, since the concept of nanomaterials are quite new to the field of toxicology, it is the emerging field of "bionanochemistry," an area concerned primarily with bioassays,

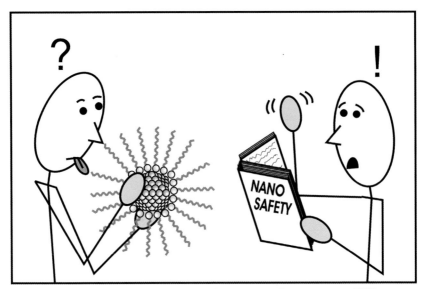

<u>Caution</u>: Nanomaterials may have a different toxicity than their bulk counterparts.

Figure 1   *Even well known compounds may present unexpected health risks when they are fashioned as nanoscale building blocks*

biolabels and bioelectronics, which will likely provide useful knowledge and technology to address such safety issues and other questions central to issues of nanotoxicity. For some very recent work on this subject see Appendix B, Cytotoxicity of Nanoparticles.

Aims of this Book

About half a century ago, Richard Feynman gave his prophetic lecture *"Plenty of Room at the Bottom."* He outlined in this talk the foundations of nanoscience, and the promise that totally synthetic constructions could eventually be built with molecular scale precision. Nanoscience research has been rapidly increasing across the globe during the past decade. It is now widely accepted by the scientific, industrial, government and business communities, that nanoscience will be of integral importance in the development of future technologies. Nanoscience is being touted as the engine that will drive the next industrial revolution.

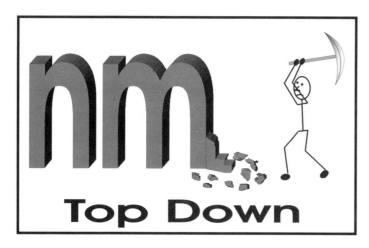

One of the hallmarks of nanoscience is its interdisciplinary nature – its practice requires chemists, physicists, materials scientists, engineers and biologists to work together in close-knit teams. Communication and collaboration between disciplines will enable the most challenging scientific problems to be tackled, those that are most pressing in the successful exploitation of nanotechnology. This is a daunting challenge for students, researchers, managers, funding agencies and politicians, as the field of nanoscience is so diverse and evolving rapidly. These difficulties are exacerbated by the fact that few textbooks on nanoscience have existed that bring together the vast range of published work in an accessible primer for students and practitioners of this topic.

"*Nanochemistry – A Chemical Approach to Nanomaterials*," has been written through the eye of chemistry in an attempt to fill this void. The content of the book has been selected and organized to establish the basic principles of nanoscience through the subject of nanochemistry. Because of the interdisciplinary approach adopted by the authors, the book should be useful to a broad readership even though the fundamental science and applications will continue to change almost on a daily basis.

Nanoscience today is a creative amalgamation of bottom–up chemistry and top–down engineering techniques. We are currently witnessing an explosion of novel ideas and strategies for making and manipulating, visualizing and interrogating nanoscale materials and structures. An aim of the book is to describe the concepts and methods developed for synthesizing a range of nanoscale building blocks with strictly controlled size, shape, bulk and surface structure and composition, and properties. A further aim is to explain how these nanoscale construction units can be organized and integrated into functional architectures using a combination of self-assembly, templating, chemical lithography and other patterning methods.

Nanochemistry will be a valuable textbook for students planning an academic or industrial research career in any area related to nanoscience and nanotechnology. It provides a global perspective of the subject of nanochemistry, written with sufficient breadth and depth to make it suitable as the basis of an advanced undergraduate course in chemistry or a graduate course in materials chemistry, engineering materials science, solid-state physics and nanoscience. Much of the subject matter presented in the book can be readily adapted to introduce chemical aspects of nanoscience to a broad audience of undergraduates in their first 3 years at university. This text should provide a readily accessible road map of nanochemistry, beginning with its roots and extending to its branches, emphasizing throughout the connection of ideas from discovery to application from within and between the science disciplines. It provides a unique perspective on nanoscience and nanotechnology, an attribute which will make it invaluable for those witnessing, participating in, and trying to remain at the leading-edge of the nanoscience explosion. These include scientists, managers in university and industry, government workers, business people, even the media.

Structure of this Book

In the first textbook on nanochemistry, Arsenault and Ozin describe the methods used by chemists to make nanoscale building blocks and the techniques that can arrange them into functional architectures. Primary building blocks have dimensions spanning nanometers to micrometers, and can be made of almost all conceivable materials. Their assembly into a particular arrangement can be driven by forces acting between them, or directed by specific interactions with a structure-guiding template or patterned substrate. Like Nature's biomineral construction units, these primary building blocks can associate into secondary structures and this organization scheme can continue until the highest level of complexity is reached. This introduces the notion of building blocks assembling over multiple length scales as a way of making materials with hierarchical structures and complex form. Hierarchy allows a material to satisfy several, sometimes conflicting, sets of conditions and demands in a single material. How this relates to Nature's way of creating structures introduces important connections between molecular recognition and epitaxy, crystal growth and form, unifying ideas in coordination, bioinorganic, materials and solid-state chemistry.

The book *Nanochemistry* begins by exploring the basic tenets of self-assembly, and continues into some insightful 19th century research of Pieter Harting on synthetic morphology, Ernst Haeckel and D'Arcy Thompson on morphogenesis and Emil Fischer on the lock-and-key principle of molecular recognition. These classic studies relate in turn to 20th century research of Richard Barrer on template directed self-assembly of microporous materials, Hientz Lowenstam on organic matrix mediated formation of biominerals and Jean-Marie Lehn on supramolecular chemistry. Ideas and methods stemming from this pioneering work provide a springboard for the development of the 21st century emerging field of nanochemistry.

Continuing on this flow of ideas, the book explores the genealogy of chemical discoveries that lead to the development of nanochemistry. The toolkit of nanochemistry is described in terms of the integration of concepts and methods in chemical lithography, templating, materials chemistry and self-assembly. The underlying theme is how different patterning and templating methods spanning the scale of microns to nanometers can be creatively fused with synthesis and self-assembly of sheet, wire, tube, rod, cluster, solid and hollow sphere and hemisphere, spiral and ring building blocks to make new materials and structures. The goal is to create functional nanostructures with perceived utility in such areas as electronics, magnetics and photonics, batteries, solar cells and fuel cells, catalysis and sensing, chemical storage and release. In fact, in a few years time it will be difficult to find products and technologies that have not been significantly advanced by nanoscience.

At the end of every chapter the reader is challenged with *"Nanofood for thought,"* a collection of questions often without clear-cut answers that have been designed to inspire creative and holistic thought and facilitate a connection of ideas about the materials and methods described throughout the book. *"Nanofood for thought"* will hopefully serve not only to enhance the readers' understanding of the underlying physicochemical principles of nanochemistry but also suggest new ways and means for the reader to approach and solve basic and directed research problems in their own area of nanoscience.

The book culminates by posing the provocative question *"What comes next?."* Some fanciful thoughts are offered on how to think about assembling the future beyond the nanochemistry presented in this textbook.

Finally, a series of practical experiment outlines in nanochemistry is explored, to enable students and researchers to roll up their sleeves in the *"Nanolab,"* make matter that matters, and invent materials of the next kind!

Contents

Chapter 10   Biomaterials and Bioinspiration   473

List of Acronyms

5CB	4-pentyl-4′-CyanoBiphenyl
AAO	Anodized Aluminum Oxide
ACC	Amorphous Calcium Carbonate
AFM	Atomic Force Microscopy
ALD	Atomic Layer Deposition
APS	AminoPropyltrimethoxySilane
ATO	Antimony Tin Oxide
BDC	1,4-Benzene-DiCarboxylate
BPEI	Branched PolyEthyleneImine
BZ	Belousov-Zhabotinsky
CVD	Chemical Vapor Deposition
CC	Colloidal Crystal
CCMV	Cowpea Chlorotic Mottle Virus
CFM	Chemical Force Microscopy
CPC	Colloidal Photonic Crystal
CPMV	CowPea Mosaic Virus
CSS	Core-Shell-Shell
DDP	Decylammonium-DihydrogenPhosphate
DEISA	Directed Evaporation-Induced Self-Assembly
DLVO	Derjaguin Landau Verwey Overbeek
DNA	DeoxyriboNucleic Acid
DOS	Density Of States
DPN	Dip-Pen Nanolithography
DPPE	DiPalmitoyl PhosphatidylEthanol
DX	Double-crossover
EBG	Electronic Band Gap
EDS	Energy Dispersive Spectroscopy
EDX	Energy Dispersive X-ray analysis
EELS	Electron Energy-Loss Spectroscopy
EG	Ethylene Glycol
EISA	Evaporation Induced Self-Assembly
EMR	ElectroMagnetic Radiation

ESA	Electrostatic Self-Assembly
ESL	Edge-Spreading Lithography
EXAFS	Extended X-ray Absorption Fine Structure
FESEM	Field-Emission Scanning Electron Microscopy
FET	Field-Effect Transistor
FTO	Fluorine-doped Tin Oxide
GLYMO	(3-GLYcidyloxypropyl)-triMethOxysilane
GMR	Giant MagnetoResistance
HMDS	HexaMethylDiSilazane
HREM	High-Resolution Electron Microscopy
HRSEM	High-Resolution Scanning Electron Microscopy
HRTEM	High-Resolution Transmission Electron Microscopy
IR	InfraRed
IRRS	InfraRed Reflection Spectroscopy
ITO	Indium-Tin Oxide
IUPAC	International Union of Pure and Applied Chemistry
LbL	Layer-by-Layer
LCD	Liquid Crystal Display
LED	Light-Emitting Diode
LFM	Lateral Force Microscopy
MALDI	Matrix-Assisted Laser Desorption Ionization
MAS NMR	Magic Angle Spinning Nuclear Magnetic Resonance
MBE	Molecular Beam Epitaxy
MCP	MicroContact Printing
MEMS	MicroElectroMechanical Systems
MFM	Magnetic Force Microscopy
MHA	16-MercaptoHexadecanoic Acid
MIMIC	MIcroMolding In Capillaries
MINIM	Metal-Insulator-Nanocluster-Insulator-Metal
MIPO	Micromolding in Inverse Polymer Opals
MOCVD	Metal Organic Chemical Vapor Deposition
MOF	Metal-Organic Framework
MOSFET	Metal-Oxide-Semiconductor Field-Effect Transistor
MPP	Microscope Projection Photolithography
MS	Mass Spectroscopy
MSA	Mesoscale Self-Assembly
MTP	MicroTransfer Printing
MWNT	Multi-Walled carbon NanoTube
NCA	N-CarboxyAnhydride
NCP	NanoContact Printing
NDR	Negative Differential Resistance
NEMS	NanoElectroMechanical Systems
NIR	Near InfraRed
NLO	Non-Linear Optical
NMR	Nuclear Magnetic Resonance
NSL	NanoSphere Lithography

NSOM	Near-field Scanning Optical Microscopy
NW	NanoWire
OCP	OctaCalcium Phosphate
OCP	Overpressure Contact Printing
ODPA	*n*-OctaDecylPhosphonic Acid
ODT	1-OctaDecaneThiol
ODT	Order-Disorder Transition
OHAp	HydrOxyApatite
OTS	*n*-OctadecylTrichloroSilane
PAA	Polyacrylic acid
PAG	PhotoAcid Generator
PAH	PolyAllylamine Hydrochloride
PANI	PolyANIline
PB	PolyButadiene
PBG	Photonic Band Gap
PCCA	Polymerized Colloidal Crystalline Array
PCR	Polymerase Chain Reaction
PDAC	PolyDiallyldimethyl Ammonium Chloride
PDLC	Polymer Dispersed Liquid Crystal
PDMS	PolyDiMethylSiloxane
PDOS	Photonic Density Of States
PEDOT	Poly(3,4-EthyleneDiOxyThiophene)
PEI	PolyEthyleneImine
PEM	PolyElectrolyte Multilayer
PEO	PolyEthyleneOxide
PET	PolyEthylene Terepthalate
PI	PolyIsoprene
P-Ink	Photonic Ink
PLA	PolyLactic Acid
PMMA	PolyMethylMethAcrylate
PMO	Periodic Mesoporous Organosilica
PMS	Periodic Mesoporous Silica
POM	Polarizing Optical Microscopy
PPDSS	Poly(PentamethylDiSilylStyrene)
PPO	PolyPropyleneOxide
PPP⁻	Poly(*P*-Phenylene sulfate)
PPQ	PolyPhenylQuinoline
PPV	PolyPhenyleneVinylidene
PPy	PolyPyrrole
PS	PolyStyrene
PSS	PolyStyreneSulfonate
PU	PolyUrethane
PVP	PolyVinylPyridine
PVP	PolyVinylPyrrolidone
PXRD	Powder X-Ray Diffraction
PXV	PolyheXylViologen

QCM	Quartz Crystal Microbalance
QSE	Quantum Size Effect
RIE	Reactive Ion Etching
RNA	RiboNucleic Acid
ROP	Ring-Opening Polymerization
SAED	Selected Area Electron Diffraction
SAM	Self-Assembled Monolayer
SANS	Small Angle Neutron Scattering
SAXS	Small Angle X-ray Scattering
SEM	Scanning Electron Microscopy
SET	Single Electron Transistor
SFLS	Supercritical Fluid-Liquid-Solid
SLS	Solvent-Liquid-Solid
SNP	Single Nucleotide Polymorphism
SP-CP	Scanning Probe Contact Printing
SPM	Scanning Probe Microscopy
SQUID	Superconducting QUantum Interference Device
STM	Scanning Tunneling Microscopy
STS	Scanning Tunneling Spectroscopy
SWNT	Single-Walled carbon NanoTube
TEG	TetraEthyleneGlycol
TEM	Transmission Electron Microscopy
T_g	glass Transition temperature
TGA	Thermal Gravimetric Analysis
TMSCl	TriMethylSilyl Chloride
TMV	Tobacco Mosaic Virus
TODE	TOpologically Directed Etching
TOF-SIMS	Time Of Flight Secondary Ion Mass Spectroscopy
TOPO	TriOctylPhosphine Oxide
^{t}RNA	transfer RiboNucleic Acid
TUBEFET	carbon nanoTUBE Field Effect Transistor
UV	UltraViolet
VLS	Vapor-Liquid-Solid
VPT	Vapor Phase Transport
WAXS	Wide Angle X-ray Scattering
XANES	X-ray Absorption Near-Edge Spectroscopy
XPS	X-ray Photoelectron Spectroscopy

Teaching (Nano)Materials

G. A. OZIN

Over more than three decades of teaching and researching solid-state (nano)materials chemistry at the University of Toronto, I have often reminded undergraduate, graduate and postgraduate students of old adages such as Linus Pauling's, "*it is what is in your head that counts not that you know where to look it up in a book.*" This is to convey the idea that creativity and excitement in the field of (nano)materials comes from the spontaneous assembly, organization and integration of information encoded at some level within the brain, from which emerges a breakthrough, the joy and anticipation of the eureka moment. A teacher in (nano)materials chemistry should therefore strive to be more than the conveyor of information, and try to enlighten students about the creativity process underpinning materials discovery, and of course what happens next.

Within the context of (nano)materials chemistry, I have learned that "*the genesis of a creative idea, first and foremost, requires a knowledge bank of all classes of solids and how to make them.*" This is to remind students that chemists are at the beginning of the materials food chain and to emphasize the importance of materials synthesis. With a solid foundation in the making of materials, I develop a framework approach with which to relate volume and surface composition and structure, length scale and dimensionality, texture and morphology of materials to their properties. Armed with this information and an appreciation of where different classes of materials find industrial applications and provide value to society, the challenge in research is to discover whether a new material is best utilized to improve the performance of an existing product or process or whether it has the potential to disrupt an existing technology or create one that never before existed.

I recall Philip Ball's quote "*a breakthrough is a discovery pregnant with promise and then the hard graft begins,*" which alerts the student to the fact (with a bit of English slang) that in the world of materials, after a euphoric report in a top journal like *Nature* or *Science*, there is a long and arduous road of research and development before real applications are reduced to practice and reach the marketplace.

When introducing solid-state materials chemistry to the student, after having laid the foundation of solid-state synthesis, structure determination by X-ray diffraction and the electronic band description of solids, I often announce, "*defects, defects, defects, there is no such thing as a perfect crystal, if it did exist it would not be terribly useful and applications would be far and few between.*" Of course, students find this upsetting at first and difficult to comprehend until they learn about the aesthetics of the defect state and that it is the imperfection of solid-state materials rather than perfection that provides them with interesting properties and ultimately their function and utility. "Perfecting imperfection" in solids and knowing which kind of imperfection to perfect to achieve a particular objective is a challenging yet important and universal concept for the student of materials chemistry to grasp.

Another message I convey to excite the imagination of students about materials chemistry is that, "*materials shape is everything and the form of materials controls function and utility.*" I follow this with the idea that "*while it has traditionally been the job of materials scientists, physicists and engineers to physically sculpt materials from the top–down it is now the role of materials chemists to find creative ways of synthetically shaping materials from the bottom–up.*" Integrating the strengths of top and bottom approaches to (nano)materials, is likely the way the field will unfold in the years ahead.

Yet another surprise in store for most students is the realization that "*the shape or habit or morphology of crystals is not necessarily Platonic,*" under crystal physics control with planar faces, sharp edges and pointed apices, a primitive unit cell, planar symmetry elements and 230 space groups. Instead, crystalline materials can also exist with curved form and a biological appearance, arising from an underlying flexi-crystallography with curvilinear symmetry elements, leading to morphogenesis of shapes akin to that found in biominerals.

Nanochemistry is an outgrowth of a career of research at the leading edge of the field which emerged as a graduate course on the subject and that required about 3–4 years to develop – it was presented for the first time in the fall of 2002. About 10 graduate students took the course for credit and around the same number audited the course. The students were inspired and excited by the novelty of the materials and new ideas presented, and based on their feedback the course was reorganized and updated in 2003. Student enrolment and auditing student numbers more than doubled and tripled, and feedback was very positive.

Because of the appeal of the course to a multidisciplinary group of students in chemistry, physics, materials science and engineering, biology, geology and nanoscience, it motivated the author to write his lecture notes in the form of a textbook. The subject matter covered is based on research papers and reviews in the field of nanochemistry and materials self-assembly, to which the authors' materials chemistry group has made pioneering contributions.

Learning (Nano)Materials

A. C. ARSENAULT

After surveying the somewhat daunting task of teaching a course on nanochemistry, it is worth mentioning the daunting task in the hands of the nanochemistry student. Meandering through the academic system, learning about neatly divided subjects and disciplines, becoming more and more specialized, one can be caught entirely off-guard by someone then telling you: "Okay, now you've got to know everything – at least a little bit of everything." This feeling can arise when starting to study nanochemistry, which intersects a confluence of scientific streams. In the face of so much new information, and realizing how little we know, it can be easy to pack up and pick a different field with a lesser breadth.

I have certainly felt this way, especially starting my graduate studies, since my previous studies had been in biological and organic chemistry. However, as time went on, I realized that I had a lot more to offer than I had anticipated. Nanoscience is inherently interdisciplinary, and the background I had gave me a unique perspective on the field. In addition to the new knowledge I acquired, I could apply my organic chemistry knowledge to the understanding of molecular building blocks, my knowledge of polymers to polymer nanochemistry, and my biological knowledge to the understanding of bionanochemistry. Starting my academic career with a taste for interdisciplinary research, I also ensured a lifelong appetite for it, essential I believe for a successful career in nanoscience.

I started my graduate studies in the group of Geoffrey A. Ozin in the fall of 2001, choosing to work jointly with a second supervisor specializing in organometallic polymers, Ian Manners. Later that year I heard about a book Geoff was writing, and proposed to him my services if none had yet done so. He wholeheartedly accepted the help, and so began this particular collaboration. A few years, a few rewrites, and a few theme changes later, we found ourselves sitting in front of a textbook of which we could be proud. This experience has been a great challenge, and as with all great challenges has given me a much greater measure of personal and professional satisfaction. In fact, I can think of no better way one can begin a research career.

About the Authors

Geoffrey A. Ozin: born 23rd August 1943 in London, England, received a B.Sc. in chemistry from Kings College University of London in 1965 and his D. Phil. degree in inorganic chemistry from Oriel College Oxford University in 1967. He was ICI Fellow at the University of Southampton from 1967 to 1969 before joining the University of Toronto in 1969. He achieved the rank of full professor in 1977, University Professor in 2001 the highest rank at the University of Toronto and has been named Government of Canada Research Chair in Materials Chemistry, a national award for his contributions to science. The significance of Professor Ozin's materials chemistry research has been recognized in various ways. He won the Brockhouse Interdisciplinary Canada Prize (with Sajeer John) in 2004 and an Alexander von Humboldt Career Award 2004, was listed as ISI Highly Cited Research in Materials Science 2003, received the Canadian Society of Chemistry E. W. R. Steacie 2002 CSC Award in Chemistry, the Royal Society of Chemistry Great Britain 2002 RSC Award in Materials Chemistry, the Canadian Institute of Chemistry 2001 CIC Medal, a Canada Council Isaac Walton Killam Research Fellowship 1995–1997, the Royal Society of Canada inducted him as a Fellow, The Royal Institution of Great Britain and University College, London University jointly elected him Honorary Professorial Fellow, and he is a Visiting Scientist at the London Center for Nanotechnology and the Max Planck Institute for Materials Research in Golm, Germany. Amongst his other honors are the Pure or Applied Inorganic Chemistry Award from the Canadian Society for Chemistry; Rutherford Memorial Medal in Chemistry from the Royal Society of Canada; Alcan Award for Inorganic Chemistry from the Chemical Institute of Canada. Others include the Coblentz Memorial Prize for Molecular

Spectroscopy from the American Spectroscopy Society; Meldola Medal for Physical-Inorganic Chemistry from the Royal Society of Chemistry Great Britain. He was a Sherman-Fairchild Fellow at California Institute of Technology and he received three University of Toronto Connaught Special Research Awards, reserved for top rank scientists amongst the 3000 faculty. The Canadian Institute for Advanced Research, CIAR elected him Associate of the new Nanoelectronics group and he is a Principle Investigator at Photonics Research Ontario and Materials Manufacturing Ontario. Two of his graduate students Hong Yang and Mark MacLachlan were awarded the 1999 and 2000 NSERC Doctoral Prize for top Ph.D. Thesis research in Canadian Natural Sciences. Recognition of outstanding and sustained research accomplishments of Professor Ozin is seen in the large number of plenary and invited lectures given at international conferences, universities and industries. He serves or has served on the international editorial advisory board of the journals, Advanced Materials, Journal of Materials Chemistry, Chemistry of Materials, Journal of Solid State Chemistry and Cluster Science. He has published around 525 articles in refereed journals, has been granted 12 US patents and 6 US patent applications, and many in different countries. Professor Ozin has also contributed significantly to teaching, research and technology through the training of a large body of undergraduate, graduate and postgraduate students in materials chemistry and through his long-standing research collaborations with industry. His close ties with industry have resulted in numerous inventions and technology transfer. Around 28 of his former coworkers hold Professorial positions in Universities in Canada and other parts of the world. A large number of other coworkers secured key scientific positions in government and industrial laboratories and one had a listed spin-off company Lumenon. One of his coworkers is a Dean of Science in a top rank US University and two others are heads of European and Canadian University chemistry departments. Graphical illustrations from 26 of his papers have been used as covers, inside covers and frontispieces of top rank scientific journals and commentaries about his research appear frequently in the scientific news media.

André C. Arsenault: born on 31st July 1979, is currently in the midst of a Ph.D. degree in the groups of Geoffrey A. Ozin and Ian Manners at the University of Toronto. He completed his Honors B.Sc. at the University of Toronto in 2001 in Biological Chemistry, and subsequently joined the graduate program in materials and polymer chemistry. After taking some undergraduate courses in materials chemistry, especially by Professor Ozin, he became convinced that the world of nanomaterials was well within his previous experience. Typically, a scientific field drives towards ever more specialized and detailed knowledge, but nanoscience is different. Rather than a coherent discipline, it is really nothing more than a length scale where you have to start really worrying about the other sciences. Nano-engineers start worrying about chemistry because of large surface areas, nano-chemists worry about the processing and integration of their synthetic units into manufacturing processes, biologists see a wealth of information arising from studying the interaction of nano-systems with organisms, and nano-physicists figure out why all these happen. He is currently the author of 11 scientific publications, the holder of one US patent, and his work has appeared several times in the news media.

References

1. J. Giles, Size matters when it comes to safety, report warns, *Nature*, 2004, **430**, 599.
2. www.etcgroup.org
3. The Royal Society and The Royal Academy of Engineering Report, *Nanoscience and Nanotechnologies*, July 2004, 35–50.

Acknowledgements

Geoffrey A. Ozin: At the outset I wish to express my deepest appreciation to my wife Linda Ozin for her support and advice throughout the writing period of this textbook. I also wish to thank my co-author André Arsenault for joining me at quite an advanced and complex stage of its assembly, in particular for his creative writing and artistic contributions as well as his dedication to ensure the book was completed on schedule, even in the midst of his busy doctoral thesis research. In addition, I want to voice very special praise to Ms Sue Mamiche, who in the middle of helping manage my research group also assisted me with the completion of this book. I am indebted to my daughter Dr. Amanda Ozin for her critical advice on the health and safety aspects of nanomaterials. The cover graphic of this book, which vividly portrays the Chemical Approach to Nanomaterials theme of the textbook, is the creation of Ludovico Cademartiri, a talented doctoral student in my group in both science and art. I also wish to acknowledge the sustained financial support of my Nanochemistry research program from the Natural Sciences and Engineering Research of Canada (NSERC) who believed in my ideas from the beginning of my career as an Assistant Professor in 1969 when this field was just emerging until the present time where there are scientific breakthroughs on a daily basis and much anticipation that Nanotechnology will likely power the next industrial revolution. The recent award of a Canada Research Chair in Materials Chemistry from NSERC also enabled my group to continue its work at the leading edge of Nanochemistry. Being elected as an Associate Member of the Canadian Institute of Advanced Research (CIAR) Nanoscience Team has had a most positive impact on the research of my group. The strong backing of all my past and present Chairs, Deans, Provosts and Presidents, as well as the professional support from the Technology Transfer Group and Innovations Foundation at the University of Toronto, has been pivotal in facilitating the research and development work undertaken by my group. I also recognize my industrial colleagues, who over the years have provided my group with the financial incentive and personal encouragement to undertake adventurous Nanochemistry research. None of this would have been possible without the contributions of all of my co-workers and collaborators, who excited and challenged me every day of my career with creative and insightful ideas and questions – they have profoundly influenced my way of thinking and writing about the wonderful world of Nanochemistry. At the close, I wish to say that each day in my academic career has been an exhilarating learning experience. I have tried to instill some of this excitement and knowledge into the pages of Nanochemistry: A Chemical Approach to Nanomaterials.

André C. Arsenault: First and foremost, I express greatest thanks to my wife Charlene Arsenault, for supporting me in this endeavor at every turn. I wish to thank my supervisor Geoffrey Ozin for his invaluable advice, scientific, professional, and personal, over my years as a graduate student. The work I have done on this book is simply one of the many testaments of the inspiration and drive he provides to his students. I would like to thank my co-supervisor, Ian Manners, whose personal guidance has furthered greatly my scientific development. The talented researchers of both the Ozin and Manners groups, past and present, all deserve thanks for the innumerable things I have learned from them. As mentioned above, very special thanks go to Sue Mamiche for her level headed advice and great help in the organization of the materials in the manuscript. Also, Ludovico Cademartiri deserves special thanks for the great artistry displayed on the front cover of this book, which truly captures the essence of this book. I wish to give special thanks to NSERC for awarding me a postgraduate scholarship and most recently a prestigious Canada graduate scholarship, as well as financial support from the University of Toronto, which have allowed me to truly dedicate myself to this book and my scientific research.

Nanofood for Thought –
Thinking about Nanochemistry, Nanoscience, Nanotechnology and Nanosafety

1. What are the four main discoveries in solid-state science that created materials chemistry as we know it today. Address the same question for nanochemistry and nanoscience.
2. How would you go in the lab and get all the information in the library of congress on the head of a pin?
3. How would you reduce to practice Richard Feynman's dream of writing all of the accumulated information in all of the books in the world in a cube of material one two-hundredth of an inch wide and proving it was all there?
4. Why is nanoscience one of the most interdisciplinary sciences and therefore challenging in terms of teaching and research?
5. How many words with the prefix nano can you think of?
6. Give scientific reasons that would satisfy both the public and a scientist why self-replicating nanobots and the gray goo are unlikely scenarios for nanoscience.
7. Free nanoparticles have raised health and environmental safety concerns because their toxicology may be distinct to that of microscale particles. Smaller particles have larger surface area and the ability to enter cells more readily and in different ways than larger ones. Another safety concern is the risk of explosion posed by particles through spontaneous combustion or ignition. Would you expect nanoscale particles to present a greater or smaller risk of explosion than microscale ones? Do you think the shape of nanoparticles, like sphere *versus* rod, would affect explosion hazard? What research would you undertake to determine the explosion characteristics of a range of nanoparticles? How would you minimize the risk of explosion of combustible nanoparticles?
8. Do you think recent developments in bionanochemistry, in which biomolecules are being integrated with nanomaterials, can help elucidate the toxicological effects of nanoparticles on human health and safety? How?
9. There is a smell of salt in the air when near the sea – is this just an expression?
10. In the hustle and bustle of a modern city, which kind of nanoparticles are you ingesting every day of your life, do you think any of these are likely to be dangerous to your health, and what do you suggest should be done about it?

11. Asbestos fibers are about 3–20 microns in length with a width of as little as 10 nm. What are the properties of this material that make it hazardous, and would you expect any effect upon reducing the length down to the nanoscale?

12. Recently, capped metal and semiconductor nanoclusters have been self-assembled with living nanostructures like virus particles, bacteria and fungi. What questions would you ask if you were working with these bionano-materials?

13. Could you imagine using biological nanomachines to dispose of nano-particles?

14. How would you go about developing a cheap and portable, sensitive and selective detector for nanoparticles?

15. We know nanoparticles are very small and can pass through most of today's filtration membranes, so how would you clean them up from the air we breathe and the water we drink?

16. Do you think we should be more concerned about nanotechnology than biotechnology?

Nanochemistry Basics

"We are like dwarfs on the shoulders of giants, so that we can see more than they."
Bernard of Chartres, 12th century

1.1 Materials Self-Assembly

When thinking about self-assembly of a targeted structure from the spontaneous organization of building blocks with dimensions that are beyond the sub-nanometer scale of most molecules or macromolecules, there are five prominent principles that need to be taken into consideration. These are: (i) building blocks, scale, shape, surface structure, (ii) attractive and repulsive interactions between building blocks, equilibrium separation, (iii) reversible association–dissociation and/or adaptable motion of building blocks in assembly, lowest energy structure, (iv) building block interactions with solvents, interfaces, templates, (v) building-blocks dynamics, mass transport and agitation.

A challenge for perfecting structures made by this kind of self-assembly chemistry is to find ways of synthesizing (bottom-up) or fabricating (top-down) building blocks not only with the right composition but also having the same size and shape. No matter which way building blocks are made they are never truly monodisperse, unless they happen to be single atoms or molecules. There always exists a degree of polydispersity in their size and shape, which is manifest in the achievable degree of structural perfection of the assembly and the nature and population of defects in the assembled system. Equally demanding is to make building blocks with a particular surface structure, charge and functionality. Surface properties will control the interactions between building blocks as well as with their environment, which ultimately determines the geometry and distances at which building blocks come to equilibrium in a self-assembled system. Relative motion between building blocks facilitates collisions between them, whilst energetically allowed aggregation de-aggregation processes and corrective movements of the self-assembled structure will allow it to attain the most stable form. Providing the building blocks are not too strongly bound in the assembly it will be able to adjust to an orderly structure. If on the other hand the building blocks in the assembly are too strongly interacting, they will be unable to adjust their relative positions within the assembly and a less

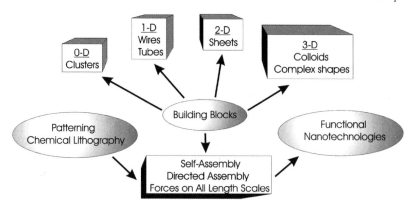

Figure 1.1   *A flowchart delineating the factors that must be considered when approaching the self-assembly of a nanoscale system*

ordered structure will result. Dynamic effects involving building blocks and assemblies can occur in the liquid phase, at an air/liquid or liquid/liquid interface, on the surface of a substrate or within a template co-assembly.

As this text describes, building blocks can be made out of most known organic, inorganic, polymeric, and hybrid materials. Creative ways of making spheres and cubes, sheets and discs, wires and tubes, rings and spirals, with nm to cm dimensions, abound in the materials self-assembly literature. They provide the basic construction modules for materials self-assembly over all scales, a new way of synthesizing electronic, optical, photonic, magnetic materials with hierarchical structures and complex form, which is the central theme running throughout this chapter. A flowchart describing these main ideas is shown in Figure 1.1.

1.2 Big Bang to the Universe

At the beginning it is said that nothing self-assembled into everything. An energetic singularity, where all symmetry and forces were one, transformed in time to all that we know today about our universe. This is the "ultimate in materials self-assembly over 'all' scales": From the smallest, densest most energetic form of matter to an expanding and seemingly infinite universe built from a diversity of materials, comprised of organic and inorganic molecules, polymers and networks, large and small, cast through the construction tool kit of chemistry, physics and biology as lifeless abiological and living biological systems.

1.3 Why Nano?

Nano-, a prefix denoting a factor of 10^{-9} has its origin in the Greek *nanos*, meaning dwarf. The term is often associated with the time interval of a nanosecond, a billionth of a second, and the length scale of a nanometer, a billionth of a meter or 10 Å. In its broadest terms, nanoscience and nanotechnology congers up visions of making, imaging, manipulating and utilizing things really small. Feynman's prescient nanoworld "on the head of a pin" inspires scientists and technologists to venture into this uncharted nano-terrain to do something big with something small. It excites investors and corporations, governments and policy makers to gamble on nanoscience breakthroughs and create new nanotechnologies.

While early theoretical concepts and experimental results for nano-size materials and devices appeared five decades ago, it is rather recent scientific developments that have inspired a resurgence of activity in the field. The stimulus for this growth can be traced to new and improved ways of making and assembling, positioning and connecting, imaging and measuring the properties of nanomaterials with controlled size and shape, composition and surface structure, charge and functionality, for use in the macroscopic real world.

1.4 What do we Mean by Large and Small Nanomaterials?

It was not so long ago in the world of molecules and materials that 1 nm (1 nm = 10 Å) was considered large in chemistry while 1 μm (1 μm = 1000 nm = 10,000 Å) was considered small in engineering physics. Matter residing in the "fuzzy interface" between these large and small extremes of length scales emerged as the science of nanoscale materials and has grown into one of the most exciting and vibrant fields of endeavor, showing all the signs of having a revolutionary impact on materials as we know them today.

In our time, "nano" has left the science reservation and entered the industrial technology consciousness and public and political perception. Indeed, bulk materials can be remodeled through bottom-up synthetic chemistry and top-down engineering physics strategies as nanomaterials in two main ways, the first by reducing one or more of their physical dimensions to the nanoscale and the second by providing them with nanoscale porosity (Figure 1.2). When talking about finely divided and porous forms of nanostructured matter, it is found that "nanomaterials characteristically exhibits physical and chemical properties different from the bulk as a consequence of having at least one spatial dimension in the size range of 1–1000 nm".

It is the "synthesis, manipulation and imaging of materials having nanoscale dimensions, the study and exploitation of the differences between bulk and

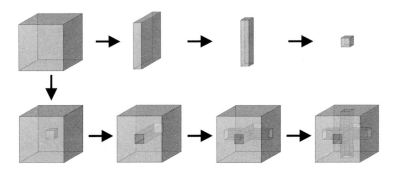

Figure 1.2 *Dividing matter to the nanoparticle and nanoporous state*

nanoscale materials, and the understanding and utilization of nanomaterials scaling laws by interdisciplinary scientists, that drive contemporary endeavors in nanoscience and nanotechnology." It is an astute awareness of the unique attributes of nanoscale matter and the tremendous opportunities for discovery and innovation that abound in the field of nanoscience, that provides much of the inspiration for researchers working in the field of nanomaterials.

1.5 Do it Yourself Quantum Mechanics

The formalism underpinning the scaling laws of materials with size tunable properties is couched in the physics language of "quantum size effects," (QSE). The Schrödinger wave equation is solved for an electron and hole in a box having either 1D, 2D or 3D and spatial dimensions of the order of the Bohr radius of the electron, hole or exciton (electron–hole pair). In this intermediate size range between molecular and bulk matter, called the nanoscale, individual energy states of molecules and continuous energy bands of solids become discrete and their energy separations display an analytic dependence on the spatial dimension of the material. This is expressed in a scheme of the type shown in Figure 1.3, which represents the "transformation of the electronic density of states (DOS) of valence and conduction bands in metals and semiconductors from continuous to discrete to individual in bulk, quantum-confined and molecular states of matter."[1] As the dimensionality of quantum-confined matter changes from that of a 2D quantum well, to a 1D quantum

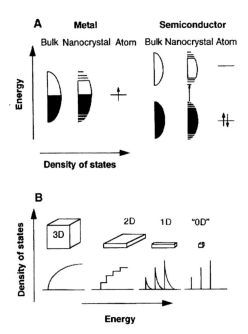

Figure 1.3 *Relation between bulk, quantum confined and molecular states of matter* (Reproduced with permission from Ref. 1)

wire and finally a 0D quantum box, the DOS correspondingly alters from the parabolic shape of the 3D bulk material, first to a 2D stepped parabola, then to a 1D inverse stepped parabola and eventually to a 0D discrete form.

The challenge in materials chemistry is to understand how the properties of nanomaterials scale with physical dimension because different properties scale in distinct ways with size. Thus, electronic band gaps scale differently to melting points and magnetization and so forth. The point is that in chemical approaches to the preparation and assembly of well-defined nanostructures into functional and integrated systems of nanocomponents, it is vital to appreciate how the properties of materials scale with size in order to target the right combination of materials compositions and length scales to achieve a desired objective.

1.6 What is Nanochemistry?

With nanoscience being the discipline concerned with making, manipulating and imaging materials having at least one spatial dimension in the size range 1–1000 nm and nanotechnology being a device or machine, product or process, based upon individual or multiple integrated nanoscale components, then what is nanochemistry? In its broadest terms, the defining feature of nanochemistry is the utilization of synthetic chemistry to make nanoscale building blocks of different size and shape, composition and surface structure, charge and functionality. These building blocks may be useful in their own right. Or in a self-assembly construction process, spontaneous, directed by templates or guided by chemically or lithographically defined surface patterns, they may form architectures that perform an intelligent function and portend a particular use.

1.7 Molecular vs. Materials Self-Assembly

The driving forces for molecule organization are quite varied and can be ionic, covalent, hydrogen, non-covalent and metal–ligand bonding interactions, which may result in structures and properties not found in the individual components. This is Jean Marie Lehn's supramolecular chemistry, the science of molecule-based assemblies.

Materials self-assembly chemistry transcends that of molecular assembly. It is distinct solid-state materials chemistry where building blocks and their assemblages are unconstrained by scale and not restricted to just chemical bonding forces. The way to view materials self-assembly over "all" scales, the subject of this text, is in terms of a map of bonding forces that operate between building blocks and over different length scales. The forces responsible for materials self-assembly at length scales beyond the molecular include capillary, colloidal, elastic, electric, magnetic and shear. In a self-organizing system of materials construction units like nanoclusters, nanorods or nanosheets, a particular architecture forms spontaneously with a structural design, which is determined by the size and shape of individual nanocomponents and map of bonding forces between them. The system proceeds towards a state of lower free energy and greater structural stability.

1.8 What is Hierarchical Assembly?

A feature of self-assembly is hierarchy, where primary building blocks associate into more complex secondary structures that are integrated into the next size level in the hierarchy. This organizational scheme continues until the highest level in the hierarchy is reached, as illustrated hypothetically in Figure 1.4. These hierarchical constructions may exhibit unique properties that are not found in the individual components. Hierarchy is a characteristic of many self-assembling biological structures and is beginning to emerge as a hallmark of materials self-assembly that encompasses multiple length scales.

1.9 Directing Self-Assembly

Self-assembly of molecules and materials may be directed by templates. Directed self-assembly of building blocks, which may involve structure-directing additives, often molecular and organic, in addition to the constituent building units, is considered to be distinct from spontaneous self-assembly. The template in this building block auto-construction process can serve to fill space, balance charge and direct the formation of a specific structure. For instance, microporous alumino-silicates called zeolites are templated by single molecules while mesoporous silicas are templated by assemblies of molecules like block copolymers or lyotropic liquid crystals. In this definition, template assembly is synonymous with co-assembly and distinct from self-assembly. Template directed assembly, may also involve the intervention of a lithographically or otherwise patterned substrate planar or curved, where spatially defined hydrophobic–hydrophilic, electrostatic, hydrogen bonding, metal–ligand or acid–base interactions between substrate and building blocks guide the assemblage into a predetermined architecture. A case in point concerns patterns

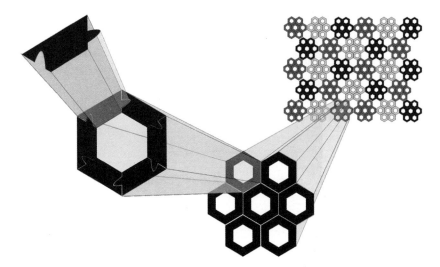

Figure 1.4   *A hypothetical hierarchical system, exhibiting distinct building rules at different length scales*

of hydrophobicity that have been used to organize non-polar polymers into microlens arrays. A lithographically defined relief pattern carved in the surface of a substrate may also be used to direct building block assembly within, an example being the alignment and packing of nanorods in substrate nanowells. The direction of the assembly process may also be driven by the involvement of a porous template that has been patterned at the nanoscale. One such template is a nanochannel membrane made of polymers, silicon and alumina used to replicate nanowires.

1.10 Supramolecular Vision

Jean Marie Lehn (Figure 1.5) had a vision of supramolecular chemistry, which unveiled a world far beyond the molecule.[2,3] It introduced to chemistry how molecular recognition can lead to the self-assembly of supramolecular materials. Lehn's construction kit consisted of complementary organic and inorganic molecules that recognize one another through lock and key types of intermolecular interactions to spontaneously form molecular assemblies. The driving forces for this molecular organization are quite varied and can be ionic, covalent, non-covalent, coordination and hydrogen bonding interactions, which may result in structures and properties not found in the individual components. Therefore, self-assembly as a route to materials has its roots firmly in organic chemistry where the ability to make molecules of almost any shape and functionality lends itself well to designing complementary interactions. However, it was realized relatively quickly that self-assembly is a very general principle, one which relies on reversible interactions and a balance of attractive and repulsive forces. Self-assembly, therefore, encompasses

Figure 1.5   *Jean-Marie Lehn, pioneer of supramolecular chemistry*
(Reproduced with permission from research group website: www.isis.
u-strasbg.fr/supra/)

all scales, with the possibility of a completely rational and predictable route to materials.[4] The self-assembly approaches to materials we attempt to illustrate in this book attest to this fact.

1.11 Genealogy of Self-Assembling Materials

It is often interesting, and always valuable, to trace the genealogy of a new branch of chemistry. One can argue that materials self-assembly had its beginnings with the observation by Pieter Harting that organics are somehow involved in the organization of inorganics in biomineral formation.[5] Harting found that the mixing of calcium ions with carbonate and phosphate under quiescent conditions in organic concoctions, such as albumen, blood, mucus and bile, produced calcareous concretions with natural form. The synthetic morphologies of Harting were published in 1873 and can be considered to predate biomimetic inorganic materials chemistry by more than a century (Figure 1.6). Harting had stumbled into organic-mediated mineralization without much knowledge of organic or inorganic chemistry, crystal nucleation, growth and tectonics.

Harting made his breakthrough at a time when there was a revolt from morphology and a movement to separate organic from inorganic chemistry, just at a point in his work when they most needed to unify. The zeitgeist, or prevailing view, of Harting's time was not conducive to the emergence of biomimetic inorganic chemistry as a field, or to templated materials synthesis for that matter, and it had to await developments in molecular recognition, self-assembly, templating, biochemistry, solid-state chemistry and biomineralization. Harting hoped that his synthetic morphology would lead to a new field just as important as Wöhler's organic chemistry.[6] Harting's aspirations lay dormant until reawakened by the discovery over a century later of inorganic liquid crystals and their mineralization to inorganics. These could be produced with morphologies and surface patterns that resembled self-made shapes and designs found in nature's biominerals.[7]

While Harting was experimenting with biomimetics in the laboratory, Ernst Haeckel was on a field trip with his friends aboard the H.M.S. Challenger during which time he discovered thousands of different kinds of sculpted silica microskeletons mineralized by single cell marine organisms known as the radiolaria.[8] His personal etcher Nitsche documented in meticulous artistic detail a myriad of filigree skeletal patterns produced by these simple organisms. Reproductions of Haeckel's exquisite drawings of radiolarian microskeletons continue to adorn the walls of museums and homes (Figure 1.7). The growth, form and biological function of the tiny lace-like radiolarian skeletons remains both an enigma and a delight to scientists and artists alike. In MacKay's recent translation of *Crystal Souls: Studies of Inorganic Life*, the last book written by Haeckel, it seems that a tenuous connection may have been made between the organic liquid crystals discovered by Reinitzer and Lehmann, and the morphogenesis of the radiolarian microskeletons.[9] Haeckel surmised that there was a relation between the organization of liquid crystals and the way silica was ordered into spectacular patterns by the radiolarian. The forethought of Haeckel in this century old work is

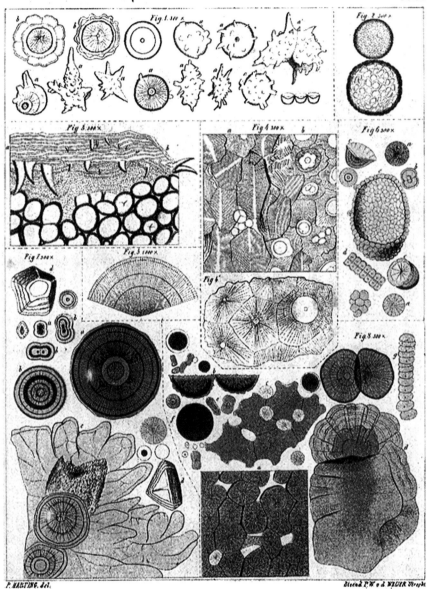

Figure 1.6 *Image of synthetic morphologies produced by Pieter Harting, hand-drawn by himself in 1872*

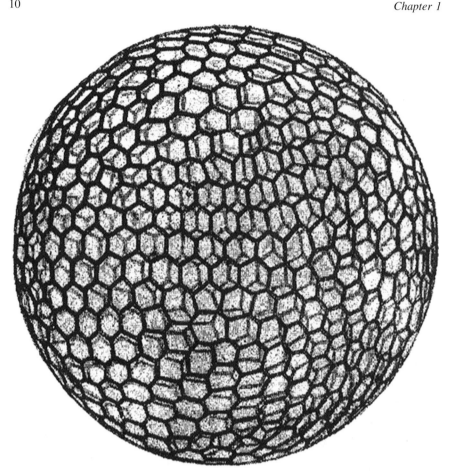

Figure 1.7   *Sketch of a radiolarian discovered by Ernst Haeckel*
(Reproduced with permission from Ref. 11)

remarkable when placed in the context of supramolecular templating, especially
the polymerization of silicate and phosphate liquid crystals to create silica and
aluminophosphate "replica" materials with complex and natural form.[10]

The connection of this early thinking about the importance of interfaces between
organics and inorganics is intriguing when placed in the light of the classic 1917 text
of D'Arcy Thompson, *On Growth and Form* (Figure 1.8).[11] He was the first to apply
physico-geometrical principles to explain the morphogenesis of biominerals. His
view of diatoms and radiolarians was that a contiguous assembly of close-packed
cells, alveoli or bubbles, functioned as a template to direct the deposition of silica.
By localizing the biosilicification process to spaces within the protoplasmic froth, a
mineralized copy of the templating interfaces could be formed. In retrospect, this
can be seen to be a paradigm for using organics to template microporous (small
molecule organics), mesoporous (micelles, liquid crystals, block-copolymers), and
macroporous (microemulsions, vesicles, colloidal crystals) materials.

Figure 1.8 *On growth and form, the groundbreaking text by D.W. Thompson* (Reproduced with permission on from Ref.11)

1.12 Unlocking the Key to Porous Solids

Emil Fischer's 1894 lock-and-key principle, molecules recognizing one another through site-selective binding, spawned fundamental studies of intermolecular forces and cooperative interactions.[12] The complementary hydrogen bonding, electrostatic and hydrophobic interactions between organic molecules underpins recognition events, self-assembly, replication and catalytic processes in biology. Nowhere is this more clear than in Watson–Crick hydrogen bonded base pairs in the DNA double helix, the complementary nature of transfer RNA in protein synthesis, and the exquisite substrate selectivity in enzyme action. In a similar vein, Richard Barrer recognized that complementary interactions involving quaternary alkylammonium cations and silicates/aluminates were able to direct the assembly of crystalline microporous alumino-silicates called zeolites (Figure 1.9).[13] These types of minerals are also naturally occurring, and their name translates to "boiling stones" since their large volume of adsorbed moisture can be released upon heating. Such microporous materials could act as hosts to selectively recognize adsorbed molecules or catalyze the reaction of organic guests based on their size and shape.[14] Barrer's organic molecule templating of zeolites can be considered to have set the stage for the development of template-based synthesis of inorganic materials, which led to the emergence of materials self-assembly chemistry. In particular, Charles Kresge showed that the length scale of templating porous solids can be increased far beyond the small molecule.[15,16] Moreover, Edith Flanigen and Robert Bedard demonstrated that the compositions of porous solids may be extended from traditional aluminosilicate zeolites and silicate molecular sieves to encompass a large portion of the periodic table.[17,18] Throughout much of this work, the underlying theme of organics directing inorganics to create composite materials remains essentially the same.

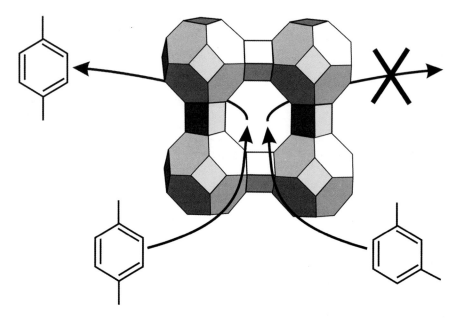

Figure 1.9   *A zeolite's crystalline aluminosilicate framework assembles around an organic template molecule providing pores after its removal. Some molecules, such as linear para-xylene, can permeate through the small pores, while more bulky molecules, such as meta-xylene, are excluded due to their size*

Self-assembly of specifically designed molecular and cluster building blocks often under exceeding gentle conditions, called soft chemistry, has also led to a diversity of open-framework solids, far beyond what is possible with microporous oxides like the zeolites and molecular sieves.[19,20] Open-framework materials (Figure 1.10)[21] now include metal phosphates, cyanides, halides, nitrides, sulfides and selenides, as well as metal interconnected heterocyclic ligands like porphyrins[22] and bipyridyls. The chemical and structural diversity of this class of materials is constantly expanding. Their open structures are aesthetically appealing because of the chemical

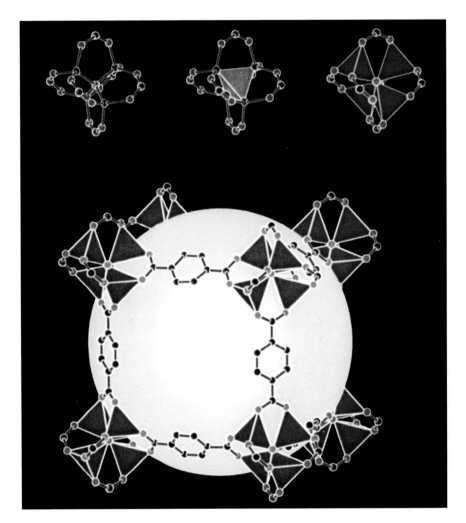

Figure 1.10 *The top of this figure shows a central tetra-zinc core used to build a framework by connecting these clusters through dicarboxylate linkages. The resulting porous solids can have very large accessible surface areas, shown as the large white sphere*
(Reproduced with permission from Ref. 21)

and stereochemical control potentially attainable in their construction. The majority use metal–ligand bonding to link the individual components into crystalline frameworks containing spacious cavities and channels.[23,24] The internal voids can account for greater than 50% of the volume of the crystal. Disordered template, water or solvent molecules may occupy the void spaces. Spaces in one framework may be filled by one or more independent framework. This creates crystalline entangled structures known as self-interpenetrating networks,[25,26] a phenomenon which reduces expected surface areas. The frameworks can be cationic, anionic or neutral, allowing size, shape and chemically selective ion exchange and adsorption. Linking of modules may facilitate guest or temperature induced angular distortions of the framework, which can have profound impacts on various properties such as electronic coupling.[27] As a result, open-framework materials display flexibility and expand in size to accommodate adsorbed guests, thereby making the material interesting for separation, catalysis and sensing applications.[28]

Both oxide and non-oxide porous frameworks offer interesting opportunities for host–guest inclusion chemistry aimed at creating composite materials.[29–31] Guests may be atomic, ionic, molecular, cluster or polymeric. The periodic array of channels and cavities within these porous solids allows assembly of novel kinds of materials, examples being metal and semiconductor clusters, wires and superlattices thereof, buckyminsterfullerene (buckyballs, a carbon allotrope with formula C_{60}) arrays, aligned chromophores, vectorial electron transport chains, and photosynthetic mimics.[32] Nanochemistry of this type is pointing the way to quantum electronic, photonic and data storage materials for advanced devices.

1.13 Learning from Biominerals – Form is Function

Hientz Lowenstam is considered to be the father of biomineralization. He first proposed his theory of organic matrix mediated mineralization in the 1970s.[33–39] A personal account of the thinking behind his discovery involves a fascinating encounter with a chiton (sea urchin) while he was sitting at a water pool in the Grand Bahama Banks. Lowenstam noticed that the chiton created chevron marks in the rocks as it fed by scraping off algae attached to the rock. He reasoned that the radula (teeth on the tongue of the chiton) must be harder than the rock in order to make marks in this hard mineral (Figure 1.11).

Returning to the laboratory, he found that the radula were comprised of magnetite, which proved the mineralization process that created them occurred intracellularly. Although it took a decade for his proposal to be accepted, it is now the basis for all mineralization processes in biological systems. In a classic 1981 *Science* magazine article, Lowenstam documented about 60 minerals that are formed biologically and classified the phyla amongst which they are distributed. Most biominerals are composed of calcium carbonate, phosphate and oxalate, silica and iron oxides and much has been reported on the biorganic and bioinorganic chemistry responsible for nucleation, growth and form of biominerals.

Pioneering biomineralization and biomimetic research from the groups of Steve Wiener and Lia Addadi,[40–44] Robert Williams,[45–47] Stephen Mann,[48–53] Galen

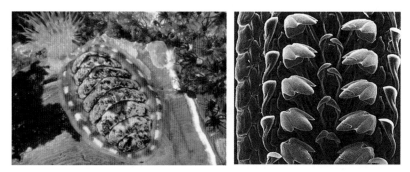

Figure 1.11 *The chiton, left, a primivite marine mollusk. Chiton radula, right, composed of biomineralized iron oxide, are hard enough to scrape algae-covered rocks* (Left image reproduced with permission from Jon Gross photography. Right image reproduced with permission from www.personal.dundee.ac. uk/~amjones/chitintr.htm)

Stucky and Daniel Morse[54–57] have provided a wealth of information on the role of the organic matrix in biological and abiological mineralization processes. Proteins have been extracted from biominerals, and amino-acid sequencing and site-selective mutagenesis have identified nucleation sites on the protein that may be responsible for mineralization.[58,59] The use of many of these extracted proteins in synthetic systems can lead to unusual crystal forms and morphologies resembling those in natural systems,[60–66] giving further indication of their role in the biomineral matrix. The picture emerging from these studies is that complementary interactions between the organic matrix and the mineralizing inorganic, involving matching of charge, stereochemistry and geometry, establish the site of nucleation, growth and generation of mineral shape.[49,67–69] In essence, the organic functions as a blueprint that carries programmed information for the synthesis of materials with natural form and a biological function.

It has been said that the outstanding problem in biology is the form problem.[70,71] In biomineralization and biomimetic chemistry, the form problem is morphogenesis, the origin and control of a shape. Biomineralization centers on the idea that organics control the nucleation, growth and form of inorganics, and it is this process that creates hierarchical composite structures with unique chemical and physical properties. Conversely, biomimetic chemistry aims to exploit the principles of biomineralization in order to synthesize novel materials where form controls function and utility (Figure 1.12).[72,73] As mentioned before, nature is frugal in her choice of inorganic materials, however, the mineralized designs and diversity of naturally occurring forms have only recently been seriously challenged by synthesis. The chemical bonding of biominerals is not unusual, and varies from amorphous covalent networks found in silica forms to multi-crystalline ionic lattices in calcareous shapes. In molecular terms, it is relatively easy to comprehend the early stages of self-organization, molecular recognition, and nucleation that precede the morphogenesis of biomineral form, however, it is not obvious how complex shapes emerge and how, in turn, they can be copied synthetically.

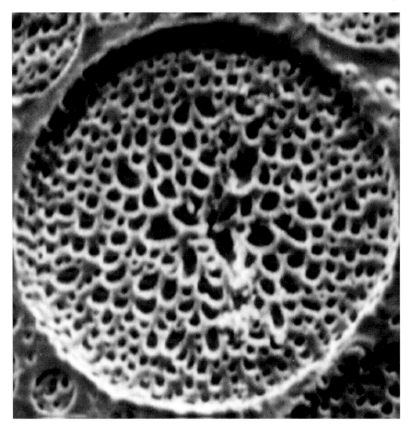

Figure 1.12   *Synthetic morphology resembling biomineralized systems*

Biological materials display hierarchical organization of structure.[74] At each level of this hierarchy, distinctive building rules are apparent. Some of these levels do not demand periodicity in the strict definition of ideal crystals, and their description requires a break from standard crystallography. The traits of biological minerals are manifested in an astonishing array of curved shapes, surface patterns and hierarchical order. These are not the characteristics of conventional crystals with their limited range of polyhedral habits and similar rules of construction from atoms to bulk.[75] Nature has learned how to replace the Platonic concept of arranging atoms in a pre-existing framework of planar symmetry elements, by the idea that local interactions between atoms can introduce long-range order and curvature into materials. It is all a matter of balance: it is energetically favorable for atoms to lay in a classical crystal unless other interactions bring the non-classical shapes to a lower free energy.

1.14 Can You Curve a Crystal?

Bernal showed how a shift from the theory of exact equivalence and long-range order in the perfect crystal, to quasi-equivalence and short-range order in biological

Figure 1.13 *Introducing curvature into crystals allows a break from traditional crystallography. The concept of minimal surfaces has been used quite famously by the architect Antoní Gaudí, who used a model built of hanging strings with weights (upside down on the left) to calculate the exact curvatures needed to support the weight of the massive Sagrada Familia church in Barcelona, right*
(Left image reproduced with permission from http://www.hawbaker.org/1/ public Right image reproduced with permission from wikipedia, the free encyclopedia, http://en.wikipedia.org)

crystals is necessary for curved shapes to emerge.[76] To expand upon the idea, the symmetry elements of the 230 space groups determine the number of ways that identical atoms can be arranged in an infinite array. Here the underlying space pattern of the primitive unit cell dictates the shape of a crystal. To deal with a system whose shape does not conform to this stereotype, it is necessary to abandon the idea that the environment of every lattice site has to be identical for order to emerge, and to consider ways of introducing curvature into crystallography and rounded form into solid-state chemistry. MacKay's "flexicrystallography" is a mathematical construct founded upon minimal surfaces with which to formalize these topological ideas.[77] The concept of minimal surfaces has also been used in architecture, and made famous by Antoní Gaudí's *Sagrada Familia* church in Barcelona (Figure 1.13).

1.15 Patterns, Patterns Everywhere

In 1952, stimulated by the visual imagery of natural shapes, Alan Turing developed a theory of how patterns can appear spontaneously in chemical reaction–diffusion systems far from equilibrium.[78] Turing's contributions proved relevant to the understanding of morphogenesis, and suggested to developmental biologists that

Figure 1.14 *Pattern formation in nature: seashell stripes*
(Reproduced with permission from Daniel Guerin Photography)

these kinds of chemical processes might be responsible for the appearance of complex patterns in nature. The appearance of regular stripes in clouds is an example of non-equilibrium pattern formation, demonstrating that order might also arise from more than well understood, chemical interactions. Since the publication of Turing's influential work there has been intense activity on the mathematical modeling of patterns in a diversity of biological and non-biological systems.[79] Excellent examples can be found in Meinhardt's book on *The Algorithmic Beauty of Sea Shells*, in which the pigment patterns found on mollusk shells are mathematically modeled using differential equations that describe various kinds of activator–inhibitor reaction–diffusion systems (Figure 1.14).[80] Ball's recent text, *The Self-Made Tapestry*,[81] is a scholarly exploration of pattern formation in the natural world, from bubbles, waves, bodies and branches to breakdown, fluids, grains and communities. Ball's obsession with materials and patterns can be considered to represent a 21st century revival of D'Arcy Thompson's classic text *On Growth and Form*.[11]

1.16 Synthetic Creations with Natural Form

As mentioned earlier, about a century ago Haeckel in his book *Crystal Souls* wrote that the properties of organic liquid crystals were somehow connected with radiolarian microskeletal forms. A century later it has been found that the polymerization of silicate and phosphate liquid crystals can produce silica and aluminophosphate morphologies with natural form.[82] By natural form, we imply the visual perception of a class of objects that share particular features typically associated with biological shapes. Features such as curvature, hierarchy and complexity are not exclusive to biological shapes, but the fact that they are observed in so many biological systems makes the synthetic systems interesting, potentially from a fundamental point of view. However, form is not sufficient. Just as an understanding of the evolution of organic soft-matter in living systems depended on a molecular level understanding of DNA and its interaction with proteins, the biomineral problem demanded at least the same level of comprehension of the involvement of inorganic ions and solids in biological systems.

Kresge's discovery of supramolecular templating of mesoporous silica materials and the authors' work on synthetic shape and patterns based upon the mineralization of inorganic liquid crystals has stimulated interest in morphosynthesis in materials chemistry.[51,67,83–85] Inorganic liquid crystals, especially the unique kind of self-organization and mineralization they exhibit, provide a new tool for enhancing the understanding of the form problem. While morphogenesis was originally motivated by Darwinian theories of evolution, morphosynthesis is driven by the evolution of a materials society. The control of morphology through biomimetic inorganic chemistry is expected to play a central role in the development of new materials, products and processes, where form controls function and utility. Motivation derives from the notion that the control of morphology is the mainstay of materials science, particularly for tuning mechanical, thermal acoustical, magnetic, electronic, optical, and catalytic properties of materials. By evaluating the multiplicity of forms we can self-assemble synthetically, we can catalog an array of parts which could be used in the designs of future nano-devices. A synthetic seashell is shown in Figure 1.15, and is one of a myriad shapes achievable through chemical means.

Kresge's synthesis of mesostructured inorganics (2–10 nm length scale) involves the polymerization of an inorganic liquid crystal formed from a templating mesophase, creating its inorganic replica. This represents a new approach to materials synthesis, a paradigm shift in materials chemistry, and ushers in an exciting new era of materials research. The synthetic methodology has been generalized to

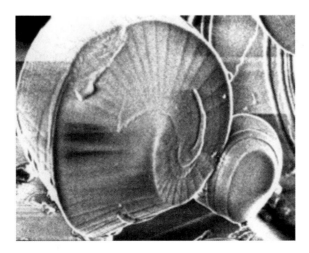

Figure 1.15 *Synthetic seashell from a liquid crystal templated aluminosilicate*

include diverse kinds of supramolecular[86] (surfactant) and macromolecular[87] (block-copolymer) templates.[88] In addition, purely inorganic liquid crystals such as niobate nanotubules can also be used as templates.[89,90] This approach has provided silica and organosilica materials with structural features in the range of 2–35 nm, modular mesostructures that assemble from cluster building blocks, compositions that pervade a good part of the periodic table, electrical properties that can be insulating, semiconducting or metallic, and morphologies and surface patterns that may span three spatial dimensions.

1.17 Two-Dimensional Assemblies

If we pass from materials self-assembly in three dimensions to that in two dimensions, we find there is also a natural progression of concepts and methods behind its development, using layer-by-layer deposition of mono- and multilayer films. This research direction can be traced through seminal work of Langmuir and Blodgett,[91] Sagiv, Nuzzo, Whitesides, Decher, Mallouk and Marks on self-assembly of controlled architecture films at air, liquid and solid interfaces. Irving Langmuir, while working at General Electric Schenectady, developed a generalized theory of adsorption for solid and liquid surfaces based on the idea of satisfying unsaturated surface valency forces.[92] Following up on the lead provided by Lord Rayleigh's theory of wave motion[93] and Agnes Pockels' "kitchen sink" surface tension measurements of oil on water,[94] he developed the Langmuir film balance to measure the properties of monolayer surface films at the boundary between air and water. He confirmed the films were one molecule thick, recorded pressure–area isotherms, and obtained the sizes and shapes of organic molecules even before the advent of X-ray diffraction. Katherine Blodgett in 1935 working with Langmuir at GE developed monolayer film transfer to different supports and built multilayer assemblies. The Langmuir-Blodgett (LB) technique lay dormant until after the Second World War, at which point a surge of interest developed as the commercial potential of all types of thin films in electronics and optics became apparent. This coincided with the advent of transistors, integrated circuits, the computer-communication revolution, and the thrust towards nanotechnology and ultimately molecular electronics.[95] One of those chiefly responsible for arousing worldwide interest in the uses of Langmuir–Blodgett assemblies was Hans Kuhn, starting in the late 1960s at the University of Marburg. He confirmed several important aspects of these films such as their stability, defect levels, and ability to make them with molecular scale patterning.[96]

Repetitive transfer of LB surfactant monolayers provided a rational approach to known thickness of LB multilayer films. Through judicious selection of the structure of the surfactant and due cognizance of interactions between surfactants, it proved possible to create a wide variety of LB multilayer film architectures with structure and composition designed for a particular function. A surfactant is an amphiphile, bearing two chemically distinct termini, and understanding the chemical composition of the substrate permits one to know both how the surfactant will orient with respect to it, as well as which end the surfactant will expose to further bind with a properly chosen counterpart. Two-dimensional supramolecular chemistry of this genre has provided access to ferromagnetic,[97] ferroelectric,[98] semiconducting,[99] dielectric,[100] non-linear optical,[101,102] electroactive[103] and photosensitive[104] multi-layers, in which the spacing between active centers in the film can be controlled with angstrom precision. Such control over the separation of chemical species was of pivotal importance for understanding certain details of the surface enhanced Raman effect[105] as well as aspects of energy and electron transfer.[106,107]

The thermal fragility inherent in organic-based LB films could be circumvented by directly anchoring organic moieties to a substrate. Pioneering experiments by Sagiv[108] demonstrated that alkoxysilanes bearing long-alkane chains were able to anchor to surface hydroxyls and spontaneously self-assemble into close-packed well-ordered films. This methodology provided a rational approach to the synthesis of mono and multilayer functional organic surfaces. Marks and others expanded upon this lead to create designer organic surfaces, noteworthy being aligned chromophoric assemblies for second harmonic generation.[109] Mallouk devised a synthetic approach to inorganic LB-like films[110] (Figure 1.16). The method involved initial attachment of α,ω-alkanehydroxyphosphonate to glass or an α,ω-alkanethiophosphonate to gold surfaces. This was followed by the sequential assembly of Zr^{4+} and α,ω-bis-alkanephosphonates to give well-ordered zirconium organophosphonate multilayer films.

Self-assembled LB films of this type display higher thermal and mechanical stability than their purely organic LB counterparts. The methodology allows precise control of film thickness and porosity, substituents on the alkane, nature of the metal ion connector and distance between active sites. Numerous device concepts based on inorganic LB films have been brought to practice, including chiral separation, second harmonic generation, vectorial electron transport and chemical sensing.

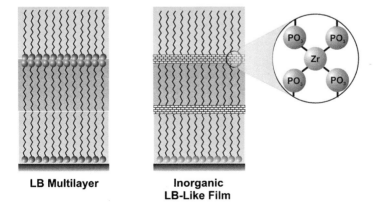

LB Multilayer   **Inorganic LB-Like Film**

Figure 1.16 *Langmuir–Blodgett films are formed through the repetitive transfer of amphiphile monolayers, left. Inorganic LB-like films can be synthesized from diphosphonates and metal ions, right*

Decher developed a simple and elegant yet extremely powerful approach to the formation of controlled architecture multilayer polymer films based upon the layer-by-layer self-assembly of oppositely charged polyelectrolytes to create an electrostatically bound superlattice[111–114] (Figure 1.17). The method begins with the anchoring of a primer layer such as 3-aminopropyltrimethoxysilane to hydroxyl groups on the surface of glass or silicon. In aqueous solution, the amine groups are protonated and the primer layer develops an overall positive charge. This facilitates the adsorption from solution of a monolayer of a polyanion, such as poly-(styrenesulfonate). The process of polyanion adsorption on the primer layer, and as a consequence the formation of an electrical double layer due to dissociation into the aqueous phase of some fraction of countercations, serves to create a sufficient excess of surface negative charge[115] to allow the subsequent adsorption of a monolayer of a polycation such as poly(allylaminehydrochloride). This method of depositing single layers of polyelectrolytes of alternating charge can be repeated several times to produce multilayer polymer films that exhibit an impressive degree of regularity in the interlayer repeat distance.[116] Decher's technique allows the construction of functional polyelectrolyte electrostatic superlattices one layer at a time in which the individual polymer layers can be electrically insulating, semiconducting or metallic and may contain photo-, redox or optically active components. In fact, further investigations have shown that this method is even more versatile than previously thought. The materials to be layered are not limited to polymer electrolytes, but effectively to any species bearing several charged groups such as colloids of various types,[117] viruses[118] and large proteins.[119] The versatility of the method for building designer multilayers is proving to be of considerable scientific and technological interest.[120]

In a creative extension of this research, Mallouk[121–123] and others[124] applied Decher's technique, layer-by-layer deposition of polyelectrolyte films of alternating charge, to the self-assembly of electrostatic superlattices comprised of exfoliated charged lamellae of layered metal phosphonates, oxides or sulfides in combination with for example, polymers, clusters, coordination compounds and bioinorganics.

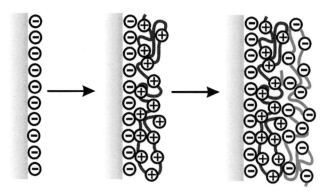

Figure 1.17 *Iterative formation of layer-by-layer electrostatic multilayers by alternately dipping in solutions of polycations and polyanions*

This bricks-and-mortar, materials self-assembly chemistry provides a simple and elegant way to assemble inorganic and organic components into controlled thickness multilayer composite films for a range of perceived device applications. For instance, by assembling monodispersed gold nanoclusters between thin oxide layers, single electron transistor (SET) and coulomb blockade behavior have been demonstrated.[125] The method may also provide access to high dielectric constant oxide thin films for the gate electrode in next generation, smaller and faster field effect transistors. Another intriguing application of designer electrostatic super-lattices involves control over the abrasion resistance of surfaces. This has been achieved by coating a surface with an ultra-hard composite nanolaminate that is constructed of alternating monolayers of colloidal zirconium oxide and poly-allylaminehydrochloride.[126] These tough organic–inorganic composite films are assembled in a slow "serial" process, one layer at a time, but with the ultimate control in composition.

1.18 SAMs and Soft Lithography

Nuzzo noticed how long chain alkanethiolates could self-assemble on gold surfaces to give well-ordered monolayer films, self-assembled monolayers, now affectionately known as SAMs.[127,128] Whitesides effectively expanded upon the SAM methodology to develop soft lithography.[129] This technique makes use of a patterned elastomer made of polydimethylsiloxane (PDMS), as a mask, stamp or mold. With this simple process, high quality patterns and structures can be created with lateral dimensions from about 5 nm to 500 mm, in two- and three-dimensions, that cannot easily be achieved by photolithography. Figure 1.18 shows square drops of colored water held in this shape by hydrophobic lines surrounding them.[130] Whitesides' strategy for patterning and shaping materials over different length scales and dimensions embraces chemical concepts of self-assembly, templating and crystal engineering, together with soft lithographic techniques of microcontact printing and micromolding. It is an alternative non-photolithographic microfabrication method to complement photolithography, where feature size is limited by optical diffraction. As well, the short wavelength radiation required for photolithographically defining smaller and smaller designs necessitates prohibitively expensive and complex facilities and technologies.

While soft lithography circumvents in theory the diffraction limits of 2D photolithography, in practice the PDMS structures need to be templated upon some fabricated pattern. Usually the pattern is cast from a photolithographically patterned master, thereby imposing onto soft lithography the same size constraints as photolithography, but smaller features can be accessed by fabricating a master through e-beam, UV or X-ray lithography. Even though very fine lithography may be expensive, the master can be repeatedly used and the cost per PDMS replica becomes quite small. Soft lithography provides access to topologically complex 3D objects for building micromachines and electrical circuits, accommodates a wide range of materials, and enables a rational surface chemistry. Soft lithography is easily accessible to chemists interested in experimenting with reliable, convenient

Figure 1.18   *Square drops of water through soft lithography*
                  (Reproduced with permission from Ref. 130)

and inexpensive methods for patterning planar or curved surfaces with organic, inorganic, polymer, liquid crystal, ceramic or biological structures. Whitesides' research illustrates the potential of soft lithographic procedures in a wide range of areas, from microelectronics to optics, microanalysis to sensors, microelectro-mechanical systems to cell biology. It is abundantly clear that the utility of soft lithography in diverse areas of chemistry, for rapid and low cost patterning and shaping of materials, will definitely increase in the near future. It is an indispensable component of the new materials self-assembly chemistry for the next millennium.

1.19 Clever Clusters

The self-assembly of organic monolayer films on planar substrates is analogous to their assembly on a monodisperse nanocluster, the curved cousin of the plane.

In these systems nanocluster nucleation, growth and stabilization is achieved through capping with organic ligands, surfactants and polymers. The idea of capping clusters with organics can be traced to synthetic work of Steigerwald and Brus that concerned the formation of narrow size distribution metal chalcogenide clusters in reverse micelles, and their stabilization by capping with surface thiolates.[131] Also Bentzon showed that the thermal decarbonylation of iron pentacarbonyl in the presence of oleic acid led to surfactant-capped iron oxide nanoclusters,[132] that had such a narrow size distribution that they spontaneously assembled into 2D and 3D capped cluster superlattices. The magnetic properties of periodic arrays of superparamagnetic and ferromagnetic, capped nanocluster films are expected to be most interesting especially in the context of high density, magnetic data storage media. In principle, every cluster in a perfect array could be individually addressed, leading to a storage density approaching the theoretical limits of magnetism.[133]

These breakthroughs set the stage for an elegant series of experiments by Bawendi and Alivisatos on the nucleation and growth of monodispersed zinc and cadmium chalcogenide nanoclusters, capped with a phosphine oxide surfactant.[134] Extraordinary control over cluster size and width of the distribution, often better than a single atomic layer, was feasible by controlling the cluster nucleation and growth using the phosphineoxide as a dual-purpose solvent and capping agent (Figure 1.19). It was also found that polydisperse samples could be size selected to an extremely narrow distribution by several cycles of precipitation from a good solvent by addition of a non-solvent. Cluster films supported on SAMs[135] and cluster supercrystals[136] could be obtained using this synthetic technique. Access to size

Figure 1.19 *Solutions of CdSe nanoparticles of different sizes under UV light, showing size-dependent emission properties*
(Courtesy of Prof. G.D. Scholes)

tunable and monodispersed capped semiconductor clusters opened up a myriad of exciting opportunities in cluster science. These include establishment of scaling laws for cluster physical properties,[137,138] such as melting points,[139] phase transitions,[140] phonon and electronic energies,[141] and magnetic relaxation times. It also enabled the development of electroluminescent cluster-based devices,[142,143] SETs,[144,145] and the study of single clusters using near field optical[146] and laser scanning confocal spectroscopy.[147,148] Significantly, a biological rather than an optoelectronic application may emerge from this kind of research. It centers on the selective fluorescent labeling of biological materials with brightly emissive and appropriately end-functionalized capped II–VI semiconductor clusters of different size.[149] Specific sites in biological specimens have been tagged with this new class of "inorganic dyes." They have the benefit of narrower emission, higher luminescent yield and greater photochemical stability than currently used organic dye labels.

Gold colloids are not new. Nearly two centuries ago, Michael Faraday was generating deeply colored gold colloids using gold tetrachloroaurate and white phosphorus.[150] The use of citric acid as a reductant was successfully applied in the 1950s,[151,152] and is still used by researchers today. However, Whyman's phase-transfer reductive synthesis[153] and Whetten's aerosol cluster beam synthesis[154] of alkanethiolate capped gold nanoclusters opened the way to a rational surface chemistry of gold nanocrystals with the study of 3D versions of SAMs on metal surfaces. This work spawned refined synthetic methods for improving the dispersity of the system to the point that alkanethiolate gold clusters are sufficiently single sized to allow them to be crystallized into well-ordered ensembles, such as cluster wires, rings, mono and multilayer films and crystals,[155] and inverse colloidal crystals using a variety of templating and patterning techniques.[156] Numerous studies have detailed the size-scaling physical properties[157] and behavior of these gold "designer molecules."[158] Some interesting properties include preferred (magic) cluster sizes and shapes, faceting of the clusters, interdigitated-bunching of the chains of the capping alkanethiolates, defects and order–disorder transitions of capped cluster supercrystals, QSEs on optical and electrochemical properties, charge-transport through a single cluster, and SET behavior.[159]

1.20 Extending the Prospects of Nanowires

Lieber, Yang and Xia, amongst others, have made great strides with the synthesis of semiconductor and metal nanowires using vapor and solution phase preparative strategies.[160] A wide range of nanowire compositions with controlled length, diameter, doping and surface structure, charge and functionality, have been reported. These nanowires are often found to grow as oriented single crystals. Diameter dependent QSEs on the electronic band gap have been observed for single crystal silicon nanowires. Interesting new chemistry and physics is beginning to emerge on prototype semiconductor nanowire devices based on single and crossed nanowire architectures. Demonstration nanowire devices like diodes, transistors, light emitting diodes, logic circuits, lasers and sensors, have been described.

Figure 1.20   *Ultra-small pitch wire arrays made by superlattice templating*
(Reproduced with permission from Ref. 161)

A grand challenge with nanowire building blocks is self-assembly into functional architectures that can form the basis for the development of nanoscale electronic, optoelectronic and photonic circuitry with exceedingly high densities of nanowires. These densities can only be achieved if the pitch between nanowires is also nanoscale, a feat that has recently been achieved by Heath, Lieber, Rogers, and Williams using bottom-up directed self-assembly and top-down microfabrication techniques investigated in Chapter 5. A crossed nanowire array made by Heath *et al.* using a creative directed self-assembly technique is shown in Figure 1.20.[161]

1.21 Coercing Colloids

Self-assembled materials have moved well beyond the 10 nm length scale with the advent of synthetic microspheres having extraordinarily narrow size distributions (less than about 2–3% polydispersity).[162] The ability to synthesize monodispersed latex and silica microspheres by heterogeneous emulsion polymerization[163] and sol–gel chemistry,[164] respectively, with diameters in the 50 nm–3 μm range provides a new class of materials with a range of exciting applications in chemistry and physics (Figure 1.21).[165] They can be crystallized into structurally well-ordered colloidal crystals, films and patterns, which can serve as templates to make inverted replica colloidal crystals with a wide range of compositions. One of the most appealing properties is that these periodic dielectric lattices are 3D photonic crystals, the optical analog of the electronic semiconductor.[166] This realization has spawned tremendous interest in them as miniature optical devices for all-optical circuits in telecommunications. Many other exciting opportunities are emerging for colloidal and inverted colloidal crystals, which take advantage of their ordered porosity and the "structural colors" that emerge from the interference of light with a microstructure, rather than absorption of light by a chromophore. Some of these opportunities include solar cells which amplify light absorption, batteries

Figure 1.21 *"Inverse opal" photonic crystal made of titania, templated by an array of*
polystyrene spheres
(Reproduced with permission from Ref. 165)

which change color as they discharge, chemical sensors with a multiple color-based readout, and paper-thin full color video displays.

Preparing colloidal crystals and films can be achieved using Stokes sedimentation,[167] electrophoresis[168] or evaporation induced self-assembly.[169] Microsphere epitaxy with a lithographically micropatterned substrate has been used to direct the crystallization of close-packed as well as non-close-packed colloidal crystal films.[169] Crystallization of microspheres as monolayer films has been exploited for lithographic patterning of surfaces[170] and for projection lithography,[171] while micromolds[172] and surface relief patterns[173] enable microsphere assembly and control over the size, morphology and orientation of colloidal crystals.

SAM and micromolding in capillary (MIMIC) soft lithographic patterning techniques have been combined with surfactant-based supramolecular templating to synthesize periodic mesoporous silica materials with micron scale designs.[175] Materials of this type exhibit hierarchical order with structural organization over nanometer and micrometer length scales.

Hierarchical synthesis of this genre has been elevated to an even higher level of structural complexity by utilizing a creative combination of micromolding together with colloidal crystal and triblock-copolymer templating (Figure 1.22).[176,177] Microchannels are used to template microspheres, the voids between them in turn template an ordered microphase separated triblock-copolymer,[178] and this polymer finally templates the formation of an ordered oxide. This strategy provides access to inorganic materials with micron scale designs of periodic macroporous crystals having walls composed of ordered mesoporous oxides such as silica, titania and niobia with nanometer scale pores. Here we have an inorganic sol–gel-block-copolymer mesophase (10 nm length scale)[179] that is infiltrated into the interstitial spaces of an organized array of latex spheres (100 nm length scale) in a decorated mold (1000 nm length scale) which yields after polymerization a composite organic–inorganic material with hierarchical construction and ordering over three distinct length scales.

Soft lithographic patterning and colloidal crystal templating over multiple length scales is producing structures with complex texturing of a type that have never been seen before in materials chemistry. Yet we cannot think too highly of ourselves, for hierarchy of this genre is the hallmark of Nature's biomineral constructions. This can be seen in the delicate ornate microskeletons of the single-cell marine organisms known as the diatoms. Nature in this case also makes use of organic molds and templates to shape and decorate silica over many length scales in a single body. In a microscopy image of the diatom *Cymbela Mexicana,* the hierarchical assembly is clearly visible (Figure 1.23). Silica frustules are sculpted in the 1000 nm size range, and these are permeated with 100 nm dimension ordered macropores and 10 nm scale periodic mesopores. Seeing these beautiful constructions is quite humbling, for it shows us our crude, albeit impressive, imitations for what they truly are. At the same time, they do show us the prospects of what we may achieve one day, and while this is not always heartening it invariably presses us to even greater scientific heights.[180]

Figure 1.22 *Hierarchically ordered oxide through a combination of microchannel, microsphere, and block-copolymer templating*
(Reproduced with permission from Ref. 176)

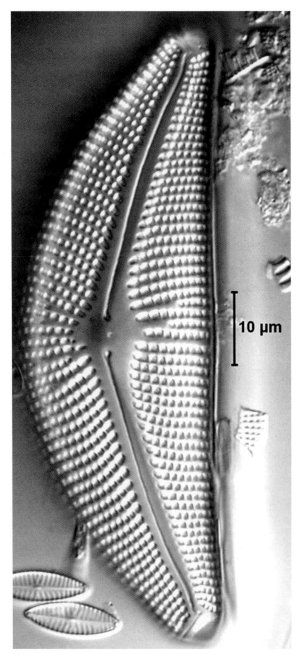

Figure 1.23 *Microstructure of siliceous diatom,* Cymbela Mexicana, *observed by microscopy*
(Reproduced with permission from www.diatom.acnatsci.org/AlgaeImage)

1.22 Mesoscale Self-Assembly

Continuing the theme of length scales it is interesting to look at a recent and fascinating example of 3D materials self-assembly of millimeter scale geodesic objects and examine why this type of research is scientifically very important and where it is likely to find utility in the future. Currently, microfabrication based on photolithography is inherently a 2D technology. New architectures envisaged for future microelectronic, optoelectronic, photonic and microelectro-mechanical devices such as photonic band-gap crystals, biomimetic structures and neural-type computers, necessitate 3D microfabrication schemes. A primitive yet pivotal 3D self-assembly demonstration experiment has amalgamated a creative combination of three elements, namely: (a) spherical templating by a drop of liquid dispersed in a non-miscible liquid (b) capillary-based self-assembly at the interface of the spherical droplet and surrounding liquid of hexagonally shaped gold platelets with a central hole having surfaces that are made selectively hydrophobic and hydrophilic by coating with alkanethiols or silica, respectively, and finally (c) welding together of the gold hexagons into a mechanically robust "micro-geodesic" dome by microelectrodeposition of silver (Figure 1.24). While the open spherical lattice generated by this strategy is comprised of structural objects with no particular function, the demonstration provides compelling proof of concept for

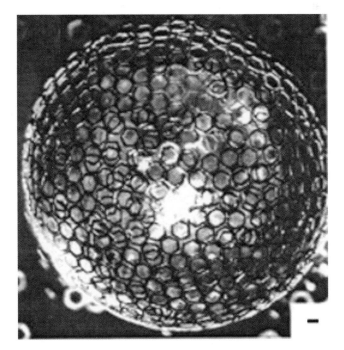

Figure 1.24 *Micron-sized self-assembled geodesic dome assembled onto a drop of water in heptane, the scale bar showing 1 μm*
(Reproduced with permission from Ref. 181)

the feasibility of constructing topologically complex 3D microstructures through self-assembly.[181] The future success of this approach necessitates the use of individual or collections of functional objects that spontaneously organize and make connections with themselves and the outside world to generate fully integrated and practical electronic, optical, chemical or mechanical systems. Exciting demonstrations from Whitesides' group, which show convincingly that this can be done, have appeared in droves in the literature and will be explored further in later chapters.

Clearly, the toolbox of materials self-assembly chemistry has been greatly expanded since the work of Harting, Haeckel and Thompson. The newly acquired ability to synthesize hierarchical materials built of individual parts with tunable compositions and dimensions with designed surface structure, charge and functionality suggests that it should be possible in the near future to self-assemble components and interconnects of fully integrated and functional electronic, optical, mechanical, fluidic and chemical systems in a parallel procedure and over all length scales.

1.23 Materials Self-Assembly of Integrated Systems

Simple, elegant and robust attributes of self-assembly are now being combined with powerful methods of inorganic and solid-state chemistry to create materials with unprecedented structures, compositions and morphologies. This is facilitating purely synthetic approaches to hierarchical systems with components integrated over multiple length scales. By combining self-assembly and microfabrication, it has become feasible to organize and connect organic, inorganic and polymeric chemical components with well-defined functions into integrated electronic, photonic, mechanical, analytical and chemical systems for a future nanotechnology.[182] Motivation is being driven by the notion that complex systems in biology, physics, engineering and chemistry are generally based on hierarchical building principles, that is, smaller units are assembled into larger ones, which in turn are organized at a higher dimension. This construction process is continued until the highest level of structural complexity in the hierarchy has been attained.

The hallmark of an integrated system is the assembly of components into an architecture that performs a certain functional task. In the cell of a higher green plant the photosynthetic chloroplast (a hierarchical solar cell) and the ATP-producing mitochondrion (a hierarchical energy converter) are perhaps two of the most impressive examples of integrated chemical systems. Within the body of a computer, atoms are assembled into insulators, semiconductors, metals, and dopants. These form gates, active layers, sources, drains, junctions, metal leads and contacts. These in turn are the building blocks that constitute the transistors, diodes and capacitors of the integrated circuits, millions of which comprise circuit boards and sub-assemblies of an integrated microelectronic system. Familiar integrated chemical systems include heterogeneous catalysts, photoelectrochemical cells, solid-state lithium batteries, hydrogen–oxygen fuel cells, instant color photographic film, sensors and chromatographic stationary phases.

The ability to self-assemble organic, inorganic and polymeric materials over all length scales and spatial dimensions has taken synthetic chemistry to a level of

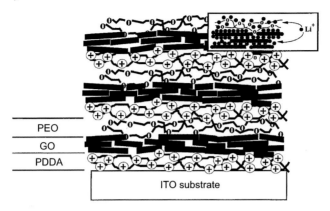

Figure 1.25 *Layer-by-layer self-assembled lithium battery*
(Reproduced with permission from Ref. 187)

structural complexity that approaches biology and engineering physics. Synthetic strategies can now be devised for organizing, patterning and connecting functional chemical components into architectures that were not previously possible. Demonstrated examples include layer-by-layer assembly of a thin film Zener diode from conducting polymers and semiconductor nanoclusters,[183] Metal-Insulator-Metal Nanocluster-Insulator-Metal (MINIM) SET,[184] a multicolor pixel voltage-controllable semiconductor nanocluster-luminescent polymer light emitting diode,[185] an all-plastic field effect transistor driven light emitting diode,[186] and a high density rechargeable graphite oxide nanoplatelet-polyelectrolyte lithium ion battery[187] (Figure 1.25). These few early examples demonstrate the power and versatility of a materials self-assembly approach to integrated systems.

In the chapters that follow, the materials self-assembly theme will be expanded upon by introducing various classes of nanoscale building blocks, and referring to pioneering case studies in the literature where building block chemistry over multiple length scales has led to new and interesting architectures. These novel architectures, often unachievable by any other means, have novel and exciting structure–property–function relations, which may lead to utility and a niche in the materials world of this new millennium.

1.24 References

1. A.P. Alivisatos, Semiconductor clusters, nanocrystals, and quantum dots, *Science*, 1996, **271**, 933.
2. J.M. Lehn, *Supramolecular Chemistry: Concepts and Perspectives*, VCH, Weinheim, Germany, 1995.
3. J.M. Lehn, Supramolecular chemistry, *Science*, 1993, **260**, 1762.
4. A. Stein, S.W. Keller and T.E. Mallouk, Turning down the heat: design and mechanism in solid-state synthesis, *Science*, 1993, **259**, 1558.
5. P. Harting, *Natuurkd. Verh. Koninkl. Acad.*, 1873, **13**, 1.
6. F. Wöhler, On the artificial production of urea, *Ann. Phys. Chem.*, 1828, **12**, 253.

7. S. Mann and G.A. Ozin, Synthesis of inorganic materials with complex form, *Nature,* 1996, **382**, 313.

8. E. Haeckel, *Art Forms in Nature*, Prestel-Verlag, Munich, 1998.

9. E. Haeckel, translated by A.L. Mackay, Crystal souls, *Forma*, 1999, **14**, 1–146.

10. S. Oliver, A. Kuperman, N. Coombs, A. Lough and G.A. Ozin, Lamellar aluminophosphates with surface patterns that mimic diatom and radiolarian microskeletons, *Nature*, 1995, **378**, 47.

11. D.W. Thompson, *On Growth and Form*, 1992, complete revised edition; Dover Publications, New York, p. 708.

12. These thoughts are outlined in Emil Fischer's 1902 Nobel speech: www.nobel.se/chemistry/laureates/1902/fischer-lecture.pdf

13. R.M. Barrer, Zeolite synthesis: an overview, *NATO ASI Ser., Ser. C: Math. Phys. Sci.*, 1988, **231** (*Surf. Organomet. Chem.: Mol. Approaches Surf. Catal.*), 221.

14. R.M. Barrer, Zeolite inclusion complexes, *Inclusion Compd.*, 1984, **1**, 191.

15. C.T. Kresge, M.E. Leonowicz, W.J. Roth, J.C. Vartuli and J.S. Beck, Ordered mesoporous molecular sieves synthesized by a liquid-crystal template mechanism, *Nature,* 1992, **359**, 710.

16. J.S. Beck, J.C. Vartuli, W.J. Roth, M.E. Leonowicz, C.T. Kresge, K.D. Schmitt, C.T.-W. Chu, D.H. Olson, E.W. Sheppard, S.B. McCullen, J.B. Higgins and J.L. Schlenker, A new family of mesoporous molecular sieves prepared with liquid crystal templates, *J. Am. Chem. Soc.*, 1992, **114**, 10834.

17. S.T. Wilson, B.M. Lok, C.A. Messina, T.R. Cannan and E.M. Flanigen, Aluminophosphate molecular sieves: a new class of microporous crystalline inorganic solids, *J. Am. Chem. Soc.*, 1982, **104**, 1146.

18. E.M. Flanigen, Zeolites and molecular sieves: a historical perspective, *Studies in Surface Science and Catalysis*, 2001, vol. 137 (*Introduction to Zeolite Science and Practice*, 2nd edn), p. 11.

19. A.K. Cheetham, G. Ferey and T. Loiseau, Open-framework inorganic materials, *Angew. Chem. Int. Ed.*, 1999, **38**, 3268.

20. O.M. Yaghi, M. O'Keeffe, N.W. Ockwig, H.K. Chae, M. Eddaoudi and J. Kim, Reticular synthesis and the design of new materials, *Nature*, 2003, **423**, 705.

21. H. Li, M. Eddaoudi, M. O'Keeffe and O.M. Yaghi, Design and synthesis of an exceptionally stable and highly porous metal-organic framework, *Nature*, 1999, **402**, 276.

22. M.E. Kosal, J.H. Chou, S.R. Wilson and K.S. Suslick, A functional zeolite analogue assembled from metalloporphyrins, *Nat. Mater.*, 2002, **1**, 118.

23. N.L. Rosi, M. Eddaoudi, J. Kim, M. O'Keeffe and O.M. Yaghi, Advances in the chemistry of metal-organic frameworks, *Cryst. Eng. Commun.*, 2002, **4**, 401.

24. O.M. Yaghi, H.L. Li, C. Davis, D. Richardson and T.L. Groy, Synthetic strategies, structure patterns, and emerging properties in the chemistry of modular porous solids, *Acc. Chem. Res.*, 1998, **31**, 474.

25. S.R. Batten, Coordination polymers, *Curr. Opin. Solid State Mater. Sci.*, 2001, **5**, 107.

26. M. Fujita, Y.J. Kwon, O. Sasaki, K. Yamaguchi and K. Ogura, Interpenetrating molecular ladders and bricks, *J. Am. Chem. Soc.*, 1995, **117**, 7287.

27. G.J. Halder, C.J. Kepert, B. Moubaraki, K.S. Murray and J.D. Cashion, Guest-dependent spin crossover in a nanoporous molecular framework material, *Science*, 2002, **298**, 1762.

28. P.H. Dinolfo and J.T. Hupp, Supramolecular coordination chemistry and functional microporous molecular materials, *Chem. Mater.*, 2001, **13**, 3113.

29. O.M. Yaghi, G.M. Li and H.L. Li, Selective binding and removal of guests in a microporous metal-organic framework, *Nature*, 1995, **378**, 703.

30. A. Stein, Advances in microporous and mesoporous solids – highlights of recent progress, *Adv. Mater.*, 2003, **15**, 763.

31. F. Schuth and W. Schmidt, Microporous and mesoporous materials, *Adv. Mater.*, 2002, **14**, 629.

32. M.W. Anderson, J.M. Shi, D.A. Leigh, A.E. Moody, F.A. Wade, B. Hamilton and S.W. Carr, The confinement of buckminsterfullerene in one-dimensional channels, *Chem. Commun.*, 1993, (6), 533.

33. H.A. Lowenstam, Goethite in radular teeth of recent marine gastropods, *Science*, 1962, **137**, 279.

34. K.M. Towe, H.A. Lowenstam and M.H. Nesson, Invertebrate ferritin: occurrence in Mollusca, *Science*, 1963, **142**, 63.

35. H.A. Lowenstam, Opal precipitation by marine gastropods (Mollusca), *Science*, 1971, **171**, 487.

36. H.A. Lowenstam, Lepidocrocite, an apatite mineral, and magnetite in teeth of chitons (Polyplacophora), *Science*, 1967, **156**, 1373.

37. H.A. Lowenstam and D.P. Abbott, Vaterite, mineralization product of the hard tissues of a marine organism (Ascidiacea), *Science*, 1975, **188**, 363.

38. H.A. Lowenstam, Minerals formed by organisms, *Science*, 1981, **211**, 1126.

39. H.A. Lowenstam and S. Weiner, On biomineralization, Oxford University Press, Oxford, UK, 1989.

40. L. Addadi and S. Weiner, Interactions between acidic proteins and crystals: stereochemical requirements in biomineralization, *Proc. Natl Acad. Sci. USA*, 1985, **82**, 4110.

41. L. Addadi and S. Weiner, Control and design principles in biomineralization, *Angew. Chem. Int. Ed. Engl.*, 1992, **31**, 153.

42. G. Failini, S. Albeck, S. Weiner and L. Addadi, Control of aragonite or calcite polymorphism by mollusk shell macromolecules, *Science*, 1996, **271**, 67.

43. L. Addadi and S. Weiner, Biomineralization: crystals, asymmetry and life, *Nature*, 2001, **411**, 753, 755.

44. L. Addadi, E. Beniash and S. Weiner, Assembly and mineralization processes in biomineralization: strategies for forming biological composite materials, *Supramol. Org. Mater. Des.*, 2002, 1.

45. R.J.P. Williams, An introduction to biominerals and the role of organic molecules in their formation, *Philos. Trans. R. Soc. London Ser. B: Biol. Sci.*, 1984, **304**, 411.

46. S. Mann, J.P. Hannington and R.J.P. Williams, Phospholipid vesicles as a model system for biomineralization, *Nature*, 1986, **324**, 565.

47. C.C. Perry, J.R. Wilcock and R.J.P. Williams, A physicochemical approach to morphogenesis: the roles of inorganic ions and crystals, *Experientia*, 1988, **44**, 638.

48. E. Dujardin and S. Mann, Bio-inspired materials chemistry, *Adv. Mater.*, 2002, **14**, 775.

49. S. Mann, The chemistry of form, *Angew. Chem. Int. Ed.*, 2000, **39**, 3392.

50. S. Mann, Biomineralization and biomimetic materials chemistry, *Biomimetic materials chemistry*, VCH, New York, NY, 1996, pp. 1–40.

51. S. Mann and G. Ozin, Synthesis of inorganic materials with complex form, *Nature*, 1996, **382**, 313.

52. S. Mann, D.D. Archibald, J.M. Didymus, T. Douglas, B.R. Heywood, F.C. Meldrum and N.J. Reeves, Crystallization at inorganic–organic interfaces: biominerals and biomimetic synthesis, *Science*, 1993, **261**, 1286.

53. S. Mann, Molecular tectonics in biomineralization and biomimetic materials chemistry, *Nature*, 1993, **365**, 499.

54. M. Fritz, A.M. Belcher, M. Radmacher, D.A. Walters, P.K. Hansma, G.D. Stucky, D.E. Morse and S. Mann, Flat pearls from biofabrication of organized composites on inorganic substrates, *Nature*, 1994, **371**, 49.

55. J.N. Cha, K. Shimizu, Y. Zhou, S.C. Christiansen, B.F. Chmelka, T.J. Deming, G.D. Stucky and D.E. Morse, Learning from biological systems: novel routes to biomimetic synthesis of ordered silica structures, *Mater. Res. Soc. Symp. Proc.*, 2000, **599** (*Mineral. Nat. Synth. Biomater.*), 239.

56. J.N. Cha, G.D Stucky, D.E. Morse and T.J. Deming, Biomimetic synthesis of ordered silica structures mediated by block copolypeptides, *Nature*, 2000, **403**, 289.

57. H.C. Lichtenegger, T. Schoeberl, M.H. Bartl, H. Waite and G.D. Stucky, High abrasion resistance with sparse mineralization: copper biomineral in worm jaws, *Science*, 2002, **298**, 389.

58. M.A. Cariolou and D.E. Morse, Purification and characterization of calcium-binding conchiolin shell peptides from the mollusk, *Haliotis rufescens*, as a function of development, *J. Comp. Physiol. B*, 1988, **157**, 717.

59. X.Y. Shen, A.M. Belcher, P.K. Hansma, G.D. Stucky and D.E. Morse, Molecular cloning and characterization of lustrin A, a matrix protein from shell and pearl nacre of *Haliotis rufescens*, *J. Biol. Chem.*, 1997, **272**, 32472.

60. N. Kroger, R. Deutzman and M. Sumper, Polycationic peptides from diatom biosilica that direct silica nanosphere formation, *Science*, 1999, **286**, 1129.

61. G. He, T. Dahl, A. Veis and A. George, Nucleation of apatite crystals in vitro by self-assembled dentin matrix protein 1, *Nat. Mater.*, 2003, **2**, 552.

62. Y.C. Liou, A. Tocilj, P.L. Davies and Z.C. Jia, Mimicry of ice structure by surface hydroxyls and water of a beta-helix antifreeze protein, *Nature*, 2000, **406**, 322.

63. A. Berman, L. Addadi and S. Weiner, Interactions of sea-urchin skeleton macromolecules with growing calcite crystals: a study of intracrystalline proteins, *Nature*, 1988, **331**, 546.

64. J.N. Cha, K. Shimizu, Y. Zhou, S.C. Christiansen, B.F. Chmelka, G.D. Stucky and D.E. Morse, Silicate in filaments and subunits from a marine sponge direct the polymerization of silica and silicones in vitro, *Proc. Natl Acad. Sci. USA*, 1999, **96**, 361.

65. G. Falini, S. Albeck, S. Weiner and L. Addadi, Control of aragonite or calcite polymorphism by mollusk shell macromolecules, *Science*, 1996, **271**, 67.

66. A.M. Belcher, X.H. Wu, R.J. Christensen, P.K. Hansma, G.D. Stucky and D.E. Morse, Control of crystal phase switching and orientation by soluble mollusc-shell proteins, *Nature*, 1996, **381**, 56.

67. G.A. Ozin, Morphogenesis of biomineral and morphosynthesis of biomimetic forms, *Acc. Chem. Res.*, 1997, **30**, 17.

68. G.K. Hunter, Interfacial aspects of biomineralization, *Curr. Opin. Solid State Mater. Sci.*, 1996, **1**, 430.

69. L.A. Estroff and A.D. Hamilton, At the interface of organic and inorganic chemistry: bioinspired synthesis of composite materials, *Chem. Mater.*, 2001, **13**, 3227.

70. B.C. Goodwin, The life of form. Emergent patterns of morphological transformation, *C.R. Acad. Sci., Ser. III*, 2000, **323**, 15.

71. B.C. Goodwin, What are the causes of morphogenesis?, *Bioessays*, 1985, **3**(1), 32.

72. B. Bensaude-Vincent, H. Arribart, Y. Bouligand and C. Sanchez, Chemists and the school of nature, *New J. Chem.*, 2002, **26**, 1.

73. M. Sarikaya, H. Fong, D.W. Frech and R. Humbert, Biomimetic assembly of nanostructured materials, *Mater. Sci. Forum*, 1999, **293** (*Bioceramics*), 83.

74. National Research Council, *Hierarchical Structures in Biology as a Guide for New Materials Technology*, National Academy Press, Washington, DC, 1994.

75. C. Giacovazzo, G. Artioli, D. Viterbo and G. Ferraris, *Fundamentals of Crystallography*, Oxford University Press, Oxford, 2002.

76. J.D. Bernal, Symmetry and genesis of form, *J. Mol. Biol.*, 1967, **24**, 379; J.D. Bernal and C.H. Carlisle, Range of generalized crystallography, *Sov. Phys. Cryst. USSR*, 1969, **13**, 811.

77. A.L. Mackay, Flexicrystallography: curved surfaces in chemical structures, *Curr. Sci.*, 1995, **69**, 151.

78. A.M. Turing, The chemical basis of morphogenesis, *Bull. Math. Biol.*, 1990, **52**, 153. (Reprinted from: A.M. Turing, Chemical basis of morphogenesis, *Trans. Roy. Soc.*, 1952, **B237**, 37.)

79. J.N. Weiss, Z.L. Qu and A. Garfinkel, Understanding biological complexity: lessons from the past, *FASEB J.*, 2003, **17**, 1.

80. H. Meinhardt, *The Algorithmic Beauty of Seashells*, Springer, Berlin, 1995.

81. P. Ball, *The Self-Made Tapestry: Pattern Formation in Nature*, Oxford Press, Oxford, 2001.

82. N. Coombs, D. Khushalani, S. Oliver, G.A. Ozin, G.C. Shen, I. Sokolov and H. Yang, Blueprints for inorganic materials with natural form: inorganic liquid crystals and a language of inorganic shape, *J. Chem. Soc.-Dalton Trans.*, 1997, 3941.

83. S. Oliver, A. Kuperman, N. Coombs, A. Lough and G.A. Ozin, Lamellar aluminophosphates with surface patterns that mimic diatom and radiolarian microskeletons, *Nature*, 1995, **378**, 47.

84. H. Yang, N. Coombs and G.A. Ozin, Morphogenesis of shapes and surface patterns in mesoporous silica, *Nature*, 1997, **386**, 692.

85. S.M. Yang, I. Sokolov, N. Coombs, C.T. Kresge and G.A. Ozin, Formation of hollow helicoids in mesoporous silica: supramolecular origami, *Adv. Mater.*, 1999, **11**, 1427.

86. J. Rathousky, M. Zukalova and A. Zukal, Supramolecular templating: a new strategy for the synthesis of mesoporous molecular sieves, *Chem. Listy*, 1997, **91**, 413.

87. C. Goltner-Spickermann, Nanocasting of lyotropic liquid crystal phases for metals and ceramics, *Colloid Chem. 1 Top. Curr. Chem.*, 2003, **226**, 29.

88. A. Stein, *The Role of Surfactants and Amphiphiles in the Synthesis of Porous Inorganic Solids*, ed. J. Texter, Marcel Dekker, New York, 2001, p. 819.

89. F. Camerel, J.C.P. Gabriel and P. Batail, Synthesis of a mesoporous composite material prepared by the self-assembly of mineral liquid crystals, *Chem. Commun.*, 2002, **17**, 1926.

90. J.C.P. Gabriel, F. Camerel, B.J. Lemaire, H. Desvaux, P. Davidson and P. Batail, Swollen liquid-crystalline lamellar phase based on extended solid-like sheets, *Nature*, 2001, **413**, 504.

91. G.G. Roberts, Langmuir–Blodgett films, *Contemp. Phys.*, 1984, **25**, 109.

92. www.nobel.se/chemistry/laureates/1932/langmuir-lecture.html

93. www.nobel.se/physics/laureates/1904/strutt-bio.html

94. M.E. Derrick, Profiles in chemistry: Agnes Pockels, 1862–1935, *J. Chem. Ed.*, 1982, **59**, 1030.

95. R.P. Feynman, famous "Plenty of room at the bottom" speech: www.its.caltech.edu/~feynman/plenty.html

96. H. Kuhn, Present status and future prospects of Langmuir–Blodgett film research, *Thin Solid Films*, 1989, **178**, 1.

97. C. Mingotaud, C. Lafuente, J. Amiell and P. Delhaes, Ferromagnetic Langmuir–Blodgett film based on Prussian blue, *Langmuir*, 1999, **15**, 289.

98. A.V. Bune, V.M. Fridkin, S. Ducharme, L.M. Blinov, S.P. Palto, A.V. Sorokin, S.G. Yudin and A. Zlatkin, Two-dimensional ferroelectric films, *Nature*, 1998, **391**, 874.

99. K.K. Kan, G.G. Roberts and M.C. Petty, Langmuir–Blodgett film metal-insulator semiconductor structures on narrow-band gap semiconductors, *Thin Solid Films*, 1983, **99**, 291.

100. C.D. Fung and G.L. Larkins, Planar silicon field-effect transistors with Langmuir–Blodgett gate insulators, *Thin Solid Films*, 1985, **132**, 33.

101. D.W. Kalina and S.G. Grubb, Langmuir–Blodgett films of non-centrosymmetric azobenzene dyes for non-linear optical applications, *Thin Solid Films*, 1988, **160**, 363.

102. I.R. Girling, S.R. Jethwa, R.T. Stewart, J.D. Earls, G.H. Cross, N.A. Cade, P.V. Kolinsky, R.J. Jones and I.R. Peterson, 2nd-order non-linear optical effects in Langmuir–Blodgett films, *Thin Solid Films*, 1988, **160**, 355.

103. J.S. Facci, P.A. Falcigno and J.M. Gold, Characterization of electroactive Langmuir–Blodgett monolayers of (ferrocenylmethyl)dimethyloctadecylammonium sulfate at gold and air water interfaces, *Langmuir*, 1986, **2**, 732.

104. M. Suzuki, Photosensitive polyimide Langmuir–Blodgett films derived from 4-(17-octadecenyl)pyridine and polyamic acid, *Thin Solid Films*, 1989, **180**, 253.

105. A. Otto, I. Mrozek, H. Grabhorn and W. Akemann, Surface-enhanced Raman scattering, *J. Phys.: Condens. Matter.*, 1992, **4**, 1143.

106. D.M. Adams, L. Brus, C.E.D. Chidsey, S. Creager, C.R. Kagan, P.V. Kamat, M. Lieberman, S. Lindsay, R.A. Marcus, R.M. Metzger, M.E. Michel-Beyerle, J.R. Miller, M.D. Newton, D.R. Rolison, O. Sankey, K.S. Schanze, J. Yardley and X.Y. Zhu, Charge transfer on the nanoscale: current status, *J. Phys. Chem. B*, 2003, **107**, 668.

107. M. Fujihara, Photoinduced electron-transfer and energy-transfer in Langmuir–Blodgett films, *Adv. Chem. Ser.*, 1994, **240**, 373.

108. L. Netzer and J. Sagiv, A new approach to construction of artificial monolayer assemblies, *J. Am. Chem. Soc.*, 1983, **105**, 674.

109. S. Yitzchaik and T.J. Marks, Chromophoric self-assembled superlattices, *Acc. Chem. Res.*, 1996, **29**, 197.

110. H. Lee, L.J. Kepley, H.G. Hong and T.E. Mallouk, Inorganic analogs of Langmuir–Blodgett films – adsorption of ordered zirconium 1,10-decanebisphosphonate multilayers on silicon surfaces, *J. Am. Chem. Soc.*, 1988, **110**, 618.

111. G. Decher, Fuzzy nanoassemblies: toward layered polymeric multicomposites, *Science*, 1997, **277**, 1232.

112. G. Decher and J.D. Hong, Build up of ultrathin multilayer films by a self-assembly process. 1. Consecutive adsorption of anionic and cationic bipolar amphiphiles on charged surfaces, *Makromol. Chem.-Macromol. Symp.*, 1991, **46**, 321.

113. G. Decher and J.D. Hong, Build up of ultrathin multilayer films by a self-assembly process. 2. Consecutive adsorption of anionic and cationic bipolar amphiphiles and polyelectrolytes on charged surfaces, *Ber. Bunsen-Ges. Phys. Chem. Chem. Phys.*, 1991, **95**, 1430.

114. G. Decher, J.D. Hong and J. Schmitt, Build up of ultrathin multilayer films by a self-assembly process. 3. Consecutively alternating adsorption of anionic and cationic polyelectrolytes on charged surfaces, *Thin Solid Films*, 1992, **210**, 831.

115. J.B. Schlenoff and S.T. Dubas, Mechanism of polyelectrolyte multilayer growth: charge overcompensation and distribution, *Macromolecules*, 2001, **34**, 592.

116. J. Schmitt, T. Grunewald, G. Decher, P.S. Pershan, K. Kjaer and M. Losche, Internal structure of layer-by-layer adsorbed polyelectrolyte films – a neutron and X-ray reflectivity study, *Macromolecules*, 1993, **26**, 7058.

117. N.A. Kotov, I. Dekany and J.H. Fendler, Layer-by-layer self-assembly of polyelectrolyte-semiconductor nanoparticle composite films, *J. Phys. Chem.*, 1995, **99**, 13065.

118. Y. Lvov, H. Haas, G. Decher, H. Mohwald, A. Mikhailov, B. Mtchedlishvily, E. Morgunova and B. Vainshtein, Successive deposition of alternate layers of polyelectrolytes and a charged virus, *Langmuir*, 1994, **10**, 4232.

119. Y. Lvov, K. Ariga and T. Kunitake, Layer-by-layer assembly of alternate protein polyion ultrathin films, *Chem. Lett.*, 1994, **12**, 2323.

120. M. Schonhoff, Self-assembled polyelectrolyte multilayers, *Curr. Opin. Colloid Interface Sci.*, 2003, **8**, 86.

121. C. Guang, H.G. Hong and T.E. Mallouk, Layered metal phosphates and phosphonates – from crystals to monolayers, *Acc. Chem. Res.*, 1992, **25**, 420.

122. S.W. Keller, H.N. Kim and T.E. Mallouk, Layer-by-layer assembly of intercalation compounds and heterostructures on surfaces – toward molecular beaker epitaxy, *J. Am. Chem. Soc.*, 1994, **116**, 8817.

123. M.M. Fang, D.M. Kaschak, A.C. Sutorik and T.E. Mallouk, A "mix and match" ionic–covalent strategy for self-assembly of inorganic multilayer films, *J. Am. Chem. Soc.*, 1997, **119**, 12184.

124. E.R. Kleinfeld and G.S. Ferguson, Stepwise formation of multilayered nanostructural films from macromolecular precursors, *Science*, 1994, **265**, 370.

125. D.L. Feldheim, K.C. Grabar, M.J. Natan and T.E. Mallouk, Electron transfer in self-assembled inorganic polyelectrolyte/metal nanoparticle heterostructures, *J. Am. Chem. Soc.*, 1996, **118**, 7640.

126. Y.J. Liu, A. Rosidian and R.O. Claus, Mechanical properties of electrostatically self-assembled Al_2O_3–ZrO_2 nanocomposites prepared at room temperature, *J. Cluster Sci.*, 1999, **10**, 421.

127. C.D. Bain, E.B. Troughton, Y.T. Tao, J. Evall, G.M. Whitesides and R.G. Nuzzo, Formation of monolayer films by the spontaneous assembly of organic thiols from solution onto gold, *J. Am. Chem. Soc.*, 1989, **111**, 321.

128. A.R. Bishop and R.G. Nuzzo, Self-assembled-monolayers: recent developments and applications, *Curr. Opin. Colloid Interface Sci.*, 1996, **1**, 127.

129. Y.N. Xia and G.M. Whitesides, Soft lithography, *Angew. Chem. Int. Ed.*, 1998, **37**, 551.

130. N.L. Abbott, J.P. Folkers and G.M. Whitesides, Manipulation of the wettability of surfaces on the 0.1-micrometer to 1-micrometer scale through micromachining and molecular self-assembly, *Science*, 1992, **257**, 1380.

131. M.L. Steigerwald, A.P. Alivisatos, J.M. Gibson, T.D. Harris, R. Kortan, A.J. Muller, A.M. Thayer, T.M. Duncan, D.C. Douglass and L.E. Brus, Surface derivatization and isolation of semiconductor cluster molecules, *J. Am. Chem. Soc.*, 1988, **110**, 3046; M.L. Steigerwald and L.E. Brus, Semiconductor crystallites – a class of large molecules, *Acc. Chem. Res.*, 1990, **23**, 183.

132. M.D. Bentzon, J. Van Wonterghem, S. Mrup, A. Thölen and C.J.W. Koch, Ordered aggregates of ultrafine iron-oxide particles – super crystals, *Philos. Mag. B*, 1989, **60**, 169.

133. D.A. Thompson and J.S. Best, The future of magnetic data storage technology, *IBM J. Res. Dev.*, 2000, **44**, 311.

134. C.B. Murray, D.J. Norris and M.G. Bawendi, Synthesis and characterization of nearly monodisperse CdE (E = S, Se, Te) semiconductor nanocrystallites, *J. Am. Chem. Soc.*, 1993, **115**, 8706.

135. V.L. Colvin, A.N. Goldstein and A.P. Alivisatos, Semiconductor nanocrystals covalently bound to metal-surfaces with self-assembled monolayers, *J. Am. Chem. Soc.*, 1992, **114**, 5221.

136. C.B. Murray, C.R. Kagan and M.G. Bawendi, Self-organization of CdSe nanocrystallites into 3-dimensional quantum-dot superlattices, *Science*, 1995, **270**, 1335.

137. P. Alivisatos, Colloidal quantum dots. From scaling laws to biological applications, *Pure Appl. Chem.*, 2000, **72**, 3.

138. G. Schmid, Large clusters and colloids – metals in the embryonic state, *Chem. Rev.*, 1992, **92**, 1709.

139. A.N. Goldstein, C.M. Echer and A.P. Alivisatos, Melting in semiconductor nanocrystals, *Science*, 1992, **256**, 1425.

140. S.H. Tolbert and A.P. Alivisatos, Size dependence of a first-order solid–solid phase-transition – the wurtzite to rock-salt transformation in CdSe nanocrystals, *Science*, 1994, **265**, 373.

141. U. Banin, Y.W. Cao, D. Katz and O. Millo, Identification of atomic-like electronic states in indium arsenide nanocrystal quantum dots, *Nature*, 1999, **400**, 542–544; L. Brus, Electronic wave-functions in semiconductor clusters – experiment and theory, *J. Phys. Chem.*, 1986, **90**, 2555; L. Brus, Size dependent development of band-structure in semiconductor crystallites, *New J. Chem.*, 1987, **11**, 123.

142. V.L. Colvin, M.C. Schlamp and A.P. Alivisatos, Light-emitting-diodes made from cadmium selenide nanocrystals and a semiconducting polymer, *Nature*, 1994, **370**, 354.

143. S. Coe, W.K. Woo, M. Bawendi and V. Bulovic, Electroluminescence from single monolayers of nanocrystals in molecular organic devices, *Nature*, 2002, **420**, 800.

144. D.L. Klein, R. Roth, A.K.L. Lim, A.P. Alivisatos and P.L. McEuen, A single-electron transistor made from a cadmium selenide nanocrystal, *Nature*, 1997, **389**, 699.

145. R.P. Andres, T. Bein, M. Dorogi, S. Feng, J.I. Henderson, C.P. Kubiak, W. Mahoney, R.G. Osifchin and R. Reifenberger, "Coulomb staircase" at room temperature in a self-assembled molecular nanostructure, *Science*, 1996, **272**, 1323.

146. M. Nirmal, B.O. Dabbousi, M.G. Bawendi, J.J. Macklin, J.K. Trautman, T.D. Harris and L.E. Brus, Fluorescence intermittency in single cadmium selenide nanocrystals, *Nature*, 1996, **383**, 802.

147. J. Tittel, W. Gohde, F. Koberling, T. Basche, A. Kornowski, H. Weller and A. Eychmuller, Fluorescence spectroscopy on single CdS nanocrystals, *J. Phys. Chem. B.*, 1997, **101**, 3013.

148. S. Empedocles and M. Bawendi, Spectroscopy of single CdSe nanocrystallites, *Acc. Chem. Res.*, 1999, **32**, 389.

149. W.J. Parak, D. Gerion, T. Pellegrino, D. Zanchet, C. Micheel, S.C. Williams, R. Boudreau, M.A. Le Gros, C.A. Larabell and A.P. Alivisatos, Biological applications of colloidal nanocrystals, *Nanotechnology*, 2003, **14**, R15.

150. M. Faraday, Experimental relations of gold (and other metals) to light, *Philos Trans. R. Soc.*, 1857, **147**, 145.

151. J. Turkevich, P.C. Stevenson and J. Hillier, A study of the nucleation and growth processes in the synthesis of colloidal gold, *Discuss. Faraday Soc.*, 1951, **11**, 55.

152. J.W. Slot and H.J. Geuze, A new method of preparing gold probes for multi-labeling cytochemistry, *Eur. J. Cell Biol.*, 1985, **38**, 87.

153. M. Brust, M. Walker, D. Bethell, D.J. Schiffrin and R. Whyman, Synthesis of thiol-derivatized gold nanoparticles in a 2-phase liquid–liquid system, *Chem. Commun.*, 1994, **7**, 801.

154. R.L. Whetten, J.T. Khoury, M.M. Alvarez, S. Murthy, I. Vezmar, Z.L. Wang, P.W. Stephens, C.L. Cleveland, W.D. Luedtke and U. Landman, Nanocrystal gold molecules, *Adv. Mater.*, 1996, **8**, 428.

155. P. Mulvaney and L.M. Liz-Marzan, Rational material design using Au core-shell nanocrystals, *Coll. Chem. 1 (Top. Curr. Chem.)*, 2003, **226**, 225.

156. A.E. Saunders, P.S. Shah, M.B. Sigman Jr., T. Hanrath, H.S. Hwang, K.T. Lim, K.P. Johnston and B.A. Korgel, Inverse opal nanocrystal superlattice films, *Nano Lett.*, 2004, 1943.

157. T. Castro, R. Reifenberger, E. Choi and R.P. Andres, Size-dependent melting temperature of individual nanometer-sized metallic clusters, *Phys. Rev. B*, 1990, **42**, 8548.

158. G. Schmid and B. Corain, Nanoparticulated gold: syntheses, structures, electronics, and reactivities, *Eur. J. Inorg. Chem.*, 2003, **17**, 3081.

159. M.M. Alvarez, R.L. Whetten, S.W. Chen, R.S. Ingram, M.J. Hostetler, J.J. Pietron, R.W. Murray, T.G. Schaaff and J.T. Khoury, Gold nanoelec-trodes of varied size: transition to molecule-like charging, *Science*, 1998, **280**, 2098.

160. Y.N. Xia, P.D. Yang, Y.G. Sun, Y.Y. Wu, B. Mayers, B. Gates, Y.D. Yin, F. Kim and Y.Q. Yan, One-dimensional nanostructures: synthesis, characteri-zation, and applications, *Adv. Mater.*, 2003, **15**, 353.

161. N.A. Melosh, A. Boukai, F. Diana, B. Gerardot, A. Badolato, P.M. Petroff and J.R. Heath, Ultrahigh-density nanowire lattices and circuits, *Science*, 2003, **300**, 112.

162. Y.N. Xia, B. Gates, Y.D. Yin and Y. Lu, Monodispersed colloidal spheres: old materials with new applications, *Adv. Mater.*, 2000, **12**, 693.

163. H. Kawaguchi, Functional polymer microspheres, *Prog. Polym. Sci.*, 2000, **25**, 1171.

164. P. Dong, Advances in preparation and application of monodisperse colloidal silica particles, *Prog. Natl Sci.*, 2000, **10**, 575.

165. J.E.G.J. Wijnhoven and W.L. Vos, Preparation of photonic crystals made of air spheres in titania, *Science*, 1998, **281**, 802.

166. J.D. Joannopoulos, P.R. Villeneuve and S.H. Fan, Photonic crystals: putting a new twist on light, *Nature*, 1997, **386**, 143.

167. R.C. Salvarezza, L. Vazquez, H. Miguez, R. Mayoral, C. Lopez and F. Meseguer, Edward–Wilkinson behavior of crystal surfaces grown by sedimentation of SiO$_2$ nanospheres, *Phys. Rev. Lett.*, 1996, **77**, 4572.

168. M. Trau, D.A. Saville and I.A. Aksay, Assembly of colloidal crystals at electrode interfaces, *Langmuir*, 1997, **13**, 6375. M. Holgado, F. Garcia-Santamaria, A. Blanco, M. Ibisate, A. Cintas, H. Miguez, C.J. Serna, C. Molpeceres, J. Requena, A. Mifsud, F. Meseguer and C. Lopez, Electrophoretic deposition to control artificial opal growth, *Langmuir*, 1999, **15**, 4701.

169. P. Jiang, J.F. Bertone, K.S. Hwang and V.L. Colvin, Single-crystal colloidal multilayers of controlled thickness, *Chem. Mater.*, 1999, **11**, 2132.

170. A. van Blaaderen, R. Ruel and P. Wiltzius, Template-directed colloidal crystallization, *Nature*, 1997, **385**, 321.

171. C.L. Haynes and R.P. Van Duyne, Nanosphere lithography: a versatile nanofabrication tool for studies of size-dependent nanoparticle optics, *J. Phys. Chem. B*, 2001, **105**, 5599.

172. J.C. Love, D.B. Wolfe, H.O. Jacobs and G.M. Whitesides, Microscope projection photolithography for rapid prototyping of masters with micron-scale features for use in soft lithography, *Langmuir*, 2001, **17**, 6005.

173. E. Kim, Y.N. Xia and G.M. Whitesides, Two- and three-dimensional crystallization of polymeric microspheres by micromolding in capillaries, *Adv. Mater.*, 1996, **8**, 245.

174. A. Arsenault, S.B. Fournier-Bidoz, B. Hatton, H. Miguez, N. Tetrault, E. Vekris, S. Wong, S.M. Yang, V. Kitaev and G.A. Ozin, Towards the synthetic all-optical computer: science fiction or reality? *J. Mater. Chem.*, 2004, **14**, 781.

175. M. Trau, N. Yao, E. Kim, Y. Xia, G.M. Whitesides and I.A. Aksay, Microscopic patterning of orientated mesoscopic silica through guided growth, *Nature*, 1997, **390**, 674.

176. P. Yang, T. Deng, D. Zhao, P. Feng, D. Pine, B.F. Chmelka, G.M. Whitesides and G.D. Stucky, Hierarchically ordered oxides, *Science*, 1998, **282**, 2244.

177. P. Yang, A.H. Rizvi, B. Messer, B.F. Chmelka, G.M. Whitesides and G.D. Stucky, Patterning porous oxides within microchannel networks, *Adv. Mater.*, 2001, **13**, 427.

178. O.D. Velev, T.A. Jede, R.F. Lobo and A.M. Lenhoff, Porous silica via colloidal crystallization, *Nature*, 1997, **389**, 447.

179. D.Y. Zhao, J.L. Feng, Q.S. Huo, N. Melosh, G.H. Fredrickson, B.F. Chmelka and G.D. Stucky, Triblock copolymer syntheses of mesoporous silica with periodic 50 to 300 angstrom pores, *Science*, 1998, **279**, 548.

180. H. Arribart and B. Bensaude-Vincent, Chemists' defiance: imitating the beauty of living things, *Recherche*, 1999, **325**, 56.

181. W.T.S. Huck, J. Tien and G.M. Whitesides, Three-dimensional mesoscale self-assembly, *J. Am. Chem. Soc.*, 1998, **120**, 8267.

182. A.J. Bard, *Integrated Chemical Systems: a Chemical Approach to Nanotechnology*, Wiley, New York, NY, 1994.

183. T. Cassagneau, T.E. Mallouk and J.H. Fendler, Layer-by-layer assembly of thin film Zener diodes from conducting polymers and CdSe nanoparticles, *J. Am. Chem. Soc.*, 1998, **120**, 7848.

184. D.L. Feldheim, K.C. Grabar, M.J. Natan and T.E. Mallouk, Electron transfer in self-assembled inorganic polyelectrolyte/metal nanoparticle heterostructures, *J. Am. Chem. Soc.*, 1996, **118**, 7640.

185. V.L. Colvin, M.C. Schlamp and A.P. Alivisatos, Light-emitting diodes made from cadmium selenide nanocrystals and a semiconducting polymer, *Nature*, 1994, **370**, 354.

186. H. Sirringhaus, N. Tessler and R.H. Friend, Integrated optoelectronic devices based on conjugated polymers, *Science*, 1998, **280**, 1741.

187. T. Cassagneau and J.H. Fendler, High-density rechargeable lithium-ion batteries self-assembled from graphite oxide nanoplatelets and polyelectrolytes, *Adv. Mater.*, 1998, **10**, 877.

Nanofood for Thought – Nanochemistry, Genealogy, Materials Self-Assembly, Length Scales

1. Discuss connections between D'Arcy Wentworth Thomson's ideas on morphogenesis of porous siliceous microskeletons of radiolaria in biology and template directed self-assembly of periodic nanoporous forms of silica in the materials chemistry laboratory.

2. In the context of nanochemistry briefly explain with examples what is meant by (a) self-assembly, (b) co-assembly, (c) template directed self-assembly, (d) colloidal assembly, (e) inverse colloidal assembly, (f) soft lithography, (g) cluster superlattice, (h) electrostatic superlattice, (i) hierarchical structure, (j) nanocomposite, (k) periodic mesostructure and (l) topotactic metamorphosis.

3. Amplify upon the statement: "Self-assembly enables the diversity of chemistry to be combined with the complexity of biology to create materials that auto-construct over 'all' length scales and this represents a new frontier in solid-state chemistry".

4. In a few sentences delineate the key scientific developments in the 20th century that enabled solid-state chemistry, as we know it today.

5. What were the key breakthroughs that enabled today's nanoscience?

6. Where did the term "self-assembly" first appear in the scientific literature?

7. Why is it important to realize that not all nanomaterials are small? Describe such a "large" nanomaterial with size-tunable properties in the micron range.

8. Which seminal contributions of Pieter Harting and Heinz Lowenstam can be cited to support the consensus that they are, respectively, the fathers of biomimetic inorganic chemistry and biomineralization?

9. Why is curvature an important concept in materials chemistry and biology? Why is curvature becoming particularly relevant in materials self-assembly chemistry?

10. What can you accomplish with inorganic Langmuir–Blodgett multilayer film that would be impossible with the organic analog?

11. Discuss the different classes of nanostructures that can be synthesized. How could each of these classes of structures be useful in nano-devices?

Soft Lithography

CHAPTER 2

Chemical Patterning and Lithography

"I'll print it, And shame the fools."
Alexander Pope (1688–1744)

2.1 Soft Lithography

Nuzzo noticed how long chain alkanethiolates could self-assemble on gold surfaces to give well-ordered monolayer films,[1–4] now affectionately known as self-assembled monolayers (SAMs). Whitesides, shown in Figure 2.1, effectively expanded upon the SAM methodology to develop the now ubiquitous process of soft lithography.[5–7] This technique makes use of a patterned elastomer such as poly(dimethylsiloxane) (PDMS), as a mask, stamp or mold.[8] With this simple process, high-quality patterns and structures can be created with lateral dimensions from about 5 nm to 500 mm.[9,10] Patterns can be two- and three-dimensional,[11,12] and often cannot easily be achieved by photolithography.[13]

Whitesides' strategy for patterning and shaping materials over different length scales and dimensions embraces chemical concepts of self-assembly, templating and crystal engineering, together with soft lithographic techniques of microcontact printing and micromolding. It is an alternative low-cost, non-photolithographic microfabrication method to complement photolithography, where the feature size is usually limited by optical diffraction. In addition, the short wavelength radiation required for photolithographically defining smaller and smaller designs necessitates prohibitively expensive and complex facilities and technologies, especially for those wanting to demonstrate proof-of-concept experiments.

In printing mode, soft lithography circumvents the diffraction limits of 2D photolithography. Some form of lithography needs to be employed to make a template, or master, for the stamp. Expensive and high-resolution methods such as e-beam or UV lithography need to be used only once, since one template can generate about a hundred stamps if well maintained. Soft lithography provides access to topologically complex 3D objects for building micro- or nanomachines, accommodates a wide range of materials and enables a rational surface chemistry.

Figure 2.1 *George Whitesides, father of soft lithography*
(Image provided by G.M. Whitesides)

It is easily accessible to chemists interested in experimenting with reliable, convenient and inexpensive methods for patterning planar[14] or curved[15,16] surfaces with organic, inorganic, polymer, liquid crystal, ceramic or biological structures.[17–24] The enormous efforts of the Whitesides' research group and other groups around the world illustrate the potential of soft lithographic procedures in a wide range of areas, from microelectronics to optics, microanalysis to sensors, microelectromechanical systems to cell biology.[25–31]

Although soft lithography is still in its infancy, it is abundantly clear that the utility of the method in diverse areas of chemistry, for rapid and low-cost patterning and shaping of materials, will definitely increase in the near future. It is an indispensable component in the toolbox of materials self-assembly over all scales in the format of directed self-assembly, a creative fusion of concepts and methods in self-assembly and lithography.

2.2 What are Self-Assembled Monolayers?

The archetype SAM is formed from the chemisorption of alkanethiols on gold substrates and their spontaneous organization into close-packed well-ordered monolayer assemblies like those illustrated in Figure 2.2.[32] Because it is so compact, the SAM shields the substrate, and the thiol endgroups dictate the surface

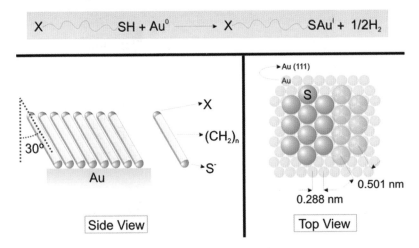

Figure 2.2 *Chemical reaction occurring during the formation of thiol self-assembled monolayers on gold (top). A side view shows SAM molecules are not standing straight up but are tilted about 30°. Sulfur atoms occupy threefold sites on the Au (111) surface, resulting in a highly ordered close-packed monolayer*

properties. The surface chemistry of alkanethiols on gold is formally that of an oxidative addition, whereby the thiol S–H bond is cleaved on a gold surface site. This leads to an Au–S covalent bond and a small amount of released hydrogen. Many diffraction, microscopy and spectroscopy diagnostics of SAM formation and structure provide detailed information about surface Au–S bonds, loss of surface hydrogen, delocalization of surface charge and non-bonded interactions between alkane chains that are specific to particular gold crystal surface orientations.[33–36] For example, the thickness and organization of a SAM depends not only on the alkane chain length but also on the tilt angle of the chain with respect to the surface[37] (Figure 2.2). Silver is an excellent substrate for SAM formation like gold, but unlike gold the molecular tilt angle is only 10°, making films of identical molecules on silver thicker by about 1.14 (cos 10/cos 30).

The spatial separation and packing of alkanethiolates on gold depends on the structure of the crystallographic gold site to which the thiolate sulfur is bonded. There are significant differences, for example, between a threefold triangular pyramidal site on a {111} basal plane and a fourfold square pyramidal site on a {100} plane. An example of a close-packed SAM visualized by AFM is shown in Figure 2.3.[38]

The real power of SAMs in diverse areas of chemistry, physics, materials science, engineering and biology stems from the ability to chemically tailor the terminal X groups of the alkanethiolate shown in Figure 2.2. This creates a rational surface organic chemistry of structurally well-defined surface monolayers, opening up a cornucopia of opportunities for controlling surface wettability, tribology, adhesion, corrosion, etch protection, chemical and electrochemical reactivity,[39–45] to name but a few possibilities. This obvious utility is compounded by the fact that using gold as a starting point, soft lithography has been elaborated to include

Figure 2.3   *Molecule resolution AFM image of a hexanethiol SAM on gold*
(Reproduced with permission from Ref. 38)

the printing of a variety of reactive inks capable of forming well-ordered
monolayers on substrates such as metals, semiconductors and insulators.[46–51]

2.3 The Science and Art of Soft Lithography

The central tool in patterning surfaces through chemistry is PDMS. This
polymer can be fashioned in the form of a soft elastomeric stamp utilized for
printing micron and submicron-scale designs of alkanethiol chemical inks on
planar and curved surfaces. A schematic illustration of the procedure for casting
PDMS replicas from a master having relief structures on its surface is shown in
Figure 2.4.

The master, fabricated by some form of lithography, is first silanized and made
hydrophobic by exposure to $CF_3(CF_2)_6(CH_2)_2SiCl_3$ vapor, then filled with liquid
PDMS prepolymer and subsequently cured. The surface treatment allows the
PDMS replica to be easily removed from the master without sticking problems.
Each master can be used to fabricate more than 50 PDMS replicas, and
representative ranges of the critical dimensions h, d, and l depicted in Figure 2.4
are 0.2–20, 0.5–200 and 0.5–200 μm, respectively. Each PDMS stamp, in turn, can
be used to print its inscribed pattern hundreds of times.

The PDMS master stamp can be used to print SAM patterns on planar and curved
surfaces in different ways and for distinct purposes. Some of the methods of
microcontact printing of alkanethiols on gold-coated silicon substrates are depicted
in Figure 2.5. A 5 nm thick layer of titanium or chromium on silicon provides
satisfactory adhesion of gold by creating a stable gradient in interfacial surface free
energy. This enables patterning by the transfer of alkanethiol ink from the surface

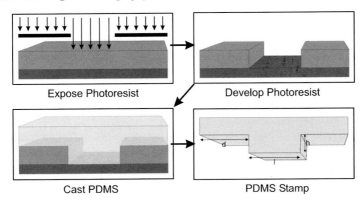

Figure 2.4 *Procedure for making a PDMS stamp. The first step is coating a thin layer of photoresist onto a silicon wafer, followed by exposure to UV light through a shadow mask. The exposed photoresist is washed away with developer (or alternatively the unexposed photoresist is washed away), and the patterned "master" treated with perfluoroalkyltrichlorosilane to reduce its stickiness. A PDMS prepolymer is then poured onto the master, cured and removed to form the PDMS stamp. Many variations on this procedure are possible, and essentially any topological feature can be replicated in this fashion*

of the PDMS stamp to the gold surface when they are placed in conformal contact. The un-inked regions of the gold can be subsequently functionalized by dipping the SAM 1 patterned substrate into a different alkanethiol, SAM 2, as indicated in Figure 2.5.[6]

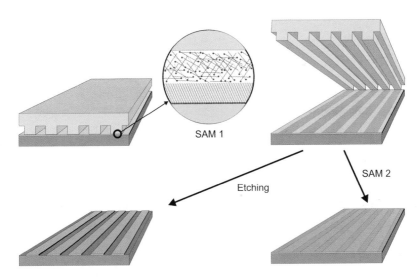

Figure 2.5 *Principles of microcontact printing with alkanethiols. A printed long chain thiol, SAM 1, forms an ordered surface layer; the unmodified regions can be either etched or modified with a different thiol, SAM 2*

These microcontact-printed SAMs have been utilized for patterning surface charge, acidity and basicity, hydrophobicity and hydrophilicity, hydrogen bonding and metal–ligand coordination, chirality and photochromicity, and applications for these chemically patterned surfaces now abound in the literature.[52] One of the most exciting uses for hydrophobic SAMs is as etch protection for making surface relief patterns in single crystal silicon wafers, since a SAM is impermeable and protects the underlying substrate from chemical reactions. Representative wet etch chemistry for selectively removing gold from unprotected SAM regions of an Si–Ti–Au substrate is given in the following equations:[43]

$$4Au + 8CN^- + O_2 + 2H_2O \rightarrow 4[Au(CN)_2]^- + 4OH^-$$

$$Au + 2CN^- + [Fe(CN)_6]^{3-} \rightarrow [Fe(CN)_6]^{4-} + [Au(CN)_2]^-$$

Subsequent removal of the Ti adhesion layer is accomplished with dilute aqueous HF and then the Si wafer, for example with [100] orientation, can be anisotropically etched using ^{i}PrOH–NaOH to create surface relief designs like those shown in Figure 2.6. In this figure a geometrically well-defined square pyramidal-shaped microwell with {111} sidewalls has been used for the confined crystallization of silica microspheres to form a planarized and [100]-oriented photonic crystal within a silicon chip – a first step towards the dream of an all-optical microphotonic crystal chip for computing with light.[53]

2.4 Patterning Wettability?

The ability to chemically tailor the terminal group of an alkanethiolate SAM enables control of the surface properties of a structurally defined organic surface.[54,55] One of the first and still most dramatic illustrations of this capability is in the chemical control of contact angle between water droplets on a hydrophobic

Figure 2.6 *Soft lithographically defined surface relief pattern anisotropically etched in silicon [100], and used for confined colloidal crystal growth*
(Reproduced with permission from Ref. 53)

and hydrophilic SAM as illustrated in Figure 1.18. Here, hydrophilic squares are wet by water, while the hydrophobic lines or "corrals" separating them confine the water to square drops.[56] In essence, contact angle and wettability of a liquid on a SAM are determined by the polarity of the endgroup on the alkanethiol. Synthetic control of contact angle and wettability provides many opportunities for the patterning and molding of a wide range of materials as will be illustrated in the following examples. Contact angle control in the form of a wettability gradient can even cause water droplets to defy gravity and run uphill! (Figure 2.7)[57]

2.5 Condensation Figures

By cooling a surface that has been patterned with hydrophobic and hydrophilic SAMs to a temperature below the dew point of air, water droplets gradually form on the polar hydrophilic surfaces to form water condensation figures like those shown in Figure 2.8.[58,59] In this illustration, the electron micrograph images on the left depict hydrophobic and hydrophilic SAM-patterned surfaces having one or more types of periodicity, while the images on the right are optical diffraction patterns obtained by shining a laser on water condensation figures formed on these SAM-patterned surfaces. The optical diffraction patterns of surfaces with more than one periodicity are distinguished in spectacular visual imagery. By effectively reflecting in reciprocal space the underlying symmetry of the SAM-patterned condensation figures, we find out about the kinetics of formation and structure of the water droplet pattern on the surface itself. Since this work, soft lithographically

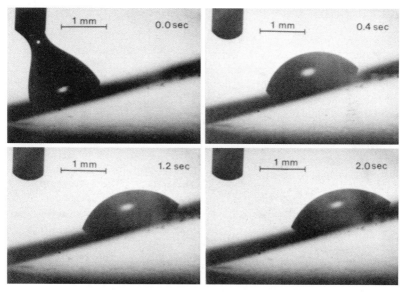

Figure 2.7 *A gradient in surface free energy can cause a water droplet to crawl uphill towards the more hydrophilic end of the surface*
(Reproduced with permission from Ref. 57)

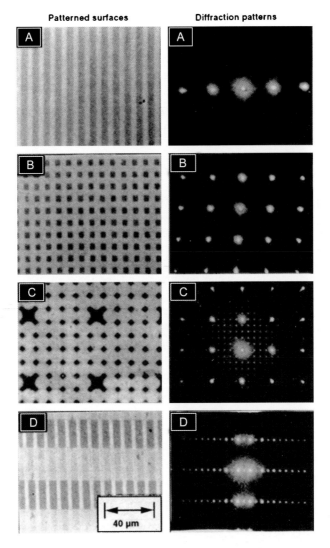

Figure 2.8 *Soft lithographically patterned water condensation figures, and the diffraction*
patterns they generate
(Reproduced with permission from Ref. 59)

defined chemically sensitive diffraction gratings, whose intensity of diffracted light
depends on the refractive index of the grating material, are proving useful in
biosensing, bioanalytics and biodiagnostics.[60,61]

2.6 Microlens Arrays

SAM patterning provides a convenient means of making arrays of microlenses.
Arrayed microlenses are widely used in a variety of applications that involve

miniaturized optical components.[62,63] They are the drivers of optical communication systems, laser printers, facsimile machines, digitized information storage or processing devices and microscope projection photolithography[64–67] (MPP). MPP consists of shining a large pattern through an array of microlenses, by using a microscope in reverse, to get millions of miniature replicas on a substrate. Microlenses also serve as diode laser correctors and fiber-optic couplers. Nature has learned long ago the advantages of microlens arrays, and uses them for instance in the bio-optics of the compound eye of an insect where many microlenses give the organism a greater periphery of vision.[68] In a 2D array of microlenses, ideally each lenslet presents a high-quality optical surface equivalent to a macroscopic, single-element lens. This facilitates a huge reduction in space as an array of microlenses replaces multiple conventional macroscopic lenses, enabling brighter light sources, shaper images, more precisely controlled laser beams and more accurate detectors.[69]

One of the first examples of the SAM-guided assembly of liquids was used to make a microlens array made of polyurethane hemispheres.[70] It was prepared by directed self-assembly, consisting of deposition of polyurethane prepolymer on the non-polar regions of a hydrophobic and hydrophilic-patterned SAM. The prepolymer selectively wets the non-polar region and forms hemi-spherical droplets (similar to condensation figures as described earlier), which were polymerized by UV curing to form the microlens replica. An example of a hexagonally ordered microlens array made by dewetting is shown in Figure 2.9.[71] Other polymeric shapes and arrays can be made by straightforward adaptation of this general approach, *e.g.*, an array of linear polymethylmethacrylate (PMMA) structures can be formed on the polar regions of a pattern of hydrophilic and hydrophobic parallel lines, with possible use as optical waveguides or microfluidic channels.

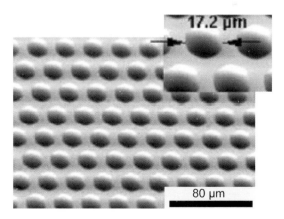

Figure 2.9 *Hexagonal microlens array generated by dip-coating onto a substrate patterned by soft lithography, imaged by SEM*
(Reproduced with permission from Ref. 71)

2.7 Nanoring Arrays

A general method for the self-assembly of nanorings with a wide range of polymeric and inorganic compositions of controlled diameters and thickness has been developed using soft lithographic techniques. It makes use of the ability of SAMs to pattern microdroplet condensation figures, with the creative twist that these droplets can function as templates for molding other materials into nanoring replicas.[72] The strategy for achieving this objective is illustrated in Figure 2.10. Key considerations for successful nanoring synthesis are the relative polarities and volatilities of the liquid forming the microdroplets and the solvent for the nanoring precursor, as well as the concentration of the precursor. The precursor solution selectively wets the regions between the microdroplets and evaporates first, depositing the precursor around the perimeter of the microdroplets. This forms the nanoring replica, after which the initial microdroplet is allowed to evaporate. The internal diameter of the nanorings is determined by the outside diameter of the microdroplets and the thickness and width of the annulus are controlled by the concentration of precursor in solution. In this example, the ring material was a polymer such as polystyrene deposited from a solution in chloroform.

This nanoring synthesis procedure could be adapted to a range of inorganic compositions, like silica or metal colloids, and has been demonstrated with the redox tunable inorganic polymer polyferrocenylsilane.[73] A nanoring array comprising a novel bimetallic polyferrocenylsilane is shown in Figure 2.11. The metallo-polymer is composed of alternating substituted silicon atoms and ferrocene groups, made by the ring opening polymerization of a sila-1-ferrocenophane. The introduction of an acetylene side arm on the silicon of the backbone allows the

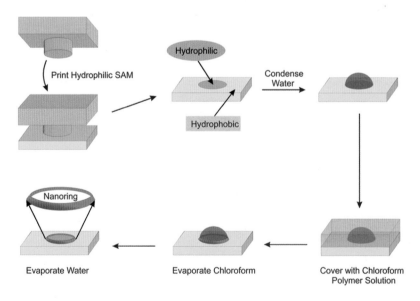

Figure 2.10 *Steps in the fabrications of soft lithographically patterned nanoring arrays*

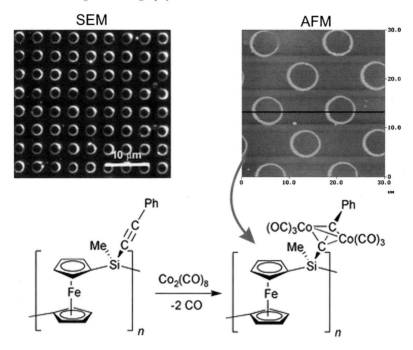

Figure 2.11 *Nanoring array of cobalt-clusterized PFS metallopolymer. On the left is a large-scale SEM image; on the right is an AFM inset. Below is shown the synthesis and structure of the polymer used for making these rings* (Reproduced with permission from Ref. 73)

coordination of a dicobaltoctacarbonyl cluster.[74] The bimetallic polyferrocenylsilane nanoring array can be pyrolyzed to a metal alloy magnetic nanocluster containing ceramic, formulated as $Fe_nCo_m/SiC/C$, in high ceramic yield and with shape retention to yield a ceramic nanoring array. Redox and magnetically tunable nanorings may prove to be interesting platforms for the development of novel kinds of electrochemical sensors, optical resonators and recording media. They could also be the critical magnetic components of a future nanoelectrical generator or metamaterial lens.

2.8 Patterning the Solid State

The most direct method of making solid-state materials is by the intimate mixing of precursors followed by their reaction at elevated temperatures.[75] Ideally the mixing should be at the smallest scale achievable in order to reduce diffusion lengths of reactants, form products in the shortest time, and perform the synthesis under the mildest possible conditions. This approach is still utilized to make some of the most important solid-state materials currently utilized in a range of high technology devices, products and processes.[76] These include perovskite, spinel, layered dichalcogenide and rutile structures that find applications as high T_c superconductors, giant magnetoresistant (GMR) recording heads, battery cathodes and solar cells.[77–81] While the understanding of the solid-state chemistry for making these materials is elegant, the "shake-and-bake" trial-and-error process of physically mixing and directly reacting solid precursors to obtain non-descript powders is rather crude and often poorly applicable to their end-use. Much effort and costs must therefore be expended for the reprocessing of these materials. Approaches employed for reducing diffusion lengths to enhance rates of reaction between solid materials include hot pressing, to reduce porosity and increase contact area, as well as making core–corona particles and superlattice reagents.

A major step forward in direct solid-state synthesis where the diffusion length is reduced to the nanoscale involves the creative use of SAM-patterned microdroplets as spatially defined reaction chambers for the mixing of precursors and subsequent thermal post-treatment steps required to form a replica pattern of product material.[82] This kind of crystal engineering is quite general for making a wide range of solid-state materials, the basic steps being shown in Figure 2.12. Because the synthesis involves a heating step to react precursors, the substrate to be patterned with hydrophobic and hydrophilic SAMs is made of a thermally stable material like silicon. The anchoring SAM in this case is an appropriately end-functionalized trichloroalkylsilane or trialkoxyalkylsilane. By dipping the SAM-patterned substrate in a solution of precursors, microdroplets containing a predetermined concentration of the precursors are formed, which on evaporation leave behind a replica pattern of size-tunable solid precursor dots. These dots can then be thermally treated to induce reaction to the desired solid-state-patterned material. The archetype examples used to reduce this intriguing idea to practice were simple or mixed nitrates of iron, cobalt and nickel, which on thermal treatment in air transform to their respective metal oxides. Further thermal treatment in hydrogen gas converts these to the corresponding metal clusters. All products are

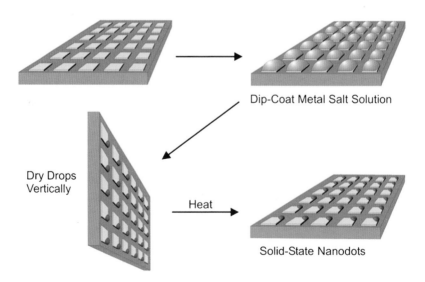

Dip-Coat Metal Salt Solution

Dry Drops Vertically

Heat

Solid-State Nanodots

Figure 2.12 *Soft lithographic patterning of classical solid-state materials*

nanoscale, smaller than the printed patch size, and in designs that mirror the original SAM pattern.

Representative optical, scanning electron and atomic force microscope images that show the patterning of nanoscale arrays of $Co(NO_3)_2$, Co_3O_4 and Co, synthesized by SAM patterning of solid-state syntheses of the aforementioned type, are displayed in Figure 2.13. The optical micrograph (dark field) of an ordered 2D array of nanoparticles of $Co(NO_3)_2$ that was fabricated on an Si/SiO_2 substrate by selective dewetting from a 0.01 M nitrate solution in 2-propanol is shown in the top panel of Figure 2.13. The surface was patterned with an array of hydrophilic $Si–SiO_2$ grids with an area of 5×5 μm^2, separated by 5 μm. In the panel below, an SEM image is shown of the patterned array in the top panel, after the nitrate had been decomposed into Co_3O_4 by heating the sample in air at 600° C for 3 h. These Co_3O_4 nanoparticles have a hemispherical shape as seen in the inset, shown in oblique view. The lower panel depicts an AFM image of the 2D array shown in the middle panel, after it had been heated and reduced in a flow of hydrogen gas at 400° C for 2 h. By changing the precursor solution concentration, as well as the wetting patch size, the cobalt nanoclusters could be easily controlled to have lateral dimensions between 70 and 460 nm and vertical dimensions between 15 and 230 nm. Clearly this method has tremendous versatility: control of the geometry and dimensions of the array, as well as the composition and dimension of nanoscale materials made by direct reaction of solid-state precursors is possible. Fifty years of "shake-and-bake" solid-state chemistry is now available for patterning by the nanoscientist!

2.9 Primed for Printing Polymers

The preceding example showed us how we can arbitrarily pattern just about any type of solid-state material by soft lithography. This technique, evaporation of a droplet on a wetting pattern, should be widely applicable to any material dissolved or suspended in solution. For instance, instead of using metal salts we could use capped nanoclusters (see Chapter 6) to obtain nanocluster superlattice nanostructures, or a dissolved polymer to obtain polymer nanostructures. The problem with polymer solutions, however, is that they can be quite viscous. Therefore, during drying a droplet of polymer solution is likely to dry as an irregular shape due to kinetic trapping. In addition, the nanostructure that is obtained is not bound to the substrate, and could hinder applications requiring robust attachment.

Polymer synthesis usually proceeds in solution, with an initiator molecule which polymerizes several monomer units into a chain. However, if instead of in solution we support an initiator on a substrate, polymer growth leads to surface-bound polymer chains.[83,84] Since they all grow out at the same rate, surface-grafted chains are mostly stretched out and a very high density of coverage can be achieved.

One example of such a surface-initiated polymer growth begins with the patterning of a gold substrate with an alkanethiol SAM, followed by exposure of the unmodified areas with a hydroxyl-terminated SAM.[85] The hydroxyl groups can serve as catalysts for the polymerization of gamma-caprolactone, when used with an

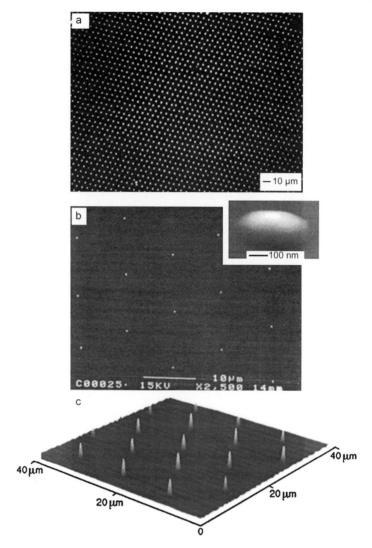

Figure 2.13   *Soft lithographically patterned solid-state materials. A and B are SEM images*
at different magnification. C is an AFM image showing spot topography
(Reproduced with permission from Ref. 82)

aluminum alkoxide co-catalyst. Polymer growth was found to proceed exclusively
from the hydroxyl-terminated patterns, to give well-defined features whose height
is dependent on the length of exposure to the monomer solution. An AFM image of
such soft lithographically defined patterns is shown in Figure 2.14. Metal catalysts
can also be used for surface-initiated polymerization,[86] starting with the patterning
of a gold substrate with an alkanethiol SAM. The unmodified regions are then
electrochemically coated with a thin film of copper,[87] after which the SAMs are
removed by electrochemical oxidation. The copper-coated regions can be used as

Figure 2.14 *AFM image of a pattern of polycaprolactone lines created by the amplification of a printed initiator pattern*
(Reproduced with permission from Ref. 85)

catalysts for the decomposition–polymerization of diazomethane, resulting in surface-bound polymethylene (identical in structure to polyethylene).

Especially exciting is the prospect of growing patterns of surface-initiated polymers by living polymerization. A living polymerization[88–90] is one in which the growing chain end remains active even after the dissolved monomer is depleted. By simply adding another monomer into the system, the chain keeps growing, adding the new monomer to the existing chain to form well-defined, narrow distribution block copolymers (see Chapter 9). This technique allows the ultimate in polymer architectural control and an almost infinite variety of highly ordered hierarchical systems. Many other systems have been investigated, and it is likely that with experimental effort, any polymer which can be grown in solution can be grown as surface-bound patterns.[91]

2.10 Beyond Molecules – Transfer Printing of Thin Films

Most of the great utilities of microcontact printing stem from the SAM. For various systems, printing of a chemical ink results in the formation of a monolayer on the substrate, which imparts it with radically different surface properties. However, microcontact printing is by no means limited to chemical inks. One could imagine printing a whole host of chemicals, materials or objects such as nanocrystals[92] from the surface of a PDMS stamp onto a substrate. The only requirement needed is a strong adhesion between the material to be transferred and the substrate. In fact, this adhesion does not need to be so strong, as long as it is greater than the adhesion of the material to the PDMS. Printing with such a stamp and its subsequent removal therefore leaves a layer of a given material on the substrate if we correctly choose the interfacial energies.

Excellent demonstrations of this concept have been made consisting of the transfer of metal films and multilayer films onto a variety of substrates.[93] The first step in this process consists of evaporating onto a PDMS stamp a thin film of metal such as gold. Thermal evaporation provides a very homogenous coating of metal, as well as a very directional one. This results in a coating being deposited onto the raised and recessed regions of the stamp, but none deposited onto the sidewalls. The pattern is then printed onto a substrate primed with a SAM bearing exposed thiol groups, which creates a strong bond between the gold layer and the substrate. If the stamp is removed, the adherent gold layer remains on the substrate. This process is not limited to single composition layers, and in fact multilayers of material have been printed using the interfacial interactions of the exposed layer for material transfer.

There are several advantages of this method *versus* the more developed process of photolithography and deposition used by the microelectronic industry. First, a pattern is printed directly, circumventing the need for photoresists, solvents and developing. Second, the contact that is made upon printing is extremely gentle. This has been found ideal for making contacts to fragile layers for use in molecular electronics.[94] In addition, the technique is ideal for making 3D structures through the printing of multiple layers. If a corrugated stamp is used with sidewalls that are not parallel to the evaporation source, the whole stamp becomes coated with metal. Upon printing, both the gold-coated raised and recessed regions along with the gold-coated sidewalls are transferred giving rise to a 3D structure in a single step.[95] A picture of a structure formed by repetition of this transfer is shown in Figure 2.15.

2.11 Electrically Contacting SAMs

Making good electrical contact with SAMs is crucial for the development of molecular electronic devices. Microcontact transfer printing of a metal overlayer on an elastomeric stamp to the surface of a chemically or physically more adherent alkanethiolate SAM, as described earlier, provides an elegant and straightforward solution to this problem. This is because of the gentle nature in which the new metal–SAM interface is created, by the transfer under ambient conditions of the metal film to the molecular monolayer. But what happens if we need to use the more conventional way of making electrical contacts to films by vapor deposition of the metal component onto the surface of the SAM? Does one observe adsorption of metal atoms like Al, Cu, Ag and Au on the surface of the SAM, or do these atomic species penetrate through the close-packed alkane layer of the SAM to the alkanethiolate–gold substrate interface?

Knowledge of this type will underpin the successful utilization of SAMs for a number of device applications where optimal electrical contact to the SAM determines the performance of the device. There are many variables in this kind of experiment that can determine its outcome including (i) alkane chain length, (ii) alkane terminal group, (iii) odd or even alkane chain length, (iv) metal atom

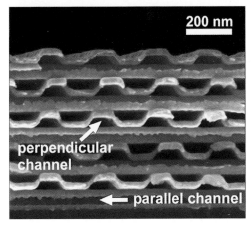

Figure 2.15 *A three-dimensional structure formed by repetitive nanotransfer printing of a series of line ridges. The top image shows two consecutive printings, while the bottom shows a sample printed nine times*
(Reproduced with permission from Ref. 95)

deposition method (*e.g.*, thermal, sputter, ablated), (v) metal type, (vi) metal–alkanethiol and metal–metal bond strengths and (vii) temperature. These are considered to be the most important factors determining interaction strengths and diffusion barriers of deposited metal atoms with the surface and internal regions of the SAM and the SAM–substrate interface, as illustrated in Figure 2.16.

This information was established using metal atom coverage-dependent time of flight secondary ion mass spectroscopy (TOF SIMS), infrared reflection spectroscopy (IRRS) and X-ray photoelectron spectroscopy (XPS).[96] Thermally deposited Al, Cu, Ag and Au atoms were found to simultaneously adsorb at the SAM surface and penetrate to the SAM–gold interface for methoxy-terminated C16 chain length alkanethiolate SAMs on gold {111}. The mechanism of adsorption involves MeO–M coordination, and the mode of penetration involves transient defects in the SAM. The reaction-diffusion behavior of Al is special because a stronger Al–AuSR bond than the Au/Ag/Cu–AuSR ones means Al atom penetration culminates after

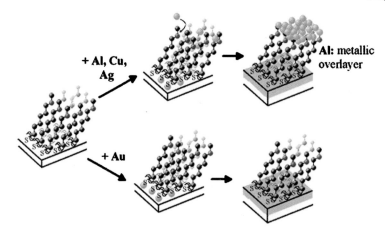

Figure 2.16 *The behavior of metals thermally deposited onto a SAM-terminated surface*
(Reproduced with permission from Ref. 96)

a ~1:1 Al:Au adlayer has been attained because of a greater diffusion barrier for
diffusion at the SAM–gold interface eliminating or reducing transient defect
formation necessary for penetration to occur. By contrast, deposited Cu, Ag and Au
atoms are able to diffuse at the SAM–gold interface enabling transient defects to
form and metal atoms to find their way to the interface at all coverage.

2.12 SAM Crystal Engineering

Nature's biomineralization of inorganic materials is a synthetic paradigm that
materials chemists have long strived to emulate. Shapes and patterns abound, where
form and composition, crystallinity or glassy disorder, orientation and alignment
can determine the function and utility of the biologically formed mineral. One of
the most impressive examples of this kind of crystal engineering in the laboratory
involves the use of SAM-patterned charge to control the nucleation and growth of
calcite single crystals in predetermined arrays and orientations on a substrate.[97,98]

The strategy used to accomplish this goal is illustrated in Figure 2.17. The idea is
to pattern surfaces with charged end-groups on the alkanethiol. SAMs are so well
ordered that the terminal charged groups can template a crystal lattice plane
complementary to the periodic SAM surface, a type of "molecular epitaxy." The
size and separation of the patches and the charge and chemical makeup of
the terminal group on the SAM together determine the pattern and rate of diffusion
of nutrients to the patches. This patterned charged template controls the rate of
nucleation of the seed crystals and crystal growth, as well as the ultimate
dimensions and separations of the crystals on the substrate. While the method can
be seen from Figure 2.18 to be capable of controlling crystal orientation, it does not
have any built-in control mechanism for establishing the relative orientation
(registry) of one crystal with respect to another.

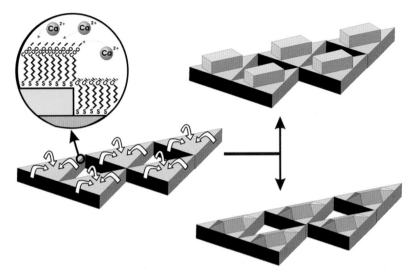

Figure 2.17 *Preferred nucleation of calcium carbonate as well as depletion of neighboring areas leads to crystal growth only on carboxylate-terminated SAM*

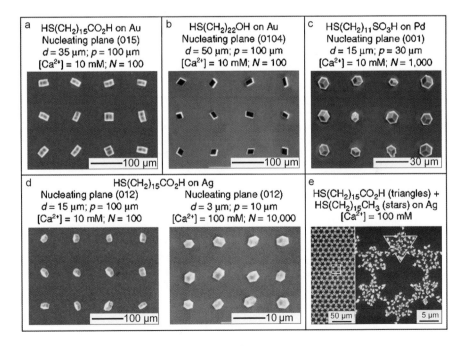

Figure 2.18 *Calcium carbonate crystals nucleating from precisely defined crystal planes by organized SAM assemblies*
(Reproduced with permission from Ref. 98)

A strategy for achieving oriented growth, patterning as well as alignment of preformed crystals through a directed self-assembly procedure has been devised. The key for aligning the crystals is the use of specifically shaped patches of SAM patterns that are both chemically and geometrically commensurate with dominant faces of the crystals, a type of "crystal epitaxy." Proof of concept that crystal engineering with this high level of perfection was possible made use of high-quality preformed hexagonal symmetry mesoporous silica crystals with a well-defined octahedral morphology.[99] The essence of the method is that triangular-shaped patches of functional SAMs are patterned to match the triangular-shaped faces of the octahedral morphology crystals. This type of "crystal epitaxy" facilitates the spatial organization, orientation and registered alignment of mesoporous silica octahedral morphology crystals, as seen from Figure 2.19. This demonstration experiment is approaching the quality and level of crystal engineering perfection that we have grown to recognize in biological mineralization. It is important to realize that such a system is in no way unique. The literature is replete with examples of extremely well-defined, faceted nanocrystals. The shape selective and crystal plane selective alignment of these building blocks on substrates should be a simple matter then trial-and-error. Perfection is now in the hands of everyone!

2.13 Learning From Nature's Biocrystal Engineering

To devise rational methods for making single crystals with micron-scale patterns is quite challenging, yet nature has achieved this in a myriad of exquisitely ornamented biominerals. An impressive example of this is seen in the SEM image shown in Figure 2.20, which displays a part of the skeleton of a brittlestar, *Ophiocoma wendtii* (Ophioroidea, Echinodermata). This entire structure is made out of a single calcite crystal! Each circular plate is a perfect lens used for some

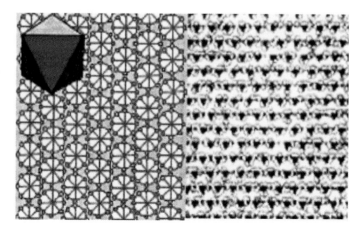

Figure 2.19 *Schematic (left) and SEM image (right) of preformed mesoporous silica octahedral morphology crystals organized on patterned, well-defined patches* (Reproduced with permission from Ref. 99)

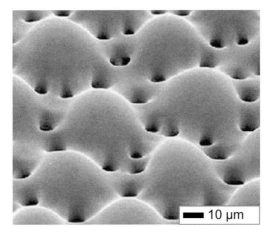

Figure 2.20 *Ornate optically functional microstructure of a brittlestar*
(Reproduced with permission from Ref. 100)

optical function, still poorly understood, while the mesh is used by the organism for its mechanical properties.[100] This is an appealing example of how the organism has biologically orchestrated calcium carbonate into a particular crystal polymorph and form to achieve a specific function. Could such a feat be copied using materials self-assembly strategies? This question has been unequivocally answered with an elegant bio-inspired template-directed demonstration of the growth of large micropatterned single calcite crystals[101] (Figure 2.21).

The procedure commences with the preparation of a photoresist micropattern based on a square grid of circular posts with feature sizes of 10 μm and an aspect ratio of 1, as shown in the top panel of Figure 2.21. The micropatterned substrate was coated with a transparent 5–10 nm thick film of gold. The tip of an AFM was then used to print a nanoscale region of a SAM of $HS(CH_2)_nA$ (A = OH, CO_2H, SO_3H) on the gold (this is called Dip-Pen Nanolithography and will be introduced in Chapter 4). This patch is an ordered region, designed to induce the nucleation of calcite in a controlled orientation. Mixtures of different alkane chain length alkanethiols terminating in phosphate, methyl and hydroxyl groups were used to derivatize the remainder of the gold surface to intentionally form a disordered organic surface that suppresses the nucleation of calcite and favors the formation of amorphous calcium carbonate (ACC). Next a thin gas-permeable polymeric film, made of PDMS, was coated on the functionalized substrate. The substrate, so patterned, functionalized and packaged, was then placed in a 1 M $CaCl_2$ solution (pH 7–9) in a closed desiccator containing $(NH_4)_2CO_3$ powder. CO_2 diffusion through the PDMS film creates a mesh of metastable ACC within the interstitial spaces of micropatterned substrate. Calcite crystal nucleation from the ACC film occurs only at the nanodomains defined by the tip of the AFM printed SAM, these seeds serving as templates for inducing and directing growth of oriented calcite over the whole substrate as illustrated in the middle panel of Figure 2.21.

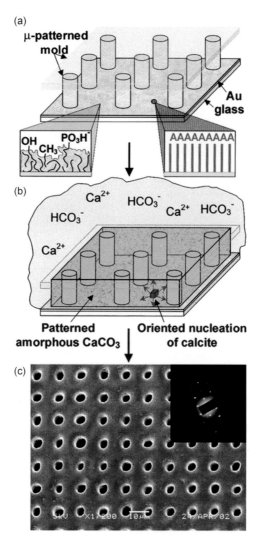

Figure 2.21 (a, b) *Steps in the formation of a patterned calcite single crystal.* (c) *SEM image of a calcite single crystal with circular pores* (Reproduced with permission from Ref. 101)

A SEM image of a single calcite crystal micropatterned in this way is shown in the lower panel of this figure. It is important to note that although the calcite mesh is porous the whole structure is a single crystal, which has been confirmed through optical birefringence measurements in a polarizing optical microscope. Patterned single crystals of this genre could spawn all sorts of interesting science and technology that was not possible before, such as enabling the topotactic growth of single crystals in single crystals such as apatite in calcite and perhaps an inroad to single crystal photonic crystals.

2.14 Colloidal Microsphere Patterns

While the synthesis of monodisperse silica and latex microspheres and their crystallization into close-packed arrays have been known for more than half a century,[102] it is only recently that a need has arisen for rational methods to arrange them into predefined patterns. When organized as films, monolayers of microspheres have attracted considerable interest, for example as templates for performing microsphere lithography[103] where voids between spheres are used as masks for deposition of material. Monolayers can also be used to project a hexagonal array of miniature patterns as miniature optical components for microsphere projection photolithography.[104] Microsphere multilayers that display optical Bragg diffraction are generating considerable interest for making planarized microphotonic crystals for all-optical devices, circuits and chips (see Chapter 7). Because the surface charge of microspheres can be systematically tailored through chemistry, it has proven possible to exploit electrostatic and capillary forces associated with charge-patterned SAMs to organize microspheres into predetermined patterns.[105] The strategy is summarized in Figure 2.22. Substrates are first chemically patterned with anionic and cationic regions to govern the deposition of charged microspheres. The direct observation of the microsphere assembly process suggests that two steps are involved: an initial patterned attachment of microspheres to the substrate *via* screened electrostatic forces, followed by an additional ordering of the microspheres by lateral capillary interactions upon drying. Using this strategy it is feasible to create 2D arrays of microspheres such as that shown in the bottom-left panel of Figure 2.22. This depicts an optical microscope image of a sample array, where the inset shows an SEM image of the structure-directing template. A high degree of microsphere focusing in a charged patch is inferred from the mapped distribution of microspheres within the printed circles. The proposed mechanism of microsphere ordering involves a balance of capillary forces between the spheres and the patch, where the ratio of the diameter of patch to sphere size controls whether ordered or disordered microspheres are obtained. The charge on the patch and microspheres in turn controls whether spheres assemble on or between the patches, with both arrangements achieved experimentally. Sphere arrays of this kind could be transferred to other substrates by microtransfer printing described earlier, and/or used as templates for making nanorings by edge spread lithography mentioned in Chapter 12.

2.15 Switching SAM Function

SAMs are useful because of the well-defined surface property they impart, but sometimes one is unable to get a molecule with a desired end-group to form a well-ordered layer. To solve this problem, one can scan across a series of different chain length molecules and vary the experimental conditions, which can become synthetically tedious. Another option is to coat a substrate with a SAM bearing a particular chemical group on its terminus, and then perform a transformation on this group to convert it to the desired functionality. The reactions available to effect this transformation are too numerous to mention, since any organic reaction available in

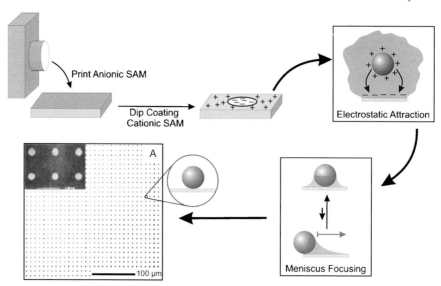

Figure 2.22 *Steps in the electrostatic patterning of an array of single microspheres. Anionic patches are printed, followed by modification of the unprinted areas with a cationic SAM. Cationic spheres are attracted to the patch through electrostatic interactions. When the solvent evaporates, the spheres are driven to the center of the patch through minimization of the meniscus area, which minimizes surface energy. The final result is an array of patches bearing single spheres, shown in a light microscopy image and SEM (inset) (Reproduced with permission from Ref. 105)*

the solution phase can theoretically be performed on a surface. The first reports to prove this idea started appearing in the early 1990s,[106–110] and it has since proven to be quite general enabling chemical, electrochemical or photochemical transformations, ligand binding and more.

One very interesting way to modify a SAM is through electrochemistry. Electrochemistry is a convenient way to supply electrons as reactants, and both the rate and the amount of electron transfer can be controlled with utmost precision. It has been shown that SAMs can be modified in this way on a very small scale by applying a voltage between an AFM tip and a substrate.[111] The voltage is localized to a small area underneath the tip, leading to local oxidation of terminal methyl groups to carboxylic acid groups. This method has since been shown to be applicable on a much larger scale. If a patterned electrode, such as a copper TEM grid, is brought into contact with a SAM layer on a conductive substrate and a voltage is applied across the layer, any area where the electrode is in contact undergoes an electrochemical oxidation.[112] For instance, a SAM, which is terminated with methyl groups, is oxidized to give terminal carboxylic acid groups. In this way, large SAM areas can be inter-converted using well-known electrochemical reactions. A schematic of how this procedure functions is shown in Figure 2.23.

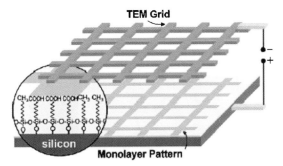

Figure 2.23 *Printing patterns of electrochemically oxidized SAMs* (Reproduced with permission from Ref. 112)

2.16 Patterning by Photocatalysis

It is well and good to convert one terminal group for another on a SAM, but what happens when we want to remove part of a SAM to have access to the bare substrate? An elegant method for performing this task has recently been reported, and relies on the ability of titania to photocatalytically decompose a wide range of organic molecules.[113] When in the presence of UV light, band gap excitation of titania creates electrons and holes where the electrons can convert water and oxygen molecules into highly reactive hydroxyl and superoxide radicals, which rapidly degrade organic molecules through oxidation. Therefore, when a SAM is supported on a titania substrate, UV irradiation causes rapid degradation of the monolayer.[114] This in itself may not be useful, but it can be adapted to create patterns on a silicon substrate.[115]

The process begins with the formation of a monolayer of trichloroalkylsilane on the surface of a silicon wafer. On a separate quartz slide, a pattern of trichloroalkylsilane is printed using a PDMS stamp, and the remaining areas are coated with titania through atomic layer deposition (ALD). ALD relies on alternating chemical reactions, which saturate the available surface bonds during each exposure step.[116] In the case of titania, if a quartz substrate is exposed alternately to titanium tetrachloride and water vapor, a film of titanium dioxide is gradually built up. The patterned quartz slide is then pressed against the SAM-coated silicon wafer, and the assembly is exposed to UV radiation. The UV decomposition of the SAM in contact with the titania pattern occurs about 20 times faster than the other areas, permitting the "sacrificial" patterning of the silicon wafer in just 2 min. The bare silicon substrate can then be further modified, and in this case is demonstrated by the selective ALD growth of zirconia. An AFM image of such a patterned silicon substrate is shown in Figure 2.24.

It is worth mentioning that there are other ways of removing SAMs from gold at the nanoscale to reveal the bare substrate. One strategy involves writing on a SAM on gold under ambient conditions with UV light from a near field scanning optical microscope (NSOM).[117,118] At sub-diffraction limit length scales this causes the written anchoring thiol group to be oxidized to a sulfonate, which is more polar and

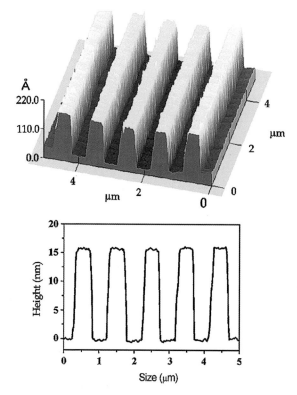

Figure 2.24 *AFM image of zirconia lines deposited onto photocatalytically patterned silicon*
(Reproduced with permission from Ref. 115)

more weakly bound to gold than the thiol group allowing it to be washed away with water thereby exposing the bare surface of the gold. Thin films of conjugated polymers can also be photo-oxidized in this way to form nanoscale patterns.[119]

2.17 Reversibly Switching SAMs

As we have seen, the most important aspect of SAMs is the utmost control they give us over surface properties. We can go one step further and manage to build various SAMs, which not only enable us to choose particular surface properties, but also allow us to change them with external stimuli. This external stimulus could be something like pressure, and such pressure-sensitive surfaces abound in industrial uses such as pressure-sensitive adhesives. We can use this surface-active behavior to create a long-lasting superhydrophobic surface. This is done by grafting a perfluoroalkyl trichlorosilane onto a stretched sheet of PDMS treated with an ozone plasma to create chemically reactive surface groups. If we then remove the stress on the elastomer sheet and allow it to relax, the areal density of fluorinated groups increases and these become much more tightly packed. The surface becomes much more hydrophobic, and provides greatly improved barrier properties to a variety of fluids.[120]

One of the simplest external triggers, light, can also be used to control the wetting characteristics of a surface.[121] A calix[4]resorcinarene is a relatively large and rigid cyclic molecule, which can be modified with, for instance, thiol groups on one face and photoisomerizable azobenzene units bearing octyl chains on the other. When these moieties are adsorbed onto a gold substrate they form a SAM with the azobenzene groups in the *trans* form, exposing on the surface low-surface energy octyl chains. The macrocyclic amphiphile is designed to ensure sufficient free volume for *trans*-to-*cis* isomerization, and irradiation with UV light results in isomerization to the *cis* isomer, exposing polar *cis*-azo groups and leading to increase in surface free energy. Blue light can then be used to photoisomerize the *cis*-azobenzene back to its *trans* form. A light gradient can easily be applied using the Gaussian intensity profile of a laser beam, and thus oil droplets and other objects can be guided on this photoresponsive surface along the induced gradient in surface free energy. The photodriven capture of a microsphere by a droplet of oil on such a responsive surface is shown in Figure 2.25.

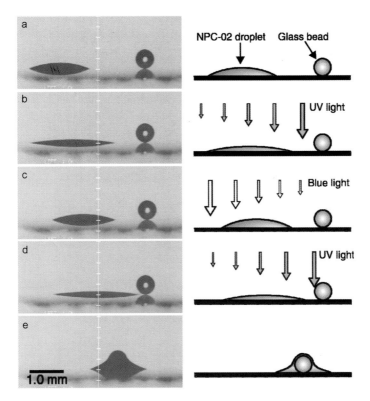

Figure 2.25 *Capture of a microsphere by an oil droplet by photoinducing an inchworm-like motion in the droplet*
(Reproduced with permission from Ref. 121)

2.18 Electrowettability Switch

An obvious goal in surface science is to achieve reversible switching of wettability with an electrical stimulus,[122] the most straightforward to apply in a device. A remarkably exciting example of tailoring the surface of a SAM to facilitate dynamic changes in interfacial properties and wettability in response to an electric potential is summarized in Figure 2.26.[123] The strategy for achieving this goal involves a novel alkanethiolate SAM on gold with a close-packed large head group based on a triphenylester as seen in the left panel of the figure. By hydrolyzing the ester to the carboxylic acid a sparsely packed alkanethiolate SAM with a small head group and flexible alkane chain is formed, as seen in the central panel of Figure 2.26. Voltage biasing the gold substrate negative or positive causes conformational switching and alteration between hydrophilic carboxyl and hydrophobic alkane surfaces, respectively, as illustrated in the right panel of the figure.

The voltage-induced switching action between hydrophilic and hydrophobic surfaces can be probed at the molecular scale by changes in methylene vibrational modes, and at the macroscopic scale by alterations in contact angle. In this way it can be shown that high and low packing density SAMs display distinct responses to an electrical potential. Intelligent SAM-patterned surfaces of this genre provide a novel means for regulating patterns of adhesion, friction and biocompatibility for utility in microfluidics, bioseparations, data storage and controlled release. Surface-active coatings have also been used, for instance, to actively control the orientation of various liquid crystals.[124]

Another chemical strategy for achieving electrowettability control of a surface involves the combination of a redox-active conducting polymer and a surfactant, like polyaniline (PANI) and dodecylbenzene sulfonic acid (DBSA), placed in contact with a polymer electrolyte such as glycerol–sodium dodecylbenzenesulfonate.[125] Electrical control of the state of charge of the PANI, through oxidation and reduction in an electrochemical cell, causes the charge-balancing DBSA to switch its orientation in the film, such that either the polar sulfonate head group or the non-polar alkane chain is exposed at the PANI–air interface where it can exert control over

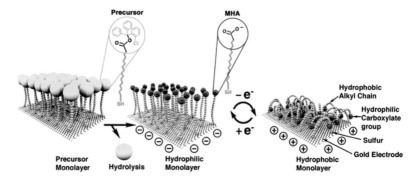

Figure 2.26   *Design of a dynamic SAM with switchable surface properties*
(Reproduced with permission from Ref. 123)

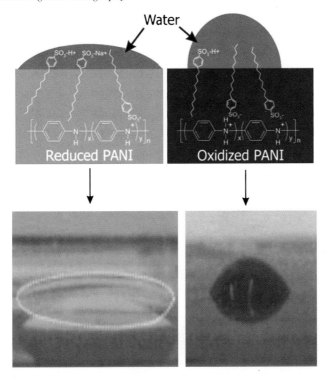

Figure 2.27 *Chemical process responsible for electrodewetting in polyaniline films (above) and optical pictures showing the effect on the contact angle of a water droplet* (Reproduced with permission from Ref. 125)

the wettability of a water droplet. This concept is illustrated in Figure 2.27 showing also a photograph of the effect on the contact angle of a drop of water on the surface of the film.

2.19 Sweet Chips

SAM-patterned substrates have gained widespread acceptance in the materials sciences for the organization of a range of functional inorganic and polymeric building blocks for applications in electronic, optical, photonic and magnetic devices. They are also acquiring prominence in the biological sciences for organizing DNA, proteins, carbohydrates and whole cells for utilization in genomic and proteomic investigations.

Biochips containing large-scale arrays of different bio-molecules anchored to micron-scale SAM patches are proving to be enormously interesting for high-throughput screening of specific interactions between DNA, proteins and carbohydrates. A recent example concerns a carbohydrate chip that enables the molecular basis of protein–carbohydrate interactions to be analyzed.[126] This information is important for understanding aspects of cell communication, adhesion and

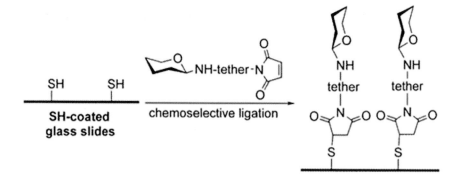

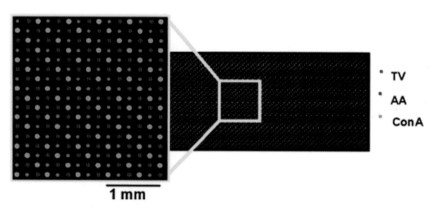

1 mm

Figure 2.28   *Chemical reactions resulting in surface-bound carbohydrate monolayers. The bottom shows an array of carbohydrate patches, which can be used for testing biological functions*
(Reproduced with permission from Ref. 126)

differentiation. Knowledge of these interactions can also provide the basis for developing new drugs to combat cell inflammation and metathesis.

This concept is illustrated in Figure 2.28 where different carbohydrate molecules have been tethered to patches on a glass slide *via* linkages of maleimide to a thiol. Using chips of this genre selective binding experiments of tethered carbohydrates, having distinct structural features, to fluorescence-tagged lectins were able to demonstrate a correlation between binding affinities for the molecule recognition events on the microarray and for the same molecules in solution. This provided strong evidence that carbohydrate chips are eminently suitable for large-scale rapid analysis of carbohydrate–protein interactions. Similar concepts and methods have been used in DNA and protein chips.

2.20 All Fall Down in a Row Lithography

Vertical microstructures can be likened to a row of undisturbed dominos. Periodic arrays of rectangular, triangular and cylindrical posts, defined in an SU-8 polymer photoresist by conventional photolithography, can behave quite similarly. That is, through the application of shear from an applied PDMS block, these vertical microstructures all fall down in a row like tipped micron-scale dominos.[127] The basic procedure is outlined in the left part of Figure 2.29.

The methodology delineated in the figure shows how sheared polymer vertical features collapse in a uniform fashion to provide a periodic 3D overlapping array, which is able to transfer intact to the shearing PDMS slab. Topologically complex microstructures of the genre accessed by this method, shown in the bottom SEM image, would be hard if not impossible to make by conventional forms of lithography.

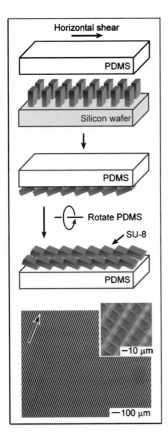
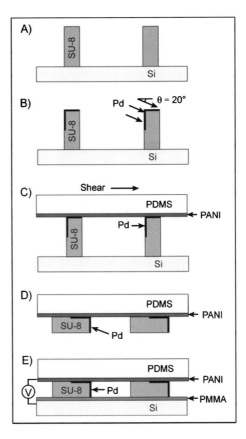

Figure 2.29 *Application of shear to an array of vertical microstructures causes them to topple like dominos. A metal coating on these structures can be used to transfer a surface charge onto a PMMA film*
(Reproduced with permission from Ref. 127)

Predictably, the repetitive patterns of microstructures obtainable by shear-induced tipping are found to depend on the direction in which the shear force is applied with respect to the original vertically standing microstructure array. These arrays have an aesthetically pleasing structure, but without modification they have very little functionality. Further steps, such as metal deposition, can be taken to create functional microstructures.

For instance, by using shadow deposition of metals like Pd onto one side of the top of the vertical microstructures and subsequently toppling these metal-modified microstructures with a PANI-coated PDMS slab, it proved possible to use the resulting electrically contacted array as a conducting patterned template for replicating charge into an electret.[128] The procedure is outlined in the right part of Figure 2.29, where the toppled microstructures are made to contact a film of PMMA coated on a silicon wafer. Application of a voltage causes a modification of the surface charge of the PMMA film, localized to the contacting Pd film having a thickness of only 15 nm. On removing the toppled microelectrode array from the PMMA film, it was possible to visualize the pattern of imprinted electrical charge in the PMMA through the surface potential image detected directly by Kelvin probe force microscopy. Microstructure toppling is a clever way to transfer the thickness control of thermal evaporation to lateral control of feature size.

2.21 References

1. R.G. Nuzzo, F.A. Fusco and D.L. Allara, Spontaneously organized molecular assemblies. 3. Preparation and properties of solution adsorbed monolayers of organic disulfides on gold surfaces, *J. Am. Chem. Soc.*, 1987, **109**, 2358.
2. M.D. Porter, T.B. Bright, D.L. Allara and C.E.D. Chidsey, Spontaneously organized molecular assemblies. 4. Structural characterization of normal-alkyl thiol monolayers on gold by optical ellipsometry, infrared-spectroscopy, and electrochemistry, *J. Am. Chem. Soc.*, 1987, **109**, 3559.
3. C.D. Bain, E.B. Troughton, Y.T. Tao, J. Evall, G.M. Whitesides and R.G. Nuzzo, Formation of monolayer films by the spontaneous assembly of organic thiols from solution onto gold, *J. Am. Chem. Soc.*, 1989, **111**, 321.
4. A.R. Bishop and R.G. Nuzzo, Self-assembled-monolayers: recent developments and applications, *Curr. Opin. Colloid Interface Sci.*, 1996, **1**, 127.
5. C.D. Bain and G.M. Whitesides, Molecular-level control over surface order in self-assembled monolayer films of thiols on gold, *Science*, 1988, **240**, 62–63.
6. Y. Xia and G.M. Whitesides, Soft lithography, *Angew. Chem. Int. Ed.*, 1998, **37**, 550.
7. Y. Xia, J.A. Rogers, K.E. Paul and G.M. Whitesides, Unconventional methods for fabricating and patterning nanostructures, *Chem. Rev.*, 1999, **99**, 1823.
8. E. Kim, Y. Xia and G.M. Whitesides, Polymer microstructures formed by molding in capillaries, *Nature*, 1995, **376**, 581.
9. T.W. Odom, V.R. Thalladi, J.C. Love and G.M. Whitesides, Generation of 30–50 nm structures using easily fabricated, composite PDMS masks, *J. Am. Chem. Soc.*, 2002, **124**, 12112.

10. A. Kumar and G.M. Whitesides, Features of gold having micrometer to centimeter dimensions can be formed through a combination of stamping with an elastomeric stamp and an alkanethiol ink followed by chemical etching, *Appl. Phys. Lett.*, 2002, **63**, 2002.

11. X.M. Zhao, Y. Xia and G.M. Whitesides, Fabrication of three-dimensional microstructures. Microtransfer molding, *Adv. Mater.*, 1996, **8**, 837.

12. H. Wu, T.W. Odom, D.T. Chiu and G.M. Whitesides, Fabrication of complex three-dimensional microchannel systems in PDMS, *J. Am. Chem. Soc.*, 2003, **125**, 554.

13. See interview with George Whitesides: www.sciencewatch.com/july-aug2002/sw_july-aug2002_page3.htm

14. L. Yan, W.T.S. Huck and G.M. Whitesides, Self-assembled monolayers (SAMs) and synthesis of planar micro- and nanostructures, *J. Macromol. Sci. – Polymer Rev.*, 2004, **C44**, 175.

15. R.J. Jackman, J.L. Wilbur and G.M. Whitesides, Fabrication of submicro-meter features on curved substrates by microcontact printing, *Science*, 1995, **269**, 664.

16. K.E. Paul, M. Prentiss and G.M. Whitesides, Patterning spherical surfaces at the two-hundred-nanometer scale using soft lithography, *Adv. Funct. Mater.*, 2003, **13**, 259.

17. G.S. Ferguson, M.K. Chaudhury, G.B. Sigal and G.M. Whitesides, Contact adhesion of thin gold films on elastomeric supports: cold welding under ambient conditions, *Science*, 1991, **253**, 776.

18. M. Husemann, D. Mecerreyes, C.J. Hawker, J.L. Hedrick, R. Shah and N.L. Abbott, Surface-initiated polymerization for amplification of self-assembled monolayers patterned by microcontact printing, *Angew. Chem. Int. Ed.*, 1999, **38**, 647.

19. W.B. Caldwell, D.J. Campbell, K.M. Chen, B.R. Herr, C.A. Mirkin, A. Mlik, M.K. Durbin, P. Dutta and K.G. Huang, A highly ordered self-assembled monolayer film of an azobenzenealkanethiol on Au(111) – electrochemical properties and structural characterization by synchrotron in-plane X-ray-diffraction, atomic-force microscopy, and surface-enhanced Raman-spectroscopy, *J. Am. Chem. Soc.*, 1995, **117**, 6071.

20. H. Yang, P. Deschatelets, S.T. Brittain and G.M. Whitesides, Fabrication of high performance ceramic microstructures from a polymeric precursor using soft lithography, *Adv. Mater.*, 2001, **13**, 54.

21. K.L. Prime and G.M. Whitesides, Self-assembled organic monolayers: model systems for studying adsorption of proteins at surfaces, *Science*, 1991, **252**, 1164.

22. G.P. Lopez, M.W. Albers, S.L. Schreiber, R. Carroll, E. Peralta and G.M. Whitesides, Convenient methods for patterning the adhesion of mammalian cells to surfaces using self-assembled monolayers of alkanethiolates on gold, *J. Am. Chem. Soc.*, 1993, **115**, 5877.

23. P.A. DiMilla, J.P. Folkers, H.A. Biebuyck, R. Haerter, G.P. Lopez and G.M. Whitesides, Wetting and protein adsorption on self-assembled mono-layers of alkanethiolates supported on transparent films of gold, *J. Am. Chem. Soc.*, 1994, **116**, 2225.

24. R.S. Kane, S. Takayama, E. Ostuni, D.E. Ingber and G.M. Whitesides, Patterning proteins and cells using soft lithography, *Biomaterials*, 1999, **20**, 2363.

25. J.C. McDonald and G.M. Whitesides, Poly(dimethylsiloxane) as a material for fabricating microfluidic devices, *Acc. Chem. Res.*, 2002, **35**, 491.

26. E.E. Endler, K.A. Duca, P.F. Nealey, G.M. Whitesides and J. Yin, Propagation of viruses on micropatterned host cells, *Biotechnol. Bioeng.*, 2003, **81**, 719.

27. C.S. Chen, M. Mrksich, S. Huang, G.M. Whitesides and D.E. Ingber, Geometric control of cell life and death, *Science*, 1997, **276**, 1425.

28. C.S. Chen, M. Mrksich, S. Huang, G.M. Whitesides and D.E. Ingber, Micropatterned surfaces for control of cell shape, position, and function, *Biotechnol. Prog.*, 1998, **14**, 356.

29. A. Brock, E. Chang, C.-C. Ho, P. LeDuc, X. Jiang, G.M. Whitesides and D.E. Ingber, Geometric determinants of directional cell motility revealed using microcontact printing, *Langmuir*, 2003, **19**, 1611.

30. R. Singhvi, A. Kumar, G.P. Lopez, G.N. Stephanopoulos, D.I.C. Wang, G.M. Whitesides and D.E. Ingber, Engineering cell-shape and function, *Science*, 1994, **264**, 696.

31. B. Kasemo, Biological surface science, *Surf. Sci.*, 2002, **500**, 656.

32. A. Ulman, Formation and structure of self-assembled monolayers, *Chem. Rev.*, 1996, **96**, 1533.

33. J.A. Rodriguez, J. Dvorak, T. Jirsak, G. Liu, J. Hrbek, Y. Aray and C. Gonzalez, Coverage effects and the nature of the metal–sulfur bond in S/Au(111): high-resolution photoemission and density-functional studies, *J. Am. Chem. Soc.*, 2003, **125**, 276.

34. J. Hautman, J.P. Bareman, W. Mar and M.L. Klein, Molecular-dynamics investigations of self-assembled monolayers, *J. Chem. Soc. Faraday Trans.*, 1991, **87**, 2031.

35. M.C. Leopold, J.A. Black and E.F. Bowden, Influence of gold topography on carboxylic acid terminated self-assembled monolayers, *Langmuir*, 2002, **18**, 978.

36. D. Roy and J. Fendler, Reflection and absorption techniques for optical characterization of chemically assembled nanomaterials, *Adv. Mater.*, 2004, **16**, 479.

37. C.T. Buscher, D. McBranch and D.Q. Li, Understanding the relationship between surface coverage and molecular orientation in polar self-assembled monolayers, *J. Am. Chem. Soc.*, 1996, **118**, 2950.

38. C.A. Alves, E.L. Smith and M.D. Porter, Atomic scale imaging of alkanethiolate monolayers at gold surfaces with atomic force microscopy, *J. Am. Chem. Soc.*, 1992, **114**, 1222.

39. M.K. Chaudhury, Adhesion and friction of self-assembled organic mono-layers, *Curr. Opin. Colloid Interface Sci.*, 1997, **2**, 65.

40. J.E. Houston and H.I. Kim, Adhesion, friction, and mechanical properties of functionalized alkanethiol self-assembled monolayers, *Acc. Chem. Res.*, 2002, **35**, 547.

41. C.M. Whelan, M. Kinsella, L. Carbonell, H.M. Ho and K. Maex, Corrosion inhibition by self-assembled monolayers for enhanced wire bonding on Cu surfaces, *Microelec. Eng.*, 2003, **70**, 551.

42. A. Kumar, H.A. Biebuyck, N.L. Abbott and G.M. Whitesides, The use of self-assembled monolayers and a selective etch to generate patterned gold features, *J. Am. Chem. Soc.*, 1992, **114**, 88.

43. Y. Xia, X.-M. Zhao, E. Kim and G.M. Whitesides, A selective etching solution for use with patterned self-assembled monolayers of alkanethiolates on gold, *Chem. Mater.*, 1995, **7**, 2332.

44. P.E. Laibinis and G.M. Whitesides, Self-assembled monolayers of n-alkanethiolates on copper are barrier films that protect the metal against oxidation by air, *J. Am. Chem. Soc.*, 1992, **114**, 9022.

45. L. Yan, C. Marzolin, A. Terfort and G.M. Whitesides, Formation and reaction of interchain carboxylic anhydride groups on self-assembled monolayers on gold, *Langmuir*, 1997, **13**, 6704.

46. P.E. Laibinis, G.M. Whitesides, D.L. Allara, Y.T. Tao, A.N. Parikh and R.G. Nuzzo, Comparison of the structures and wetting properties of self-assembled monolayers of n-alkanethiols on the coinage metal surfaces, copper, silver, and gold, *J. Am. Chem. Soc.*, 1991, **113**, 7152.

47. L.B. Goetting, T. Deng and G.M. Whitesides, Microcontact printing of alkanephosphonic acid on aluminum: pattern transfer by wet chemical etching, *Langmuir*, 1999, **15**, 1182.

48. J.C. Love, D.B. Wolfe, R. Haasch, M.L. Chabinyc, K.E. Paul, G.M. Whitesides and R.G. Nuzzo, Formation and structure of self-assembled monolayers of alkanethiolates on palladium, *J. Am. Chem. Soc.*, 2003, **125**, 2597.

49. T.L. Breen, P.M. Fryer, R.W. Nunes and M.E. Rothwell, Patterning indium tin oxide and indium zinc oxide using microcontact printing and wet etching, *Langmuir*, 2002, **18**, 194.

50. S.R. Wasserman, Y.T. Tao and G.M. Whitesides, Structure and reactivity of alkylsiloxane monolayers formed by reaction of alkyltrichlorosilanes on silicon substrates, *Langmuir*, 1989, **5**, 1074.

51. Y. Xia, M. Mrksich, E. Kim and G.M. Whitesides, Microcontact printing of octadecylsiloxane on the surface of silicon dioxide and its application in microfabrication, *J. Am. Chem. Soc.*, 1995, **117**, 9576.

52. J.J. Gooding, F. Mearns, W.R. Yang and J.Q. Liu, Self-assembled monolayers into the 21(st) century: recent advances and applications, *Electroanalysis*, 2003, **15**, 81.

53. G.A. Ozin and S.M. Yang, The race for the photonic chip: colloidal crystal assembly in silicon wafers, *Adv. Funct. Mater.*, 2001, **11**, 95.

54. C.D. Bain and G.M. Whitesides, Correlations between wettability and structure in monolayers of alkanethiols adsorbed on gold, *J. Am. Chem. Soc.*, 1988, **110**, 3665.

55. N.L. Abbott, C.B. Gorman and G.M. Whitesides, Active control of wetting using applied electrical potentials and self-assembled monolayers, *Langmuir*, 1995, **11**, 16.

56. N.L. Abbott, J.P. Folkers and G.M. Whitesides, Manipulation of the wettability of surfaces on the 0.1-micrometer to 1-micrometer scale through micromachining and molecular self-assembly, *Science*, 1992, **257**, 1380.

57. M.K. Chaudhury and G.M. Whitesides, How to make water run uphill, *Science*, 1992, **256**, 1539.

58. G.P. Lopez, H.A. Biebuyck, C.D. Frisbie and G.M. Whitesides, Imaging of features on surfaces by condensation figures, *Science*, 1993, **260**, 647.

59. A. Kumar and G.M. Whitesides, Patterned condensation figures as optical diffraction gratings, *Science*, 1994, **263**, 60.

60. R.C. Bailey, J.M. Nam, C.A. Mirkin and J.T. Hupp, Real-time multicolor DNA detection with chemoresponsive diffraction gratings and nanoparticle probes, *J. Am. Chem. Soc.*, 2003, **125**, 13541.

61. J.B. Goh, P.L. Tam, R.W. Loo and M.C. Goh, A quantitative diffraction-based sandwich immunoassay, *Anal. Biochem.* 2003, **313**, 262.

62. M. Hutley, R. Stevens and D. Daly, Microlens arrays, *Phys. World*, 1991, **4**, 27.

63. D. Daly, *Microlens Arrays*, CRC Press, Boca Raton, FL, 2000.

64. M.-H. Wu and G.M. Whitesides, Fabrication of diffractive and micro-optical elements using microlens projection lithography, *Adv. Mater.*, 2002, **14**, 1502.

65. M.-H. Wu, C. Park and G.M. Whitesides, Fabrication of arrays of microlenses with controlled profiles using gray-scale microlens projection photolithography, *Langmuir*, 2002, **18**, 9312.

66. H. Wu, T.W. Odom, and G.M. Whitesides, Generation of chrome masks with micrometer-scale features using microlens lithography, *Adv. Mater.*, 2002, **14**, 1213.

67. J.C. Love, D.B. Wolfe, H.O. Jacobs, and G.M. Whitesides, Microscope projection photolithography for rapid prototyping of masters with micron-scale features for use in soft lithography, *Langmuir*, 2001, **17**, 6005.

68. W.H. Miller, G.D. Bernard and J.L. Allen, The optics of insect compound eyes, *Science*, 1968, **162**, 760.

69. N.F. Borrelli, R.H. Bellman, J.A. Durbin and W. Lama, Imaging and radiometric properties of microlens arrays, *Appl. Opt.*, 1991, **30**, 3633.

70. H.A. Biebuyck and G.M. Whitesides, Self-organization of organic liquids on patterned self-assembled monolayers of alkanethiolates on gold, *Langmuir*, 1994, **10**, 2790.

71. M.H. Wu and G.M. Whitesides, Fabrication of two-dimensional arrays of microlenses and their applications in photolithography, *J. Micromech. Microeng.*, 2002, **12**, 747.

72. G. Lu, W. Li, J.M. Yao, G. Zhang, B. Yang and J.C. Shen, Fabricating ordered two-dimensional arrays of polymer rings with submicrometer-sized features on patterned self-assembled monolayers by dewetting, *Adv. Mater.*, 2002, **14**, 1049.

73. S.B. Clendenning, S. Fournier-Bidoz, A. Pietrangelo, G.C. Yang, S.J. Han, P.M. Brodersen, C.M. Yip, Z.H. Lu, G.A. Ozin and I. Manners, Ordered 2D arrays of ferromagnetic Fe/Co nanoparticle rings from a highly metallized metallopolymer precursor, *J. Mater. Chem.*, 2004, **14**, 1686.

74. A. Berenbaum, M. Ginzburg-Margau, N. Coombs, A.J. Lough, A. Safa-Sefat, J.E. Greedan, G.A., Ozin and I. Manners, Ceramics containing magnetic Co–Fe alloy nanoparticles from the pyrolysis of a highly metallized organometallic polymer precursor, *Adv. Mater.*, 2003, **15**, 51.

75. A.R. West, *Solid State Chemistry and Its Applications*, Wiley, New York, 1984.

76. P.F. McMillan, High pressure synthesis of solids, *Curr. Opin. Solid State Mater. Sci.*, 1999, **4**, 171.

77. Y.D. Tretyakov and E.A. Goodilin, Chemical principles of the metal-oxide superconductors preparation, *Uspekhi Khimii*, 2000, **69**, 3.

78. A. Simon, Superconductivity and chemistry, *Angew. Chem. Int. Ed.*, 1997, **36**, 1789.

79. J.I. Martin, J. Nogues, K. Liu, J.L. Vicent and I.K. Schuller, Ordered magnetic nanostructures: fabrication and properties, *J. Magn. Magn. Mater.*, 2003, **256**, 449.

80. B. Raveau, A. Maignan, C. Martin and M. Hervieu, Colossal magnetoresistance manganite perovskites: relations between crystal chemistry and properties, *Chem. Mater.*, 1998, **10**, 2641.

81. K. Kang, S.H. Dai and Y.H. Wan, Progress in synthetic methods, of cathode material $LiM_xMn_{2-x}O_4$ for lithium ion batteries, *J. Inorg. Mater.*, 2001, **16**, 586.

82. Z. Zhong, B. Gates, Y. Xia and D. Qin, Soft lithographic approach to the fabrication of highly ordered 2D arrays of magnetic nanoparticles on the surfaces of silicon substrates, *Langmuir*, 2000, **16**, 10369.

83. O. Prucker and J. Ruhe, Synthesis of poly(styrene) monolayers attached to high surface area silica gels through self-assembled monolayers of azo initiators, *Macromolecules* 1998, **31**, 592.

84. O. Prucker and J. Ruhe, Polymer layers through self-assembled monolayers of initiators, *Langmuir*, 1998, **14**, 6893.

85. M. Husemann, D. Mecerreyes, C.J. Hawker, J.L. Hedrick, R. Shah and N.L. Abbott, Surface-initiated polymerization for amplification of self-assembled monolayers patterned by microcontact printing, *Angew. Chem. Int. Ed.*, 1999, **38**, 647.

86. W.F. Guo and G.K. Jennings, Directed growth of polymethylene films on atomically modified gold surfaces, *Adv. Mater.*, 2003, **15**, 588.

87. W.F. Guo and G.K. Jennings, Use of underpotentially deposited metals on gold to affect the surface-catalyzed formation of polymethylene films, *Langmuir*, 2002, **18**, 3123.

88. X.P. Chen and K.Y. Qiu, Study of "living"/controlled radical polymerization, *Prog. Chem.*, 2001, **13**, 224.

89. T. Ando, M. Kamigaito and M. Sawamoto, Recent development of transition metal-catalyzed living radical polymerization – design and development of the metal complexes, *Kobunshi Ronbunshu*, 2002, **59**, 199.

90. N. Hadjichristidis, S. Pispas, H. Iatrou and M. Pitsikalis, Linking chemistry and anionic polymerization, *Curr. Org. Chem.*, 2002, **6**, 155.

91. D.J. Dyer, Patterning of gold substrates by surface-initiated polymerization, *Adv. Funct. Mater.*, 2003, **13**, 667.

92. Q.J. Guo, X.W. Teng, S. Rahman and H. Yang, Patterned Langmuir–Blodgett films of mondisperse nanoparticles of iron oxide using soft lithography, *J. Am. Chem. Soc.*, 2003, **125**, 630.

93. Y.L. Loo, R.L. Willett, K.W. Baldwin and J.A. Rogers, Interfacial chemistries for nanoscale transfer printing, *J. Am. Chem. Soc.* 2002, **124**, 7654.

94. Y.L. Loo, D.V. Lang, J.A. Rogers and J.W.P. Hsu, Electrical contacts to molecular layers by nanotransfer printing, *Nano Lett.*, 2003, **3**, 913.

95. J. Zaumseil, M.A. Meitl, J.W.P. Hsu, B.R. Acharya, K.W. Baldwin, Y.L. Loo and J.A. Rogers, Three-dimensional and multilayer nanostructures formed by nanotransfer printing, *Nano Lett.*, 2003, **3**, 1223.

96. A.V. Walker, T.B. Tighe, O.M. Cabarcos, M.D. Reinard, B.C. Haynie, S. Uppili, N. Winograd and D.L. Allara, The dynamics of noble metal atom penetration through methoxy-terminated alkanethiolate monolayers, *J. Am. Chem. Soc.*, 2004, **126**, 3954.

97. J. Aizenberg, A.J. Black and G.M. Whitesides, Oriented growth of calcite controlled by self-assembled monolayers of functionalized alkanethiols supported on gold and silver, *J. Am. Chem. Soc.*, 1999, **121**, 4500.

98. J. Aizenberg, A.J. Black and G.M. Whitesides, Control of crystal nucleation by patterned self-assembled monolayers, *Nature*, 1999, **398**, 495.

99. Z.R.R. Tian, J. Liu, H.F. Xu, J.A. Voigt, B. Mckenzie and C.M. Matzke, Shape-selective growth, patterning, and alignment of cubic nanostructured crystals via self-assembly, *Nano Lett.*, 2003, **3**, 179.

100. J. Aizenberg, A. Tkachenko, S. Weiner, L. Addadi and G. Hendler, Calcitic microlenses as part of the photoreceptor system in brittlestars, *Nature*, 2001, **412**, 819.

101. J. Aizenberg, D.A. Muller, J.L. Grazul and D.R. Hamann, Direct fabrication of large micropatterned single crystals, *Science*, 2003, **299**, 1205.

102. Y. Xia, B. Gates, Y. Yin and Y. Lu, Monodispersed colloidal spheres: old materials with new applications, *Adv. Mater.*, 2000, **12**, 693.

103. J.C. Halteen and R.P. Van Duyne. Nanosphere lithography: a materials general fabrication process for periodic particle array surfaces, *J. Vac. Sci. Tech. A*, 1995, **13**, 1553.

104. J.C. Love, D.B. Wolfe, H.O. Jacobs and G.M. Whitesides, Microscope projection photolithography for rapid prototyping of masters with micron-scale features for use in soft lithography, *Langmuir*, 2001, **17**, 6005.

105. J. Aizenberg, P.V. Braun and P. Wiltzius, Patterned colloidal deposition controlled by electrostatic and capillary forces, *Phys. Rev. Lett.*, 2000, **84**, 2997.

106. N. Tillman, A. Ulman and J.F. Elman, Oxidation of a sulfide group in a self-assembled monolayer, *Langmuir*, 1989, **5**, 1020.

107. N. Balachander and C.N. Sukenik, Monolayer transformation by nucleophilic-substitution – applications to the creation of new monolayer assemblies, *Langmuir*, 1990, **6**, 1621.

108. C.S. Dulcey, J.H. Georger, V. Krauthamer, D.A. Stenger, T.L. Fare and J.M. Calvert, Deep UV photochemistry of chemisorbed monolayers – patterned coplanar molecular assemblies, *Science*, 1991, **252**, 551.

109. D.G. Kurth and T. Bein, Surface-reactions on thin-layers of silane coupling agents, *Langmuir*, 1993, **9**, 2965.

110. W.J. Dressick, C.S. Dulcey, J.H. Georger, G.S. Calabrese and J.M. Calvert, Covalent binding of Pd catalysts to ligating self-assembled monolayer films for selective electroless metal-deposition, *J. Electrochem. Soc.*, 1994, **141**, 210.

111. R. Maoz, S.R. Cohen and J. Sagiv, Nanoelectrochemical patterning of monolayer surfaces: toward spatially defined self-assembly of nanostructures, *Adv. Mater.*, 1999, **11**, 55.

112. S. Hoeppener, R. Maoz and J. Sagiv, Constructive microlithography: electrochemical printing of monolayer template patterns extends constructive nanolithography to the micrometer–millimeter dimension range, *Nano Lett.*, 2003, **3**, 761.

113. A. Fujishim and K. Honda, Electrochemical photolysis of water at a semiconductor electrode, *Nature*, 1972, **238**, 37; A. Mills and S. LeHunte, An overview of semiconductor photocatalysis, *J. Photochem. Photobiol. A-Chem.*, 1997, **108**, 1.

114. J.P. Lee, H.K. Kim, C.R. Park, G. Park, H.T. Kwak, S.M. Koo and M.M. Sung, Photocatalytic decomposition of alkylsiloxane self-assembled mono-layers on titanium oxide surfaces, *J. Phys. Chem. B*, 2003, **107**, 8997.

115. J.P. Lee and M.M. Sung, A new patterning method using photocatalytic lithography and selective atomic layer deposition, *J. Am. Chem. Soc.*, 2004, **126**, 28.

116. T. Suntola, Atomic layer epitaxy, *Thin Solid Films*, 1992, **216**, 84.

117. E. Betzig, J.K. Trautman, T.D. Harris, J.S. Weiner and R.L. Kostelak, Breaking the diffraction barrier – optical microscopy on a nanometric scale, *Science*, 1991, **251**, 1468.

118. E. Betzig and J.K. Trautman, Near-field optics – Microscopy, spectroscopy, and surface modification beyond the diffraction limit, *Science*, 1992, **257**, 189.

119. P.K. Wei, J.H. Hsu, B.R. Hsieh and W.S. Fann, Surface modification and patterning of conjugated polymers with near-field optical microscopy, *Adv. Mater.*, 1996, **8**, 573.

120. J. Genzer and K. Efimenko, Creating long-lived superhydrophobic poly-mer surfaces through mechanically assembled monolayers, *Science*, 2000, **290**, 2130.

121. K. Ichimura, S.-K. Oh and M. Nakagawa, Light-driven motion of liquids on a photoresponsive surface, *Science*, 2000, **288**, 1624.

122. B.S. Gallardo, V.K. Gupta, F.D. Eagerton, L.I. Jong, T.V.S. Craig, R.R. Shah and N.L. Abbott, Electrochemical principles for active control of liquids on submillimeter scales, *Science*, 1999, **283**, 57.

123. J. Lahann, S. Mitragotri, T.N. Tran, H. Kaido, J. Sundaram, I.S. Choi, S. Hoffer, G.A. Somorjai and R. Langer, A reversibly switching surface, *Science*, 2003, **299**, 371.

124. Y.-Y. Luk and N.L. Abbott, Surface-driven switching of liquid crystals using redox-active groups on electrodes, *Science*, 2003, **301**, 623.

125. J. Isaksson, C. Tengstedt, M. Fahlman, N. Robinson and M. Berggren, A solid-state organic electronic wettability switch, *Adv. Mater.*, 2004, **16**, 316.
126. S. Park, M.R. Lee, S.J. Pyo and I. Shin, Carbohydrate chips for studying high-throughput carbohydrate–protein interactions, *J. Am. Chem. Soc.*, 2004, **126**, 4812.
127. B.D. Gates, Q.B. Xu, V.R. Thalladi, T.B. Cao, T. Knickerbocker and G.M. Whitesides, Shear patterning of microdominos: a new class of procedures for making micro- and nanostructures, *Angew. Chem. Int. Ed.*, 2004, **43**, 2780.
128. H.O. Jacobs and G.M. Whitesides, Submicrometer patterning of charge in thin-film electrets, *Science*, 2001, **291**, 1763.

Nanofood for Thought – Soft Lithography, SAMs, Patterning

1. How may SAM methods be utilized to accomplish the following materials targets: (a) micron-scale checker board array of oriented CdS and CdSe crystals; (b) 50 nm wide grid lines of gold; (c) rows of aligned carbon nanotubes; (d) chemically sensitive switchable optical grating; (e) mesoporous silica grating; (f) micron dimension hexagonal pattern of argentothiol nanoclusters; (g) alternating strips of polystyrene and polyvinylpyridine; (h) copper microelectronic circuit; (i) millimeter diameter glass fiber coated with separated rings of micron dimension gold discs; (j) triangular drops of water; (k) liquid droplet spatio-temporal control at the sub-millimeter scale; (l) microgeodesic dome; (m) homeotropic or planar anchoring of the same thermotropic liquid crystal on a planar substrate; (n) an electronic ink based on a microsphere on a surface; (n) tetrahedral disposed chemically smart patches on the surface of a microsphere?
2. Describe a materials chemistry approach to make nanorods (i) roll uphill and water droplets (ii) form a checkerboard pattern of blue and green colored squares.
3. How would you go about chemically patterning and characterizing the surface of a SAM with different amide functionalities? Where could this methodology prove valuable technologically?
4. Explain how you would employ a SAM to control the optical birefringence properties of a nematic liquid crystal and how you would use the knowledge to build a flat panel liquid crystal display with a wide viewing angle. What benefits does this method offer over the old fashioned "rubbing" technique for controlling the director fields of thermotropic liquid crystals?
5. How could you use soft lithography to make a micron-scale square grid of capped ferromagnetic nanoclusters and what would you use it for?
6. In the context of advanced crystal engineering can you think of a materials self-assembly technique for creating a micron-scale checkerboard pattern of oriented calcite nanocrystals on a silicon substrate?
7. Can you imagine a synthetic approach for creating a micron-scale pattern of submicron diameter holes running through a single crystal of calcite?
8. How and why would you make a micron-scale checkerboard pattern of 10 nm sized $NiFe_2O_4$ nickel ferrite spinel nanocrystals on a silicon substrate?
9. How could the wettability of a SAM surface be altered in a systematic way without changing the SAM?
10. What innovative experiments could you perform with a SAM on the tip of an SPM and on a planar substrate?
11. Can you imagine some creative experiments that you could perform if you were able to bring together with nanoscale precision SAM 1 and SAM

2-coated substrates such that they just touched? How could such a junction be achieved?

12. Devise an experiment that you could perform on mixing two different alkanethiols on a gold substrate when the alkanethiols have (a) different alkane chain lengths; (b) different chain terminal functional groups; (c) different chain internal functional groups.

13. How does an alkanethiol chemisorbed on Au (111) differ from Au (100) and Ag (111) and how can one make use of these differences?

14. What controls the tilt angle of an alkanethiol SAM on gold?

15. Delineate the types of defects that exist in an alkanethiol SAM on a gold surface and explain how they will influence the use of SAM in perceived applications.

16. Describe how defects in SAMs may be intentionally created, and how such controlled defects can be used for nanoscale patterning.

17. How would you organize a collection of monodispersed octahedral morphology micron size crystals in a micron-scale hexagonal pattern on a substrate, where all the crystals have the same orientation?

18. In the context of coordination chemistry and metal–ligand bonding conceive of a chemical approach that enables functional materials to be patterned at the (a) micron; (b) submicron; and (c) nanometer scale on the surface of a SAM.

19. Try to develop a surface organometallic chemistry of a SAM that would be sufficiently original, important and potentially technologically relevant to secure your tenure as a new faculty member in a university chemistry department in your country.

20. How would you measure the stability constant of an α,ω-alkanedithiol SAM with respect to its coordinating ability to a metal ion, and how would you expect it to compare with the classical solution phase value?

21. How many chemical and physical methods can you think of that could be successfully used to remove alkanethiols from spatially selected areas of a SAM on gold? Why would you want to do such a thing?

22. How could you enhance the thermal stability of a SAM?

23. What approaches could you use to control the diffusion rate of an alkanethiol on gold?

24. Which volatile products would you expect to detect by laser desorption mass spectroscopy of a SAM made of two different alkanethiols on gold, and what information would this knowledge provide about the nature of the interaction, between the alkanethiol and gold as well as the distribution of the two different alkanethiols in the monolayer film?

25. It is not common knowledge that alkanethiols form SAMs on the surface of liquid mercury. How would you expect these SAMs to differ from those formed on thermally or sputter-deposited gold on a silicon substrate? What would you expect to happen if you microcontact printed a SAM pattern on the surface of mercury? Devise strategies to probe the evolution of such SAM patterns in real time.

26. Can you think of a soft lithography way of making a chemically sensitive 1D grating out of a polymer gel? Why would you want to do this?

27. Can you imagine how to manipulate the phenomenon of electrically controlled dewettability of a liquid droplet on a substrate to build a polychrome pixilated display? What do you think would be the pros and cons of such a dewettability display with respect to electrophoretics and electrochromics currently battling for control of the color display market?

28. Do you think there is a spin-off business opportunity in microcontact-printed or micromolded nailpolish?

Building Layer by Layer

CHAPTER 3

Layer-by-Layer Self-Assembly

*Shrek: "No! Layers! Onions have layers! Ogres have layers! Onions have layers.
You get it? We both have layers."
Donkey: "Oh, you both have layers. Oh. You know, not everybody like onions."
Shrek (2001)*

3.1 Building One Layer at a Time

In the previous chapter we took a look at soft lithography, which relies mainly on the formation of a well-ordered surface monolayer to control surface properties, and to pattern materials. We have looked therefore at how a surface layer can influence subsequent steps in a material process, but little attention has been paid to the layers themselves. A thin layer can have very interesting and desirable properties. For instance, if the layer is conductive it provides a two-dimensionally conductive path, while occupying a minimum of volume. As well, if a layering process is repeated several times, the buildup can lead to three-dimensional materials. If we can control the vertical layering process as well as the lateral distribution of layers, we can fabricate a three-dimensional material with control over shape, dimensions, and composition over all length scales. This chapter will investigate this possibility, and describe synthetic routes to a host of planar functional materials.

3.2 Electrostatic Superlattices

Iler discovered in 1966 that certain surfaces modified with cationic surfactants bind monolayers of negatively charged colloidal silica or latex particles.[1] The anionic layers can then bind cationic colloids like boehmite or gelatin, a cationic colloidal biopolymer.[2] Alternate immersions in suspensions of opposite charge gave multilayer films one layer at a time.

Decher in 1991, extended the concept to layer-by-layer (LbL) growth of soluble organic polyelectrolytes.[3–6] The procedure is illustrated in Figure 3.1, and can be generalized to any species bearing multiple charges. The first, or primer, layer serves to impart a surface with a charge. This primer may consist of an

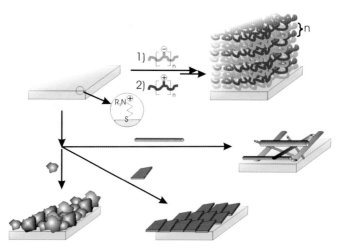

Figure 3.1 *Schematic diagram showing the buildup of electrostatic multilayers of soluble polyelectrolytes (above) as well as the generalization of the procedure to charged objects such as clusters, sheets and rods (below)*

aminoalkylchloro- or alkoxysilane, which anchors to hydroxyl groups on a silica or silicon surface,[7] or an aminoalkylthiol that can chemisorb on gold to form close-packed self-assembled monolayers (see Chapter 2). At neutral pH the exposed amine functionality is protonated, thereby exposing a positively charged surface to an aqueous solution. This surface is able to electrostatically bind a water-soluble anionic polyelectrolyte from an aqueous solution, which is held tightly through multiple electrostatic interactions. In contrast, a molecule bearing a few charges on it would only be weakly bound. A polyelectrolyte monolayer forms, and this layer overcompensates for the positive charge of the alkylammonium primer head-group, thereby leaving an excess negative charge at the surface.[8] This occurs first of all because of charge screening in the solution, and because polyelectrolytes are macromolecules: once a certain number of repeat units have compensated for all the charges in the underlying layer, there are still a few charged repeat units dangling above. This fortuitous situation enables the sequential deposition process to be repeated with a cationic and then an anionic polyelectrolyte solution. The deposition characteristics of a set of polyelectrolytes can be readily modified by, for instance, varying the pH, salt concentration, or polyelectrolyte charge density.[9–12] The method provides a purely synthetic pathway to LbL self-assembled poly-electrolyte multilayer films of alternating charge, with a multitude of compositions. This is an extremely general, electrostatically driven self-assembly procedure and essentially any species bearing multiple ionic charges can be layered, as shown in Figure 3.1. This technique does not require complicated equipment but can be automated, and layers can be built up to any desired thickness at will, and can even be made erasable.[13] Species such as polymers,[6] proteins,[14] polynucleic acids,[15] viruses,[16] polymetallic clusters,[17] colloids,[18–20] dyes,[21] to name a few, can be employed to build up multicomponent films in an LbL fashion with nanometer scale precision. This technique is so general and gentle it could even be used in theory to assemble multilayers of living, breathing cells!

In the above self-assembly strategy, electrostatic interactions are responsible for LbL growth of electrolytes and a wide variety of complex multilayer structures are readily accessible with control of layer thickness, composition, and function. To achieve polyelectrolyte multilayer self-assembly it is not necessary to have specific functionality of film components because charge self-regulation causes a monolayer of polyanion or polycation to be deposited in each step. The subtle question of whether these layers interpenetrate can be probed by techniques such as heavy atom labeling,[22] neutron scattering,[23,24] Förster energy transfer,[25] X-ray reflectivity[26] or direct imaging TEM.[27]

3.3 Organic Polyelectrolyte Multilayers

An example that effectively demonstrates the power of the LbL approach is seen in the self-assembly of redox active organic polyelectrolyte multilayers (Figure 3.2).[28,29] In this case, LbL solution phase self-assembly of a cationic polyviologen with an anionic sulfonated polyelectrolyte creates a redox active multilayer film with controlled thickness. When deposited on a conducting substrate, the result is a multilayer film with electrochemically addressable polyviologens that has electrochromic, ion transport, and electrocatalytic functionality.

Detailed studies of these multilayer films have demonstrated (i) redox reversible electron hopping between colorless(2+)–purple(1+) viologen units in increasing thickness layer pairs on an Au substrate by cyclic voltammetry, (ii) active potential control of coloring effect by viologen reduction on ITO-glass monitored by UV–Vis absorption, (iii) ion transport of Ca^{2+} from $CaCl_2$ solution to balance the reduced charge, and (iv) an electrocatalytic advantage in a H_2-air fuel cell with the over potential for O_2 reduction at ITO-glass electrode reduced by 200 mV. Such a merging of different functionalities into one very thin film is a hallmark of the LbL process.

3.4 Layer-by-Layer Smart Windows

LbL formation of electrostatic superlattices, as illustrated in this chapter, is an extremely versatile materials platform. From the standpoint of film formation, this technique offers unprecedented control over the film thickness as well as

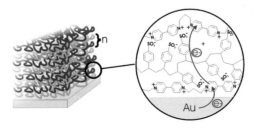

Figure 3.2 *Schematic of electron hopping between polyviologen layers in an LbL film*

composition with an almost unlimited choice of polyelectrolytes and stacking sequence. The example above illustrated how the electrostatic multilayer technique can be used to synthesize electroactive films. One of the ways this redox activity can be harnessed is through the optical changes in the material upon oxidation or reduction. This effect is termed electrochromism, and arises when either the oxidized or reduced state of a material has strong absorptions in the visible spectrum, leading to intensely colored materials. There is a wide range of electrochromic polymers available commercially, and many of these can be obtained as polyelectrolytes or charged nanoscale colloids, allowing their implementation in electrostatic multilayers.

Recent examples in the literature have shown that polyelectrolyte films can indeed be used as electrochromics, and in some cases outperform the best electrochromic systems available.[30–32] In a particular example, electrochromic films are assembled from polyhexylviologen (PXV) as the polycation, and negatively charged poly(3,4-ethylenedioxythiophene):poly(styrene sulfonate) (PEDOT:PSS) colloids as the polyanion, onto conductive indium–tin oxide (ITO) glass. At a negative potential, the colorless dicationic PXV undergoes a reduction to its deep-blue colored radical cation, and the PEDOT:PSS also becomes colored due to an undoping of the conductive state. Both of these polymers are commercially available, and since both are electroactive and colored in their reduced states, the color density is maximized. An optical picture of such an LbL smart window is shown in Figure 3.3. As can be

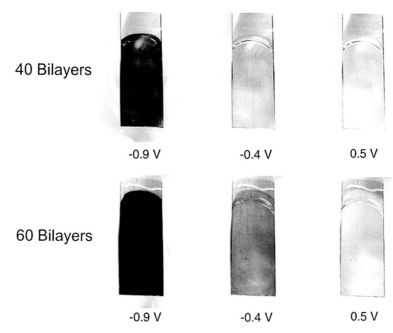

40 Bilayers

    −0.9 V   −0.4 V   0.5 V

60 Bilayers

    −0.9 V   −0.4 V   0.5 V

Figure 3.3   *Optical performance for PEDOT:PSS-PXV multilayers of 40 and 60 bilayers at different potential*
(Reproduced with permission from Ref. 31)

clearly seen, the optical contrast is quite high, making this process a leading contender for the fabrication of next generation display devices. Although the example shown only switches a single color on and off, this principle should hold for other colors if other electrochromic polymers are obtained as polyions.

3.5 How Thick is Thin?

There are several established methods for probing the thickness of polyelectrolyte multilayers, as well as their buildup *in situ*. The first of these methods is optical spectroscopy,[5] where the measurement is usually easy but sometimes hard to interpret. For this experiment the multilayer films must be grown upon a transparent substrate such as glass, and must contain a moiety which absorbs light at a wavelength range of interest. By simply monitoring the optical absorption as a function of number of layers, we can get an idea of how much material is being deposited. It is most convenient to use some sort of dye with a well-defined high-intensity absorption band in the visible region, but UV absorptions from aromatic rings, conjugated systems, or metal atoms can also be used. To accurately measure film thickness the observation area as well as the extinction coefficient of the absorbing group must be precisely known, which is why optical spectroscopy is usually corroborated with other analytical techniques.

Ellipsometry is another technique used to measure the property of thin films, which has become essential to the understanding of self-assembled multilayers.[33] Ellipsometry uses polarized light to probe the dielectric properties of a thin film, and can determine a film thickness, refractive index, as well as its roughness and large-scale molecular orientation. Since it is a change in polarization which is effected, ellipsometry can probe the properties of layers much thinner than the wavelength of light, down to atomic layers. A significant amount of theoretical fitting must be done to extract the required parameters from the experimental data, but several fitting software packages are available and the data obtained is quite reliable.

Finally, quartz crystal microbalance (QCM) measurements are often performed on these systems to exactly determine the amount of polymer adsorbed in each layering step.[34] Quartz is in the family of piezoelectric materials, and if subjected to an alternating electric field a quartz crystal cut in a particular orientation will contract and expand with a characteristic oscillation frequency. This frequency is extremely sensitive to the mass of material deposited onto the surface of the crystal. Minute increases in mass of a surface layer are translated into a decrease in the oscillation frequency (tuning fork analogy), allowing one to measure in real time the film deposition process. While QCM is widely applicable to many systems, care must be taken to avoid experimental error. If the film being probed is viscoelastic (rubbery), this will dampen the crystal oscillation, and the measured frequency will be lower than for a non-viscoelastic film of the same mass.

3.6 Assembling Metallopolymers

High molecular weight polyferrocenylsilanes with redox active iron sites in the backbone are a new class of inorganic polymers.[35,36] They are synthesized from

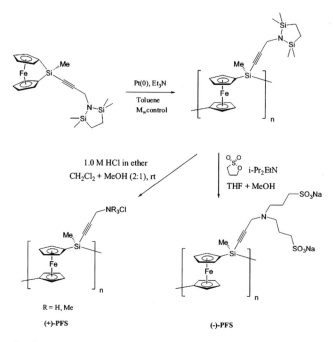

Figure 3.4 *Synthesis of water-soluble polyferrocenylsilanes*

the thermal,[37] anionic[38] or transition metal[39] catalyzed ring-opening polymerization of strained ring-tilted [1]silaferrocenophanes.[40] Because of the diverse types of functional groups that can be bound to the cyclopentadienyl rings and silicon atoms in the polymer backbone, a wide range of polymer compositions and architectures become accessible with novel properties and potential applications abound. The ability to incorporate charged functional groups into polyferrocenylsilanes, as illustrated in Figure 3.4, has enabled the synthesis of water-soluble cationic and anionic ferrocenylsilane polyelectrolytes.[41,42] These can be included in the toolbox of LbL self-assembly for making hybrid organic–inorganic polyelectrolyte multilayers as well as all-inorganic versions thereof. Because of the reversible ferrocene/ferrocenium (Fc/Fc$^+$) redox cycling behavior and hopping semiconductivity of oxidized polyferrocenylsilanes,[43] as well as their ability to act as precursors to shape retentive magnetic ceramics,[44,45] polyferrocenylsilane electrolytes provide access to tailored multilayer films on both planar and curved primed substrates with potentially interesting semiconductive, magnetic, chemical- and bio-sensing, drug delivery, photonic and lubrication properties.

3.7 Directly Imaging Polyelectrolyte Multilayers

While it is true that many diagnostic techniques have been successfully applied to probe the thickness of polyelectrolyte multilayers like ellipsometry, optical and

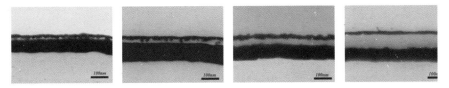

Figure 3.5 *Direct visualization of electrostatic multilayers of 5, 10, 20 and 30 bilayers of PFS–PSS*
(Reproduced with permission from Ref. 27)

vibrational spectroscopies, and QCM, these are rather indirect methods. To directly visualize the multilayers, two kinds of electron microscopy methods have been employed. One involves heavy metal atom or cluster labeling of layers in the film and imaging by transmission electron microscopy (TEM).[46] The other approach involves sputter deposition of a thin gold layer on the top surface of the multilayer and imaging the cross-section by TEM.[27] An example that vividly illustrates this sandwiching and imaging method for 5, 10, 20, and 30 polystyrenesulfonate–polyferrocenylsilane (PSS–PFS) multilayers is depicted in Figure 3.5. The sandwich thickness monotonically increases by an amount corresponding to that obtained by indirect methods of the kind mentioned above. Some diffusion and aggregation of gold atoms into the surface region of the multilayers is seen for the thinner films, presumably due to the high energy of gold clusters generated by the argon ion sputtering process. This agglomeration problem could be avoided by using the gentler technique of microtransfer printing of a top gold layer on the multilayer rather than the more severe gold sputtering method (see Chapter 2). With high resolution TEM it might prove possible to probe the nature of the organization of the polymer chains in these PSS–PFS multilayers by direct visualization of the backbone iron atoms in individual layers.

3.8 Polyelectrolyte–Colloid Multilayers

The work on organic and inorganic polyelectrolyte multilayers leads naturally to electrostatic superlattices assembled by sequential adsorption of polyelectrolytes and various kinds of colloids bearing multiple ionic charges, which include inorganic nanosheets[47] and nanoclusters.[2] Many naturally occurring or synthetic materials have layered structures.[48,49] These consist typically of sheets having strong *intra*sheet covalent bonds, but relatively weak *inter*sheet bonds (between layers) that can be broken to suspend individual sheets in solution, called exfoliation. Utilization of exfoliated single-sheet silicates, metal oxides, metal phosphates and metal chalcogenides in the layering electrostatic self-assembly process creates a new class of multilayer polyelectrolyte–inorganic hybrid materials.[47,50]

The general approach is illustrated in Figure 3.6. The nanolaminate component may be endowed with desired mechanical, electrical, optical, redox, and magnetic properties. Examples of such hybrid multilayers include exfoliated α-Zr(HPO$_4$)$_2$-

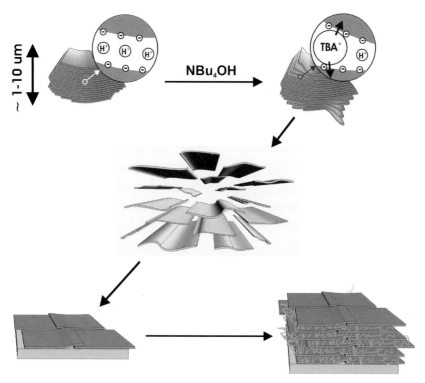

Figure 3.6 *Exfoliation of anionic layered materials with bulky tetrabutylammonium cations leads to a stable colloidal suspension, which can be layered onto positively charged substrates and built into multilayers*

H_2O[51,52] and $K_4Nb_6O_{17}$-$3H_2O$ nanosheets[53] interlaced with polyallylaminehydrochloride (PAH). As well, Keggin ion nanoclusters $Al_{13}O_4(OH)_{12}(H_2O)_{24}^{7+}$ [54,55] and bio-inorganics like cytochrome C^{56} have been layered with PAH.

Multilayer designs of this genre provide access to a myriad of functional superlattices with a diversity of perceived uses that include passivating or corrosion protection layers,[57] ultra-thin insulating field-effect transistor (FET) gate,[58] high dielectric constant films for microelectronics,[59] recognition elements in chemical sensors,[60,61] non-linear optics (NLO),[62] photoluminescent[20] and electroluminescent[63] devices, batteries,[68] molecular diodes,[65] (photo)electron-transfer multilayers,[66] and templates for the growth of zeolite films.[67]

As illustrated in Figure 3.6, a typical process begins with the formation of a 4-aminobutyldimethoxysilane or 2-mercaptoethylamine priming layer on Si or Au, respectively. The reaction scheme shown is for the sequential adsorption of exfoliated alpha-zirconium phosphate (α-ZrP) polyanions and organic oligo- or polycations as shown in the top panel. The α-ZrP sheets are exfoliated by treating the bulk material with tetrabutylammonium hydroxide, and the exchanging of protons between layers by bulky tetrabutylammonium cations pushes the layers apart. The technique of ellipsometry is most effective for monitoring the thickness change of the deposited film *versus* time for adsorption of α-ZrP onto 2-MEA/Au. The ellipsometric thickness, for alternate 15 min, adsorptions of α-ZrP and PAH on 2-MEA/Au, follows a well-defined linear behavior as with other soluble polyelectrolytes.

3.9 Graded Composition LbL Films

Films with spatially graded compositions fashioned at the nanometer scale may provide enhanced electrical, optical, and mechanical properties compared to homogeneous analogs. There are several well-known top–down strategies for making controlled thickness and composition gradient structures, like MOCVD and MBE methodologies used in semiconductor-based high electron mobility transistors and quantum cascade lasers. It is only recently, however, that bottom–up approaches have risen to the challenge of making graded composition films for a variety of purposes.

In the context of LbL electrostatic self-assembly it is straightforward to introduce compositional gradients into films by simply controlling the amount and identity of the constituents that comprise each bilayer. A recent example that colorfully demonstrates the simplicity and effectiveness of this approach is the creation of an LbL quantum dot polyelectrolyte multilayer film that displays a luminescent "nanorainbow,"[68] seen by confocal fluorescent microscopy in Figure 3.7. The colors emanate from a gradient of luminescent, capped semiconductor nanoclusters whose diameter monotonically increases from the bottom to the top of the film. The smaller the cluster, the wider its band gap and the shorter the wavelength (higher energy) of the emitted luminescence. The films were assembled in an LbL manner from aqueous dispersions of anionic thioglycolic acid capped CdTe nanoclusters and cationic poly(diallyldimethylammonium chloride) (PDAC). This is a nice example which exploits the synergism between the size-control achievable for the optical properties of semiconductor quantum dots with spatial control of LbL electrostatic self-assembly.

Another example comprises a graded composition silver nanocluster-polyelectrolyte film, which functions through the existence of a refractive index gradient as a dielectric mirror. These films were made by hydrogen reduction of an

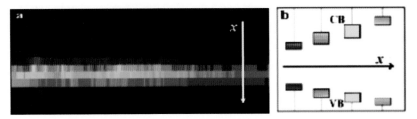

Figure 3.7 *Confocal microscope image of a cross-section of an LbL film made of green, yellow, orange and red CdTe quantum dots. On the right is shown the diameter-dependent band gap in these dots, which leads to the different fluorescence colors*
(Reproduced with permission from Ref. 68)

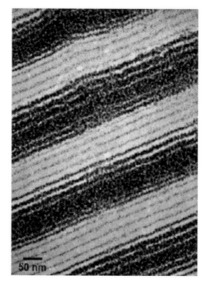

Figure 3.8   *TEM of a graded composition LbL multilayer where the dark lines correspond*
to layers permeated with silver clusters
(Reproduced with permission from Ref. 70)

LbL-assembled poly(allylamine hydrochloride)-poly(acrylic acid) multilayer, in which silver ions had been exchanged with the protons of the acrylic acid groups.[69] The silver atoms so formed underwent aggregation to silver nanoclusters in spatially predefined regions of the film to generate a silver refractive index gradient, an example of which is shown in Figure 3.8. These tailored gradients have been shown to behave as one-dimensional diffraction gratings,[70] and the exquisite control over the individual layers in this process results in a considerable command over optical properties.

    LbL gradient films with diverse compositions and architectures may be formed on both planar and curved surfaces – accomplishments that may not be feasible by top–down means. Gradient structures made by LbL bottom–up methods are expected to find a myriad of uses, as antireflection coatings for instance.[71] In areas like photonics the coupling of light into a photonic crystal requires a smooth change in refractive index (apodization) to avoid excessive reflective light losses, and in optics an array of lenses needs a coating with a graded refractive index to prevent parasitic scattering losses.

3.10 LbL MEMS

Flexible cantilever arrays for chemical sensing applications are usually made by top–down silicon micro-machining methods. An alternative bottom–up approach for making cantilever arrays employs an LbL electrostatic self-assembly strategy, using a multilayer composition designed to enable actuation of the cantilevers and stability to repetitive movement.[72]

Materials of choice for construction of the cantilevers must provide the required flexibility. This was achieved by utilizing nanometer thickness, alternating charge layers of cationic poly(diallyldimethyl ammonium chloride) PDAC and anionic delaminated montmorillonite clay sheets. Actuation of the cantilevers was realized through a magnetic over-layer of iron oxide magnetic nanocrystals deposited on the surface of the cantilevers. Through the sequence of photo-lithographic patterning and LbL deposition steps summarized at the top of Figure 3.9, it proved possible to create an array of clay–polymer–magnetite ultra-thin cantilevers each individually anchored to a silicon substrate and displaying synchronized movements in response to the application of an external magnetic field.

The process begins by spin-coating a layer of a positive photo resist on a silicon wafer. A channel is then photo-lithographically created in the resist and then the entire photo resist is exposed to UV light except for the areas that will define the cantilevers. LbL deposition of the polymer–clay–magnetite nanocrystal multilayer over the entire area of the photo resist follows. The final step involves removal of

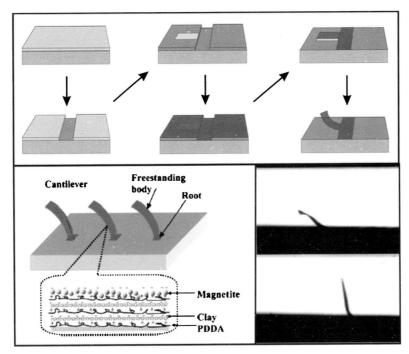

Figure 3.9 *Top, fabrication steps in the fabrication of magnetic cantilevers: first a channel is created in a coating of photo-resist, then the whole substrate is exposed by UV light, except for a perpendicular line which remains insoluble. Next, the LbL process is carried out over the whole sample, and the undeveloped photo-resist dissolved. Finally, acetone is used to dissolve the exposed resist, liberating the cantilever. Bottom left, structure of the LbL magnetic cantilevers. Bottom right, optical images of the cantilever before and after applying a magnetic field*
(Reproduced with permission from Ref. 72)

the UV exposed photo resist with developer and then unexposed photo resist with acetone to create freestanding cantilevers that show the desired response to a changing magnetic field.

LbL self-assembled ultra-thin cantilever arrays may be tailored with a wide range of compositions, sizes, and shapes to create functional microelectromechanical systems tailored for diverse applications such as sensors and actuators.

3.11 Trapping Active Proteins

One of the most exciting new collaborations of materials chemistry is with the biological sciences. Advances in the chemistry, physics, and engineering of materials drive the incredible explosion in biological research with new tools and finer probes. By the same token, when new molecules or assemblies are discovered in biology, they can be emulated by materials scientists or even incorporated into synthetic constructions. The incorporation of intact and functioning biological molecules and assemblies into synthetic materials is of great current interest, and can be achieved readily using the LbL electrostatic self-assembly strategy.[73–77] Most soluble biological molecules, such as enzymes and DNA, bear multiple charged groups and can be deposited in a polyelectrolyte multilayer by co-depositing with a polyelectrolyte of opposite charge. Since these molecules are bound only by charged groups on the outside of the molecule they can often be layered with the retention of biological activity. For instance, the electrostatic LbL strategy has been used to form multilayers containing P450, a common enzyme which converts olefins to epoxides.[78] This enzyme uses hydrogen peroxide as an oxidant, and adds an oxygen atom across the double bond of camphor derivatives. Other substrates can fit inside the cavity of this enzyme such as benzene or styrene derivatives, and their conversion by this enzyme into epoxides which react with DNA make them toxic to organisms. However, this enzymatic non-specificity can be used to our advantage. If the P450 multilayer films are exposed to styrene and hydrogen peroxide, the enzyme cleanly converts this to styrene oxide. Hydrogen peroxide serves to re-oxidize the catalytic iron center in P450, a step that can be circumvented by doing this directly by electrochemistry. Correspondingly, it was found that applying an oxidative potential re-oxidizes the enzymatic center after it has oxidized a styrene molecule, effectively making this an electroenzymatic transformation. Many other enzymes rely on changes in oxidation state to carry out their catalytic cycle, such as nitrogenase (fixes atmospheric N_2) and the oxygen evolving complex (generates O_2 in green plants), and these could all in theory be implemented into electrostatic multilayers.

3.12 Layering on Curved Surfaces

An important extension of the LbL method on planar substrates involves the formation of multilayers of oppositely charged nanoscale organic and inorganic polyelectrolytes, sheets and clusters onto the curved surfaces of spheres and rods of larger dimension. This is another illustration of the power of the technique, since a surface of any shape can be coated just as effectively as a planar one. In this way,

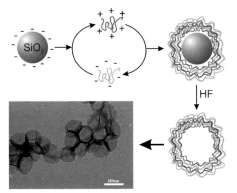

Figure 3.10 *Process used for making hollow capsules by LbL. Bottom left shows a TEM image of hollow PFS–PSS spheres made using this process*

it is possible to prepare, for example, surface functionalized microspheres[79] and hollow capsules thereof[80–82] by removing the templating microsphere. This technique has been extended to coating spheres with biological molecules,[83] ceramics,[84] and more. In addition, the coating procedure can be extended to just about any type of colloid, including oxides, polymers,[80] metals,[85] and even crystals of molecular drugs such as acetaminophen. The general process is illustrated in Figure 3.10 where it can be seen that alternating positively and negatively charged polyelectrolytes can be deposited as controlled thickness multilayers on a negatively charged silica microsphere. The sign of the surface charge of the composite microsphere-multilayer is typically determined by zeta potential measurements, which measures the mobility of a particle in an electric field, and can be directly visualized by the attraction or repulsion of smaller negatively charged silica spheres[86] imaged by scanning electron microscopy (SEM).

Surface tailored multilayer microspheres can be quite varied in layer thickness and composition[87] and may be interesting candidates as beads in combinatorial synthesis, targeted chemical delivery and release,[88] and building blocks for tunable colloidal photonic crystals. As mentioned above, sacrificial chemical removal of the templating microsphere can be utilized as a route for making hollow multilayer capsules.[89] A PFS–PSS example obtained by aqueous HF etching of the silica microsphere is shown in the lower left panel of Figure 3.10. By screening the charge on the polyelectrolyte layers with dissolved electrolytes such as sodium chloride, or modifying the charge on the backbone by redox tuning the polymer backbone Fe^{2+}/Fe^{3+} sites, it may be possible to tune the separation between polymer layers in the hollow capsule, or its porosity, and hence gain command over the diffusion of solution species into and out of the capsules. Research by a number of groups has shown that the control over loading and release of species inside these hollow capsules is indeed possible.[90–92] Through this chemical command over diffusion of solution species, polyelectrolyte multilayer hollow microspheres could prove useful for controlled encapsulation and release of chemicals like pharmaceuticals, pesticides, herbicides, and fragrances.[93–95]

3.13 Crystal Engineering of Oriented Zeolite Film

The LbL approach exemplified above is appealing because of its versatility for chemically tailoring the surfaces of planar and curved surfaces with a wide variety of nanoscale dimension polymeric and inorganic materials. An elegant application of this approach utilizes positively and negatively charged, size monodisperse, surface functionalized zeolite crystals with well-defined crystal habit as building blocks. Zeolites are very important industrial materials used for catalysis, separation, detergents and waste remediation amongst other things (see Chapter 8). However, for many applications these materials are required as membranes or thin films.[96] The LbL strategy by definition is thus a great prospect for building functional zeolite thin films.

The LbL electrostatic superlattices produced are comprised of oriented zeolite crystals and polyelectrolytes.[97] An illustration of the entire concept is presented in Figure 3.11. Protonated amino groups and deprotonated carboxylic acid groups can

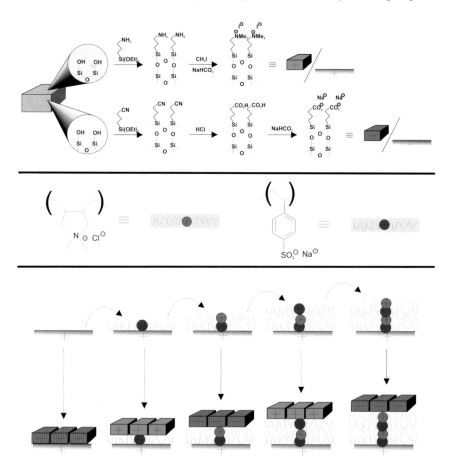

Figure 3.11 *Functionalization of zeolite nanocrystal and glass surfaces (above). Organic polyelectrolytes used in this work (middle). A sample of possible stacking designs for zeolite–polyelectrolyte composites (bottom)*

be selectively anchored to the surface silanols on the zeolite crystal external surface as shown in the figure. The modification affects only the surface since the reactants are size-excluded from the zeolite micropores. These anchoring groups provide positively and negatively charged surfaces to the zeolite crystals when dispersed in water for the LbL process on glass substrates surface functionalized in a similar fashion.

Armed with cationic and anionic water-soluble polyelectrolytes, single size and shape zeolite crystals with positively or negatively charged functionalized surfaces, and a glass substrate primed with either a positive or negative anchoring layer, it is possible to self-assemble a wide range of hybrid multilayer architectures comprised of charge alternating layers of polymers and zeolites. As shown, nearly close-packed and oriented zeolite crystal layers can be interleaved with polyelectrolyte layers. The zeolite layer component is "inorganic and hard" while the polymer layer is "organic and soft." Stacking designs can be quite varied, for example, alternating zeolite and polyelectrolyte layers, bilayers or trilayers as depicted in Figure 3.11 and realized experimentally in Figure 3.12. Preferred orientation of the zeolite crystals in these hybrid stacks is proven by powder X-ray diffraction (PXRD) by the observation of only specific reflections. This is diagnostic of particular crystal planes being parallel to the substrate surface, and thus only certain planes interact with the X-ray beam. In a regular powder, grains are randomly oriented and X-ray reflections are observed for all lattice planes. This LbL self-assembled multilayer polyelectrolyte–zeolite nanocomposite has an organic–inorganic hybrid composition and layer stacking architecture reminiscent of that found in biomineralized nacre[98] in the abalone shell shown in Figure 3.13,[99] which is comprised of alternating layers of protein and oriented calcite crystals. The fracture toughness of nacre is about 1000× that of calcite itself because of the ability of the soft protein layers to dissipate the impact energy of a predator attack

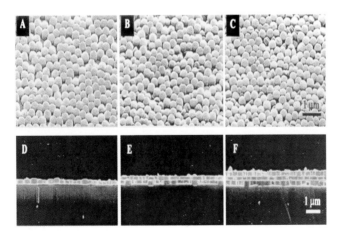

Figure 3.12 *Top view, above and side view, below, of zeolite–polyelectrolyte multilayers with 1, 2 and 3 zeolite crystal layers*
(Reproduced with permission from Ref. 97)

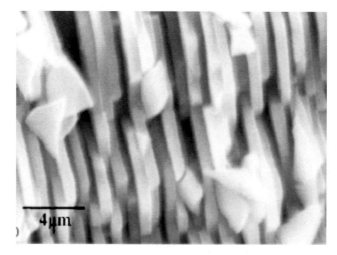

Figure 3.13 *SEM micrograph showing the layered structure of abalone shell*
(Reproduced with permission from Ref. 99)

on the shell (see Chapter 9). This is a clear-cut case where form determines function!

3.14 Zeolite-Ordered Multicrystal Arrays

When we look at electrostatic superlattices fabricated by LbL deposition, these are rarely well ordered. Rather, the layers resemble a tangled mess of strands where a particular polymer chain can penetrate into several adjacent layers. Order can be induced by incorporating colloidal sheets into the process, and these act as physical barriers to prevent interpenetration. Surprisingly enough, however, ordered multilayers can arise from molecular precursors if these are picked judiciously. For instance, thin polyurethane films can be grown in an LbL fashion by exposing a substrate alternately to 1,4-phenylene diisocyanate and 2-butyne-1,4-diol.[100] By investigating these films by angle-resolved IR reflectance, it was found that they consisted of oriented polymer chains. Because the polymer is so well aligned, it can be used as a template to achieve vectorial control of the nucleation and growth of zeolites. Researchers placed the multilayer polyurethane films into a hydrothermal reactor with tetraethyl orthosilicate (TEOS, silica precursor), tetrapropylammonium hydroxide base catalyst (the zeolite cavity template) and water. After subsequent SEM and X-ray reflection measurements, it was found that the substrate was neatly tiled with an array of zeolite crystals, 95% of which had their *c*-axes aligned with the substrate. Different modifications were performed on this system, and by varying the polyurethane precursors as well as the sol–gel mixture composition, crystals of different sizes, shapes, and orientations could be produced as ordered arrays on a substrate. An SEM image of an array of zeolite crystals with their *a*-axes oriented perpendicular to the glass substrate is shown in Figure 3.14.

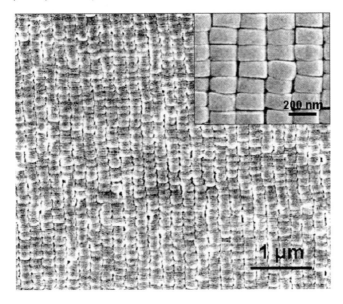

Figure 3.14   *SEM image of an ordered zeolite nanocrystal array templated by an ordered polymer multilayer*
(Reproduced with permission from Ref. 100)

3.15 Crosslinked Crystal Arrays

Surface functionalization of single size and shape crystal building blocks enable them to self-assemble, organize and bind to suitably functionalized substrates. In the case of zeolite A crystals discussed above, micron scale cubes were arranged as close-packed monolayers and multilayers, however, their adhesion to the substrate was found to be weak because of the small area to large mass ratio. An elegant way to surmount this mechanical stability problem involves chemically crosslinking the crystals through imine or urethane linkages as illustrated in Figure 3.15.[101] The procedure begins with anchoring of $(MeO)_3Si(CH_2)_3NH_2$ to the surface of the zeolite A crystals. This step is crystal surface specific as the tethering molecule is too large to enter the eight-ring pore. A glass plate is then primed with the same tethering agent and reacted with either terepthaldicarboxyaldehyde or 1,4-diisocyanatobutane in toluene. Next, a toluene dispersion of the surface functionalized crystals is allowed to undergo evaporation induced self-assembly on the chemically derivatized glass plates where they are linked *via* their mutually contacting surfaces through imine or urethane chemical crosslinkages. The amino groups on the five remaining non-crosslinked zeolite A crystal surfaces were then linked by treatment with terepthaldicarboxaldehyde or 1,4-diisocyanatobutane in toluene with the intention that lateral crosslinking of adjacent crystal surfaces will ensue as indicated in Figure 3.15. Proof that this scheme was operational is seen by the effect of ultrasound of the crystal array compared to one that had not been subject to this crystal crosslinking procedure. The success of this process can be vividly visualized

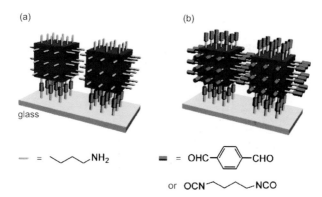

Figure 3.15 *Schematic depicting the crosslinking of amine-terminated zeolite crystals by dialdehydes or diisocyanates*
(Reproduced with permission from Ref. 101)

in the optical microscope images of the before and after crosslinked crystal arrays, Figure 3.16, demonstrating the robust attachment deriving from the crosslinking procedure. This result speaks well for generally enhancing the mechanical stability of self-assembled microcrystal and nanocrystal building blocks on substrates for a range of advanced materials applications.

3.16 Layering with Topological Complexity

The LbL strategy is very versatile, capable of coating planar and curved substrates alike as seen for the coating of dispersed microspheres. However, a sphere need not be the endpoint on the complexity axis. Many other materials display highly ordered, periodic curved surfaces, which may also be coated with electrostatic multilayers. A great example of one of these materials is a colloidal crystal. A colloidal crystal is formed through the self-assembly of monodisperse building blocks into a close-packed structure, and will be looked at in more detail in Chapter 7. If we use colloids that are on the order of the wavelength of light, very interesting optical diffraction properties can be obtained as a result of light interference with the regular microstructure. If we take a crystal formed from spherical building blocks, the material consists of 74% spheres and 26% free space. This free space in turn is organized as highly curved periodic void spaces between individual colloidal building blocks. It was found that even this constrained space could be coated with polyelectrolytes in an LbL fashion.[102] The process can be conveniently monitored by visible spectroscopy, since the coating of the material inside the voids manifests itself as a gradual red shift in the Bragg diffraction peak of the {111} lattice planes. The procedure used is illustrated in Figure 3.17, which also shows the results of the LbL optical monitoring, right, and an SEM of the final composite with an idealized schematic of the layer structure.

This method is ideal for fabrication of functional optical materials. Colloidal crystals can be deposited in a very ordered fashion, and these high-optical quality

Before Cross-linking   **After Cross-linking**

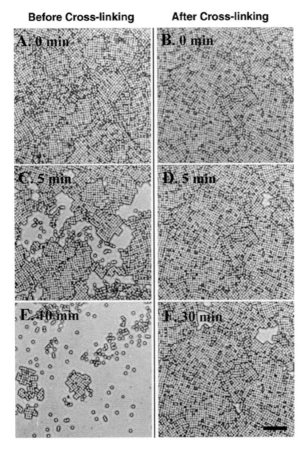

Figure 3.16 *Optical microscopy images of chemically crosslinked and non-crosslinked zeolite monolayers after sonication for the given periods of time in toluene* (Reproduced with permission from Ref. 101)

materials can then be coated with a given number of layers to achieve a desired optical function. This is in contrast to the method entailing the LbL coating of spheres in solution followed by their organization into ordered arrays. Colloidal crystals obtained by this method can have poor optical functionality, as evidenced by the low resolution of optical features such as the Bragg diffraction peak.[103] This highlights the importance of "structural transfer": to make materials as perfect as possible, we should strive to use process conditions that incorporate steps known to give perfect materials. The order of performed steps is very important, and should be well planned to achieve the highest possible structural integrity.

3.17 Patterned Multilayers

The LbL technique can produce films with a multitude of properties. However, applications by and large require the organization of materials into functional

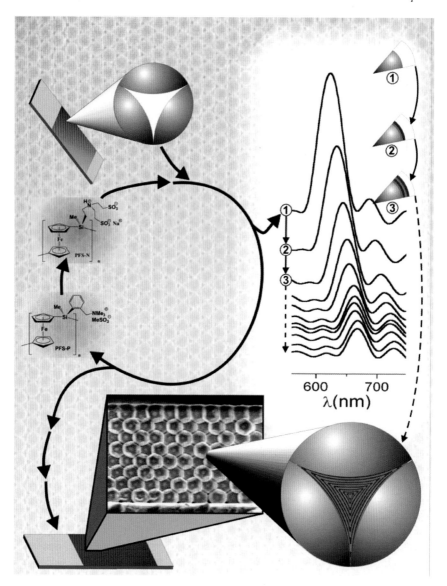

Figure 3.17 *Procedure for the coating of preformed colloidal crystal films with LbL polyelectrolyte electrostatic superlattices, and the gradual shift in the optical properties observed*

patterns. Therefore, to truly take advantage of the LbL technique we must be able to produce these multilayers with predetermined patterns, such as those used for electronic circuits. Photolithography is the common way of patterning thin films of materials, but relies on very efficient photochemistry and harsh solvent development steps. Patterning polyelectrolyte multilayers by photolithography may consequently pose great experimental challenges, which could hamper their introduction into devices.

A strategy for patterning electrostatic multilayers by soft lithography has been developed.[104,105] The process begins with the microcontact printing of a carboxylic acid SAM onto a gold surface. The remaining areas are then derivatized with an oligo-ethylene glycol (EG) terminated SAM. This patterned substrate is then dipped iteratively into solutions of PDAC and poly(styrene sulfonate) (PSS), both of which are completely ionized at any given pH value making them "strong" polyelectrolytes. This PDAC/PSS multilayer builds up exclusively on the anionic carboxylate-terminated SAM, since deposition onto the EG surface would disrupt energetically favorable hydrogen bonding of the EG oxygen with the water solvent. After this deposition is complete, the substrate is switched to deposition baths containing branched poly(ethyleneimine) (BPEI) and poly(acrylic acid) (PAA) in order to form BPEI/PAA multilayers. The BPEI molecules interact with the EG SAM through multisite hydrogen bonding and hydrophobic interactions, which causes this multilayer to grow on the EG terminated monolayer. Especially interesting is the fact that the BPEI/PAA pair does not grow at all on the preformed PDAC/PSS multilayer. It is presumed that BPEI is adsorbed onto the terminal PSS layer, but then is removed during the PAA treatment due to the energetically favorable formation of a BPEI/PAA complex in solution. It was found that this selectivity was maintained by using poly(allylamine hydrochloride) (PAH) instead of BPEI. To clearly demonstrate the selectivity of this approach, multilayers were grown using fluorescent molecules. The soft lithographically patterned substrate was first coated with a multilayer of PDAC and poly(p-phenylene sulfate) (PPP$^-$), a fluorescent polymer emitting in the blue region, which deposited selectively onto the carboxylate-terminated SAM. The substrate was then coated with a multilayer of PAH and a sulfonated ruthenium phenanthroline dye Ru(phen')$_3$, which deposited selectively on the EG terminated SAM. Striking fluorescence images illustrating the success of this procedure are shown in Figure 3.18. Recently, a technique of micro-transfer printing has been elaborated, which greatly simplifies the formation of such patterned multilayers.[106]

These patterned multilayers have also been used for the templated deposition of two different sizes of microspheres onto a substrate.[107] This illustrates the power of this technique for the orthogonal assembly of nanostructures, and will likely be expanded to further levels of complexity in the near future.

3.18 Non-Electrostatic Layer-by-Layer Assembly

In this chapter, almost all the examples given deal with the iterative layering of polyelectrolytes of opposite charge. The buildup of multilayers is accomplished by the cooperative action of a large number of relatively weak interactions (relative to covalent bonds), in this case electrostatic bridges. These bridges can be replaced by interactions such as hydrogen bonding,[108] or ligand–receptor interactions,[109] and even covalent bonds as in the case of the ordered zeolite arrays mentioned above. A hydrogen bond is weaker than an electrostatic bond but this difference could be

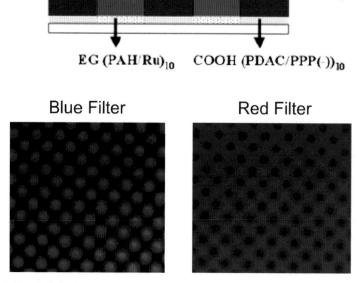

EG (PAH/Ru)$_{10}$ COOH (PDAC/PPP(-))$_{10}$

Blue Filter   **Red Filter**

Figure 3.18 *Side-by-side assembly of red and blue fluorescent materials. Fluorescence images using a blue or red filter clearly displays the formed patterns* (Reproduced with permission from Ref. 105)

reduced by using a so-called hydrogen-bonding motif. The most famous of these is found in double-stranded DNA, where adenine and cytosine bases make strong and selective multiple hydrogen bonds with thymine and guanine bases, respectively. In an extrapolation of this concept, one could imagine making DNA multilayers by hydrogen bonding. If polycytosine is exposed to a positively charged substrate, it will form a monolayer due to electrostatic attraction with the negatively charged phosphate-containing backbone. If this layer is followed by alternating exposure to solutions of polyguanine and polycytosine, layers could build up based on the strong G–C triple hydrogen bonds. Many other polymer pairs can be deposited by hydrogen bonding, and these can be dissolved by changing the pH, or conversely be crosslinked by UV irradiation and made more robust.[110]

3.19 Low-Pressure Layers

The LbL process for fabricating materials has certainly picked up steam as it is found that more and more species can be deposited from solution by soft chemical approaches. However, materials engineers and physicists have been using a similar strategy for many years. Probably, one of the most commonly used techniques is molecular beam epitaxy (MBE).[111–113] In this type of setup, a vacuum chamber is loaded with sources for each of the elements required in the layer of material deposited. In ultrahigh vacuum, material is deposited onto a substrate by heating the appropriate sources with a resistor, and a thin film of alloy of the selected elements is deposited whose thickness is proportional to the exposure time. In this process, very homogenous, crystalline material is deposited. It is, therefore, very important to ensure that materials are deposited epitaxially onto crystalline substrates with a similar lattice constant to avoid interfacial strains.

Another widely used technique is chemical vapor deposition (CVD).[114–116] In this procedure, the elements we want to deposit are incorporated into a volatile molecule. For instance, if we need a source of phosphorus we can use phosphine gas (PH_3), or if we need a copper precursor we can use a copper acetylacetonate (acac) complex. The volatile precursor, or mixture of precursors, is flowed in gas form through a low-pressure reactor. The pressure is typically adjusted to get an optimal gas-phase concentration of precursor, which can be diluted with a variety of buffer gases such as nitrogen or argon. The mixture of gases then comes into contact with a substrate, which is heated to a temperature that causes the decomposition of the precursors. This breaks down the precursor molecules to their elemental forms, and the volatilizing groups are eliminated as waste gases. There are many variations on this technique such as using a plasma discharge or other energy source instead of heat to decompose the precursors, or tuning the deposition selectivity by using hot-walled or cold-walled reactors. The CVD process is quite versatile and can be extended to a great variety of solid materials, making it a very valuable coating procedure. Although both MBE and CVD techniques are primarily from the engineering field, commercially available reactors are becoming increasingly affordable and in some cases a home-built apparatus can be relatively straightforward to produce. The bottom line is that these are indispensable procedures, which should be in the toolbox of all materials chemists.

3.20 Layer-by-Layer Self-Limiting Reactions

Very much related to CVD is the technique of atomic layer deposition (ALD).[117] Instead of being carried out in the gas-phase, however, ALD is usually carried out in solution although a gas-phase process may often be more convenient. ALD begins by treating a surface with a reactant, which will saturate all available surface sites. For instance, if we expose a silica surface to a solution of silicon tetrachloride in methylene chloride, the Si–Cl bonds will react with all available surface hydroxyl groups to form a covalently bound monolayer. The surface will then be capped with Si–Cl groups instead of Si–OH groups. If this surface is then dipped into water-saturated methylene chloride, the water will react with the surface Si–Cl bonds thereby regenerating a hydroxyl-capped surface. If these steps are repeated, a layer of amorphous silica is produced. This LbL approach can be repeated with a great variety of reactants provided they meet the requirements of surface capping and activation for subsequent steps. In a previous example, organized polymer films were used to template the formation of ordered zeolite nanocrystals. The buildup of these organized polymer layers by iterative saturating reactions is a perfect example of ALD.

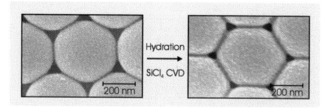

Figure 3.19 *Silica ALD onto colloidal crystal of silica microspheres*
(Reproduced with permission from Ref. 118)

As mentioned, this process can be carried out by using reactant vapors, which are flowed through a reactor. During each exposure the surface states are saturated, leading to buildup of films of controlled thickness. An image of a silica colloidal crystal before (left) and after (right) coating with a thin film of silica by ALD is shown in Figure 3.19,[118] demonstrating that this process can be quite useful even in very constrained environments.

3.21 References

1. R.K. Iler, Multilayers of colloidal particles, *J. Colloid Interf. Sci.*, 1966, **21**, 569.
2. W.F. Harrington and P.H. von Hippel. The structure of collagen and gelatin, *Adv. Protein Chem.*, 1961, **16**, 1.
3. G. Decher and J.D. Hong, Buildup of ultrathin multilayer films by a self-assembly process. 1. Consecutive adsorption of anionic and cationic bipolar amphiphiles on charged surfaces, *Makromol. Chem. Macromol. Symp.*, 1991, **46**, 321.
4. G. Decher and J.D. Hong, Buildup of ultrathin multilayer films by a self-assembly process. 2. Consecutive adsorption of anionic and cationic bipolar amphiphiles and polyelectrolytes on charged surfaces, *Ber. Bunsen-Ges.*, 1991, **95**, 1430.
5. G. Decher, J.D. Hong and J. Schmitt, Buildup of ultrathin multilayer films by a self-assembly process. 3. Consecutively alternating adsorption of anionic and cationic polyelectrolytes on charged surfaces, *Thin Solid Films*, 1992, **210**, 831.
6. G. Decher, Fuzzy nanoassemblies: toward layered polymeric multicomposites, *Science*, 1997, **277**, 1232.
7. R.K. Iler, *The Chemistry of Silica: Solubility, Polymerization, Colloid and Surface Properties and Biochemistry of Silica*, Wiley, New York, 2001.
8. J.B. Schlenoff and S.T. Dubas, Mechanism of polyelectrolyte multilayer growth: charge overcompensation and distribution, *Macromolecule*, 2001, **34**, 592.
9. D. Yoo, S.S. Shiratori and M.F. Rubner, Controlling bilayer composition and surface wettability of sequentially adsorbed multilayers of weak polyelectrolytes, *Macromolecule*, 1998, **31**, 4309.

10. S.S. Shiratori and M.F. Rubner, pH-dependent thickness behavior of sequentially adsorbed layers of weak polyelectrolytes, *Macromolecule*, 2000, **33**, 4213.
11. S.L. Clark, M.F. Montague and P.T. Hammond, Ionic effects of sodium chloride on the templated deposition of polyelectrolytes using layer-by-layer ionic assembly, *Macromolecule*, 1997, **30**, 7237.
12. M.A. Plunkett, P.M. Claesson, M. Ernstsson and M.W. Rutland, Comparison of the adsorption of different charge density polyelectrolytes: a quartz crystal microbalance and X-ray photoelectron spectroscopy study, *Langmuir*, 2003, **19**, 4673.
13. S.A. Sukhishvili and S. Granick, Layered, erasable, ultrathin polymer films, *J. Am. Chem. Soc.*, 2000, **122**, 9550.
14. J.D. Hong, K. Lowack, J. Schmitt and G. Decher, Layer-by-layer deposited multilayer assemblies of polyelectrolytes and proteins: from ultrathin films to protein arrays, *Prog. Colloid Polym. Sci.*, 1993, **93**, 98.
15. Y. Lvov, G. Decher and G. Sukhorukov, Assembly of thin films by means of successive deposition of alternate layers of DNA and poly(allylamine), *Macromolecule*, 1993, **26**, 5396.
16. Y. Lvov, H. Haas, G. Decher, H. Moehwald, A. Mikhailov, B. Mtchedlishvily, E. Morgunova and B. Vainshtein, Successive deposition of alternate layers of polyelectrolytes and a charged virus, *Langmuir*, 1994, **10**, 4232.
17. F. Caruso, D.G. Kurth, D. Volkmer, M.J. Koop and A. Muller, Ultrathin molybdenum polyoxometalate-polyelectrolyte multilayer films, *Langmuir*, 1998, **14**, 3462.
18. J.H. Fendler and F.C. Meldrum, The colloid-chemical approach to nanostructured materials, *Adv. Mater.*, 1995, **7**, 607.
19. E.R. Kleinfeld and G.S. Ferguson, Stepwise formation of multilayered nanostructural films from macromolecular precursors, *Science*, 1994, **265**, 370.
20. N.A. Kotov, I. Dekany and J.H. Fendler, Layer-by-layer self-assembly of polyelectrolyte-semiconductor nanoparticle composite films, *J. Phys. Chem.*, 1995, **99**, 13065.
21. K. Ariga, Y. Lvov and T. Kunitake, Assembling alternate dye-polyion molecular films by electrostatic layer-by-layer adsorption, *J. Am. Chem. Soc.*, 1997, **119**, 2224.
22. S. Joly, R. Kane, L. Radzilowski, T. Wang, A. Wu, R.E. Cohen, E.L. Thomas and M.F. Rubner, Multilayer nanoreactors for metallic and semiconducting particles, *Langmuir*, 2000, **16**, 1354.
23. J. Schmitt, T. Grunewald, G. Decher, P.S. Pershan, K. Kjaer and M. Losche, Internal structure of layer-by-layer adsorbed polyelectrolyte films – a neutron and X-ray reflectivity study, *Macromolecule*, 1993, **26**, 7058.
24. I. Estrela-Lopis, S. Leporatti, S. Moya, A. Brandt, E. Donath and H. Mohwald, SANS studies of polyelectrolyte multilayers on colloidal templates, *Langmuir*, 2002, **18**, 7861.
25. J.W. Baur, M.F. Rubner, J.R, Reynolds and S. Kim, Foerster energy transfer studies of polyelectrolyte heterostructures containing conjugated polymers: a means to estimate layer interpenetration, *Langmuir*, 1999, **15**, 6460.

26. G. Decher, Y. Lvov and J. Schmitt, Proof of multilayer structural organization in self-assembled polycation polyanion molecular films, *Thin Solid Films*, 1994, **244**, 772.
27. J. Halfyard, J. Galloro, M. Ginzburg, Z. Wang, N. Coombs, I. Manners and G.A. Ozin, Direct visualization of layer-by-layer self-assembled multilayers of organometallic polymers, *Chem. Commun.*, 2002, **16**, 1746.
28. J.B. Schlenoff, D. Laurent, H. Ly and J. Stepp, Redox-active polyelectrolyte multilayers, *Adv. Mater.*, 1998, **10**, 347.
29. J.B. Schlenoff, H. Ly and M. Li, Charge and mass balance in polyelectrolyte multilayers, *J. Am. Chem. Soc.*, 1998, **120**, 7626.
30. D. DeLongchamp and P.T. Hammond, Layer-by-layer assembly of PEDOT/ polyaniline electrochromic devices, *Adv. Mater.*, 2001, **13**, 1455.
31. D.M. DeLongchamp, M. Kastantin and P.T. Hammond, High-contrast electrochromism from layer-by-layer polymer films, *Chem. Mater.*, 2003, **15**, 1575.
32. D.M. DeLongchamp and P.T. Hammond, High-contrast electrochromism and controllable dissolution of assembled Prussian blue/polymer nanocomposites, *Adv. Funct. Mater.*, 2004, **14**, 224.
33. A. Tronin, Y. Lvov and C. Nicolini, Ellipsometry and X-ray reflectometry characterization of self-assembly process of polystyrenesulfonate and polyallylamine, *Colloid Polym. Sci.*, 1994, **272**, 1317.
34. F. Caruso, K. Niikura, D.N. Furlong and Y. Okahata, Ultrathin multilayer polyelectrolyte films on gold: construction and thickness determination. 1, *Langmuir*, 1997, **13**, 3422.
35. K. Kulbaba and I. Manners, Polyferrocenylsilanes: processable, metal-containing polymeric materials, *Polym. News*, 2002, **27**, 43.
36. I. Manners, Putting metals into polymers, *Science*, 2001, **294**, 1664.
37. D.A. Foucher, B.Z. Tang and I. Manners, Ring-opening polymerization of strained, ring-tilted ferrocenophanes: a route to high-molecular-weight poly(ferrocenylsilanes), *J. Am. Chem. Soc.*, 1992, **114**, 6246.
38. R. Rulkens, A.J. Lough and I. Manners, Anionic ring-opening oligomerization and polymerization of silicon-bridged [1]ferrocenophanes: characterization of short-chain models for poly(ferrocenylsilane) high polymers, *J. Am. Chem. Soc.*, 1994, **116**, 797.
39. Y. Ni, R. Rulkens, J.K. Pudelski and I. Manners, Transition metal catalyzed ring-opening polymerization of silicon-bridged [1]ferrocenophanes at ambient temperature, *Macromol. Rapid Commun.*, 1995, **16**, 637.
40. A.G. Osborne and R.H. Whiteley, Silicon bridged [1] ferrocenophanes, *J. Organomet. Chem.*, 1975, **101**, C27.
41. F. Jakle, Z. Wang and I. Manners. Versatile and convenient routes to functionalized poly(ferrocenylsilanes), *Macromol. Rapid Commun.*, 2000, **21**, 1291.
42. Z. Wang, A. Lough and I. Manners, synthesis and characterization of water-soluble cationic and anionic polyferrocenylsilane polyelectrolytes, *Macromolecule*, 2002, **35**, 7669.
43. R. Rulkens, A.J. Lough, I. Manners, S.R. Lovelace, C. Grant and W.E. Geiger, Linear oligo(ferrocenyldimethylsilanes) with between two and nine

ferrocene units: electrochemical and structural models for poly(ferrocenylsilane) high polymers, *J. Am. Chem. Soc.*, 1996, **118**, 12683.

44. B.Z. Tang, R. Petersen, D.A. Foucher, A. Lough, N. Coombs, R. Sodhi and I. Manners, Novel ceramic and organometallic depolymerization products from poly(ferrocenylsilanes) via pyrolysis, *Chem. Commun.*, 1993, **6**, 523.

45. M.J. MacLachlan, M. Ginzburg, N. Coombs, T.W. Coyle, N.P. Raju, J.E. Greedan, G.A. Ozin and I. Manners, Shaped ceramics with tunable magnetic properties from metal-containing polymers, *Science*, 2000, **287**, 1460.

46. S. Joly, R. Kane, L. Radzilowski, T. Wang, A. Wu, R.E. Cohen, E.L. Thomas and M.F. Rubner, Multilayer nanoreactors for metallic and semiconducting particles, *Langmuir*, 2000, **16**, 1354.

47. E.R. Kleinfeld and G.S. Ferguson, Stepwise formation of multilayered nanostructural films from macromolecular precursors, *Science*, 1994, **265**, 370.

48. M. Zikmund and K. Hrnciarova, Anionic clays. Structure, synthesis, applications, *Chemicke Listy*, 1997, **91**, 169.

49. H. Shioyama, The interactions of two chemical species in the interlayer spacing of graphite, *Synth. Met.*, 2000, **114**, 1.

50. S.W. Keller, H.N. Kim and T.E. Mallouk, Layer-by-layer assembly of intercalation compounds and heterostructures on surfaces – toward molecular beaker epitaxy, *J. Am. Chem. Soc.*, 1994, **116**, 8817.

51. H.N. Kim, S.W. Keller, T.E. Mallouk, J. Schmitt and G. Decher, Characterization of zirconium phosphate polycation thin films grown by sequential adsorption reactions, *Chem. Mater.*, 1997, **9**, 1414.

52. D.M. Kaschak, J.T. Lean, C.C. Waraksa, G.B. Saupe, H. Usami and T.E. Mallouk. Photoinduced energy and electron transfer reactions in lamellar polyanion/polycation thin films: toward an inorganic "leaf", *J. Am. Chem. Soc.*, 1999, **121**, 3435.

53. M.M. Fang, C.H. Kim, G.B. Saupe, H.N. Kim, C.C. Waraksa, T. Miwa, A. Fujishima and T.E. Mallouk, Layer-by-layer growth and condensation reactions of niobate and titanoniobate thin films, *Chem. Mater.*, 1999, **11**, 1526.

54. M.T. Pope and A. Muller. Polyoxometalate chemistry – an old field with new dimensions in several disciplines, *Angew. Chem. Int. Ed.*, 1991, **30**, 34.

55. I. Ichinose, H. Tagawa, S. Mizuki, Y. Lvov and T. Kunitake. Formation process of ultrathin multilayer films of molybdenum oxide by alternate adsorption of octamolybdate and linear polycations, *Langmuir*, 1998, **14**, 187.

56. Y. Lvov, K. Ariga, I. Ichinose and T. Kunitake, Assembly of multicomponent protein films by means of electrostatic layer-by-layer adsorption, *J. Am. Chem. Soc.*, 1995, **117**, 6117.

57. T.R. Farhat and J.B. Schlenoff, Corrosion control using polyelectrolyte multilayers, *Electrochem. Solid State Lett.*, 2002, **5**, B13.

58. T.F. Guo, S.C. Chang, S. Pyo and Y. Yang, Vertically integrated electronic circuits via a combination of self-assembled polyelectrolytes, ink-jet printing, and electroless metal plating processes, *Langmuir*, 2002, **18**, 8142.

59. J.D. Mendelsohn, C.J. Barrett, V.V. Chan, A.J. Pal, A.M. Mayes and M.F. Rubner, Fabrication of microporous thin films from polyelectrolyte multilayers, *Langmuir*, 2000, **16**, 5017.

60. G.B. Sukhorukov, M.M. Montrel, A.I. Petrov, L.I. Shabarchina and B.I. Sukhorukov, Multilayer films containing immobilized nucleic acids. Their structure and possibilities in biosensor applications, *Biosens. Bioelectron.*, 1996, **11**, 913.

61. X. Yang, S. Johnson, J. Shi, T. Holesinger and B. Swanson, Polyelectrolyte and molecular host ion self-assembly to multilayer thin films: an approach to thin film chemical sensors, *Sens. Actuators B-Chem.*, 1997, **45**, 87.

62. A. Laschewsky, B. Mayer, E. Wischerhoff, X. Arys, P. Bertrand, A. Delcorte and A. Jonas, A new route to thin polymeric, non-centrosymmetric coatings, *Thin Solid Films*, 1996, **285**, 334.

63. H. Hong, D. Davidov, M. Tarabia, H. Chayet, I. Benjamin, E.Z. Faraggi, Y. Avny and R. Neumann, Blue to red electroluminescence from self-assembled films, *Synth. Met.*, 1997, **85**, 1265. M.Y. Gao, C. Lesser, S. Kirstein, H. Mohwald, A.L. Rogach and H. Weller, Electroluminescence of different colors from polycation/CdTe nanocrystal self-assembled films, *J. Appl. Phys.*, 2000, **87**, 2297.

64. T. Cassagneau and J.H. Fendler, High density rechargeable lithium-ion batteries self-assembled from graphite oxide nanoplatelets and polyelectrolytes, *Adv. Mater.*, 1998, **10**, 877.

65. T. Cassagneau, T.E. Mallouk and J.H. Fendler, Layer-by-layer assembly of thin film Zener diodes from conducting polymers and CdSe nanoparticles, *J. Am. Chem. Soc.*, 1998, **120**, 7848.

66. D.M. Kaschak and T.E. Mallouk, Inter- and intralayer energy transfer in zirconium phosphate poly(allylamine hydrochloride) multilayers: an efficient photon antenna and a spectroscopic ruler for self-assembled thin films, *J. Am. Chem. Soc.*, 1996, **118**, 4222.

67. G.S. Lee, Y.J. Lee and K.B. Yoon, Layer-by-layer assembly of zeolite crystals on glass with polyelectrolytes as ionic linkers, *J. Am. Chem. Soc.*, 2001, **123**, 9769.

68. A.A. Mamedov, A. Belov, M. Giersig, N.N. Mamedova and N.A. Kotov, Nanorainbows: graded semiconductor films from quantum dots, *J. Am. Chem. Soc.*, 2001, **123**, 7738.

69. T.C. Wang, R.E. Cohen and M.F. Rubner, Metallodielectric photonic structures based on polyelectrolyte multilayers, *Adv. Mater.*, 2002, **14**, 1534.

70. A.J. Nolte, M.F. Rubner and R.E. Cohen, Creating effective refractive index gradients within polyelectrolyte multilayer films: molecularly assembled rugate filters, *Langmuir*, 2004, **20**, 3304.

71. J. Hiller, J.D. Mendelsohn and M.F. Rubner, Reversibly erasable nanoporous anti-reflection coatings from polyelectrolyte multilayers, *Nat. Mater.*, 2002, **1**, 59.

72. F. Hua, T. Cui and Y.M. Lvov, Ultrathin cantilevers based on polymer-ceramic nanocomposite assembled through layer-by-layer adsorption, *Nano Lett.*, 2004, **4**, 823.

73. G. Decher, B. Lehr, K. Lowack, Y. Lvov and J. Schmitt, New nanocomposite films for biosensors – layer-by-layer adsorbed films of polyelectrolytes, proteins or DNA, *Biosens. Bioelect.*, 1994, **9**, 677.

74. Y. Lvov, K. Ariga, I. Ichinose and T. Kunitake, Assembly of multicomponent protein films by means of electrostatic layer-by-layer adsorption, *J. Am. Chem. Soc.*, 1995, **117**, 6117.

75. F. Caruso, K. Niikura, D.N. Furlong and Y. Okahata, Assembly of alternating polyelectrolyte and protein multilayer films for immunosensing. 2, *Langmuir*, 1997, **13**, 3427.

76. F. Caruso and C. Schuler, Enzyme multilayers on colloid particles: assembly, stability, and enzymatic activity, *Langmuir*, 2000, **16**, 9595.

77. N. Jessel, F. Atalar, P. Lavalle, J. Mutterer, G. Decher, P. Schaaf, J.C. Voegel and J. Ogier, Bioactive coatings based on a polyelectrolyte multilayer architecture functionalized by embedded proteins, *Adv. Mater.*, 1999, **15**, 692.

78. X.L. Zu, Z.Q. Lu, Z. Zhang, J.B. Schenkman and J.F. Rusling, Electro-enzyme-catalyzed oxidation of styrene and cis-beta-methylstyrene using thin films of cytochrome P450cam and myoglobin, *Langmuir*, 1999, **15**, 7372.

79. G.B. Sukhorukov, E. Donath, H. Lichtenfeld, E. Knippel, M. Knippel, A. Budde and H. Möhwald, Layer-by-layer self-assembly of polyelectrolytes on colloidal particles, *Colloid Surf. A: Physicochem. Eng. Aspects*, 1998, **137**, 253.

80. E. Donath, G.B. Sukhorukov, F. Caruso, S.A. Davis and H. Möhwald, Novel hollow polymer shells by colloid-templated assembly of polyelectrolytes, *Angew. Chem. Int. Ed.*, 1998, **37**, 2201.

81. F. Caruso, R.A. Caruso and H. Möhwald, Nanoengineering of inorganic and hybrid hollow spheres by colloidal templating, *Science*, 1998, **282**, 1111.

82. F. Caruso, Hollow capsule processing through colloidal templating and self-assembly, *Chem. Eur. J.*, 2000, **6**, 413.

83. F. Caruso and H. Mohwald, Protein multilayer formation on colloids through a stepwise self-assembly technique, *J. Am. Chem. Soc.*, 1999, **121**, 6039.

84. F. Caruso, R.A. Caruso and H. Mohwald, Nanoengineering of inorganic and hybrid hollow spheres by colloidal templating, *Science*, 1998, **282**, 1111.

85. S.M. Marinakos, J.P. Novak, L.C. Brousseau, A.B. House, E.M. Edeki, J.C. Feldhaus and D.L. Feldheim, Gold particles as templates for the synthesis of hollow polymer capsules. Control of capsule dimensions and guest encapsulation, *J. Am. Chem. Soc.*, 1999, **121**, 8518.

86. F. Caruso, H. Lichtenfeld, M. Giersig and H. Mohwald, Electrostatic self-assembly of silica nanoparticle – polyelectrolyte multilayers on polystyrene latex particles, *J. Am. Chem. Soc.*, 1998, **120**, 8523.

87. F. Caruso, Nanoengineering of particle surfaces, *Adv. Mater.*, 2001, **13**, 11.

88. A.A. Antipov, G.B. Sukhorukov, E. Donath and H. Mohwald, Sustained release properties of polyelectrolyte multilayer capsules, *J. Phys. Chem. B*, 2001, **105**, 2281.

89. G.B. Sukhorukov, E. Donath, S. Davis, H. Lichtenfeld, F. Caruso, V.I. Popov and H. Mohwald, Stepwise polyelectrolyte assembly on particle surfaces: a novel approach to colloid design, *Poly. Adv. Technol.*, 1998, **9**, 759.

90. D.E. Bergbreiter, Self-assembled, sub-micrometer diameter semipermeable capsules, *Angew. Chem. Int. Ed.*, 1999, **38**, 2870.

91. F. Caruso, W.J. Yang, D. Trau and R. Renneberg, Microencapsulation of uncharged low molecular weight organic materials by polyelectrolyte multilayer self-assembly, *Langmuir*, 2000, **16**, 8932.

92. A.J. Chung and M.F. Rubner, Methods of loading and releasing low molecular weight cationic molecules in weak polyelectrolyte multilayer films, *Langmuir*, 2002, **18**, 1176.

93. J. Whelan, Nanocapsules for controlled drug delivery, *Drug Discovery Today*, 2001, **6**, 1183.

94. A.J. Khopade and F. Caruso, Stepwise self-assembled poly(amidoamine) dendrimer and poly(styrenesulfonate) microcapsules as sustained delivery vehicles, *Biomacromolecule*, 2002, **3**, 1154.

95. H. Ai, S.A. Jones and Y.M. Lvov, Biomedical applications of electrostatic layer-by-layer nano-assembly of polymers, enzymes, and nanoparticles, *Cell Biochem. Biophys.*, 2003, **39**, 23.

96. F. Mizukami, Application of zeolite membranes, films and coatings, *Porous Mater. Environ. Friendly Process Stud. Surf. Sci. Catal.*, 1999, **125**, 1.

97. G.S. Lee, Y.-J. Lee and K.B. Yoon, Layer-by-layer assembly of zeolite crystals on glass with polyelectrolytes as ionic linkers, *J. Am. Chem. Soc.*, 2001, **123**, 9769.

98. Z.Y. Tang, N.A. Kotov, S. Magonov and B. Ozturk, Nanostructured artificial nacre, *Nat. Mater.*, 2003, **2**, 413.

99. J.J. Wang, Y.Z. Xu, Y. Zhao, Y.P. Huang, D.J. Wang, L. Jiang, J.G. Wu and D.F. Xu, Morphology and crystalline characterization of abalone shell and mimetic mineralization, *J. Cryst. Growth*, 2003, **252**, 367.

100. J.S. Lee, Y.-J. Lee, E.L. Tae, Y.S. Park and K.B. Yoon, Synthesis of zeolite as ordered multicrystal arrays, *Science*, 2003, **301**, 818.

101. J.S. Park, Y.J. Lee and K.B. Yoon, Marked increase in the binding strength between the substrate and the covalently attached monolayers of zeolite microcrystals by lateral molecular cross-linking between the neighboring microcrystals, *J. Am. Chem. Soc.*, 2004, **126**, 1934.

102. A.C. Arsenault, J. Halfyard, Z. Wang, V. Kitaev, G.A. Ozin, I. Manners, A. Mihi and H. Miguez, Tailoring photonic crystals with nanometer-scale precision using polyelectrolyte multilayers, *Langmuir*, 2005, **21**, 499.

103. G. Kumaraswamy, A.M. Dibaj and F. Caruso, Photonic materials from self-assembly of "tolerant" core-shell coated colloids, *Langmuir*, 2002, **18**, 4150.

104. S.L. Clark, M.F. Montague and P.T. Hammond, Ionic effects of sodium chloride on the templated deposition of polyelectrolytes using layer-by-layer ionic assembly, *Macromolecule*, 1997, **30**, 7237.

105. X.-P., Jiang, S.L. Clark and P.T. Hammond, Side-by-side directed multilayer patterning using surface templates, *Adv. Mater.*, 2001, **13**, 1669.

106. J. Park and P.T. Hammond, Multilayer transfer printing for polyelectrolyte multilayer patterning: direct transfer of layer-by-layer assembled micro-patterned thin films, *Adv. Mater.*, 2004, **16**, 520.

107. H.P. Zheng, I. Lee, M.F. Rubner and P.T. Hammond, Two component particle arrays on patterned polyelectrolyte multilayer templates, *Adv. Mater.*, 2002, **14**, 569.

108. W.B. Stockton and M.F. Rubner, Molecular-level processing of conjugated polymers. 4. Layer-by-layer manipulation of polyaniline via hydrogen-bonding interactions, *Macromolecule*, 1997, **30**, 2717.

109. M. Schutte, D.G. Kurth, M.R. Linford, H. Colfen and H. Mohwald, Metallosupramolecular thin polyelectrolyte films, *Angew. Chem. Int. Ed.*, 1998, **37**, 2891.

110. S.Y. Yang and M.F. Rubner, Micropatterning of polymer thin films with pH-sensitive and cross-linkable hydrogen-bonded polyelectrolyte multilayers, *J. Am. Chem. Soc.*, 1999, **124**, 2100. J.Y. Chen and W.X. Cao, Fabrication of a covalently attached self-assembly multilayer film via H-bonding attraction and subsequent UV-irradiation, *Chem. Commun.*, 1999, **17**, 1711.

111. J.C. Bean, Growth techniques and procedures, *Germanium Silicon: Phys. Mater. Semicond. Semimetals*, 1999, **56**, 1.

112. D.E. Savage, F. Liu, V. Zielasek and M.G. Lagally, Fundamental mechanisms of film growth, *Germanium Silicon: Phys. Mater. Semicond. Semimetals*, 1999, **56**, 49.

113. M.A. Herman and H. Sitter, *Molecular Beam Epitaxy: Fundamentals and Current Status*, vol. 7, Springer Series in Materials Science, Springer, Berlin, 1997.

114. H.O. Pierson, *Handbook of Chemical Vapor Deposition: Principles, Technology, and Applications*, 2nd edn, William Andrew Publishing, Norwich, New York, LLC, 1999.

115. K.L. Choy, Chemical vapor deposition of coatings, *Prog. Mater. Sci.*, 2003, **48**, 57.

116. G.B. Stringfellow, Fundamental aspects of organometallic vapor phase epitaxy, *Mater. Sci. Eng. B- Solid State Mater. Adv. Tech.*, 2001, **87**, 97.

117. T. Suntola, Atomic layer epitaxy, *Thin Solid Films*, 1992, **216**, 84.

118. H. Miguez, N. Tetreault, B. Hatton, S.M. Yang, D. Perovic and G.A. Ozin, Mechanical stability enhancement by pore size and connectivity control in colloidal crystals by layer-by-layer growth of oxide, *Chem. Commun.*, 2002, **22**, 2736.

Nanofood for Thought – Designer Monolayers, Multilayers, Materials Flatland

1. Think of a chemistry approach to pattern polyelectrolyte multilayers on a substrate at the micron scale and choose the pattern, polymers and number of layers to evoke a useful function.

2. What methods could provide molecular scale information about the nature of the interface between the polymers comprising an LbL polyelectrolyte multilayer stack?

3. What properties of a polyelectrolyte in an LbL polyelectrolyte multilayer determine the thickness of the individual polymer monolayers?

4. Sexithiophene type organic semiconductors are proving useful in all-organic field effect transistors (FETs), yet its electron mobility is too low to make a practical device out of this material. Why is this so and how could you chemically modify sexithiophene and use self-assembly to improve the electrical properties of the materials for making organic FETs?

5. Changing the environment around a polyelectrolyte multilayer can cause it to swell or shrink. Amplify upon this statement and think of a way that you could usefully exploit this property for a polyelectrolyte multilayer in the form of a film, hollow capsule, or grating.

6. How could you employ LbL electrostatic self-assembly to tune the properties and evoke a useful function from (a) a silica colloidal crystal (b) a gold nanorod (c) a silicon nanochannel membrane (d) a single-wall carbon nanotube (e) DNA (f) a virus?

7. Think up a strategy for developing a material that can control the release of perfume and why would you want to do such a thing?

8. Devise a method for doing end-selective LbL on (a) one end (b) both ends of a gold nanorod and what materials opportunities might such constructs enable?

9. How could a binary TiP/PAH/ZrP/PAH/HfP/PAH electrostatic superlattice be synthesized one-layer-at-a-time? What single technique can establish the morphology and distribution of components in such a structure at the nanoscale? MP represents metal phosphonate and PAH denotes poly(allylammonium)-chloride?

10. Invent a purely synthetic method for preparing micron scale (a) hollow sphere of zirconia (b) hollow sphere of a three-layer PAH/ZrP electrostatic superlattice.

11. How would you assemble a poly(diallyldimethylammonium)-chloride sodium decatungstate ($Na_4W_{10}O_{32}$) electrostatic superlattice? What property–function relations could you envision for a multilayer film with this architecture and composition?

12. Dream up a purely synthetic approach to a multilayer film containing an assembly of aligned chromophores that display efficient second harmonic generation. What advantages would these films have over poled polymer analogs?

13. Think of a materials self-assembly pathway to a multilayer polyferrocenylsilane-polystyrenesulfonate hollow capsule of around 500 nm diameter and how would you prove you were successful?

14. Conceive a way of making a self-assembling zeolite membrane and why this is an interesting construct?

15. Which chemical strategies would you employ to synthesize polypyrrole between the corrugated layers of MoS_2 and which materials diagnostics would you employ to quantify the structural and chemical outcome of the process? For what type of application could such a polypyrrole–molybdenum disulfide find utility?

16. Devise a materials self-assembly strategy for patterning at the micron scale the intercalation compound formed between a primary alkylamine $CH_3(CH_2)_nNH_2$ and tantalum disulfide TaS_2. This nanocomposite is a superconductor whose critical temperature T_c decreases with the number of methylene groups n in the alkane chain for the range $n = 2$–18. Which experiments would you perform to discover the structure, composition and redox state of the guest and host for different values of n, and try to rationalize the trend in the T_c values?

17. Detail a list of the different classes and examples of building blocks that could be used to form electrostatic multilayers.

18. Layer-by-layer electrostatic assembly can be carried out on many different planar or curved surfaces. Describe different types of topologically complex or unconventional surfaces that could be coated by LbL, and how such a procedure would be useful.

Patterning:
can
Chemistry
go
Smaller?

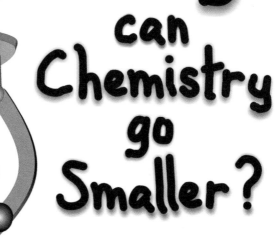

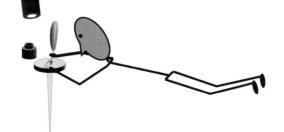

Nanocontact Printing and Writing – Stamps and Tips

Why cannot we write the entire 24 volumes of the Encyclopedia Britannica on the
head of a pin?
Richard Feynman (1918–1988)

4.1 Sub-100 nm Soft Lithography

Since the discovery that alkanethiols self-assemble on gold to form well-ordered monolayers[1,2] and that alkanethiols can be faithfully transferred from a patterned elastomeric polydimethylsiloxane (PDMS) stamp to a gold substrate, termed microcontact printing (MCP)[3–5] (see Chapter 2), to form replica patterns at the 10000–100-nm scale, there has been a determined effort to develop even finer tools to reduce the print scale below 100 nm. The opportunities for organizing matter at the sub-100-nm scale using nanocontact printing include making metal contacts sufficiently minute for electrical connections to molecules and polymers in electronic devices, to create nanocluster and nanowire electronic, photonic and sensory circuits, and to assemble nanocrystal and nanorod arrays for biodiagnostics. As well, if experience in the microelectronic industry has taught us anything it is that the feature size of devices steadily decreases to satisfy the demand for more power, efficiency, and density. It follows that almost anything fabricated by soft lithography or similar chemical means, if successful and commercialized, will need to be further miniaturized. Thus, finding ways to create smaller feature sizes than we are currently able to produce is, and will remain, an important goal of nano-technology research.

4.2 Extending Microcontact Printing

In the last few years, major strides have been made towards the realization of practical ways to perform nanocontact printing.[3] A suite of procedures that has been utilized to reduce the dimensions of size features of SAMs created by MCP

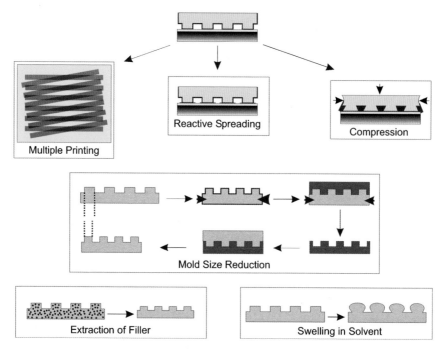

Figure 4.1 *Summary of a few of the methods available to reduce the feature size in microcontact printing*

has been demonstrated.[6] To take a microscale contact-printing mode and apply it to the nanoscale of below 100 nm, a number of methods have been employed that are variations on now standard soft lithographic techniques. These techniques are summarized graphically in Figure 4.1. (i) MCP with a PDMS stamp under lateral compression in the plane[7] increases the dimensions of the raised features on the patterned side of the stamp, while printing with a stretched stamp increases the dimension of the raised regions in the direction of stretching while reducing those perpendicular to this direction (ii) Replica molding against a PDMS mold under mechanical compression creates a permanent replica of the metastable strained mold. (iii) MCP with a PDMS stamp under substantial pressure perpendicular to the surface deforms the raised regions of the stamp and enlarges the printed area, (iv) MCP with a PDMS stamp that has been swollen with a solvent increases the area of printed features and the inverse technique, removing an inert filler from a PDMS stamp (effectively deswelling it), correspondingly decreases the feature size, (v) MCP using a PDMS stamp with controlled reactive spreading[8] causes a steady increase of printed feature area with respect to time, (vi) MCP with multiple impressions from a PDMS stamp on the same surface can reduce feature sizes in printed patterns as well as create novel patterns through Moire-type interference effects. An image of a series of lines reduced in dimensions due to reactive spreading is shown in Figure 4.2.

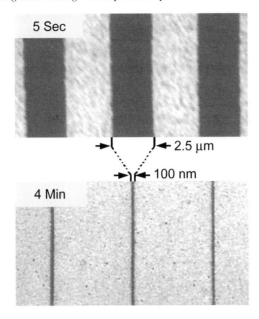

Figure 4.2 *Line patterns printed on gold, reduced in size by leaving the stamp on the substrate and allowing the ink to diffuse for a given amount of time* (Reproduced with permission from Ref. 6)

4.3 Putting on the Pressure

Deformation of the elastomeric stamp for MCP can be either a shortcoming or a boon. If care is not taken with the dimensions of the surface relief pattern in the stamp, then unwanted deformations can reduce the fidelity of the pattern transfer of a chemical-ink to a substrate. On the other hand, designed deformations of the stamp, like compression, can be efficaciously utilized to reduce the feature sizes obtained by MCP. By putting pressure on the top of a stamp new patterns may be obtained, a method that has been dubbed overpressure contact printing.[9]

This idea originates from the observation that defect-free contact printing with PDMS stamps is restricted to a certain range of height-to-width ratios. If the ratio of the surface relief pattern is outside a range of 0.2–2, the roof of the surface relief pattern can contact the substrate. Too high a compressive stamping force can have the same unwanted effect. Thinking "laterally", what if we intentionally let the roof of the stamp come down onto the substrate? Could one form new patterns?

This concept is expressed graphically in Figure 4.3, where depending on the extent of the deformation different printed patterns can emerge from the same stamp. A case in point is the printing of discs *versus* rings at low *versus* high compressive force, as indicated in the figure. Evidence that this idea actually works in practice involved the overpressure contact printing of micron scale discs and

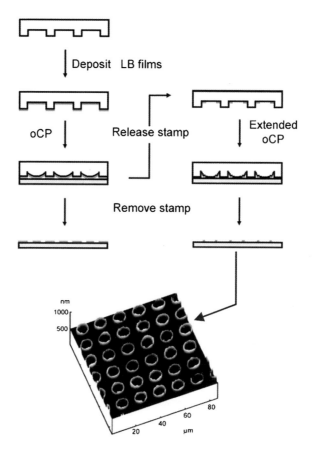

Figure 4.3 *Overpressure microcontact printing. Applying an overpressure during microcontact printing can cause the upper regions of the stamp to contact the substrate. If a PDMS stamp is coated with nanocrystals using the Langmuir–Blodgett technique this can result in nanodisks, left, or nanorings, right, the latter shown in an AFM image at the bottom*
(Reproduced with permission from Ref. 9)

rings of ferromagnetic Pt@Fe$_2$O$_3$ core–shell nanoparticles with topographic and magnetic imaging using atomic force microscopy (AFM) and magnetic force microscopy (MFM). The nano-particles were transferred to a PDMS stamp by the Langmuir–Blodgett technique (see Chapter 6) and the stamp was then brought into contact with a gold-coated silicon substrate at low then high compressive force to achieve either discs or rings, respectively. A representative AFM image of the rings obtained in this way is displayed in the bottom part of Figure 4.3. As it has been recently reported that core–shell Pt@Fe$_2$O$_3$ nanoparticles can be thermally transformed into contiguous ferromagnetic FePt alloy films by exposure to a hydrogen–argon mixture, this overpressure MCP approach also provides an entry to ferromagnetic discs and rings made of FePt. This development bodes well for sub-micron magnet materials research targeting high-density data storage as well as separation and manipulation of biomolecules.

4.4 Defect Patterning – Topologically Directed Etching

Another strategy for reducing the size features generated using MCP utilizes nanoscale defect patterns intentionally designed in SAMs. These defects form spontaneously at metal interfaces, created for example by a nanoscale thick film of Ag deposited on an Au substrate. They can have interesting reactivity, and promote nucleation and growth of materials like calcium carbonate relative to ordered regions of the SAM.[10] Defects in SAMs have also been found less protective as etch resists on gold than ordered regions of the SAM,[11] since the disordered chains shield the substrate poorly. Patterning nanoscale disorder in SAM resists, called topologically directed etching (TODE), enables patterning of the underlying substrate with sub-100-nm structures defined by the edges of larger scale features.[12] TODE is illustrated in Figure 4.4. It relies on SAM patterned regions of disorder at edges and corners of topographically patterned materials like Au, Ag, Si, SiO_2 or Al. This enables the achievement of sub-100-nm negative structures, whose thickness is limited by the edge resolution of the larger pattern. The SAM defect regions have a greater chemical reactivity, and in addition to etching are also more prone to displacement by another thiol. It is noteworthy that the use of less well protecting short-chain SAMs and the exchange of the defect region with a better protecting long-chain SAMs enables inversion of the TODE process and the obtainment of sub-100-nm positive structures. This is a nice example of "perfecting imperfection", a concept that will appear in other chapters of this textbook involving ways and means of making "smart defects" to achieve a specific function.

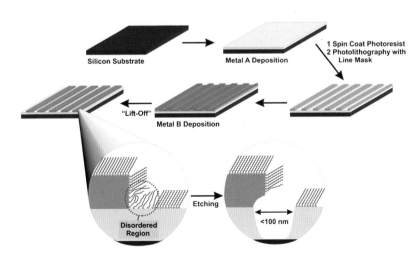

Figure 4.4 *Topologically directed etching*

4.5 Below 50 nm Nanocontact Printing

Another interesting way to reduce the size of printed features involves printing with a stamp, cast from V-grooves anisotropically etched in silicon as opposed to casting from a master with orthogonal walls.[13] This methodology has been further improved, and impressively small features have recently been achieved through the combined use of V-shaped "sharp and hard" PDMS stamps. The contact area of the stamp is below 50 nm, and using high molecular weight dendrimer inks which have low mobility avoids diffusive smearing of the patterns.[14] The methodology is schematized in Figure 4.5. The PDMS stamp is a composite construction made of a 2–3 mm layer of soft PDMS supporting a 30 μm film of hard PDMS having a higher modulus and thus being more resistant to contact deformations.[15] As an aside, this resistance to deformation has recently been tested, and it was found that "hard" PDMS could replicate features as small as 2 nm.[16] The V-shaped printing features of the composite stamp is shown on the bottom right of Figure 4.5 and were able to print dendrimer lines on a silicon substrate with a width below 50 nm. Unlike printing nanoscale patterns with the tip of a scanning probe microscope (see Section 6) in a sequential writing mode, the nanocontact printing method enables the patterning of large areas of substrates simultaneously and it is

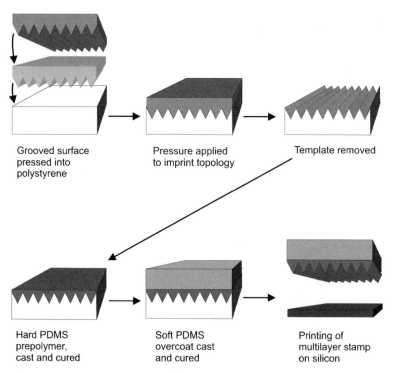

Figure 4.5 *Procedure for fabricating "sharp and hard" PDMS stamps for nanocontact printing*

anticipated that the use of higher density V-stamps with features below 10 nm will facilitate single molecule, single strand polymer, or single protein patterns to be deposited. One of the drawbacks of this method, however, is the generation of V-groove masters from which the PDMS stamps are cast. If these are initially generated from a silicon wafer, the patterns cannot have arbitrary line orientation since anisotropic etching relies on preferential dissolution along crystal planes and can only be performed along particular crystallographic orientations.[17]

4.6 Nanocontact Writing – Dip Pen Nanolithography

It has been demonstrated that the feature size of soft-lithographically printed or written SAM patterns can be reduced from the micron to the sub-100-nm nanoscale by directly writing alkanethiols on gold using the fine tip of a scanning probe microscope such as an AFM.[18,19] The method is called dip pen nano-lithography (DPN), and works under ambient laboratory conditions. It operates by delivering collections of alkanethiol molecules from the tip *via* a water meniscus, formed at the tip–substrate interface, in a positive printing mode. DPN is not limited to writing nanopatterns of alkanethiols on gold, and can be adapted to suit a wide variety of surface chemistries.[20] The writing principle of DPN is illustrated in Figure 4.6. By chemically matching the ink to be printed with the surface chemistry of the substrate it has proven feasible to directly write patterns as fine as about 10 nm of a diversity of molecules, polymers, biomolecules, and materials on disparate substrates like metals, semiconductors, and dielectrics. The fabrication of arrays of line structures with deliberately designed 12–100 nm gaps is shown in Figure 4.7.[21] These structures were made by using DPN to pattern the etch resist, 16-mercaptohexadecanoic acid, on Au/Ti/SiO$_x$/Si substrates. Wet-chemical etching was then used to remove the exposed gold. Such small-scale structures could be useful for surface enhancement in Raman spectroscopy, plasmon optical coupling, nanoparticle-based biodiagnostics, nanogaps for making electrical measurements on nanoscale matter and even molecules. The DPN

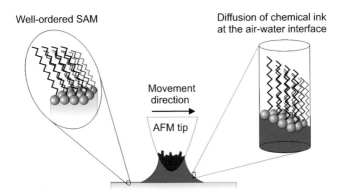

Figure 4.6 *Principle behind DPN*

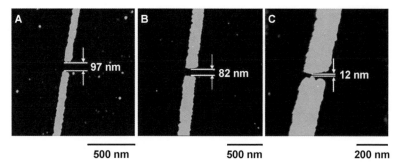

Figure 4.7 *Nanolines with nanogaps made by DPN*
(Reproduced with permission from Ref. 21)

method has been adapted and used successfully as a multitip nanoscale plotter as well as a combinatorial screening tool to probe a nanoscale library of alkanethiols on gold for elucidation of substituent and feature size effects on the exchange behavior of alkanethiols.

4.7 DPN of Silicon

Because silicon is the most important semiconductor it is particularly significant that DPN has been adapted to writing nanoscale etch protection patterns[22] enabling, for example, formation of nanoscale surface relief patterns in silicon.[23] A schematic representation of the deposition and multistage etching procedure for achieving this objective is shown in Figure 4.8. The process begins with the deposition of ODT onto an Au surface of a Au/Ti/Si multilayer substrate using DPN. Exposing the silicon is performed by selectively etching Au with ferri-ferrocyanide-based etchant, and selective Ti/SiO$_2$ etching and Si passivation

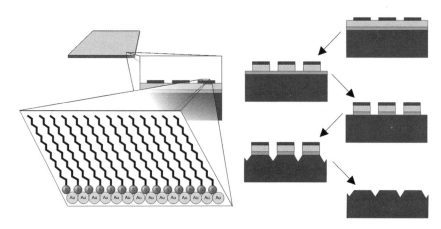

Figure 4.8 *Schematic of the nanoscale anisotropic etching of silicon*

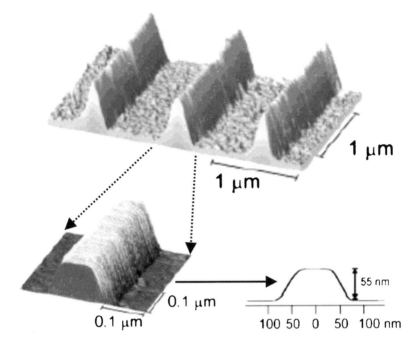

Figure 4.9 *DPN generated nanometer lines in silicon*
(Reproduced with permission from Ref. 23)

with HF. Finally, the Si is anisotropically etched selectively with basic KOH/^{i}PrOH etchant, exposing {111} crystal planes, and passivated with HF which creates surface Si–H bonds. Nanometer scale lines with a height of 55 nm and a width of about 100 nm made in this way are presented as AFM topographic images in Figure 4.9.

4.8 DPN on Glass

Thus far we have shown how to perform DPN on gold or other noble metals, as well as a variety of substrates coated with these metals. However, there is in theory an almost unlimited choice of reactions at surfaces one could utilize to pattern with DPN. One of the most well-studied surfaces is that of glass, or amorphous silica. It can be easily modified with a variety of reagents such as trialkoxyalkyl silanes, or trichloroalkyl silanes. It was found that these types of reactive silanes can also be used to write patterns on glass using DPN,[24] opening the door to direct writing on a very important surface. Alkoxysilanes are relatively reactive, and ambient humidity can cause these molecules to polymerize in solution and be rendered useless for surface modifications. It was found that if the atmospheric humidity was kept low and controlled, this type of pre-polymerization did not occur and reproducible patterns could be obtained. In Figure 4.10 some line patterns are shown that have been written on glass. On the left is the lateral force microscopy image showing

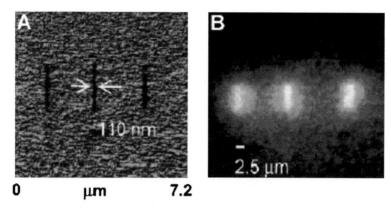

0   **µm**   **7.2**

Figure 4.10 *Nanoscale lines written on glass visualized by lateral force microscopy, left, and fluorescence, right*
(Reproduced with permission from Ref. 24)

the difference in surface properties, and on the right are some fluorescence images from lines labeled with a fluorescent tag using the ligand–receptor interaction of biotin–streptavidin. As more types of surfaces and surface chemistries are explored, DPN is likely to emerge as a very general procedure for the patterning of materials.

4.9 Nanoscale Writing on Semiconductor Nanowires

The practical realization of nanowire-based electronic and photonic devices and circuits, as described in Chapter 5, requires a suite of materials chemistry strategies for controlling nanowire diameter and length, composition and doping, crystal orientation and surface functionality, and for making longitudinal and axial heterostructures, p–n junctions and electrical contacts. It is also important to be able to modify semiconductor nanowires in a site-selective manner and with nanometer scale precision in order to control their chemical and morphological properties as well as their electrical and optical behavior.

A creative means of achieving this objective is by DPN, whereby nanoscale-induced chemical modification of the surface of the nanowire can be achieved.[25] For example, electrochemical DPN has been used to deposit aqueous KOH on a GaN nanowire and locally oxidize it to Ga_2O_3, which can subsequently be etched in HCl. The oxidation $2GaN+6OH^-+6h^+ \rightarrow Ga_2O_3+3H_2O+N_2$ appears in AFM images as nanoscale bumps on the GaN nanowire, Figure 4.11, whereas the etching $Ga_2O_3+6HCl \rightarrow 2GaCl_3+3H_2O$ is manifest as nanoscale indentations. The size of the bumps and indents scales with the voltage applied in the electrochemical oxidation step as well as the humidity. It seems that the strength of the electric field at the voltage-biased AFM tip controls the spatial extent of oxidation by promoting oxide ion transport through the Ga_2O_3 surface layer while humidity determines the extent of the water meniscus that delivers the KOH to the nanowire. Current–voltage profiles of DPN written heterostructured GaN–Ga_2O_3 nanowires shows

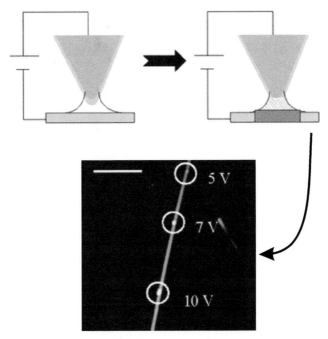

Figure 4.11 *DPN can be used to locally modify a nanowire, with the voltage-dependence of this modification shown in the AFM image at the bottom* (Reproduced with permission from Ref. 25)

them to have an electronic transport barrier and hence are electrically distinct to the parent GaN nanowire. Electrochemical DPN could also be used to locally write and reduce aqueous $AuCl_4^-$ to gold nanoclusters on a GaN nanowire for the catalytic V–L–S growth of nanowire branches from the body of the parent nanowire. With DPN it should be possible to write nanoscale morphological and chemical features on a range of nanostructures like spheres, barcoded rods, and polyhedra.

4.10 Sol–Gel DPN

An important direction in the use of DPN involves the use of nanoscale patterning of classical solid-state materials. Proof-of-concept demonstration of this target involves the delivery of metal oxide sol–gel precursors, using a triblock polypropyleneoxide–polyethyleneoxide–polypropyleneoxide (Pluronic) solvent carrier, to create nanoscale patterns of lines and dots[26] shown in Figure 4.12. These features are made of oxides of tin, aluminum, and silicon, the latter showing the expected polymerization-induced contraction effects of thermally treating the as-deposited nanolines. Atmospheric moisture and the water meniscus at the tip–substrate interface serve to hydrolyze the sol–gel precursors in a controlled fashion to the desired metal oxide. The ability to deliver sol–gel precursors or other open

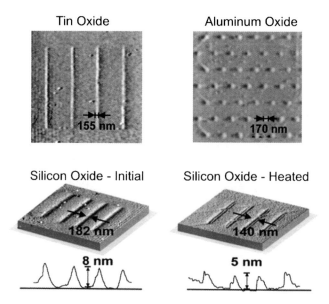

Figure 4.12   *AFM topographic images of DPN printed tin dioxide, aluminum sesquioxide and silicon dioxide both as deposited and after thermal treatment* (Reproduced with permission from Ref. 26)

framework building blocks (see Chapter 8) by DPN, means that by incorporating structure-directing templates of different kinds in the synthesis mixture, plus choosing the right pH, substrate temperature and post-treatment, it now becomes feasible to write at the scale of nanometers a periodic table of microporous and mesoporous materials. This could let a genie full of holes out of the DPN bottle!

4.11 Soft Patterning of Hard Magnets

The ability to pattern substrates with sol–gel mixtures using DPN lets one make use of the plethora of functional materials derived from the sol–gel process.[27] Important among these are the magnetic metal oxides, such as oxides of iron (maghemite, Fe_2O_3, and magnetite, Fe_3O_4) and cobalt. By incorporating mixed metals into this system various magnetic properties such as hardness, the susceptibility of demagnetization by an applied field, can be easily tuned by adjusting the metal ratios. One of these "hard" magnetic materials is barium hexaferrite, $BaFe_{12}O_{19}$ (BaFe), which is of interest as a material for magnetic data storage. The direct patterning of nanoscale BaFe features has recently been accomplished using DPN.[28] The sol–gel mixture used consisted of a 12:1 mixture of iron nitrate and barium carbonate, dissolved in ethylene glycol, which stabilizes the oxide precursors through coordination. An AFM tip is dipped into this mixture and scanned across a silicon substrate, and the resulting patterns are annealed at 450–950 °C. The thermal treatment converts the precursors to ferromagnetic BaFe, whose shapes dictated by the writing process are preserved. Nanoscale magnetic lines and bars

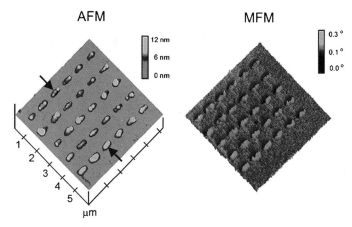

Figure 4.13 *Array of magnetic bars produced by DPN, and imaged by topographical AFM and MFM confirming the presence of the nanobars and their magnetic properties*
(Reproduced with permission from Ref. 28)

were produced, confirmed by AFM and MFM measurements. An MFM image of an array of magnetic bars is shown in Figure 4.13, where a dark or light region indicates a magnetic field and the presence of permanent magnets on the surface. DPN has also been used to pattern preformed magnetic nanoparticles on surfaces.[29] These techniques can produce features smaller than currently achievable through photolithography, and may well revolutionize the data storage industry.

4.12 Writing Molecular Recognition

An important goal in nanotechnology and bionanotechnology is using DNA to build materials. Specifically, we can take advantage of the well-known specificity of T–A and C–G nucleotide base pairing in DNA to guide nanoscale building blocks functionalized with a particular DNA strand to complementary DNA surface sites. This strategy, known as orthogonal assembly[30–32] for its ability to assemble components in a parallel fashion, offers significant opportunities for the preparation of multicomponent nanostructures with utility in biological diagnostics and catalysis, nanoelectronics and nanophotonics. The use of DNA at the nanoscale has been enabled by the direct writing of nanopatterns by DPN.[33] The concept of this kind of orthogonal assembly is mapped out in Figure 4.14, showing the DNA nanostructures and nanoparticles functionalized with complementary DNA strands.[34] In this scheme the DNA strand sequences are terminated by an amine group, and are bound to the surface through an amide linkage with the 1,16 mercaptohexadecanoic acid (MHA) patterns, denoted as the gray ellipsoids. The DNA sequences illustrated in Figure 4.14 are simple sequences used for illustration purposes. In order to achieve a higher strand selectivity, the DNA sequences used by the authors of this proof-of-concept experiment were longer and more complex.

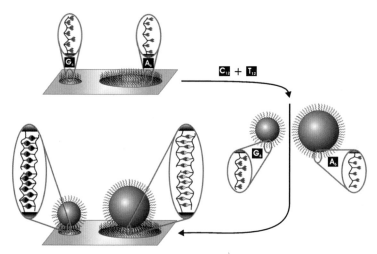

Figure 4.14 *A strategy for the orthogonal assembly of DNA-functionalized nanoparticle building blocks on DPN generated patterns*

The procedure begins with the surface patterning of Au with MHA by DPN. Then the area surrounding the MHA pattern is modified with 1-octadecanethiol (ODT) by immersing the substrate in a 1 mM ethanolic solution of ODT for 5 min. This passivates the bare Au with minimal ODT exchange with the MHA patterned regions. Then an alkylamine-modified DNA (the G_6 segment in Figure 4.14) is coupled in solution to the MHA pattern by the formation of an amide bond, and imaged with AFM as seen in Figure 4.15 top left panel. Next a second MHA pattern with smaller features (200 nm), offset from the first DNA pattern, is generated by DPN. The MHA is written on top of the ODT layer, and replaces the areas modified with ODT under normal writing conditions as evidenced by lateral force microscopy.[35] This step allows the second pattern of MHA to be placed in near-perfect registration with first pattern without chemically modifying the first one. This is followed by addition of a second alkylamine oligonucleotide (the T_6 segment in Figure 4.14) to be immobilized on the second pattern of MHA, and this pattern is imaged by AFM as seen in the bottom left panel of Figure 4.15.

DNA strands with specific oligonucleotide sequences have now been placed at specific sites on the nanopatterned gold substrate. These DNA recognition sites are able to guide the assembly of randomly dispersed nanoparticles, each functionalized with different oligonucleotides that are complementary to the ones anchored to the substrate. The result of the successful DNA-mediated orthogonal assembly is seen in the AFM images and line scans on the right side of Figure 4.15. The advantage of DNA is its high selectivity, which increases with the chain length. If one were to use a 100 base pair sequence for this experiment, in principle it would be possible to generate 4^{100} unique codes able to bind nano-objects functionalized with complimentary strands. We are sure to see in the coming years the explosion of DNA-based assembly for constructing systems with previously unthinkable complexity.

First Pattern   Adsorbed Microspheres

Second Pattern

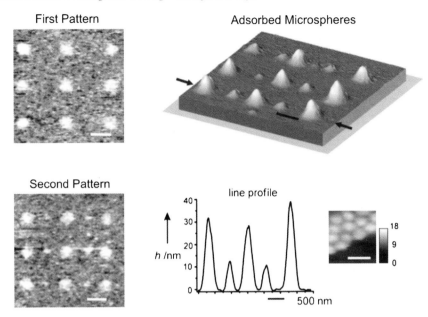

line profile

500 nm

Figure 4.15 *AFM images of the different steps in the orthogonal assembly process. A first pattern is drawn, followed by a second, after which labeled microspheres can be adsorbed selectively onto either pattern*
(Reproduced with permission from Ref. 34)

4.13 DPN Written Protein Recognition Nanostructures

In the same spirit as DNA recognition nanostructures described in Section 12, it is also possible to create DPN protein patterns[36] and protein recognition nano-structures.[37] The key ingredients of the process are a polyethyleneglycol (PEG) coated cantilever used to reduce adventitious protein adhesion and a gold-coated DPN tip covered with thioacetic acid to deliver proteins like immunoglobulin (IgG) or lysozyme (LYZ) to a gold substrate in a direct DPN writing mode. A schematic illustration of the process is shown in Figure 4.16.

IgG and LYZ nanodot arrays created in this way can be imaged by AFM, an IgG array being shown in the left part of Figure 4.17. This pattern is then exposed to solutions of the anti-IgG coated gold nanoparticles, and the outcome imaged again by AFM, shown on the right of the figure. Protein recognition at the nanoscale is clearly diagnosed by the observed increase in the width and height of the nanodots through nanoparticle binding.

4.14 Patterning Bioconstructions

It is evident that DPN is ideal for the chemical nanopatterning of biomolecules such as DNA and proteins. The complexity can be increased even further by the

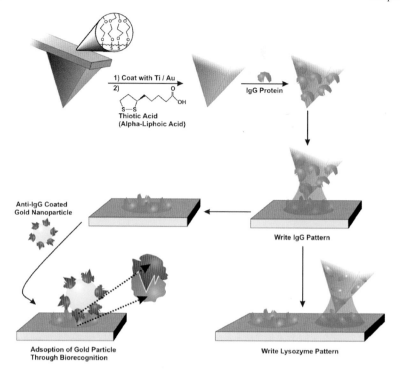

Figure 4.16 *Protein recognition nanostructures formed via direct DPN*

patterning of large biological entities such as virus capsids. It was recently reported that arrays of genetically engineered cowpea mosaic virus (CPMV) could be formed on a substrate using the surface chemistry control afforded by DPN.[38] The CPMV capsule is made of a self-assembled array of identical protein molecules

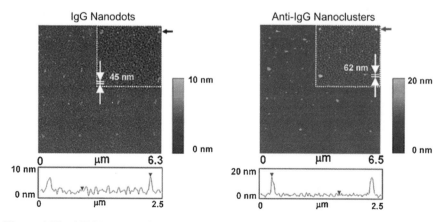

Figure 4.17 *AFM imaging of nanoscale protein recognition*
(Reproduced with permission from Ref. 37)

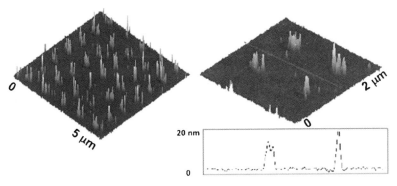

Figure 4.18   *Mutant CMV capsids patterned by DPN, and imaged by AFM*
(Reproduced with permission from Ref. 38)

arranged as a faceted polyhedron. Through DNA modification techniques,[39] a single amino acid in the protein chain can be replaced for another. The modified DNA is incorporated into a living organism, which transcribes the code into messenger RNA (mRNA), from which the ribosome molecular machine[40,41] makes many protein translations of the code. In this manner, the apexes of a self-assembled CPMV capsule were decorated with sulfur atoms by replacing an amino acid that resides at an apex with a sulfur-bearing cysteine residue. The results are highly ordered reactive sites, or "sticky spots", which can be coupled to molecules or surfaces through selective chemical bond formation.

On a gold substrate nanoscopic features of a mixed disulfide were written by DPN, and presented a low density of maleimide groups, which react with thiols. The remainder of the substrate was coated with a bio-inert SAM terminated with an oligo ethylene glycol unit.[42–45] Upon exposure of this patterned substrate to the engineered virus capsid, CPMV capsules reacted only with the written pattern, giving site-specific localization of the virus. Patterns of CPMV are shown in an AFM image in Figure 4.18. This method is ideal for immobilizing proteins in a very soft manner, preventing as much as possible denaturation of the protein leading to function loss.

4.15 Eating Patterns – Enzyme DPN

Enzymatic nanolithography is a newly developed method that utilizes the tip of a scanning probe microscope to deliver an enzyme to specific points on a bio-molecule surface, the goal being to control biochemical reactions locally and with nanometer scale precision.

Proof-of-concept of enzyme DPN was demonstrated for a thiol terminated, DNA oligonucleotide SAM on a gold substrate with DPN delivery of the enzyme DNase I.[46] By exposing the adsorbed enzyme patterned oligonucleotide SAM surface to an aqueous buffered solution of Mg(2+), the enzyme is activated to cause local digestion of the oligonucleotide, shown schematically in Figure 4.19. Rinsing of the surface with de-ionized water and drying in a nitrogen gas stream deactivates the

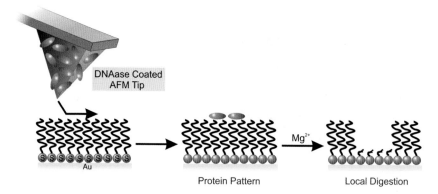

Figure 4.19 *Nanoscale writing of DNAase for local digestion of oligonucleotide monolayers*

enzyme with respect to further digestion of the oligonucleotide SAM, thereby enabling AFM imaging of the result of the enzyme DPN action. The clear-cut result is a nanoscale pattern with trenches where digestion was allowed to take place.

The control of enzyme catalyzed biochemical reactions with nanoscale precision spatial resolution provides many opportunities for tailoring the structure and properties, function and utility of biologically active surfaces at a length scale that is comparable to the dimensions of many biomolecules. Bionanodegradable electronic and optical devices are now a real possibility!

4.16 Electrostatic DPN

As we have seen in Chapter 3, thin films can be readily built up using the electrostatic attraction between oppositely charged polyelectrolytes and a charged surface. This simple technique can be combined with the power of DPN to electrostatically generate nanoscale polyelectrolyte lines.[47] One simply dips the AFM tip into a solution of polyelectrolyte and scans the tip over an oppositely charged surface, resulting in the adsorption of a thin polyelectrolyte layer. In order to make the system functional, the polyelectrolyte in question can be a doped conducting polymer such as polyaniline, polythiophene, or polypyrrole. This opens the possibility of directly writing a diversity of electrical devices and interconnects, in contrast to material-costly, subtractive fabrication methods.

4.17 Electrochemical DPN

As we have seen, DPN functions through the transport of ink molecules over the water meniscus. This meniscus forms naturally through the condensation of a certain volume of water between the AFM tip and the substrate, given a particular tip–substrate separation and atmospheric humidity. There is an interesting analogy, in that DPN is like a conventional Langmuir–Blodgett trough turned on its head, with the monolayer of amphiphile on water being dragged across the substrate instead of

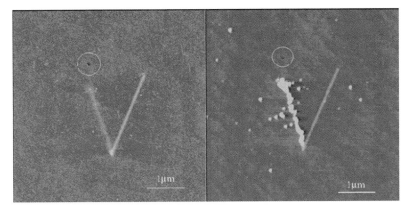

Figure 4.20 (a) *Platinum and* (b) *silicon oxide lines, deposited by electrochemical DPN* (Reproduced with permission from Ref. 52)

the substrate being pulled through the water–air interface. However, instead of only using the water–air interface for deposition, the meniscus itself can be used as a nano-chemical reactor to conduct reactions such as electrochemical depositions.

Using electrochemistry in conjunction with scanning probe microscopy is not new, and by applying a voltage between the tip and the substrate while in a liquid electrolyte one can easily conduct high resolution nano-electrochemistry, and spatially localized etching or electrodeposition of such materials as metals or semiconductors.[48–51] However, to truly use the nano-capabilities of the AFM, the liquid medium can be entirely replaced by the water meniscus. This has been successfully used to grow arbitrary patterns of metals, semiconductors, and conducting polymers[52,53] upon both conducting and insulating substrates. This expands the toolbox of DPN to a variety of materials, which cannot be deposited *via* the standard writing mode, with examples of lines drawn out of platinum metal and silicon dioxide shown in Figure 4.20. The platinum lines are made by electrochemically reducing a platinum salt at −4 V, while silica lines are made by electrochemical oxidation of a silicon surface at +10 V.

4.18 SPM Nano-Electrochemistry

The nano-electrochemical lithographic capability of a scanning probe microscopy tip has also been effectively applied to the top surface of a SAM where terminal amino groups have been written at the nanoscale to direct the self-assembly of gold nanoclusters to specific anchoring sites on the surface of the SAM.[54]

In this nanopatterning approach, depicted in Figure 4.21, a conductive nanotip is used to inscribe a local electrochemical oxidation of terminal methyl groups to carboxyls on the surface of an *n*-octadecyltrichlorosilane SAM tethered to a silicon substrate. The tip written carboxyls are then utilized to guide, through site-selective hydrogen-bonding interactions, the self-assembly of vinyl-terminated nonadecyl-trichlorosilane molecules to form a patterned bilayer architecture. These vinyl

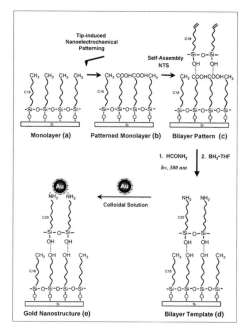

Figure 4.21 *Scheme delineating the electrochemical modification of a SAM and its decoration with gold nanoparticles. Below is shown a sketch and AFM image of Pablo Picasso's World Without Weapons*
(Reproduced with permission from Ref. 54)

groups are then photo-reacted with formamide to create terminal amide groups in their place, which are subsequently reduced with borane–tetrahydrofuran to obtain the desired patterned terminal amine functionality needed to selectively capture and pattern gold nanoclusters.

The pattern forming gold-amine co-assembly step is achieved in an aqueous environment through site-selective electrostatic interactions between protonated terminal amine groups and citrate capped gold nanoclusters. An example of a complex nano-architecture built in this way is illustrated in Figure 4.21. The 1962 Picasso sketch depicting his vision of a world without weapons was programmed into the SPM drive mechanism, electrochemically nano-etched as carboxyl terminal groups in the SAM, replicated as terminal amine groups and finally captured in gold at the nanoscale.

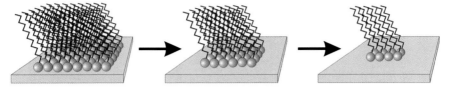

Figure 4.22   *Electrochemical "whittling" of nanoscale SAM patterns*

4.19 Beyond DPN – Whittling Nanostructures

While there may be a limit to how small a nanostructure can be directly written by DPN, even though many parameters can be adjusted to affect feature size, it is intriguing to think about whether it is possible to reduce the spatial extent of a nanostructure after it has been written on a substrate. In fact, this has been achieved by the electrochemical "whittling" of nanoscale SAMs.[55] The essence of the approach is illustrated in Figure 4.22, which depicts the observation that the peripheries of 16-mercaptohexadecanoic acid generated by DPN can be removed in a controlled fashion by the process of electrochemically induced desorption.[56] When electrochemistry is performed on a MHA monolayer, it is found that at about -1 V the thiolate ligands (RS^-) reductively desorb from the substrate as thiols (RSH) due to the reduction of a surface Au($+1$) to Au(0). At defect sites, such as edges present on DPN generated patterns, desorption is found to occur at a less negative voltage. It seems that the larger the free volume surrounding the SAM and the greater the ion accessibility to edge sites, the more facile is this process relative to the same defect-free, unpatterned SAM.[57] In practice, the gold substrate is held at a potential of -750 mV *versus* Ag/AgCl for designated periods of time in 0.5 M aqueous KOH solution. This means the size of the MHA nanostructures on the substrate could be reduced in a controlled fashion. Structures as small as 30 nm could be generated on polycrystalline gold substrates, imaged in Figure 4.23. By observing the temporal evolution of different kinds of SAM nanostructures by AFM it has been determined that the spatial extent of electrodesorption of alkanethiolates from gold is directly proportional to time.

4.20 Combi Nano – DPN Combinatorial Libraries

Combinatorial chemistry has left the fold of pharmaceutical discovery and has entered the realm of materials chemistry.[58] A great deal has now been written about the philosophy of combinatorial materials chemistry, the essence of the approach involving parallel synthesis and rapid screening of large libraries of materials to provide a straightforward, efficient and cost-effective means of enhancing the chances of discovering new and improved performance materials. Large data sets obtained from combinatorial libraries are also useful for evaluating theoretical predictions of how systematic changes in the structure and composition of materials affects the relations between property and function and can aid to pinpointing utility. Some examples of materials made and evaluated through combinatorial

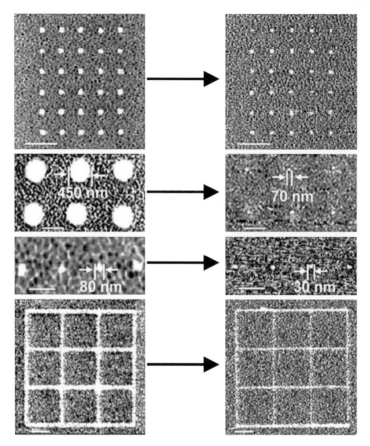

Figure 4.23   *DPN patterns before and after electrochemical desorption*
(Reproduced with permission from Ref. 55)

strategies include superconducting[59] and giant magnetoresistant[60] perovskites, blue
phosphors,[61] heterogeneous catalysts for cyclohexane dehydrogenation,[62] photocatalysts for the oxidation of chlorinated aromatics in water,[63] electrocatalysts for a
methanol fuel cell anode,[64] thin film dielectric layers for microelectronics,[65] and
anodes for a lithium solid state battery.[66] All of these examples involved materials
libraries patterned on the micron or larger length scales. DPN offers enormous
potential to expand the area density of samples in a combinatorial library by
reducing the size and spacing of samples to the nanoscale in a direct multipen
writing mode.

Proof-of-concept that combi nano is an achievable objective is demonstrated
by a SAM exchange study based on combinatorial DPN.[67] Four different
end-functionalized alkanethiolates, 1–4 as shown in Figure 4.24, were utilized in
this study. DPN was used to make combinatorial libraries of alkanethiolate SAMs
of particular area, shape, and composition. Another alkanethiol molecule is then

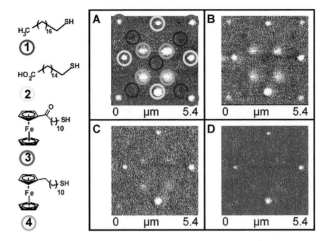

Figure 4.24 *Components used for combinatorial DPN exchange study, and visualization of SAM exchange rates by lateral force microscopy after scanning the surface with SAM 1: 2(A), 10(B), 25(C), and 35(D) times*
(Reproduced with permission from Ref. 67)

delivered to the different sites in the library, and by means of lateral force scanning probe microscopy (LFM) the kinetics of the SAM exchange process can be imaged as a function of the tip-water meniscus contact time and number and type of alkanethiol deliveries.

LFM images of four different alkanethiolate SAM nanodot arrays are shown in Figure 4.24, from which one can discern distinct alterations in the rates of exchange of the arrays as the nanotip delivers on top of them another alkanethiol. Combinatorial SAM exchange studies provide a straightforward and versatile method for generating libraries of planar nanostructures and studying their exchange processes *via* DPN, under one set of environmental conditions. It allows the study of adsorbate–adsorbate exchange at the nanoscale and can probe the effect of defects at edge and bulk regions in the SAM nanostructures.

It is likely that the integration of the strength of DPN with the power of combinatorial chemistry will lead to revolutionary advances in materials science. The discovery of new or improved organic, inorganic, polymeric, and biological materials will be facilitated because of the massive increase of sample density enabled by direct DPN writing at the nanoscale. However, combinatorial DPN may not be the panacea it might seem, since it will only be viable provided rapid screening techniques can be adapted to handle the nanoscale, as well as the tremendous increase in the number of samples to be analyzed.

4.21 Nanoplotters

How does one get around the slow serial low throughput DPN methodology? Build an "ink jet nanoprinter" comprised of individually controlled ink delivery

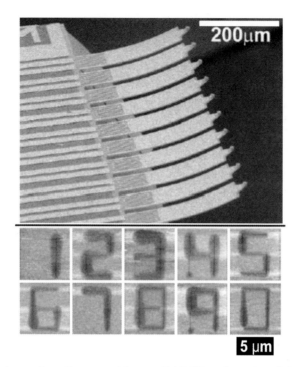

Figure 4.25   *Array of cantilevers used for parallel DPN, and a series of characters written all at once using this array*
(Reproduced with permission from Ref. 68)

cantilevers. Sounds unlikely, but indeed such a device has been reduced to practice and shown to enable parallel DPN.[68,69]

The demonstration model could simultaneously create 10 distinct SAM nano-patterns distributed in distinct spatial regions of a substrate. The pivotal break-through involved microfabrication of an array of 10 cantilevers with bimorphic construction, each spaced by 100 and 300 μm in length and electrically driven through differential thermal expansion of the bimorphs, Figure 4.25. The array could simultaneously scribe 10 different octanethiol nanoscale patterns on a gold surface, as seen in the bottom of the figure.

Practical realization of parallel DPN means that almost any number of materials of the kind described in this chapter, organics, inorganics, polymers, biomolecules, can now be written in a parallel fashion on essentially any surface and at the nanoscale, a development that bodes well for a universal direct-write scanning probe lithography!

4.22 Nanoblotters

While the development of arrays of independently addressable cantilever nanotips, as described earlier, has definitely increased the writing throughput capability of

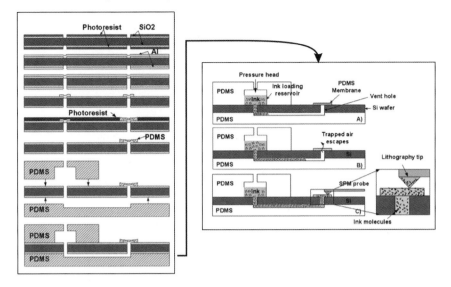

Figure 4.26 *Procedure used to make a nano-inkwell covered with a thin sheet of PDMS, left. On the right we see how this nano-inkwell operates* (Reproduced with permission from Ref. 70)

DPN, for it to be viable in a parallel and multiplexed operational mode each nanotip still requires access to a particular nano-inkwell in an array of different inks. Further, each ink needs to be sealed yet accessible in its well, allowing transfer of the individual inks to the different tips without evaporative loss, spillover, or cross contamination.

A clever solution to this problem combines pressure-induced microfluidic delivery of multiply addressable inks within channels of a specially designed PDMS chip in order to fill and replenish multiple nano-inkwells with the desired ink.[70] Noteworthy, each of these nano-inkwells is capped with a semi-permeable PDMS membrane. By simply contacting the membrane with the nanotip, ink diffuses through the membrane to wet the tip preparing it for writing, very similar to a conventional ink blotter.

The architecture of the inking chip and the method of delivering the ink to the tip is shown in Figure 4.26, as well as illustrating the multistep method of fabricating the inking chip. Every step in the production of the inking chip involves standard methods in photolithography, deposition, wet and dry etching, sacrificial layering, micromolding, and lift-off. This breakthrough in nano-inkwell design and micro-fluidic delivery of inks to nanotips is expected to greatly enhance the capability of DPN and scanning probe contact printing (SP-CP) for a myriad of applications.

4.23 Scanning Probe Contact Printing (SP-CP)

Imagine the benefits of "hybridizing" the power of MCP, delivering chemical inks from an elastomeric stamp for parallel patterning a substrate at the micron scale, with the strength of DPN, for writing with a chemically inked fine tip to serially pattern a substrate at the nanometer scale.

To experimentally realize this intermediate kind of soft lithography, dubbed SP-CP, why not make the tip of a scanning probe microscope out of PDMS? Demonstration that this inventive idea works well in a size range intermediate between those most easily achievable by MCP and DPN, first involved the creation of a flexible polyimide cantilever with an attached elastomeric tip using a multi-step microfabrication strategy. The second step was the inking and writing of an alkanethiol pattern on a gold substrate using this soft tipped cantilever.[71]

Each step in the production process to make the cantilever-tip assembly involves standard methods in photolithography, anisotropic etching, sacrificial layering, micromolding and lift-off, illustrated in Figure 4.27. Depending on the extent of etching of the anisotropically etched square pyramidal pit in the silicon substrate, the replicated PDMS tip can have either a pointed or flat shape. The shape of the inked PDMS tip provides control over whether written patterns are comprised of lines of printed dots or continuous lines. In addition, the residence

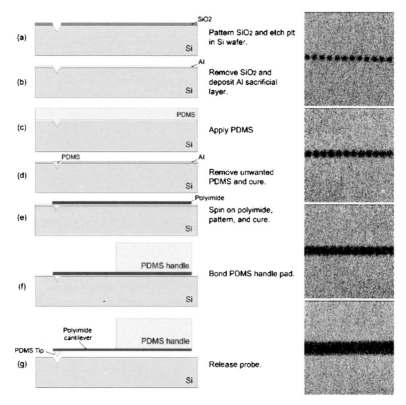

Figure 4.27   *Procedure used to make a PDMS-tipped cantilever for scanning probe contact printing, left. The right images show a series of AFM images of dots drawn at increasing residence times*
(Reproduced with permission from Ref. 71)

time of the tip on the substrate surface determines the spatial extent of diffusion of the ink on the substrate, which enables command over the size of the printed features. An example of the effect of tip residence time on written features is seen on the right of Figure 4.27, where the transition between dots and lines can be readily appreciated.

SP-CP conveniently bridges the capabilities of lower resolution parallel MCP and higher resolution serial DPN. It provides new opportunities for organizing, interrogating, and utilizing macromolecules and biomolecules at length scales comparable to their molecular dimensions.

4.24 Dip Pen Nanolithography Stamp TIP – Beyond DPN CP

Hybridizing the power of contact printing CP with the strength of DPN in the form of scanning probe contact printing SP-CP, as described above, enables structures of around 330 nm to be accessed. However, this pales in comparison with 15 nm resolution features – the best achieved by DPN to date. The observed resolution limitation of SP-CP is related to the fact that the cantilever and tip were both comprised of polymer materials, which have deleterious effects on the performance of the AFM and the achievable sharpness of the PDMS tip, respectively. A creative solution to both of these problems is to simply dip a standard hard Si_3N_4 AFM tip into a PDMS precursor followed by a thermal cure, coating it with a thin film of polymer.[72] This innovation enables sub-100-nm patterns to be written using a wide range of inks ranging from alkanethiols and inorganic salts to dendrimers and biomolecules, a feat that compares well with DPN. These soft tips may also be of great use in the AFM imaging of soft materials since hard tips can often cause disruption and destruction on the nanoscale.

4.25 Best of Both Worlds

It is clear from the examples presented in this chapter that bottom–up nanocontact printing with stamps and tips enables sub-100 nm chemical lithography. Meanwhile, patterning at this length scale by top–down methods, like ultraviolet and electron beam lithography, is pretty routine these days. Therefore, a fair question to ask is why not pattern SAMs using these methods rather than resorting to chemical lithography? The simplest answer is that the unexposed regions of SAMs being patterned in these ways, especially if they are fragile biomolecules, may be disrupted or contaminated by the adverse conditions that they experience under ultrahigh vacuum and on being exposed to ultraviolet photons or electrons.

A creative way to circumvent these deleterious effects is to integrate the best aspects of high-resolution lithography with the best features of SAMs.[73] A strategy to accomplish this objective is sketched in Figure 4.28. It essentially involves electron beam patterning of a polymeric resist PMMA on gold. The exposed gold regions can then be coated with the first alkanethiol SAM(A). Acetone removes the

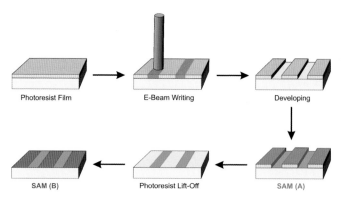

Figure 4.28 *Strategy for the soft patterning of substrates using electron beam lithography*

remaining PMMA to expose fresh gold regions, which can then be coated with the second alkanethiol SAM(B). Patterns as small as 40 nm can be achieved using this technique with an edge resolution of 3.5 nm. Because the SAM is never exposed to the electron beam this method should be applicable to the sub-100 nm patterning of delicate biomolecule monolayers like proteins, lipid bilayers, DNA, vesicles and so forth.

4.26 The Nanogenie is out of the Bottle

Armed with these straightforward protocols for performing top–down chemical nanopatterning of sub-100 nm features, and the self-assembly chemistry methods described earlier, the floodgates are now open wide to address a large number of unanswered questions and problems in nanoscience and bionanoscience. Directed self-assembly strategies that fuse nanoscale patterning with self-assembly provide a straightforward means of organizing and connecting molecules, polymers and building blocks made of diverse materials into integrated chemical, physical and biological systems at length scales not previously possible, in order to achieve a range of functional nanoscale devices never seen before.

4.27 References

1. C.D. Bain, E.B. Troughton, Y.T. Tao, J. Evall, G.M. Whitesides and R.G. Nuzzo, Formation of monolayer films by the spontaneous assembly of organic thiols from solution onto gold, *J. Am. Chem. Soc.*, 1989, **111**, 321.
2. A.R. Bishop and R.G. Nuzzo, Self-assembled-monolayers: recent developments and applications, *Curr. Opin. Colloid Interface Sci.*, 1996, **1**, 127.
3. Y. Xia and G.M. Whitesides, Soft lithography, *Angew. Chem. Int. Ed.*, 1998, **37**, 550.
4. Y. Xia, J.A. Rogers, K.E. Paul and G.M. Whitesides, Unconventional methods for fabricating and patterning nanostructures, *Chem. Rev.*, 1999, **99**, 1823.

5. G.M. Whitesides, E. Ostuni, S. Takayama, X. Jiang and D.E. Ingber, Soft lithography in biology and biochemistry, *Ann. Rev. Biomed. Eng.*, 2001, **3**, 335.

6. Y. Xia and G.M. Whitesides, Extending microcontact printing as a microlithographic technique, *Langmuir*, 1997, **13**, 2059.

7. Y. Xia and G.M. Whitesides, Reduction in the size of features of patterned SAMs generated by microcontact printing with mechanical compression of the stamp, *Adv. Mater.*, 1995, **7**, 471.

8. Y. Xia and G.M. Whitesides, Use of controlled reactive spreading of liquid alkanethiol on the surface of gold to modify the size of features produced by microcontact printing, *J. Am. Chem. Soc.*, 1995, **117**, 3274.

9. Q. Guo, X. Teng and H. Yang, Overpressure contact printing, *Nano Lett.*, 2004, **4**, 1657.

10. J. Aizenberg, A.J. Black and G.M. Whitesides, Controlling local disorder in self-assembled monolayers by patterning the topography of their metallic supports, *Nature*, 1998, **394**, 868.

11. X.-M. Zhao, J.L. Wilbur and G.M. Whitesides, Using two-stage chemical amplification to determine the density of defects in self-assembled monolayers of alkanethiolates on Gold, *Langmuir*, 1996, **12**, 3257.

12. A.J. Black, K.E. Paul, J. Aizenberg and G.M. Whitesides, Patterning disorder in monolayer resists for the fabrication of sub-100-nm structures in silver, gold, silicon, and aluminum, *J. Am. Chem. Soc.*, 1999, **121**, 8356.

13. J.L. Wilbur, E. Kim, Y. Xia and G.M. Whitesides, Lithographic molding. A convenient route to structures with sub-micrometer dimensions, *Adv. Mater.*, 1995, **7**, 649.

14. H.-W. Li, B.V.O. Muir, G. Fichet and W.T.S. Huck, Nanocontact printing: a route to sub-50-nm-scale chemical and biological patterning, *Langmuir*, 2003, **19**, 1963.

15. Information on Sylgard PDMS elastomers available at: www.dowcorning.com

16. B.D. Gates and G.M. Whitesides, Replication of vertical features smaller than 2 nm by soft lithography, *J. Am. Chem. Soc.*, 2003, **125**, 14986.

17. S. Hirai, Model of anisotropic etching process for single crystal silicon, *Int. J. Jpn Soc. Prec. Eng.*, 1999, **33**, 283.

18. R.D. Piner, J. Zhu, F. Xu, S. Hong and C.A. Mirkin, "Dip-pen" nanolithography, *Science*, 1999, **283**, 661.

19. D.S. Ginger, H. Zhang and C.A. Mirkin, The evolution of dip-pen nanolithography, *Angew. Chem. Int. Ed.*, 2004, **43**, 30.

20. H. Zhang and C.A. Mirkin, DPN-generated nanostructures made of gold, silver, and palladium, *Chem. Mater.*, 2004, **16**, 1480.

21. H. Zhang, S.W. Chung and C.A. Mirkin, Fabrication of sub-50-nm solid-state nanostructures on the basis of dip-pen nanolithography, *Nano Lett.*, 2003, **3**, 43.

22. A. Ivanisevic and C.A. Mirkin, "Dip-pen" nanolithography on semiconductor surfaces, *J. Am. Chem. Soc.*, 2001, **123**, 7887.

23. D.A. Weinberger, S. Hong, C.A. Mirkin, B.W. Wessels and T.B. Higgins, Combinatorial generation and analysis of nanometer- and micrometer-scale silicon features via "dip-pen" nanolithography and wet chemical etching, *Adv. Mater.*, 2000, **12**, 1600.

24. H. Jung, R. Kulkarni and C.P. Collier, Dip-pen nanolithography of reactive alkoxysilanes on glass, *J. Am. Chem. Soc.*, 2003, **125**, 12096.
25. B.W. Maynor, J.Y. Li, C.G. Lu and J. Liu, Site-specific fabrication of nanoscale heterostructures: local chemical modification of GaN nanowires using electrochemical dip-pen nanolithography, *J. Am. Chem. Soc.*, 2004, **126**, 6409.
26. M. Su, X. Liu, S.-Y, Li, V.P. Dravid and C.A. Mirkin, Moving beyond molecules: patterning solid-state features via dip-pen nanolithography with sol-based inks, *J. Am. Chem. Soc.*, 2002, **124**, 1560.
27. C.J. Brinker, *Sol–Gel Science: The Physics and Chemistry of Sol–Gel Processing*, Academic Press, New York, 1990.
28. L. Fu, X.G. Liu, Y. Zhang, V.P. Dravid and C.A. Mirkin, Nanopatterning of "hard" magnetic nanostructures via dip-pen nanolithography and a sol-based ink, *Nano Lett.*, 2003, **3**, 757.
29. X. Liu, L. Fu, S. Hong, V.P. Dravid and C.A. Mirkin, Arrays of magnetic nano particles patterned via "dip-pen" nanolithography, *Adv. Mater.*, 2002, **14**, 231.
30. P.E. Laibinis, J.J. Hickman, M.S. Wrighton and G.M. Whitesides, Orthogonal self-assembled monolayers: alkanethiols on gold and alkane carboxylic acids on alumina, *Science*, 1989, **245**, 845.
31. J.J. Hickman, P.E. Laibinis, D.I. Auerbach, C. Zou, T.J. Gardner, G.M. Whitesides and M.S. Wrighton, Toward orthogonal self-assembly of redox active molecules on platinum and gold: selective reaction of disulfide with gold and isocyanide with platinum, *Langmuir*, 1992, **8**, 357.
32. A. Ivanisevic, J.-H. Im, K.-B. Lee, S.-J. Park, L.M. Demers, K.J. Watson and C.A. Mirkin, Redox-controlled orthogonal assembly of charged nanostructures, *J. Am. Chem. Soc.*, 2001, **123**, 12424.
33. L.M. Demers, D.S. Ginger, S.-J. Park, Z. Li, S.-W. Chung and C.A. Mirkin, Direct patterning of modified oligonucleotides on metals and insulators by dip-pen nanolithography, *Science*, 2002, **296**, 1836.
34. L.M. Demers, S.-J. Park, T.A. Taton, Z. Li and C.A. Mirkin, Orthogonal assembly of nanoparticle building blocks on dip-pen nanolithographically generated templates of DNA, *Angew. Chem. Int. Ed.*, 2001, **40**, 3071.
35. A. Ivanisevic, K.V. McCumber and C.A. Mirkin, Site-directed exchange studies with combinatorial libraries of nanostructures, *J. Am. Chem. Soc.*, 2002, **124**, 11997.
36. K.-B. Lee, S.-J. Park, C.A. Mirkin, J.C. Smith and M. Mrksich, Protein nanoarrays generated by dip-pen nanolithography, *Science*, 2002, **295**, 1702.
37. K.-B. Lee, J.-H. Lim and C.A. Mirkin, Protein nanostructures formed via direct-write dip-pen nanolithography, *J. Am. Chem. Soc.*, 2003, **125**, 5588.
38. J.C. Smith, K.B. Lee, Q. Wang, M.G. Finn, J.E. Johnson, M. Mrksich and C.A. Mirkin, Nanopatterning the chemospecific immobilization of cowpea mosaic virus capsid, *Nano Lett.*, 2003, **3**, 883.
39. D. Voet and J.G. Voet, *Biochemistry*, 2nd edn, Wiley, New York, 1995.
40. M. Fromont-Racine, B. Senger, C. Saveanu and F. Fasiolo, Ribosome assembly in eukaryotes, *Gene*, 2003, **313**, 17.

41. D.N. Wilson and K.H. Nierhaus, The ribosome through the looking glass, *Angew. Chem. Int. Ed.*, 2003, **42**, 3464.

42. C. Palegrosdemange, E.S. Simon, K.L. Prime and C.M. Whitesides, Formation of self-assembled monolayers by chemisorption of derivatives of oligo (ethylene glycol) of structure $HS(CH_2)_{11}(OCH_2CH_2)_m$–OH on gold, *J. Am. Chem. Soc.*, 1991, **113**, 12.

43. G.P. Lopez, M.W. Albers, S.L. Schreiber, R. Carroll, E. Peralta and G.M. Whitesides, Convenient methods for patterning the adhesion of mammalian-cells to surfaces using self-assembled monolayers of alkanethiolates on gold, *J. Am. Chem. Soc.*, 1993, **115**, 5877.

44. G.P. Lopez, H.A. Biebuyck, R. Harter, A. Kumar and G.M. Whitesides, Fabrication and imaging of 2-dimensional patterns of proteins adsorbed on self-assembled monolayers by scanning electron-microscopy, *J. Am. Chem. Soc.*, 1993, **115**, 10774.

45. M. Mrksich, C.S. Chen, Y.N. Xia, L.E. Dike, D.E. Ingber and G.M. Whitesides, Controlling cell attachment on contoured surfaces with self-assembled monolayers of alkanethiolates on gold, *Proc. Natl Acad. Sci. USA*, 1996, **93**, 10775.

46. J. Hyun, J. Kim, S.L. Craig and A. Chilkoti, Enzymatic nanolithography of a self-assembled oligonucleotide monolayer on gold, *J. Am. Chem. Soc.*, 2004, **126**, 4770.

47. J.H. Lim and C.A. Mirkin, Electrostatically driven dip-pen nanolithography of conducting polymers, *Adv. Mater.*, 2002, **14**, 1474.

48. O.E. Husser, D.H. Craston and A.J. Bard, High-resolution deposition and etching of metals with a scanning electrochemical microscope, *J. Vac. Sci. Technol. B: Microelectron. Nanometer Struct.*, 1988, **6**, 1873.

49. D. Mandler and A.J. Bard, High resolution etching of semiconductors by the feedback mode of the scanning electrochemical microscope, *J. Electrochem. Soc.*, 1990, **137**, 2468.

50. A.J. Bard, F.R.F. Fan, D.T. Pierce, P.R. Unwin, D.O. Wipf and F. Zhou, Chemical imaging of surfaces with the scanning electrochemical microscope, *Science*, 1991, **254**, 68.

51. S. Manne, P.K. Hansma, J. Massie, V.B. Elings and A.A. Gewirth, Atomic-resolution electrochemistry with the atomic force microscope – copper deposition on gold, *Science*, 1991, **251**, 183.

52. Y. Li, B.W. Maynor and J. Liu, Electrochemical AFM "dip-pen" nano-lithography, *J. Am. Chem. Soc.*, 2001, **123**, 2105.

53. B.W. Maynor, S.F. Filocamo, M.W. Grinstaff and J. Liu, Direct-writing of polymer nanostructures: poly(thiophene) nanowires on semiconducting and insulating surfaces, *J. Am. Chem. Soc.*, 2002, **124**, 522.

54. S.T. Liu, R. Maoz and J. Sagiv, Planned nanostructures of colloidal gold via self-assembly on hierarchically assembled organic bilayer template patterns with in-situ generated terminal amino functionality, *Nano Lett.*, 2004, **4**, 845.

55. Y. Zhang, K. Salaita, J.-H. Lim and C.A. Mirkin, Electrochemical whittling of organic nanostructures, *Nano Lett.*, 2002, **2**, 1389.

56. M.M. Walczak, D.D. Popenoe, R.S. Deinhammer, B.D. Lamp, C. Chung and M.D. Porter, Reductive desorption of alkanethiolate monolayers at gold: a measure of surface coverage, *Langmuir*, 1991, **7**, 2687.

57. S. Imabayashi, M. Iida, D. Hobara, Z.Q. Feng, K. Niki and T. Kakiuchi, Reductive desorption of carboxylic-acid-terminated alkanethiol monolayers from Au(111) surfaces, *J. Electroanal. Chem.*, 1997, **428**, 33.

58. E.J. Amis, X.-D. Xiang and J.-C. Zhao, Combinatorial materials science: what's new since Edison? *MRS Bull.*, 2002, **27**, 295.

59. X.-D. Xiang, X. Sun, G. Briceno, Y. Lou, K.-A. Wang, H. Chang, W.G. Wallace-Freedman, S.-W. Chen and P.G. Schultz, A combinatorial approach to materials discovery, *Science*, 1995, **268**, 1738.

60. G. Briceno, H. Chang, X. Sun, P.G. Schultz and X.-D. Xiang, A class of cobalt oxide magnetoresistance materials discovered with combinatorial synthesis, *Science*, 1995, **270**, 273.

61. E. Danielson, J.H. Golden, E.W. McFarland, C.M. Reaves, W.H. Weinberg and X.D. Wu, A combinatorial approach to the discovery and optimization of luminescent materials, *Nature*, 1997, **389**, 944.

62. S.M. Senkan and S. Ozturk, Discovery and optimization of heterogeneous catalysts by using combinatorial chemistry, *Angew. Chem. Int. Ed.*, 1999, **38**, 791.

63. C. Lettmann, H. Hinrichs and W.F. Maier, Combinatorial discovery of new photocatalysts for water purification with visible light, *Angew. Chem. Int. Ed.*, 2001, **40**, 3160.

64. E. Reddington, A. Sapienza, B. Gurau, R. Viswanathan, S. Sarangapani, E.S. Smotkin and T.E. Mallouk, Combinatorial electrochemistry: a highly parallel, optical screening method for discovery of better electrocatalysts, *Science*, 1998, **280**, 1735.

65. R.B. van Dover, L.F. Schneemeyer and R.M. Fleming, Discovery of a useful thin-film dielectric using a composition-spread approach, *Nature*, 1998, **392**, 162.

66. A. Bonakdarpour, K.C. Hewitt, T.D. Hatchard, M.D. Fleischauer and J.R. Dahn, Combinatorial synthesis and rapid characterization of $Mo_{1-x}Sn_x$ ($0 \leqslant x \leqslant 1$) thin films, *Thin Solid Films*, 2003, **440**, 11.

67. A. Ivanisevic, K.V. McCumber and C.A. Mirkin, Site-directed exchange studies with combinatorial libraries of nanostructures, *J. Am. Chem. Soc.*, 2002, **124**, 11997.

68. D. Bullen, S.W. Chung, X.F. Wang, J. Zou, C.A. Mirkin and C. Liu, Parallel dip-pen nanolithography with arrays of individually addressable cantilevers, *Appl. Phys. Lett.*, 2004, **84**, 789.

69. S.H. Hang and C.A. Mirkin, A nanoplotter with both parallel and serial writing capabilities, *Science*, 2000, **288**, 1808.

70. K.S. Ryu, X. Wang, K. Shaikh, D. Bullen, E. Goluch, J. Zou and C. Liu, Integrated microfluidic inking chip for scanning probe nanolithography, *Appl. Phys. Lett.*, 2004, **85**, 136.

71. X.F. Wang, K.S. Ryu, D.A. Bullen, J. Zou, H. Zhang, C.A. Mirkin and C. Liu, Scanning probe contact printing, *Langmuir*, 2003, **19**, 8951.

72. H. Zhang, R. Elghanian, N.A. Amro, S. Disawal and R. Eby, Dip pen nanolithography stamp tip, *Nano Lett.*, 2004, **4**, 1649.

73. D. Stamou, C. Musil, W.P. Ulrich, K. Leufgen, C. Padeste, C. David, J. Gobrecht, C. Duschl and H. Vogel, Site-directed molecular assembly on templates structured with electron-beam lithography, *Langmuir*, 2004, **20**, 3495.

Nanofood for Thought – Sharper Chemical Patterning Tools

1. If you could make an LbL electrostatic superlattice using DPN what would be the first experiment that you would do that would not have previously been possible using other lower spatial resolution forms of lithography?
2. Could you imagine how to use DPN for nanoscale etching of silicon and silica and utilize the method for making nanoelectromechanical machines NEMS that outsmart microlectromechanical machines MEMS.
3. Dream about how you could make the nanoscale fine tip of a DPN instrument out of anatase and how you might employ it for developing a photocatalytic DPN experiment on SAM films of different type.
4. Imagine the different kinds of tips one could use as fine quills in an AFM-DPN, especially those available through self-assembly chemical approaches. How could such tips be installed, and what advantages would they provide?
5. Where might combinatorial chemistry *via* DPN impact materials self-assembly?
6. Describe the methods for nanoscale patterning of surfaces through chemistry, and why such approaches may become technologically invaluable in the future.
7. DPN allows the patterning at predetermined locations due to the combined imaging–writing capabilities. Describe types of surfaces that would be impossible to pattern by photolithography or soft lithography that could be patterned at the nanoscale with DPN.
8. Using DPN, how could one go about making a string of 10 silica microspheres with 10 different, decreasing diameters?
9. Detail all possible binding interactions that can be used to effect DPN.
10. How can DPN be used to fabricate a next-generation magnetic storage device?
11. Describe how DPN could be used to make a PDMS stamp used for microcontact printing at the nanoscale.
12. Dream up a creative nanoscience project that focuses upon the use of a scanning probe microscope with a tip made of polydimethylsiloxane? What could you do with this capability that would prove impossible by DPN?
13. Devise a new kind of scanning probe nanolithography project that exploits the knowledge that silver nanoclusters photocatalyze the reduction of organonitro to organoamine groups, and that amines couple to carboxylic acids to form amides and amines initiate the electrochemical formation of polyaniline.
14. Knowing that esters can be acid hydrolyzed to carboxylic acids, invent a scanning chemical probe nanolithography that writes voltage controllable wettability patterns.
15. Discuss how and why it would be interesting to develop a scanning photocatalytic probe nanolithography, SPPN.

Nanowires

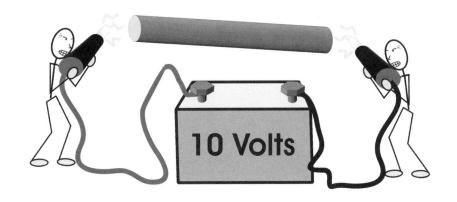

CHAPTER 5

Nanorod, Nanotube, Nanowire Self-Assembly

"God hangs the greatest weights upon the smallest wires"
Sir Francis Bacon (1561–1626)

5.1 Building Block Assembly

In the opening of this book the paradigm of materials self-assembly over "all" scales was introduced in terms of a novel construction process for making materials. The building blocks in this process are not constrained by volume or surface composition or structure, physical size or shape and are spontaneously driven to a specific architecture by forces that act over multiple length scales or directed to these architectures by templates. Building blocks can be molecules or macromolecules, nanometer scale clusters, sheets, wires and tubes, micrometer scale spheres and rods, and a variety of millimeter scale polyhedral shapes. Structure directing templates for making, organizing and interconnecting the building blocks can be nanometer scale porous hosts, micrometer scale lithographic patterns, and channels in polymer, aluminum oxide (alumina) and silicon membranes or their replicas.

In this chapter we will examine different ways of making nanoscale dimension rods, tubes and wires with strict chemical control over their length and diameter, and describe how they can be made to function as minute sensors, light emitting diodes (LED), lasers and bioanalytical devices, as well as the tiniest possible diodes and transistors in nanoelectronic circuitry.[1–3]

Note that the nomenclature nanorod and nanowire is a little arbitrary but the literature examples suggest they be distinguished by their diameter where nanorods are larger than nanowires with a demarcation line around 20 nm – see later.

5.2 Templating Nanowires

One of the most straightforward and general methods of making nanoscale wires and high-aspect ratio structures is by filling a template bearing nanoscale cylindrical holes with a given material.[4,5] If the holes are completely filled solid nanorods

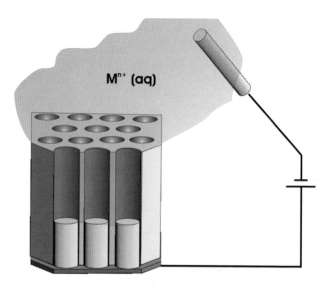

Figure 5.1 *A templating approach can produce nanowires or nanotubes from porous templates. Illustrated is the formation of metal nanowires by reduction of a metal salt into the cylindrical pores of a thin membrane*

and nanowires result, while a partial filling with a continuous coating gives rise to nanotubes. The formation of metal nanowires by an electrochemical reduction process is illustrated in Figure 5.1. If a template with very homogenous pores can be obtained, this technique reliably provides size-controlled nanorods and nanowires. Porous templates meeting these requirements abound in the materials world. Some natural minerals such as microporous zeolites are fashioned with one-, two-, or three-dimensional arrays of pores, which can be filled with other materials.[6] As well, lipid bilayers, which every living organism uses for compartmentalization and encapsulation, are dotted with one-dimensional channels created by transmembrane proteins (porins) which could potentially serve the same purpose. Chemically synthesized templates such as liquid crystal templated mesoporous aluminosilicates,[7–9] electrochemically synthesized templates such as anodically etched aluminum or silicon, self-organized systems such as microphase separated block copolymers,[10,11] and templates fabricated by engineering physics, such as track-etched polymer membranes, can all be used to make nanowire replicas.[12] As well, porous membranes can be doubly templated, giving rise to copies of the membrane out of different materials.[13] We will attempt to illustrate the templating approach to nanorods and nanowires with some chosen examples, while stressing that we are only covering a very small portion of templating possibilities.

5.3 Modulated Diameter Gold Nanorods

If one takes a p^+-doped silicon wafer (high concentration of hole carriers) and applies an oxidizing potential (anodization) in aqueous hydrofluoric acid (HF), what is formed is nanoporous silicon with a random network of pores running perpendicular to the membrane surface.[14–16] The diameter of the pores can be tuned from microns to nanometers by the choice of electrochemical current or voltage, and if porous enough the nanoscale silicon that is left can be luminescent because of spatial and quantum confinement of charge carriers, [17,18] much as isolated silicon nanocrystals luminesce if they are small enough.[19] Porous silicon in many forms has been found to be ideal for sensing applications, given the wealth of chemistry available to its surface.[20]

By lithographically patterning the surface of the silicon wafer with a regular array of nanowells, we create points where the electric field is focused during electrochemical oxidation.[21] The etching process begins at the tips of the nanowells to form nanochannels that continue to grow in a diameter self-limiting and perpendicular fashion with respect to the surface of the wafer. The result is a nanochannel silicon membrane with the pores as perfectly ordered as the inscribed lithographic pattern,[22] a process which can also be adapted to form perfect pore structures during the anodic oxidation of aluminum.[23] Indenting the starting material to be anodized allows one not only to change the lattice spacing of the nanochannels, but also their geometry.[24] If during the anodic oxidation the current is modulated, then a silicon membrane is formed that contains a parallel array of modulated diameter nanochannels, like that shown in Figure 5.2.[25,26] In this case the current and its modulation profile determine the dimensions and periodic architecture of the nanochannels. The higher the current, the greater is the relative width of the anodized channel. Such periodic, straight and modulated, nanochannel

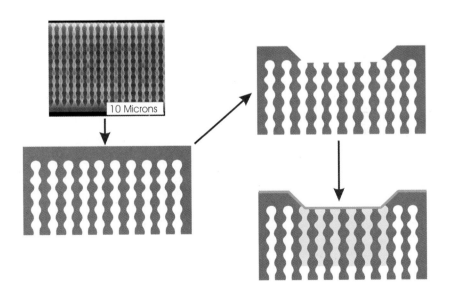

Figure 5.2   *SEM of a modulated diameter pore in silicon formed by anodic oxidation with a
sinusoidal current profile, top left, and the method used to replicate this
membrane as modulated diameter metal rods*
(Reproduced with permission from Ref. 27)

membranes are of considerable interest as two-dimensional and three-dimensional photonic crystals, respectively (see Chapter 7). To use this nanochannel silicon membrane as a template to direct the formation, for example, of modulated diameter gold nanorods, the back surface of the membrane is subjected to a local thermal oxidation and aqueous HF removal of the silica so formed. The newly exposed silicon on the backside of the membrane is anisotropically etched with ^{i}PrOH/KOH and the exposed regions of the silicon coated with sputter deposited gold. The gold layer acts as a catalyst for the electroless deposition of gold uniformly within the modulated diameter nanochannels of the membrane.[27] The steps in this process are depicted in Figure 5.2. The silicon membrane is then sacrificially etched and the released gold nanorod dispersion is collected and studied by electron and optical microscopy. It is found that the modulated diameter of the templated gold nanorods faithfully replicates the modulated current profile used in the anodic oxidation to make the modulated nanochannel silicon membrane, as seen in Figure 5.3. The observed correlation between the gold nanorod reflectivity and modulation profile shown on the right-hand side of this figure most likely originates from the size and shape dependence of the frequency of the plasmon resonance of gold.[28,29]

5.4 Modulated Composition Nanorods

The strategy of utilizing controlled diameter and length nanoscale channels in a membrane as a template to direct the deposition and growth of nanorods is quite general. The method can be adapted to create nanorods having a composition rather than a dimension profile. Modulated composition nanorods have been called "nanobarcodes," due to their striped appearance, suggesting that they could be utilized to tag such things as molecules in analytical chemistry and biology.[30] Because a characteristic diagnostic of a metal is its optical reflectivity, different segments of a barcode metal nanorod can be distinguished in an optical microscope.

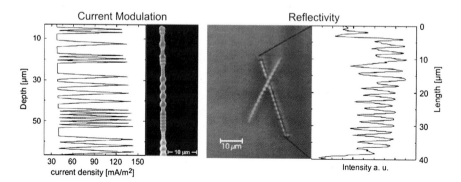

Figure 5.3   *Relationship between current density and microstructure of a modulated-diameter gold rod, left, microstructure and spatially resolved reflectivity from a modulated diameter gold rod, right*
(Reproduced with permission from Ref. 27)

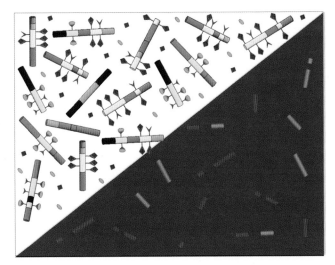

Figure 5.4 *Array of nanorods where the white segments have been tagged with receptors for a red fluorescent tag, and the yellow segments for green fluorescent tags. The bottom half shows how such a collection of rods would look like as a fluorescence image*

If a spectrometer is interfaced to the optical microscope, the shape of the selected area spectra can tell us exactly which metal is being looked at.

To use barcoded metal nanorods as an analytical screening tool, probe molecules must be anchored to specific metal segments of the nanorod, as illustrated in Figure 5.4. In this figure, there are two types of receptors anchored on different metal segments on the rods. Nanorods are then dispersed in the analyte containing the target molecules, which bear a luminescent label. The molecular-recognition binding event can then be tracked by observing diagnostic barcode fluorescence signals from the nanorods. Such a fluorescence image might look like the bottom right half of the figure. Mixtures of barcode nanorods with different stripe patterns, each type bearing a different probe molecule or receptor, could then be used as supports for DNA hybridization and immunoassays.

Metal barcode nanorods can be made by controlled coulometry electrochemical deposition of different metal segments within the nanochannels of an alumina membrane.[31] Also, any other type of material which can be electrodeposited, conductive polymers, semiconductors, or alloys, can be plated as nanowires.[32] Anodic oxidation of lithographically embossed aluminum sheet, analogous to silicon anodization described earlier, in phosphoric or oxalic acid electrolyte causes the growth of a regular array of constant diameter nanochannels perpendicular to the surface of the aluminum.[33,34] It is possible to obtain ordered arrays of size-tunable holes without the embossing step,[35] although the order of the pores is less pronounced and the ordered pore regions are polycrystalline since they are randomly nucleated in the early stages of the anodization.[36] The residual aluminum and aluminum oxide barrier layer are removed by treatment with mercuric chloride and phosphoric acid, respectively, to create the desired nanochannel alumina membrane. The proposed chemistry for anodic oxidization of aluminum and consequent alumina pore formation in phosphoric acid electrolyte is expressed in the equation

$$2Al + 3PO_4^{3-} \longrightarrow Al_2O_3 + 3PO_3^{3-}$$

The mechanism for diameter self-limiting channel growth is believed to begin with the initiation of alumina formation at the apices of the lithographically defined nanoindentations. Localized and continuous growth of oxide ensues in these regions, with concomitant dissolution and redeposition of aluminum oxide to the regions circumscribing the nanoindentations. The aluminum oxide deposited in these regions serves to create the nanochannels, as illustrated in Figure 5.5. There are two processes which give rise to ordered channels. During the etching process the applied voltage is concentrated at the tips of the growing tubes, and these sites of high field mutually repel each other to achieve an energetically favorable hexagonal packing. In addition, there is a volume expansion when converting aluminum to alumina, and the regions of high local mechanical stress repel each other as well.[37] By controlling the voltage and temperature conditions in the anodic oxidation it is possible to control the patterns of electric and stress fields to create periodic arrays of nanochannels with controlled diameters anywhere in the range of 20–500 nm, an example of which is shown in Figure 5.6.

Multi-metal nanobarcode rods can be synthesized in this type of membrane as illustrated in Figure 5.7. A thin metal film is first evaporated onto one side of the membrane, and serves as the electrode to initiate electrochemical deposition of metals within the nanochannels.[38] The metal deposition begins at the metal contacts located at the bottom of the nanochannels and the length of metal segment that grows scales directly with the number of coulombs of electricity passed. The bath can then be switched to a second metal salt and another segment grown, a process

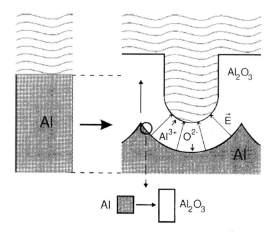

Figure 5.5 *Mechanism for pore formation in anodized aluminum*
(Reproduced with permission from Ref. 34)

Figure 5.6 *Anodic alumina membrane with ordered channels templated by imprinted indentations*
(Reproduced with permission from Ref. 33)

which can be repeated to build up a nanobarcode rod with a predetermined profile of metal segments. The metal backing is then dissolved, using nitric acid if this metal is silver for instance, and sodium hydroxide is used to remove the alumina template. This sequence of steps yields a dispersion of barcode nanorods, like that shown in Figure 5.8, where differences in optical reflectivity of metals enable barcode segments to be directly imaged.[30]

5.5 Barcoded Nanorod Orthogonal Self-Assembly

A demonstration experiment that shows the feasibility of segment specific anchoring of selected molecules to a barcode metal nanorod, called orthogonal self-assembly, is illustrated in Figure 5.9.[39] In this scheme, butylisonitrile binds indiscriminately to the Au and Pt segments. Aminoethanethiol has a greater affinity for gold than butylisonitrile[40] and thus displaces it from the Au segments. The result

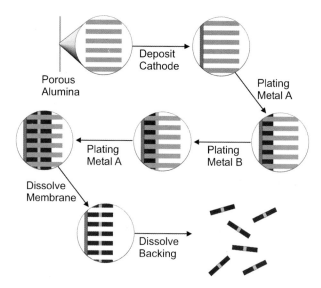

Figure 5.7 *Synthesis of multi-metal rods by electrodeposition*

is the coating of the two metals, both exposed to the same experimental steps, with two distinct monolayers of chemical species.

Validation that orthogonal self-assembly has occurred is demonstrated by anchoring a fluorescent rhodamine B isothiocyanate dye to the amino end group of the aminoethanethiol, *via* a thiourea linkage. Microscopic optical fluorescence

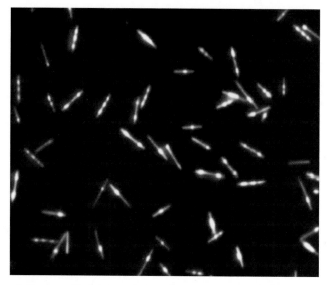

Figure 5.8 *Collection of multi-metal nanorods imaged in optical microscopy by the different reflectivities of different metals*
(Reproduced with permission from Ref. 30)

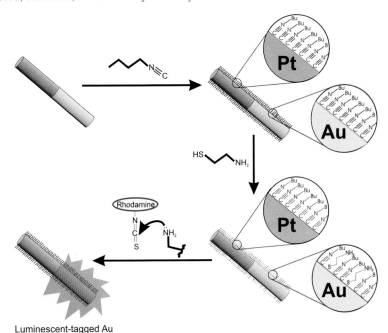

Luminescent-tagged Au

Figure 5.9 *Orthogonal assembly on nanorods. Butyl isonitrile is bound to Pt and Au surfaces. Aminoethanethiol displaces isonitriles on gold but not on platinum. Rhodamine isocyanate is reacted with terminal amino groups to fluorescently labeled gold segments*

images confirm that regions where the butylnitrile (Pt) and aminoethanethiol (Au) are anchored on the nanorod actually correspond to the nanorod Pt–Au segment profile. The observed fluorescence intensity profile of the nanorods, showing fluorescence only at the nanorod tips for 2–8–2 segment, Au–Pt–Au nanorods, confirms that orthogonal self-assembly was successfully implemented.

Barcode metal nanorod self-assembly enables bioassays such as the so-called "DNA sandwich" hybridization. This is when a target DNA strand is "sandwiched" between two capture strands whose ends are complementary to the target.[41] The example in Figure 5.10 shows that a 24-nucleotide DNA oligomer (the analyte) binds to a 12-nucleotide complementary capture sequence anchored to the nanorod and is later detected by addition of a 12-nucleotide probe sequence that is complementary to the remaining unbound 12 nucleotides and carries a $3'$ fluorescent tag, which caps and labels the DNA sandwich.[30] Both fluorescence and reflectivity images shown in the top left and right panels of Figure 5.10 serve to identify the barcode nanorods. To demonstrate the specificity of the binding event, a key control experiment omits the analyte. The very low fluorescence intensity image from the barcode nanorods (bottom left panel Figure 5.10) shows lack of binding of the analyte. Because DNA hybridization can be selectively detected on single optically encoded barcode nanorods, the way is now open for multiplexed hybridization assays. The selective detection of DNA hybridization on single optically encoded particles is an important first step towards ultrahigh information

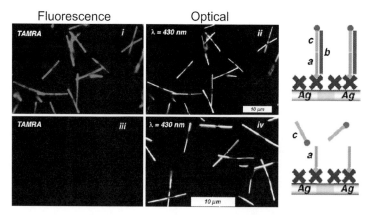

Figure 5.10 *DNA sandwich hybridization assay on metal barcode nanorods. The top shows the fluorescence observed when the target strand is present, while the bottom shows no fluorescence in the absence of the target*
(Reproduced with permission from Ref. 30)

content diagnostics. Entirely new types of DNA hybridization and immunoassays will surely be enabled by these metal barcode nanorods.

5.6 Self-Assembling Nanorods

The ability to anchor functional molecules to selective locations of a nanorod provides a number of interesting opportunities for self-assembly of nanorods into predetermined functional architectures, like end-to-end, side-to-side and crossbar arrangements. With soft-lithographically patterned substrates, nanorods can be directed to self-assemble into predetermined designs. A strategy for DNA end-functionalization and detection of gold nanorods is depicted in Figure 5.11.[42] The key step is the selective binding of thiolated DNA to the exposed ends of the gold nanorod encased in an alumina membrane. The membrane is then dissolved in NaOH to create a suspension of DNA end-functionalized gold nanorods to which rhodamine labeled complementary DNA is end-coupled to the nanorods. The success of this multistep procedure is recognized in diagnostic fluorescence microscopy images, which display fluorescence only from the ends of the gold nanorods as shown in Figure 5.11.[43]

Gold nanorods with complementary DNA recognition sites on their extremities should provide access to self-assembled nanorod architectures like dimers, trimers and chains. Similar approaches can be conceived to create selectively side-functionalized DNA and thereby a means of positioning nanorods with respect to each other in side-to-side, end-to-end and end-to-side arrangements. By this means, DNA side-functionalized gold nanorods have been directed to self-assemble on soft lithographically patterned gold surface sites that had been functionalized with complementary DNA.[44] In a similar vein, gold nanorods have been coated with ionically end-functionalized alkanethiolates and electrostatically self-assembled

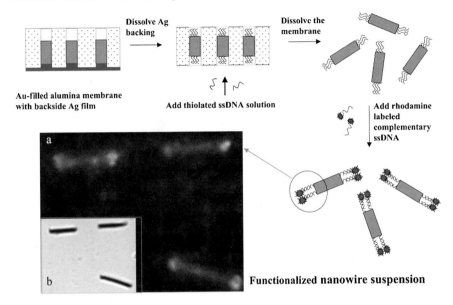

Figure 5.11   *Gold nanorod DNA end-functionalization with fluorescence detection*
(Reproduced with permission from Ref. 43)

on soft lithographically patterned surface sites or into surface relief patterns,[45] using
the strategy illustrated in Figure 5.12.

5.7 Magnetic Nanorods Bunch Up

One of the benefits of templating to create nanorods is that any material, which can
be electrodeposited, can constitute a segment of the structure. This list includes
metals, alloys, semiconductors, as well as conducting polymers, and consequently
nanorods can have a variety of segment-specific physical properties. If we focus on
metal rods, a useful property which can be exploited is magnetism. Magnetic
metals such as iron, cobalt, or nickel can be easily deposited into template
membranes.[46–49] Due to the small size of the segments, magnetic properties can be
direction- and size-dependent. For instance, if the length of the magnetic segments
is greater than their width they are preferentially magnetized along the rod axis. If
however, they are disks with their length shorter than their diameter, they are
magnetized perpendicular to the rod axis.

   This principle has recently been used to assemble nanorods with magnetic
segments into well-ordered bundles.[50] To achieve this, metal rods are electroplated
in alumina membranes in the standard fashion described above, using gold
interspersed with short disk-shaped nickel segments. The rods are then released
from the alumina with KOH solution, and dispersed in solution. When these rods
are exposed to a magnetic field, the nickel segments become permanently
magnetized perpendicular to the rod axis. Upon removal of the magnetic field, the
magnetic moments of nickel segments in different rods align head-to-tail, causing

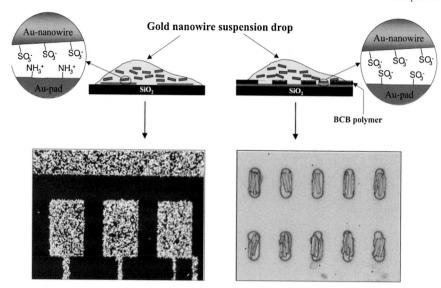

Figure 5.12 *Directed self-assembly of gold nanorods on soft lithographically patterned gold substrates. On the left, nanowires are patterned using electrostatic interactions, while on the right they are trapped into lithographically inscribed surface patterns*
(Reproduced with permission from Ref. 43)

a strong lateral attraction between rods. The result is magnetically self-assembled rod bundles, as shown in Figure 5.13, which do not form without the necessary contribution of the magnetic segments.

5.8 Magnetic Nanorods and Magnetic Nanoclusters

When a cylindrical magnetic nanorod segment is disk-shaped, its magnetization axis lies perpendicular to the rod axis, as in the previous example. It also follows that if the segment is longer than its width, its easy axis of magnetization is parallel to the nanorod axis.[51] The properties of magnetic structures electrodeposited in membranes can therefore be tuned over a wide range[52] by tuning the aspect ratio of the magnetic block as well as its composition. Nanorods of pure magnetic metals[53] such as nickel, iron and cobalt, as well as magnetic metal alloys and magnetic multilayers[54] have been synthesized. Magnetic properties of such wires can be probed by super-conducting quantum interference device (SQUID) magnetometry as a collection of free standing wires or as an array still embedded in the template membrane.[55] This technique relies on the extreme sensitivity to magnetic fields of an insulating junction between two superconductors, termed Josephson junctions. Individual nanowires can also be probed by magnetic force microscopy[56,57] (MFM) when coated onto a substrate. MFM is essentially an atomic force microscope (AFM) with a magnetic tip,[58,59] which allows the measurement of topography as well as magnetic fields through repulsive and attractive magnetic interactions. Another technique that can be

Figure 5.13 *Nanorod bundle caused by attractive interactions between disk-shaped magnetic blocks*
(Reproduced with permission from Ref. 50)

used to visualize magnetic fields is electron holography, or Lorentz microscopy.[60,61] This technique is performed in a transmission electron microscope, where the bending of the electron beam by local magnetic fields is visualized as interference patterns tracing the outline of magnetic field lines.[62]

Since electrodeposited wires can be made to have a large magnetic anisotropy, it can be straightforward to assemble them with a magnetic field. In Figure 5.14 is

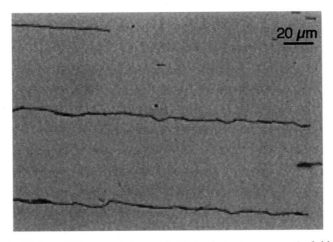

Figure 5.14 *Chains of Ni nanorods assembled into chains in a magnetic field*
(Reproduced with permission from Ref. 63)

shown an image of nickel nanorods, which were assembled into chains using a magnetic field parallel to a substrate.[63] They assemble in this way due to a head-to-tail stacking of magnetic dipoles in the wires. Magnetic fields are easy to apply, and could be used to organize these conductive magnetic wires over very large areas for future nanocircuits. Even more exciting would be the use of magnetic wires in conjunction with localized micro-magnetic fields, produced for instance by soft lithography,[64] allowing the predetermined assembly of complex functional systems.

The possibilities of magnetic field assembly using nanowires are increased even further, by considering the interaction between neighboring magnetic segments in a striped wire. If two segments of a magnetic metal such as nickel are separated by a non-magnetic block, such as gold, the field lines for the nickel blocks will interact with each other. The smaller the separation between nickel segments, the stronger the magnetic interaction, and the more intense the magnetic field gradient at this point. This is because the field strength is strong directly between the nickel tips, but drops off very rapidly as you travel away from the magnetic junction. Such magnetic field gradients tend to attract small magnetic particles, which are drawn to the region of highest field strength. This principle was demonstrated by assembling iron oxide nanoparticles onto nanowires with magnetic and non-magnetic stripes.[65] An illustration of the process is shown in Figure 5.15. The top panel shows the

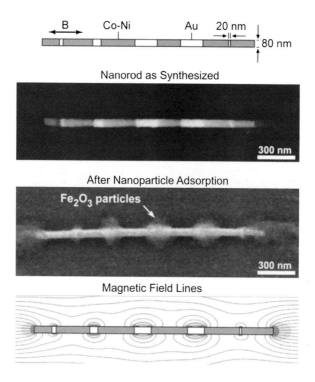

Figure 5.15 *Attraction of magnetic nanoparticles to diamagnetic segments in a magnetic barcoded nanorod*
(Reproduced with permission from Ref. 65)

structure of the nanowires used, followed by SEM images of a bare and nano-particle coated wire. The bottom panel shows the direction of magnetic field lines in such a wire. Many methods could be used to align and organize these magnetic wires. Further binding of nanoparticles at predetermined positions on the wire could create a whole range of systems with hierarchical functionality.

5.9 An Irresistible Attraction for Biomolecules

In this chapter, we have seen the power of selectively functionalized nanorods for the detection of DNA and proteins, and in Chapter 6 we will see that nanoclusters can also be used for this purpose. However, is it possible to adapt these nano-structures to separate and purify biomolecules?

An interesting strategy to achieve this goal involves the use of tri-segment Au–Ni–Au nanorods as an affinity template for histidine-tagged proteins.[66] Here, the ferromagnetic nickel segment of the nanorod serves two functions: (i) to selectively capture the labeled proteins and (ii) in an applied magnetic field, magnetically separate them from a mix of undesirable biomolecules. Subsequent chemical release of the histidine-tagged protein from the nickel segment of the magnetically collected nanorods achieves the targeted separation and purification. This strategy is summarized in the top part of Figure 5.16 for the separation of histidine-tagged proteins from untagged ones, as well as an extension of the approach to the separation of antibodies that recognize the histidine-tagged protein from ones that do not. The method works well for both situations, but for the sake of illustration we will elaborate on the former.

First, it is important to note that the capping gold segments on the nanorod were found to be a necessary part of this system. Their presence avoided etching of the nickel during the dissolution of the nanochannel alumina template used to electrosynthesize the barcode nanorods. Moreover, to minimize deleterious protein binding to the nanorods as well as to avoid nanorod aggregation as much as possible, the gold segments were coated with a thiol-terminated polyethyleneglycol. As a first step, specific anchoring of fluorescein-labeled polyhistidine to the bare nickel segment of the nanorod was demonstrated by confocal fluorescence spectroscopy. Then, the efficacy of the method for the magnetic separation of molecules was tested, by mixing nanorods having a bare nickel segment with a solution of fluorescein-labeled polyhistidine and then magnetically separating the nanorods from the solution. The separation was demonstrated visually (and quantified by optical spectroscopy) by a striking color change of the solution from bright green to colorless (Figure 5.16). In a control experiment, using Au nanorods functionalized with the same thiol terminated polyethyleneglycol there was essentially no change in the green color of the fluorescein-labeled polyhistidine solution.

This is clearly a convincing demonstration that barcode magnetic nanorods can be utilized to separate and purify proteins from a biological mixture. It is interesting to note that a nickel-based stationary phase in a chromatography format has traditionally been used to achieve this end. The advantages of the nanorod strategy can be seen to relate to the simplicity and selectivity of the protein capture

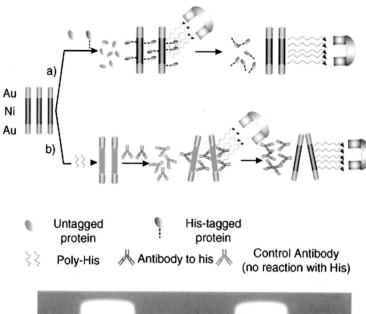

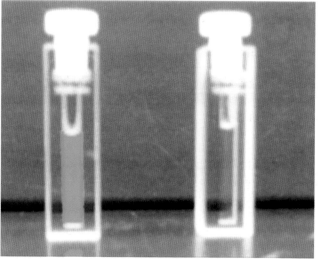

Figure 5.16 *Scheme depicting the separation of biomolecules by magnetically tagged nanowire barcodes, top. The bottom optical image shows a solution before, left, and after, right, separation of fluorescently labeled protein by the magnetic barcodes*
(Reproduced with permission from Ref. 66)

and subsequent separation–purification steps, the intrinsic high separation efficiency of the method because of the high surface area of the nanorods, and the fact that the process is a dispersion rather than a column chromatography-based separation. The high degree of tailorability amenable to the nanorod approach, through modulation of barcode composition and surface functionalization, bodes well for the future of this method for combined detection, separation and purification of diverse kinds of biomolecules.

5.10 Hierarchically Ordered Nanorods

The behavior of nanoscale building blocks can be quite difficult to predict. Quantum size effects (QSE), increased surface area, decreased influence of gravity, can make nanoscale materials very different from bulk materials, and can make it challenging to gauge their response in a given system. However, in some cases theoretical work intended for quite different systems can be used with great success to describe novel nanoscale building blocks.

For instance, surfactants and amphiphiles are molecules that are quite well understood and ubiquitous in everyday life.[67] In general terms an amphiphile is a molecule whose two ends have different properties, such as hydrophilic and hydrophobic. This allows a surfactant to bridge two phases, and homogenize normally phase-separated mixtures such as oil and water. When a surfactant with a hydrophilic head and hydrophobic tail is dissolved in water, it will minimize its energy and maximize its entropy by packing its tail side-by-side with other surfactants in a lamella. The assembly of amphiphiles has been studied extensively,[68–71] and can be likened to the close packing of geometrical shapes. For instance, if an amphiphile has a large head and a long thin tail, its geometrical shape is best described as a solid cone. The packing of cones into a close-packed array leads to a lamellae with high curvature, which may join with itself to form a sphere. When a molecule has the shape of a truncated cone, it will form lamellae with a smaller curvature, while molecules with a cylindrical shape will assemble into flat sheets. The geometrical aspects of surfactant molecules are typically characterized by a packing parameter, which approximates how this surfactant will assemble into ordered arrays. Such ordered arrays can be quite complex, especially when considering large amphiphiles such as diblock copolymers.

The packing parameter theory, which likens molecules to geometrical objects, can be extended to nanoscale solids, which *are* geometrical objects. Using this analogy, nanorods have recently been used as mesoscopic amphiphile mimics, which assemble into similar ordered phases.[72] Utilizing nanorods as such molecule-mimetics begins with the electrodeposition into a porous alumina template to form one-dimensional rods. Gold is used to form the rigid, hydrophilic parts of the rod, while electrodeposited polypyrrole (Ppy) is used to form the soft, hydrophobic parts of the rod. When the rods are released from the membrane with aqueous NaOH, it is found that the Ppy segments have a smaller diameter than the gold portions. A two-segment nanorod having a gold and Ppy sides thus has a tapered shape similar to surfactant molecules. As well, the hydrophobic Ppy blocks experience an attraction because of hydrophobic interactions when suspended in water, much like hydrophobic surfactant tails.

A model taking these attractive forces and geometrical constraints into account was developed, and various types of arrays realized experimentally by varying the block lengths and sequences. An image of a hierarchical sphere composed

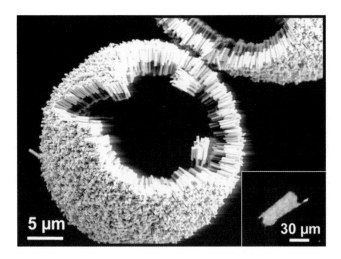

Figure 5.17   *A hollow sphere self-assembled from Au–PPy nanorods. The inset shows an*
optical picture of a curved self-assembled nanorod sheet
(Reproduced with permission from Ref. 72)

of a curved lamella of close-packed Au–Ppy rods is shown in Figure 5.17. It was
found that by increasing the length of the Au block relative to the Ppy block, the
nanorod assumed the shape of a cone with increasing focal angle. By tuning these
effects, self-assembled structures with predetermined curvature could be obtained,
much like surfactant systems. The influence of block ratios on the curvature of the
self-assembled architecture is shown graphically in Figure 5.18.

5.11 Nanorod Devices

By juxtaposing the roles of an alumina nanochannel membrane both as a template
for synthesizing barcode nanorods as well as a nanoreaction chamber for the
selective end-coating of nanorods with molecules or materials, it is possible to

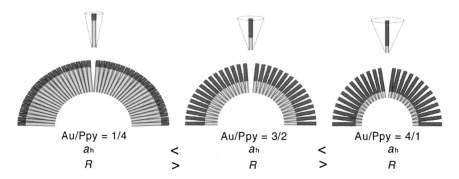

Figure 5.18   *Building rules for self-assembly using mesoscopic nanorod amphiphiles*
(Reproduced with permission from Ref. 72)

devise protocols for making in-wire and on-wire nanorod architectures that can function as nanoelectronic devices. These nanorod devices can be organized by directed self-assembly methods that utilize microfluidics[73] and electrostatics[74] and guided into position across probe electrodes to measure their electrical properties.

A scheme for making an in-wire nanorod device in which two metal segments sandwich a central SAM or other material is shown in Figure 5.19. A nanorod device with a functionalized SAM in-wire junction has been shown to display negative differential resistance[75,76] (visualized as a spike of increased current in the current–voltage curve), which bodes well for its use in nanoelectronic circuitry.

In a similar vein, nanorods can be coated using LbL electrostatic self-assembly (ESA) (see Chapter 3) to form polyelectrolyte and metal oxide multilayer coated nanorods such as a $Au(ZnO/PSS)_n ZnO/Au$ on-wire nanorod device.[77] In-wire nanorod device architectures of this type have been shown to function as field effect transistors (FET),[78] making them possible candidates for nanoscale electronic circuitry. It is important to note that the coating and wire materials illustrated here are by no means exhaustive. Methods to coat the starting alumina template with a variety of compounds abound in the literature,[79,80] and electrodeposition within these pre-coated membranes followed perhaps by further chemical deposition of material and template dissolution can lead to a "periodic table" of concentric coaxial, striped nanorods. Likely only our imagination can tell us what uses these structures may have in commercial devices.

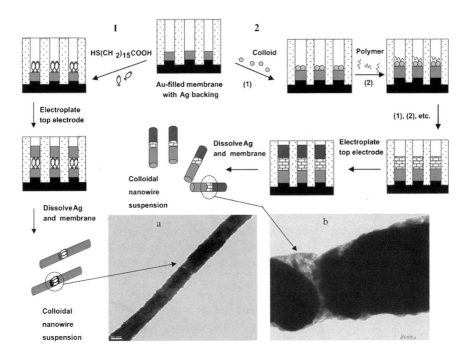

Figure 5.19 *Schemes for preparing in-wire nanorod devices*
(Reproduced with permission from Ref. 43)

5.12 Nanotubes from Nanoporous Templates

It should be apparent by now that well-defined templates offer countless opportunities for making replicas out of almost any desired material, just as templates with linear pores allow the formation of nanorods of myriad compositions. We have also seen that rods can have both axial and radial compositional profiles. The level of complexity can be raised once more if instead of filling the whole template with a material, only a uniform coating is deposited. Once the template is dissolved, what is left are hollow nanotubes, made out of almost any type of material.[81]

There are several ways of coating the inside of porous templates. One of the simplest ways consists of letting a fluid spontaneously wet the inside of the template pores.[82,83] The fluid used can be a molten polymer, a polymer solution or a sol–gel preparation,[84] and is coated onto the template using capillary forces resulting from small diameter channels with a large available surface. The nanotube is obtained by solidification of the precursor fluid, be it by cooling, heating, waiting, or adding a reactant or catalyst. This method is quite easy to apply experimentally, but it is often difficult to control the nanotube wall thickness.

Another strategy is to use a layer-by-layer growth procedure to coat the inside of the pores. This can include chemical vapor deposition (CVD) of gas-phase species,[85–87] solution phase atomic layer deposition (ALD), or layer-by-layer electrostatic assembly,[88] to name a few. The standard deposition is simply carried out in the presence of the bare membrane, and after a certain number of cycles a nanotube of predetermined wall thickness can be formed. The membrane is dissolved to obtain freestanding tubes, or a material is first deposited inside the remaining void space to obtain coaxially coated nanowires upon membrane dissolution. It must be noted that this process is rarely as simple as it seems; because of the small size of the channels, surface effects are extremely important, and diffusion of species into channels can easily be hindered.

Optimization of procedures to obtain homogenous nanotubes is always important, since only homogenous tubes can be used for most demonstration and commercial applications. Materials deposited as nanotubes can be further modified by a chemical transformation, such as high-temperature pyrolysis. For instance, if we pyrolyze in an inert atmosphere a tube made of organic polymer, a carbon (graphite or amorphous carbon) nanotube could result. In previous chapters, we have discussed polyferrocenylsilanes (PFS) polymers, which can be converted to magnetic ceramics upon heating due to the presence of iron in the polymer chain.[89] Tubes could be made of PFS, then these tubes heated to make magnetic ceramic replicas. Such a magnetic membrane might be of use, for instance, in filtering magnetic particle tagged biomolecules.

The synthesis of nanotubes formed using porous templates is illustrated for the case of silica. Silica nanotubes can be formed from an alumina membrane,[90] using cycles of liquid-phase surface reactions. First, the membrane is dipped into $SiCl_4$, which reacts with the surface hydroxyl groups in the membrane pores. Excess reactant is washed out with CCl_4 then the membrane is reacted with water. After washing out the water the steps are repeated, leading to a buildup of a SiO_2 shell on the membrane with the elimination of HCl. Thus formed nanotubes can be released into solution by dissolution of the alumina membrane in concentrated aqueous base, or prior electrodeposition into the coated membranes can result in silica-insulated metallic wires. In Figure 5.20, we can see in the top panel a silica nanotube produced by this method, and on the bottom panel a silica-coated gold nanorod formed by electroplating inside a silica-coated membrane. By including a

Figure 5.20 *Silica nanotube and silica-coated gold nanorod*
(Reproduced with permission from Ref. 90)

structure-directing template into the silica nanotube forming synthesis mixture (see Chapter 8) one can imagine perforating the channel walls of the nanotubes with periodic arrays of micropores or mesopores to get holey nanotubes!

5.13 Layer-by-Layer Nanotubes from Nanorods

In Chapter 3, we saw the power and versatility of the layer-by-layer electrostatic multilayer technique. It was noted that this type of growth can proceed on templates with quite different topologies, as well as colloids in suspension. The fabrication of metal nanorods by electrodeposition into porous membranes is relatively straightforward, and the resulting colloidal rods can be surface modified by electrostatic multilayers. The subsequent removal of the nanorod leaves a flexible, closed-ended nanotube replica of the nanorod.[91] This process begins with the formation of nickel nanorods using electrodeposition into a polycarbonate track-etched membrane. Once in solution, these nanorods are coated with alternating layers of poly(diallyldimethyl ammonium chloride) (PDAC) and poly(styrene sulfonate) (PSS), washing the rods by centrifugation in pure water after each coating step. The core–shell rods are then treated with a mild solution of HCl for 24 h, leading to nickel dissolution and the formation of polyelectrolyte multilayer nanotubes. By coating instead a polyelectrolyte ceramic precursor, hollow ceramic tubules were also produced. Figure 5.21 shows a TEM image of a ceramic nanotube produced in this fashion. This technique should be applicable to a wide variety of

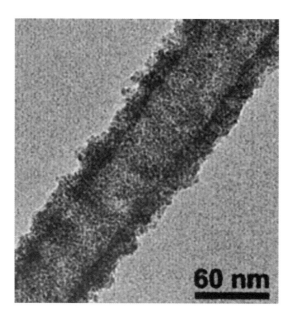

Figure 5.21   *TEM image of a ceramic nanotube prepared by layer-by-layer deposition onto a metal nanowire*
(Reproduced with permission from Ref. 91)

materials provided they can be coated onto nanorods in solution, and not affected by the core-etching step. One can imagine a creative extension on this theme, performing LbL on a barcoded nanorod. Selective etching of particular segments could lead to a whole new world of "flexible nanorods," which bend on command at the LbL nanojoints simply by applying external chemical or physical stimuli, enabling the exciting prospect of nanorod nanomechanics!

5.14 Synthesis of Single Crystal Semiconductor Nanowires

While the nanochannel template directed synthesis of nanorods works well for diameters from 500 nm down to about 20 nm, it does not work for smaller diameter nanorods because of a lower limit set on nanochannel diameters accessible by anodic oxidation of aluminum. Access to this lower range of diameters turns out to be especially important because it is at these length scales that quantum confinement effects are expected to emerge in nanowires.[92] Another drawback of the nanochannel template method is the fact that deposited nanowires tend to be polycrystalline, making their electronic, optical, or other properties less well-defined and more difficult to predict. Smaller diameter nanowires would facilitate the assembly of smaller, more densely packed nanocircuitry for electronics, optoelectronics and photonics. Note that the term nanowire is used for this sub-20 nm length scale to distinguish them from supra-20 nm nanorods, as mentioned at the beginning of this chapter.

5.15 Vapor–Liquid–Solid Synthesis of Nanowires

One of the most successful ways of synthesizing oriented single crystal semiconductor nanowires with control over their diameter, length and composition including doping as well as longitudinal and axial structures, is a catalytic nanocluster vapor–liquid–solid (VLS) growth process[93–98] shown schematically in Figure 5.22. A catalyst droplet first melts, becomes saturated with precursors, and the precipitating elements extrude out of the catalyst droplet as a single-crystal nanowire.

The physicochemical underpinnings of the synthesis of Si nanowires rests on VLS catalytic anisotropic growth on a liquid Fe–Si nanocluster of controlled diameter. The equilibrium pseudo-binary Fe–Si phase diagram is known, such that the composition and temperature that favors liquid Fe–Si in equilibrium with solid Si can be readily identified. This region of composition-temperature phase space favors VLS growth of Si nanowires. The diameter of the Si nanowire is controlled by the diameter of the Fe–Si nanoclusters formed in the laser ablation of a Fe–Si target in a buffer gas flow system in a temperature-controlled furnace. The conditions in the furnace control the time that Fe–Si liquid nanoclusters survive, which determines the length of the Si nanowires that grow.

The Si nanowire growth process involves laser ablation of the Fe–Si target, which creates a dense hot vapor that condenses into nanoclusters as the Fe and Si

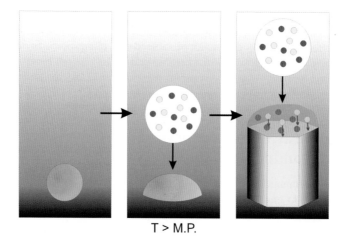

T > M.P.

Figure 5.22 *The vapor–liquid–solid growth mechanism for synthesizing crystalline nanowires*

species cool through collisions with the buffer gas. The furnace temperature is controlled to maintain the Fe–Si liquid state and Si nanowire growth begins only when the Fe–Si liquid nanocluster becomes supersaturated with Si. Nanowire growth proceeds as long as the Fe–Si nanoclusters remain in the liquid state and Si nutrient is available.

The Si nanowire grows anisotropically as a single crystal with the {111} lattice planes oriented perpendicular to the nanowire growth axis (Figure 5.23). As mentioned, the nanowire diameter is determined by the nanocluster diameter and the length is controlled by the growth period. Nanowire growth terminates when the nanowires pass out of the hot zone of the laser ablation-furnace reactor. In the case of silicon nanowires, a sheath of silica coats the surface of the wire and the catalytic Fe–Si nanocluster is attached to the end of the wire. Doping is achieved by controlling the precise stoichiometry of the target, which is made slightly rich in phosphorus for n-Si and boron for p-Si nanowires. Alternatively, controlled

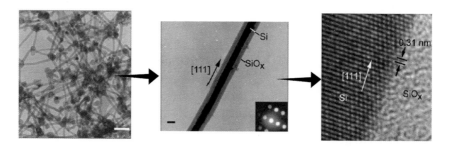

Figure 5.23 *Oriented single crystal silicon nanowires showing the catalytic Fe–Si nanocluster attached to the end of the wire and an amorphous silica sheath* (Reproduced with permission from Ref. 93)

amounts of PH_3 and BH_3 can be introduced into the gas phase during the nanowire growth process, with similar effects. Epitaxy at the nanocluster–nanowire interface seems to be responsible for oriented single crystal nanowire growth – see later.

In this example, the catalyst droplet is created *in situ* through the condensation of a laser-ablated Fe–Si target. The catalyst can also be provided separately, as a capped gold nanocluster (see Chapter 6) for instance.[99] These nanoclusters can be deposited onto a surface or alternatively grown from a gold nanometer thickness film by thermally induced dewetting, and if the substrate and nanowire have epitaxial matching of particular lattice planes the wires may grow out of the clusters perfectly orthogonal to the substrate.[100]

5.16 What Controls Nanowire-Oriented Growth?

The catalytic nanocluster in VLS growth of say silicon nanowires is actually a liquid nanodroplet supersaturated with silicon precursors like silane, from which nucleate and grow oriented single crystal silicon nanowires with a diameter that is controlled by the diameter of the nanodroplet. But what factors determine nanowire orientation?

This is a tough question because the answer likely involves surface energetics at the solid–liquid interface of the nanowire emerging from the nanodroplet, which involves complex issues like size-dependence of nanodroplet supersaturation and surface free energy, plus the effect of amorphous surface oxides and defects.

Despite these challenges some insight has recently been obtained from the observation that VLS of SiH_4/H_2 gaseous mixtures on gold nanocluster catalysts generates a narrow size distribution of single crystal silicon nanowires with diameters as low as 3 nm, devoid of a sheath of surface silicon oxide and capped with Si–H bonds instead, displaying well-defined surface facets and, especially relevant to the question posed, orientation dominated by growth for the smallest 3 nm and for the largest 13 nm ones.[101]

These observations beg the question, how does the nature of the nanocluster–nanowire interface relate to the direction of oriented nanowire growth? HRTEM images of the interface observed for small and large silicon nanowires shown in Figure 5.24 clearly show V-shaped morphology for the former and a planar one for the latter. It seems therefore that the lowest energy {111} interface controls the initial nucleation and growth process where two {111} planes merge to create a <100> growth axis for the smaller silicon nanowires, rather than a <111> growth axis interface. Clearly, surface energetics plays a key role in determining nanowire growth directions in VLS syntheses. This analysis does not, however, throw any light on the more difficult matter of how the barrier for nucleation and supersaturation scale with nanodroplet size, an excellent test for future studies!

5.17 Supercritical Fluid–Liquid–Solid Synthesis

In a creative departure from metal nanocluster seeded VLS growth of oriented single crystal semiconductor nanowires, alternative solution phase synthetic

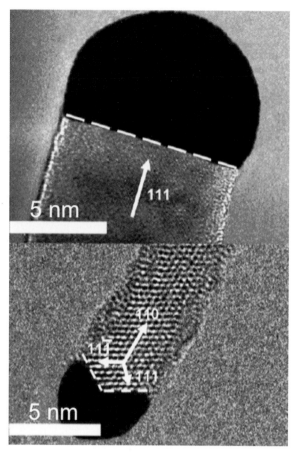

Figure 5.24 *TEM images of silicon nanowires growing in the 111 and 110 directions, terminated with an alloy catalyst droplet*
(Reproduced with permission from Ref. 101)

approaches to nanowires have been sought to expand the choice of precursors, enhance the opportunities for doping and surface functionalization, and boost production rates. For low eutectic temperature metal–semiconductor combinations like Ga/GaAs, In/InN and In/GaAs, VLS growth of semiconductor nanowires in conventional organic solvents, dubbed SLS, works well at around 200 °C. However, the method is not viable for many higher eutectic temperature systems (350–500 °C) required for gold nanocluster seeded growth of Si, Ge and GaAs nanowires. Interestingly, such temperatures are achievable by working under pressure above the critical point but below the degradation point of the organic solvent used for nanowire growth. The first proof-of-principle demonstration of supercritical fluid–liquid–solid (SFLS) synthesis focused on GaAs nanowires and was based on the reaction of (^{t}Bu)$_3$Ga and As(SiMe$_3$)$_3$ precursors catalyzed by 7 nm dodecanethiol capped gold nanoclusters in supercritical hexane at 500 °C

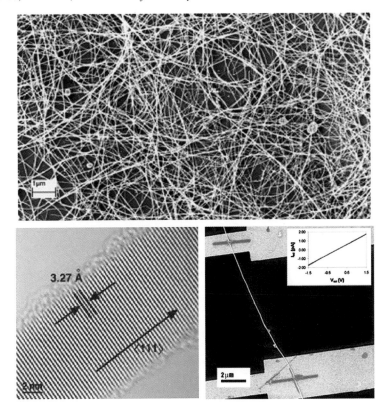

Figure 5.25 *Supercritical growth of nanowires. The top image shows a collection of GaAs nanowires grown using this method, while the bottom left image shows lattice fringes by high resolution TEM. A current–voltage plot could be obtained by evaporating electrodes onto the nanowire tips, bottom right*
(Reproduced with permission from Ref. 102)

and 370 atm.[102] Yields of ⟨111⟩ oriented single crystal GaAs nanowires were high and accompanying unwanted particulates were sparse (Figure 5.25). Electrical and XPS measurements on isolated GaAs nanowires showed that *in situ* doping had occurred in the synthesis, most likely through incorporation of Si from the As(SiMe$_3$)$_3$ precursor. Because of the potentially greater variety of materials chemistry precursors for SLS and SFLS solution phase syntheses of nanowires, it is possible that these approaches will emerge as a general route to nanowires with diverse compositions, doping and surface functionalization.

5.18 Nanowire Quantum Size Effects

VLS synthesis can be used to make single crystal silicon nanowires with diameters, tailored by the choice of catalytic nanocluster size, in the range of 1–7 nm. Note that one has to be smaller than the ~5 nm exciton size for silicon to see QSEs in

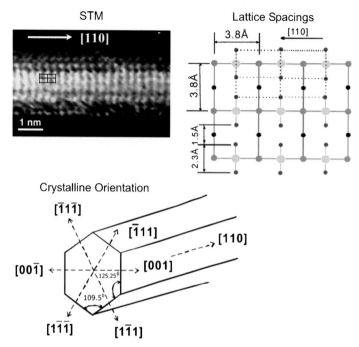

Figure 5.26 *STM of Si nanowire 001 facet, along with structural models fitting the STM data*
(Reproduced with permission from Ref. 103)

any type of silicon nanostructure. By removing the surface silicon oxide with aqueous HF, it has been proven possible to create stable, faceted, hydrogen terminated silicon nanowire surfaces, enabling atomically resolved scanning tunneling microscopy (STM) imaging (Figure 5.26).[103] As well, scanning tunneling spectroscopy measurements using an STM have been used to probe the electronic density of states (DOS) of the silicon nanowires and the corresponding electronic band gaps (E_g) as a function of the diameter of the nanowires. The experimental E_g are found to be those expected by theory, increasing with decreasing nanowire diameters. Gaps range from 1.1 eV for 7 nm to 3.5 eV for 1.3 nm, demonstrating QSE in silicon nanowires. Possible applications of single crystal silicon nanowires are UV LEDs and lasers, providing the wide band gap provides a direct electronic transition with allowed electric dipole and short radiative lifetime emission, as predicted for such small diameter structures.

While both quantum confined nanoclusters and nanowires show QSE in their electronic properties, their anisotropy make nanowires quite different. If we consider light emission, a luminescent nanocluster emits plane-polarized light while a nanorod exhibits linearly polarized light.[104] A whole host of direction-dependent electronic effects can be utilized, but this relies heavily on being able to reproducibly align single nanowires or collections of nanowires in a particular orientation.

5.19 Zoo of Nanowire Compositions and Architectures

The VLS synthesis provides a rational way of making semiconductor wires with nanometer diameters and micron lengths – in fact, the growth process can be considered akin to a living polymerization. This is to say, multi-block nanowires with blocks of controlled length can be grown by changing the precursor composition during growth. Compositions can be quite varied as seen by the synthesis of intrinsic and doped forms of Si, Ge, and SiGe, compound semiconductors such as GaN,[105] GaP, GaAs[106] and InP,[107] high-T_c superconductors,[108] oxides,[109] and more. The requirement in this type of synthesis is a thorough understanding of the temperature–pressure-composition phase diagram for the nanowire precursor elements and the metal catalyst droplet. Once at the proper temperature and pressure,[110] nanowires of different compositions can be grown, by controlling the composition of the gas-phase precursors. Various nanowire architectures have also been made by creative modifications of the VLS catalytic nanocluster growth process. One strategy enabled the growth of nanowire superlattices, made of GaAs/GaP or Si/SiGe, in a block-by-block growth process, with the formation of epitaxial (lattice matching) junctions between heterocomponents.[111] Nanowires have also been grown in a diode configuration comprised of a p-Si/n-Si epitaxial junction. Also, by modifying the VLS growth conditions epitaxial core–shell and core–multishell nanowire hetero-structures have been synthesized. As well, this growth mechanism can be adapted to growing nanowires in solution. For instance, it was reported that silicon nanowires could be grown under supercritical solvent conditions, using gold nanocrystals as catalyst particles and elemental silicon as the Si source.[112] The nanowire precursors can be pure elements, which are thermally evaporated or laser ablated, or inorganic complexes used for CVD which decompose at the synthesis conditions.

These nanowire building blocks have enabled the assembly of the smallest electronic, optoelectronic and photonic devices, such as the one shown in Figure 5.27.[113] Some of these will be described in the following sections and critically appraised in terms of whether nanodevices and nanocircuits based on nanowires can ever deliver on their promise of a smaller, denser and faster electronics and optics.

5.20 Single-Source Precursors

The VLS synthesis of nanowires can give highly homogenous products, adequate for any proof-of-concept demonstration. However, the implementation of these nanowires in commercial devices such as computers will require nanowires with an even greater degree of perfection. When a VLS synthesis is performed using laser ablation, it can be difficult to ensure total homogeneity of the nanowires. There is some heterogeneity in the atomic clusters produced by ablation, which especially affect the synthesis of nanowires containing two or more elements and may lead to structural defects. An elegant solution to this problem is to use a precursor that contains all the desired elements so that mixing is accomplished at the molecular scale. The synthesis of CdS and ZnS nanowires, have recently been reported using this strategy.[114] In the case of CdS, the cadmium and sulfur are incorporated into

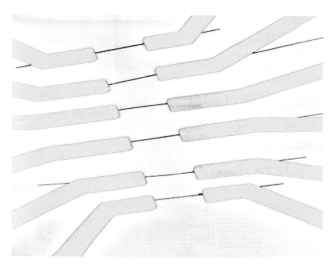

Figure 5.27   *The incredible shrinking circuit*
(Reproduced with permission from Ref. 113)

a cadmium complex bearing two bidentate sulfamide ligands, $Cd(S_2NEt_2)_2$. The ZnS nanowires are made with the analogous complex, $Zn(S_2NEt_2)_2$. The precursors are volatilized by heating at the entrance of the furnace, then are dissolved and decomposed in the gold catalyst droplets at high temperature. A 25-nm diameter CdS nanowire produced in this way is shown in a TEM image in Figure 5.28, which also shows the narrow band photoluminescence of these wires due to quantum confinement of charge carriers.

5.21 Manipulating Nanowires

For nanowires to achieve their true potential for many applications, means must be found to efficiently organize, manipulate and direct them into a diverse set of well-defined architectures. Experiments on single nanowires are necessary for probing their basic properties, but the real challenge is making arrays and connection to these arrays at the nanoscale. Several methods have been developed to put nanowires where they are needed, or organize them into ordered arrays which could be later connected into circuits.

An obvious spot we would like to place a nanowire is spanning the length of two electrodes. In this way, its electrical characteristics can be directly probed and used in a device. Aligning wires with patterned electrodes could potentially be an inefficient and time-consuming process. In order to direct nanowires between desired electrodes, an electric-field based alignment method has been used.[115] When a voltage is applied between two micropatterned electrodes, charge carriers of opposite sign are accumulated at opposite electrodes and an electric field is produced. If a nanowire in solution approaches this field, charges within the wire become polarized to mirror the applied electric field. This creates an attraction

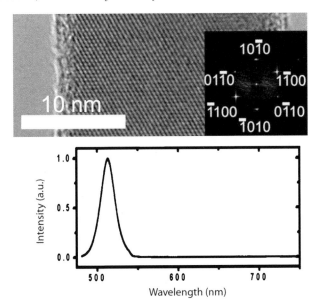

Figure 5.28 *CdS nanowire produced with a single-source precursor, above. High-quality CdS results in a sharp luminescence peak, below*
(Reproduced with permission from Ref. 114)

between the electrode pair and the nanowire, which bridges the gap between them. Once a nanowire is between electrodes, the electric field is quenched, eliminating the adsorption of a second nanowire in the gap. The assembly of a collection of 200-nm diameter nanowires across interdigitated gold electrodes overcoated by a thin insulating layer is shown in Figure 5.29. Metal spots can be evaporated onto the insulator surface to focus the electric field, and achieve a greater selectivity.

A microfluidic approach has also been used to orient nanowires[116] and direct them into functional arrays.[117] The first step consists of producing a PDMS stamp with a series of parallel rectangular grooves on its surface (see Chapter 2). The channels are then aligned under a microscope with electrodes which have been previously patterned on a substrate. A drop of nanowire suspension is flowed into the microchannels by capillary forces, and solvent evaporation aligns the wires at the edges of the channels. This method is less selective than the electric field alignment, but can be performed over a larger area in a parallel fashion.

For many applications, large two-dimensional arrays of packed nanowires are required. An organized array is ideal for further patterning using photolithography to produce functional elements. As well, certain properties arising from interwire electronic coupling can be probed and possibly used in applications. An interesting way to produce organized nanowire layers is through the Langmuir–Blodgett transfer technique. Traditionally, this method was used to form ordered monolayers of a molecular amphiphile at a water–air interface, and to transfer this layer onto substrates. However, it can be adapted to hydrophobized silver wires, which are

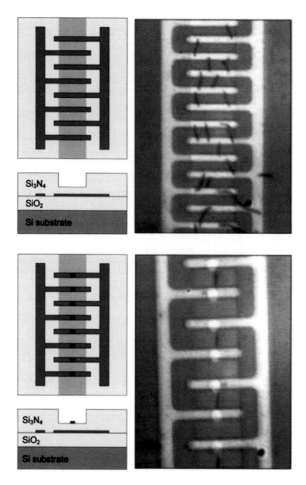

Figure 5.29 *Electric field-induced assembly of 200 nm gold nanorods across electrodes with, above, and without, below, small metal islands used to increase the effect of the electric field*
(Reproduced with permission from Ref. 115)

also found to float at a water–air interface. Compression of the interface, which increases the surface density of nanowires cause the wires to bump into each other and align into ordered phases. This ordered monolayer can then be transferred onto a substrate by slowly withdrawing it through the interface, and by keeping the surface pressure constant.[118]

The Langmuir–Blodgett technique using nanowires has an interesting analogy in the lumber industry, during logging of timber along riverways.[119] Trees are cut and cleaned and then floated onto the surface of a river with a processing plant downstream. When the river is wide the logs float with a random orientation, but when a narrow river segment is encountered both the river flow and congestion by neighboring logs causes them to align in the flow direction. In a Langmuir–Blodgett

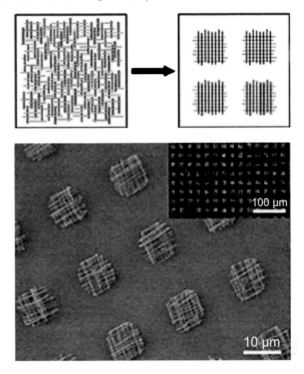

Figure 5.30 *Nanowire crossbar arrays produced by the Langmuir–Blodgett technique combined with photolithography*
(Reproduced with permission from Ref. 121)

apparatus, a computer-controlled barrier mimics the narrowing riverbank, compressing the nanowires and aligning them. Different functional systems have been achieved using this method, such as substrates for sensitive surface-enhanced Raman spectroscopy of molecules[120] using close-packed silver nanowire arrays. Multiple steps of Langmuir–Blodgett assembly and deposition, followed by photolithography, have also resulted in hierarchical crossbar arrays.[121] A SEM micrograph of such an array is shown in Figure 5.30, illustrating the potential of this procedure for a variety of nanowire-based devices.

5.22 Crossed Semiconductor Nanowires – Smallest LED

Having access to n- and p-doped direct electronic band gap semiconductor nanowires like InP, it is possible to devise materials self-assembly strategies for making a crossed n-InP/p-InP LED.[122,123] One approach to nanowire architectures of this type is summarized in Figure 5.31. In this scheme, n-InP and p-InP nanowires are organized as a crossed junction between four orthogonal gold microelectrodes patterned on a silicon substrate. This is achieved by electric field directed self-assembly, described earlier, where a voltage is applied first to one pair and then the other pair of electrodes. As mentioned above, the electric field localized between the electrodes induces an electric dipole in the nanowire, and this polarized nanowire assembles selectively across a pair of oppositely charged gold electrodes through electrostatic interactions. Once a nanowire has bridged two electrodes, the electric field is quenched by the passing of current, and thus only a single nanowire is deposited. By first depositing the p-InP and then the n-InP nanowires, we create an electrically addressable crossed p-InP/n-InP nanowire junction as shown in Figure 5.31. Another method for making this crossed nanowire architecture, as mentioned above, involves the microfluidic organization of an ethanolic suspension of doped InP nanowires by evaporation induced self-assembly (EISA)[124] within the microchannels of a PDMS elastomeric stamp held in conformal contact with the silicon substrate and registered with a pair of gold micro-contacts on the substrate surface, as shown in Figure 5.32. The process is repeated with the microchannels registered orthogonal to the original nanowire direction to create the electrically addressable crossed pair of p-InP/n-InP nanowires.

By applying a voltage across the p–n InP nanowire junction, electrons and holes recombine radiatively at the point where the n-InP and p-InP segments meet to create a nanoscale LED. The key for success in this endeavor is to achieve an abrupt and uncontaminated junction between the n- and p-doped InP segments to promote the efficient radiative recombination of electrons and holes injected into the p–n junction, thereby showing light emission from the smallest possible LED. It has since been demonstrated that the efficiency of electron–hole radiative recombination and the quantum yield of emitted light can be greatly improved by improving the quality of the p–n junction using a core–shell–shell (CSS) nanowire axial heterostructure, demonstrated for n-GaN/InGaN/p-GaN photolumi-nescent and electroluminescent blue LEDs.[125] Since the wires are as thin as can possibly be without losing their valuable physical properties, the greatest challenge is now to make arrays of closely spaced junctions. This is the so-called "pitch problem," or how to pack all these light sources into truly the smallest possible area – see later.

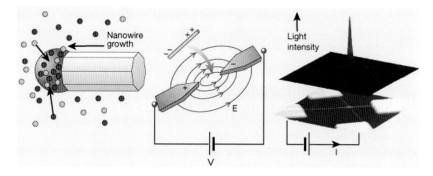

Figure 5.31 *VLS catalytic nanocluster synthesis, electric field driven assembly, and point luminescence of a crossed n–p semiconductor nanowire LED* (Reproduced with permission from Ref. 123)

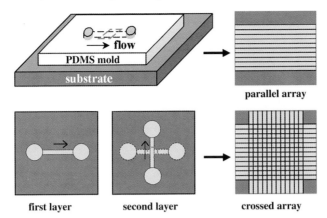

Figure 5.32 *Microfluidic EISA of nanowires into parallel and crossed arrays* (Reproduced with permission from Ref. 124)

5.23 Nanowire Diodes and Transistors

By having access to semiconductor nanowires able to transport electrons and holes, the possibility arises that they could function as building blocks for self-assembling nanoelectronic devices and circuits without having to resort to complexity and costliness of semiconductor microfabrication facilities. To evaluate the feasibility of this potentially revolutionary proposal, boron- and phosphorus-doped silicon nanowires made by VLS catalytic nanocluster synthesis methods were successfully self-assembled into semiconductor nanodevices.[126]

One example was a passive diode structures consisting of crossed p- and n-type nanowires exhibit rectifying transport similar to planar p–n junctions, as depicted in Figure 5.33. Another example was an active bipolar n–p–n transistor, built from heavily and lightly n-doped nanowires crossing a common p-type wire base, shown in the figure along with the circuit configuration. These transistors have been shown to exhibit common base and emitter current gains as large as 0.94 and 16 for heavily and lightly doped nanowires, respectively.

Materials self-assembly of semiconductor nanowire building blocks into pivotal nanoelectronic device architectures bodes well for a bottom–up approach for manufacturing devices and circuits of nanoscale dimensions – more on this exciting prospect later.

5.24 Nanowire Sensors

Silica is renowned for the ability of surface silanol groups to anchor a myriad of functional molecules to its surface *via* the coupling of Si–Cl and Si–OR groups of alkylchlorosilanes and alkylalkoxysilanes.[127] This is a theme that reoccurs throughout this text. As a sheath of silica coats the surface of VLS as-synthesized single crystal silicon nanowires, the possibility therefore exists to anchor probe molecules to the surface silanols to create highly sensitive, real-time, electrically

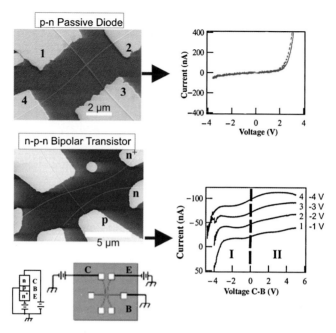

Figure 5.33 *A p–n silicon nanowire nanodiode, and a n^+–p–n silicon nanowire bipolar nanotransistor with their electrical characteristics* (Reproduced with permission from Ref. 126)

based nanowire sensors for the detection of chemical and biological species. This has been accomplished for boron-doped silicon nanowires using the strategy summarized in Figure 5.34. In the example illustrated, an aminoethylsilane was anchored to SiOH surface groups of a silicon nanowire that had been self-assembled across two contact electrodes, and this utilized as a nanoscale FET (nanoFET) for pH sensing.[128] The pH-dependent conductance observed for the nanoFET was found to be linear over a large dynamic range, and the underlying principle for sensing can be understood in terms of the change in surface charge during protonation and deprotonation. At high pH, the amine is neutral and the residual surface silanols are deprotonated. For a p-doped nanoFET, an increase of surface negative charge will attract additional holes into the p-channel of the FET and, therefore, enhance its conductance as seen in the figure. At low pH, the reverse is true, both the amines and silanols are protonated and the increase in surface positive charge will repel holes from the p-channel and decrease the conductance of the FET. There are all sorts of possibilities for nanoFET sensors, such as anchoring a biotin to the surface amine groups to detect streptavidin down to a picomolar concentration range. Almost any type of molecule can be grafted onto silica, allowing the possibility for sensors detecting almost anything. Sensing DNA or RNA, proteins and antibodies, viruses, bacteria and other cell types, various other biomolecules, heavy metal ions, are just a few of the possibilities in the nanotechnologists' toolbox of the future.

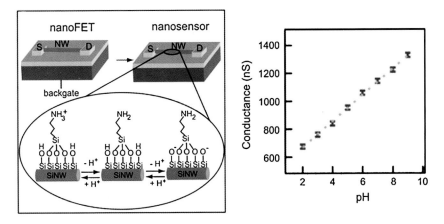

Figure 5.34 *The underlying mechanism and electrical characteristics of a NanoFET nanowire sensor*
(Reproduced with permission from Ref. 128)

5.25 Catalytic Nanowire Electronics

The ability to synthesize controlled diameter and length catalytically interesting nanowires and organize them into nanowire electronic devices, like nanowire diodes and transistors, provides a unique opportunity to tinker with surface catalyzed chemical reactions using solid-state electronic circuitry. An elegant example of this has been demonstrated using a SnO_2 nanorod FET, with a mechanism applicable to smaller diameter nanowires. The gate voltage applied to a NW-FET determines the electron concentration within the nanowire, which in turn determines the chemical reactivity and selectivity of the tin dioxide surface with respect to oxygen adsorption–desorption processes and carbon monoxide catalytic oxidation to carbon dioxide.[129]

To amplify upon this idea, cassiterite-SnO_2 nanowires were made by the topotactic oxidation of β-Sn nanowires, the latter being made by electrochemical growth of tin in anodized aluminum oxide nanochannel membranes, as mentioned previously. Because the nanowire diameter is smaller than the Debye length, that is the width of the surface space charge region, it is proposed that flat band conditions rather than Schottky junctions between grains determine the conductance of the SnO_2. It is the concentration of localized electron states in the electronic band gap near the conduction band edge associated with surface oxygen vacancies that determine the nanowire conductance. Electrons in these shallow door levels can be thermally excited into the conduction bands; hence, conductivity will increase with any process that enhances the number of vacancies. This study shows that both surface chemistry and gate voltage can accomplish this change of electron concentration and hence the conductance characteristics of the nanowire.

The effect of nanowire surface chemistry is dramatically seen in the conductance change observed on switching from a N_2 to an O_2–N_2 gas flow over a 300 °C SnO_2 NW-FET (Figure 5.35 left). In the first case, nitrogen causes desorption of oxygen

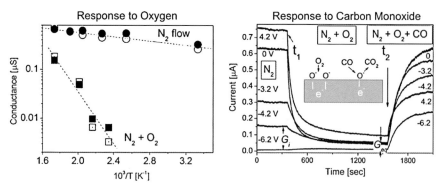

Figure 5.35 *The electrical response of a tin dioxide nanowire FET to exposure by oxygen,
left, and carbon monoxide, right*
(Reproduced with permission from Ref. 129)

from the nanowire surface making more oxygen vacancies. This boosts the number
of electrons in shallow donor levels that can be excited into the conduction band
with a concomitant growth in the steady-state conductance. Oxygen admittance
reverses the situation by tying up and stabilizing electrons at oxygen vacancy sites,
making their excitation to the conduction band more difficult with a concurrent
drop in nanowire conductance. Gate voltage can be seen to affect surface kinetics
from the time-dependent changes in conductance on switching from N_2 to O_2–N_2 to
CO–O_2–N_2 atmospheres in contact with the SnO_2 NW-FET (Figure 5.35, right).
The effect of CO, as expected, is opposite to O_2: the former depletes surface oxygen
while the latter adds to it, serving to respectively increase and diminish available
donor electrons and nanowire conductivity.

While the detailed interpretation of the kinetics of these surface processes is quite
complicated, the main observation is that surface electrons at vacancy sites are less
available for chemistry at negative gate voltages as nanowire electrons are forced
into the source–drain region, diminishing oxygen adsorption and oxidation rates.

5.26 Nanowire Heterostructures

Many semiconductor applications depend on the ability to make single crystal thin
film epitaxial heterojunctions and superlattices for electronic and optoelectronic
devices, such as light-emitting and resonant-tunneling diodes, high-electron mobility transistors and quantum cascade lasers. If semiconductor nanowire analogs are
to be realized, it is essential to be able to synthesize them as single crystals,
interfacially lattice matched, and with longitudinally and axially ordered hetero-
junction and superlattice architectures.

These have both been reduced to practice by the development of modified VLS
synthetic methods that permit nanowire growth either in a block-by-block mode
with a predetermined composition profile along the nanowire axis (Figure 5.36)
as well as in a layer-by-layer growth process around the body of the nanowire.

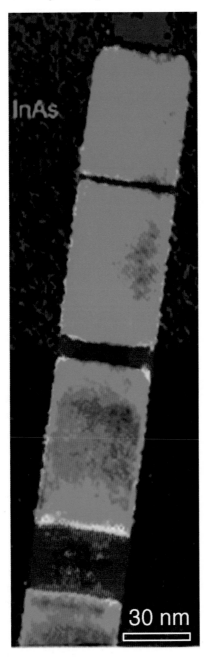

Figure 5.36 *Longitudinal InP/InAs nanowire superlattice – As and P EDX composition profile shown in green and red color code*
(Reproduced with permission from Ref. 130)

In the following sections, the synthesis and some diagnostics of these novel nanowire constructs will be described.

5.27 Longitudinal Nanowire Superlattices

Nanowire epitaxial heterojunctions and superlattices can be made by programmed switching of the vapor phase constituents in a catalytic nanocluster, VLS "block-by-block" growth process. This can be performed with catalytic gold nanoclusters, using a molecular-beam epitaxy (MBE) apparatus to switch the elemental sources, in order to form InAs/InP nanowire superlattices as seen in Figure 5.36.[130] It can also be performed on supported gold clusters made by dewetting a thin film of gold heated on a silicon substrate used for making a Si/SiGe superlattice (Figure 5.37).[131] The scheme shows different stages of the block-by-block nanowire growth process: (1) alloying between Au thin film and Si species in substrate/vapor; (2) growth of pure Si block when laser is off, only Si species deposit into the alloy droplet; (3) growth of SiGe alloy block when the laser is on, both Si and Ge species deposit into the liquid droplet; (4) growth of Si/SiGe superlattice structure by periodically turning the laser beam on and off. Longitudinal Si/SiGe nanowires form orthogonally to the silicon substrate surface and show single crystal epitaxial growth with the {111} planes oriented perpendicularly to the nanowire axis. A well-defined Si/SiGe composition profile is established by SEM, TEM and EDX diagnostics, shown in Figure 5.38.

The nanowire block-by-block growth can be performed on gas phase catalytic gold nanoclusters, formed by laser ablation of a gold target, used to make a longitudinal GaP/GaAs nanowire superlattice with a well-defined composition profile along the wire axis.[132] Energy dispersive X-ray analysis (EDX) spectra of a 20 nm diameter (GaP/GaAs)$_3$ nanowire superlattice, shown in Figure 5.39 top panel, display a clear periodic modulation of the composition along the complete length of nanowire, displaying three uniform segments of GaP interdispersed

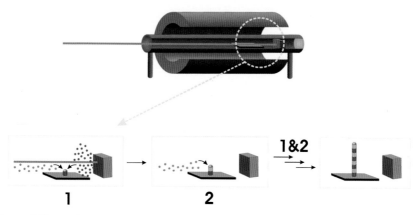

Figure 5.37   *Hybrid pulsed laser ablation CVD process for the synthesis of semiconductor nanowires with longitudinal ordered heterostructures*

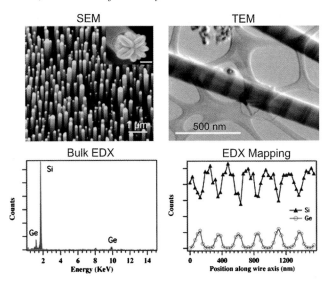

Figure 5.38 *VLS–CVD synthesized Si/SiGe nanowires showing growth orthogonal to substrate seen in the SEM, and well-defined composition profiles recorded by TEM and EDX*
(Reproduced with permission from Ref. 131)

with three of GaAs. Photoluminescence images of this (GaP/GaAs)$_3$ nanowire superlattice are only obtained by excitation with light polarized parallel and not perpendicular to the nanowire axis, seen in the middle left panel of the figure. This demonstrates one of the benefits of anisotropic nanostructures, in this instance polarization photoselection of emitted light because of the existence of electric dipole selection rules operating for the electronic states of the nanowire. The three bright regions correspond to the three direct band gap GaAs segments, whereas the three interlaced dark regions correspond to the indirect band gap GaP segments.

Impressive examples showing the type of complexity that can be achieved by this synthesis technique are seen in the PL image of a 40-nm diameter (GaP)$_5$–(GaAs)$_5$–(GaP)$_5$–(GaAs)$_5$–(GaP)$_{10}$–(GaAs)$_5$–(GaP)$_{20}$–(GaAs)$_5$–(GaP)$_{40}$–(GaAs)$_5$–(GaP)$_5$ superlattice (Figure 5.39, bottom panel). The numbers in subscript are growth times in seconds for each layer. The inset shows relative lengths of GaAs (blue)–GaP (red) layers with a scale bar of 5 μm, middle right panel. In the lower panel of this figure, an amazing PL image of a 21-layer superlattice, (GaP/GaAs)$_{10}$GaP beautifully depicts a group of four equally spaced spots on the left, two in the middle with larger gaps, and another set of four with equal spacing on the end. The superlattice is 25 μm in length.

5.28 Axial Nanowire Heterostructures

To achieve this objective VLS catalytic nanocluster growth is used to make nanowires of predetermined length from the first precursor. By lowering the flow

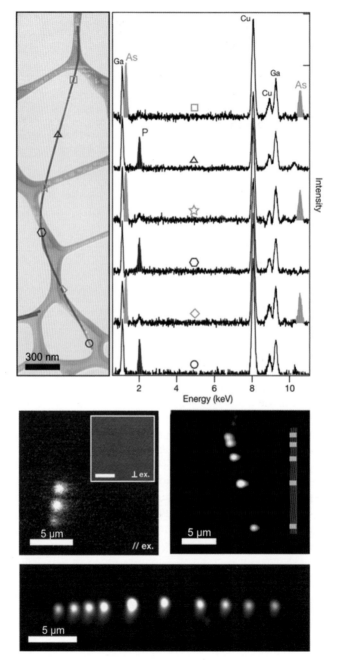

Figure 5.39 *EDX and PL profiles from GaP/GaAs longitudinal nanowire superlattices* (Reproduced with permission from Ref. 132)

rates in the reactor, a second precursor can be preferentially deposited axially and in an epitaxial fashion on the preformed nanowire.[133] This can be repeated with a third precursor to make multishell nanowire heterostructures. The process is summarized in Figure 5.40. EDX and TEM composition profile and structure analysis of Si shell (blue) on Ge core (red) nanowires confirm the single crystal nature and epitaxial matching of the core and shell regions of the axial nanowire heterostructure. Nanoelectronic and nanophotonic devices based on this axial nanowire heterostructure architecture have begun to appear in the literature.[125]

5.29 Nanowires Branch Out

Nanocluster catalyzed VLS growth of semiconductor nanowires has proven to be a powerful and general synthetic method for systematically controlling the composition, doping, diameter, length and heterostructure of axially and co-axially epitaxially modulated compositions. A creative extension of this nanowire synthetic strategy enables the subsequent growth of nanowire branches emanating from specific points on the body of the parent nanowire.[134]

The procedure commences by first placing catalytic nanoclusters from solution at specific points along the length of the parent nanowire. This step is followed by VLS growth of nanowires from the newly deposited nanoclusters. This site-selective nanocluster deposition procedure can be repeated to create further branch points to generate a particular hyper-branched construction designated for a specified function (Figure 5.41).

Proof-of-concept of the methodology was demonstrated for silicon and gallium nitride nanowires. In the case of gold nanocluster catalyzed VLS growth of silicon nanowires, the branches were grown from gold nanoclusters placed on the parent silicon nanowire and shown by high-resolution TEM to grow as single crystal nanowires in an epitaxial fashion from the body of the parent single crystal silicon nanowire (Figure 5.42). This multistep nanocluster catalyzed VLS growth process allows branched and hyper-branched semiconductor nanowires to be assembled with a predetermined composition and diameter, density and doping. One can envision the rational development of branched architectures with n–p diode and n–p–n transistor junctions, which if reduced to practice would represent an important step towards the future development of nanowire electronics in three dimensions.

5.30 Coaxially Gated Nanowire Transistor

The targeted nanowire transistor device with a coaxially gated architecture is shown schematically in the top panel of Figure 5.43. The inset shows the cross-section of the as-grown nanowire, starting with a p-doped Si core (blue, 10 nm), with subsequent layers of i-Ge (red, 10 nm), SiO_x (green, 4 nm), and p-Ge (5 nm). The source (S) and drain (D) electrodes are contacted to the inner i-Ge core, while the gate electrode (G) is in contact with the outer p-Ge shell and electrically isolated from the core by the SiO_x layer.

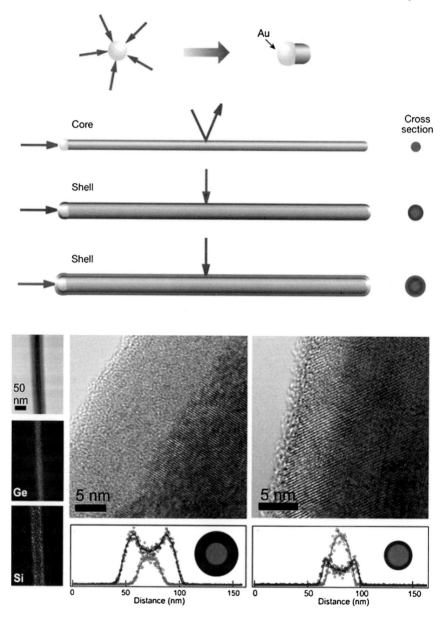

Figure 5.40   *Strategy for synthesizing epitaxial core–shell and core–multishell nanowire heterostructures, as well as TEM image and EDX elemental mapping composition profiles of c–Ge–c–Si axial nanowire heterostructures* (Reproduced with permission from Ref. 133)

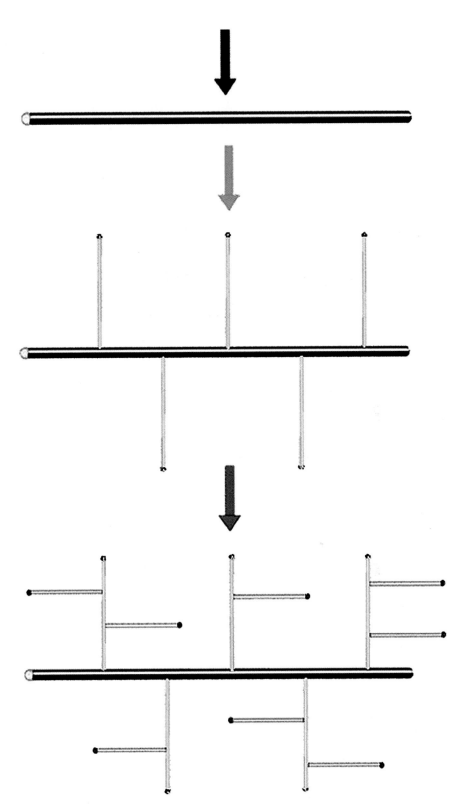

Figure 5.41 *Schematic illustrating the rational growth of dendritic nanowires* (Reproduced with permission from Ref. 134)

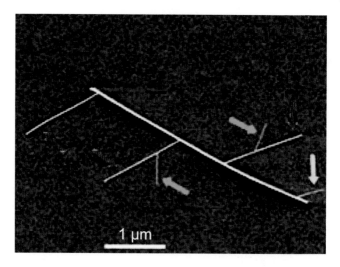

Figure 5.42 *A branched nanowire showing first generation (red arrows) and second generation (blue and yellow arrows) branches* (Reproduced with permission from Ref. 134)

The central panel of Figure 5.43 shows an SEM of a coaxial transistor that has been self-assembled between microelectrodes and selectively etched to achieve the desired architecture. This entails e-beam deposition of source and drain electrodes after etching the Ge (30% H_2O_2, 20 s) and SiO_x layers (buffered aqueous HF, 10 s) to expose the core layers. The etching of these outer layers is shown clearly in the inset and is indicated by the arrow. The gate electrodes were defined in a second step without any etching before contact deposition. The minute size of this device can be appreciated by inspection of the 500 nm scale bar. It can also be appreciated, though not mentioned by the authors, that the device as a whole is not very small and spans roughly 5 μm. Clearly, making nanoparts assemble into nanoscale, functioning devices is a necessary goal for nanotechnology.

In the bottom panel of Figure 5.43, the gate response of the coaxial transistor at a source–drain voltage of 1 V shows a maximum transconductance of 500 nAV^{-1}. Charge transfer from the p-Si core to the i-Ge shell produces a highly conductive and gateable channel. It is clear from reports on longitudinal and axial nanowire superlattices that this class of heterostructures offers tremendous promise for applications in nanoscale electronics, optoelectronics and photonics.

5.31 Vertical Nanowire Field Effect Transistors

One of the biggest challenges for nanometer width single-crystal nanowires is finding easy ways to take advantage of their impressively small cross-section. If this were somehow possible, through a creative fusion of bottom–up synthesis and top–down microfabrication to make for instance, vertically oriented nanowire architectures, then it might be possible to achieve ultrahigh density nanoelectronic

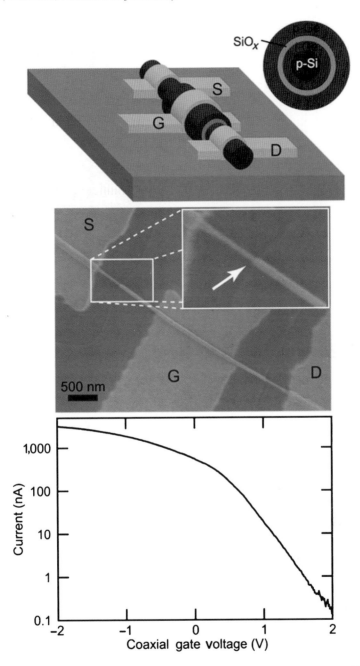

Figure 5.43 *Coaxially gated nanowire transistor design, top, SEM image, middle, and electrical characteristics, bottom*
(Reproduced with permission from Ref. 133)

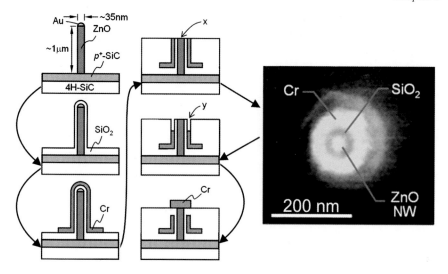

Figure 5.44 *Scheme depicting the fabrication of nanowire-based vertical field-effect transistors, as well as a top-view SEM of one stage of the process* (Reproduced with permission from Ref. 135)

circuitry with terascale capability. Quite simply, this is because the footprint of a standing up device is less than one laying down, examples of which were described earlier. The scheme outlined in Figure 5.44 provides an elegant means of achieving this end.[135] This strategy exploits the bottom–up ease of growing vertically organized wurzite ZnO nanowires on epitaxially matched heavily doped SiC, and the facility to process them from the top–down into vertical nanowire FETs. The main attractions of this approach are two-fold (i) the elimination of complex lithographic and etching steps that would be required if top–down methods were utilized to make the vertical nanowire component of the FET and (ii) the exclusion of tricky and time-consuming manipulation steps if bottom–up nanowires were used for the FET.

In the depicted vertical architecture nanowire FET the highly conductive n^+ or p^+ doped SiC substrate is chosen to feed electrons or holes into the device *via* the bottom end of the ZnO nanowire. The small epitaxial mismatch ($\sim$5.5%) between ZnO {0001} and hexagonal SiC {0001} facilitates growth of the ZnO nanowires in an orthogonal fashion with respect to the surface of the SiC substrate, and at the same time enables good electrical contact between the two materials. Following growth of the nanowire array, CVD of silica and then ion beam deposition of chromium provide uniform continuous coatings on the nanowires and function as the thin oxide and metal layers of the desired vertically oriented Metal-Oxide-Semiconductor Field-Effect Transistor (MOSFET) device. The technique of chemical mechanical polishing was subsequently used to define the length of the ZnO channel of the FET and to remove the catalytic gold nanocluster from the end of the nanowire. Making the drain contact to the top end of the ZnO nanowire required a selective wet etch to remove some chromium cladding, filling in of the tubular cyclindrical recess so formed with silica, and finally topping off of the silica layer with a chromium gate electrode. The successful implementation of this procedure was supported by a series of electron microsocopy snapshots of each of the aforementioned steps. A top-view FE-SEM image of the architecture before the chromium etching step is shown in Figure 5.44, showing clearly the conformal coating of silica and chromium on the ZnO nanowire core.

The device characteristics of both the n- and p- vertical nanowire FETs were established and compared to the performance of the best carbon nanotube FETs and silicon MOSFETs as well as thin film transistors based on ZnO. It was clear from the preliminary results obtained, with obvious extensions to single crystal silicon and germanium nanowires, that the vertical nanowire FET based on ZnO could one day rival state-of-the-art MOSFETs in device performance but with a 10% smaller footprint.

5.32 Integrated Metal–Semiconductor Nanowires – Nanoscale Electrical Contacts

We have seen earlier how spatially directed electron beam deposition can create electrical contacts to semiconductor nanowires, a key step in the assembly of nanoscale devices and nanocircuits. Contacts made in this way, however, are considerably larger than the nanowires themselves, defeating somewhat the objectives and advantages of devices and circuits fashioned at the nanoscale. A purely synthetic attempt to solve the problem of size incompatibility between contact and nanowire is to devise a solution-phase electroless deposition method that can localize gold exclusively on the tips of nanowires, which will be covered in Chapter 6. A more direct approach would be to devise a method whereby pre-selected spatial regions of a semiconductor nanowire can be transformed to the metallic state.[136]

This interesting idea has been achieved by the solid-state reaction at 550 °C of single crystal silicon nanowires with nickel coated by thermal evaporation, thereby forming metallic nickel silicide nanowires. Any excess nickel on the nanowire surface could be removed by wet etching, and the final step consisted of thermal annealing at 600 °C in N_2:H_2 90:10 forming gas. The nickel silicide nanowires made in this way were found to be single crystalline and metallic, with ideal resistivities ~ 10 $\mu\Omega$ cm and very high breakdown current densities $<10^8$ A cm^{-2}.

This methodology could also be combined with lithographic patterning, using masks to deposit nickel and spatially define the semiconducting Si and metallic NiSi regions with atomically precise interfaces in a contiguous single-crystal nanowire. As a proof-of-concept, NiSi/Si heterostructures and superlattices could be made in this way, as illustrated schematically and in optical images in Figure 5.45. Using this technique, a FET comprised of an integrated NiSi/p-Si/NiSi metal/p-doped semiconductor/metal heterostructured nanowire was demonstrated, in which nanoscale electrical contacts to the source and drain are defined by the NiSi metallic termini. This device can be seen in the bottom right part of the figure.

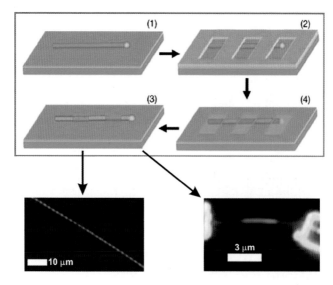

Figure 5.45 *Scheme depicting the formation of metallic NiSi regions on a Si nanowire by depositing Ni through a shadow mask. The bottom shows optical images of two heterostructures made by this method*
(Reproduced with permission from Ref. 136)

The demonstrated ability to create high-quality nanoscale conducting interconnects within a semiconducting nanowire device speaks well for the future of an integrated nanoelectronics.

5.33 Photon-Driven Nanowire Laser

The combined use of vapor phase transport (VPT) and VLS growth has been effectively used for the synthesis and self-assembly of an array of oriented single crystal ZnO nanowires (wurtzite form).[137,138] These form parallel arrays orthogonal to a sapphire substrate (Figure 5.46) that function as the tiniest UV lasers[100,139] (Figure 5.47).

The synthesis begins with a sealed quartz tube temperature gradient reactor, containing a mixture of ZnO and C powder at the hot end of the tube (905 °C) and a sapphire substrate with a nanoscale thickness layer of Au at the cold end of the tube (880 °C). At this temperature the Au film does not wet sapphire and de-wets this substrate to form an array of surface-confined gold nanoclusters. Carbo-thermal reduction of ZnO by graphitic carbon generates the reactive vapor species ZnCO, which in the concentration gradient of the tube reactor is transported in a VPT mode to the cooler end of the tube where it dissolves in the Au nanoclusters. At saturation loading of ZnCO in the gold, ZnO begins to precipitate out from the nanocluster and nucleates and grows in a VLS mode as an oriented single crystal wurtzite polymorph nanowire.

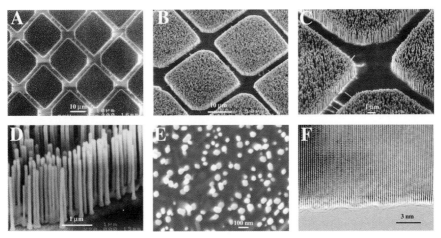

Figure 5.46 *VPT–VLS synthesis of oriented single crystal wurtzite ZnO nanowires orthogonally and parallel aligned on a sapphire substrate. Panels A–C show wide area SEM images of the patterned nanowire growth, D shows the orthogonal orientation of the nanowires, and E a magnified top view of the nanowire arrays. F shows a high-resolution TEM with characteristic ZnO lattice planes clearly visible*
(Reproduced with permission from Ref. 100)

The slower rate of growth of the {0001} crystal faces is responsible for the observed ⟨0001⟩ ZnO nanowire morphology (Figure 5.46). The diameter of the nanowire is in turn determined by the diameter of the Au nanocluster. Lateral van der Waals interactions between ZnO nanowires and crystal plane direction causes them to grow in a self-organized parallel aligned and orthogonal fashion from the surface of the sapphire substrate. Sapphire is used in this case because of the excellent lattice match between the {110} crystal planes of sapphire and the {0001} crystal planes of ZnO (less than 0.08% mismatch), leading to truly epitaxial growth of one material on the other.

The as-synthesized nanowires act as a resonance cavity for light with two naturally faceted hexagonal end-faces, which function as Fresnel reflecting mirrors (Figure 5.47). Higher refractive index ZnO ($n = 1.9$) is surrounded by lower refractive index sapphire ($n = 1.8$) and air ($n = 1$), forming natural laser cavities which serve to confine the light by total internal reflection within the nanowire optical cavity. Using 266 nm laser excitation, a threshold-intensity is found that transforms electron–hole recombination broad band background photoluminescence into end facet emitting ultraviolet narrow band laser action centered at 366 nm. Quantum confinement of electrons and holes serves to enhance radiative recombination and the efficiency of laser action in ZnO nanowires. Other materials have also been used in this respect, and UV-blue lasing has also been observed in GaN nanowires.[140] This is important since it is predicted that electrical pumping of GaN nanowire lasers should be much more facile than doing so with ZnO.

Not only can these UV lasers form parallel arrays on flat substrates, but they can also be grown in antennae-like arrays. By tuning the synthesis conditions,

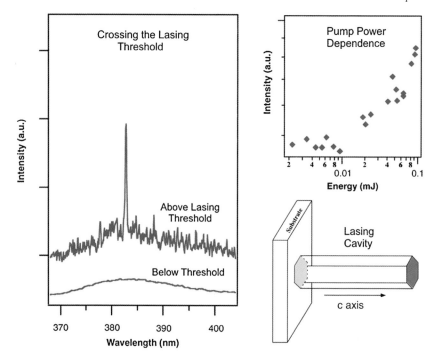

Figure 5.47 *Single crystal ZnO UV-nanolaser optical properties above and below the lasing threshold, and the dependence of the luminescence output on the pump-laser power. Also shown is the architecture of the lasing cavity formed by the two facets of the ZnO nanowire*
(Reproduced with permission from Ref. 100)

comb-like dendritic ZnO nanowires could be fabricated bearing smaller ZnO wires jutting out from a larger ZnO stem.[141] A comb-like structure synthesized in this way is shown in Figure 5.48. If the architecture of these branched constructs can be placed under synthetic control, one can imagine literally growing an all-optical nanowire photonic circuit where the ZnO nanowires have the dual function of laser light sources and waveguides!

5.34 Electrically Driven Nanowire Laser

Optical and electrical measurements have been made on single-crystal cadmium sulfide nanowires with faceted ends and show that these structures can also function as Fabry–Perot optical cavities with mode spacing inversely related to nanowire length, and a threshold for lasing characterized by optical modes with instrument-limited linewidths.[142] CdS nanowires in this experiment are first synthesized by VLS growth, and then treated with ultrasound. The ultrasonic treatment causes the nanowire tips to break, cleaving them to atomically flat planes. They are then deposited onto a heavily doped silicon substrate, which will be the anode for

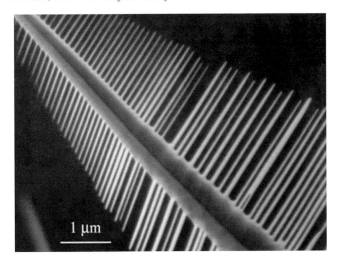

Figure 5.48 *SEM image of a comb-like, single crystal ZnO nanostructure*
(Reproduced with permission from Ref. 141)

electrically pumping the nanowire. A thin gold electrode is then deposited as
a top-contact onto the nanowire, separated by a very thin film of alumina to prevent
shorting of the circuit. An image showing the structure of the device as well as a
luminescence image depicting lasing from the nanowire tip is shown in Figure 5.49.

UV ZnO and multicolor visible CdS nanowire lasers of controlled diameter could
find a myriad of applications including optical computing and circuitry,[143] near-
field optical lithography, laser based surgery with unprecedented resolution, highly
integrated chemical and biological sensors, information storage and lab-on-chip
microanalysis. While nanowires may be linear, research in this area is sure to grow
in an exponential fashion!

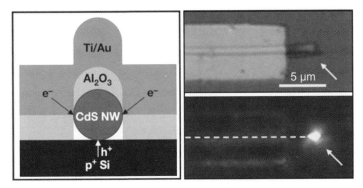

Figure 5.49 *Structure of an electrically pumped nanowire laser, left. Top-view SEM image*
of the device, as well as spatially resolved luminescence
(Reproduced with permission from Ref. 142)

5.35 Nanowire UV Photodetectors

ZnO nanowires have been shown to function as one of the tiniest photoconductors with a response that depends on the ambient atmosphere, suggesting utility as a gas-sensitive optical switch – a nanosensor.[144] In Figure 5.50 are shown the observed *I–V* results for a single 60-nm diameter ZnO nanowire that has been assembled across four Au electrode current collectors and photoexcited with 365 nm laser light. The bottom curve shows the dark current and top curve the photocurrent response of the ZnO nanowire. The sensitivity of the photo-response to light exposure at 532 and 365 nm shows that band gap excitation is required to observe a photoelectric response. By playing with diameter variations and different surface functionalization of ZnO nanowires one can imagine the construction of ZnO nanowire nano-noses for use as molecule discriminating nanosensors.

5.36 Simplifying Complex Nanowires

Many complex metal oxides display interesting and important properties like high-T_c superconductivity, giant magnetoresistance, ferroelectricity, ferromagnetism and catalysis, with myriad uses in, for example, resistance-free power transport, ultra-dense magnetic data storage, microelectronics and spintronics.[145–147] These materials are by and large made in bulk or film forms. If they could be fashioned as nanowires, implementation in devices could be more straightforward, and the emergence of new chemistry and physics might uncover a multitude of

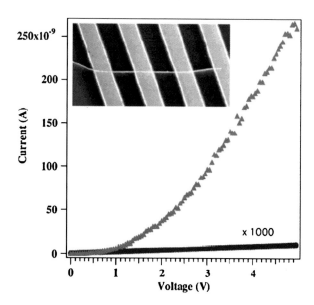

Figure 5.50 *SEM image of a ZnO nanowire photodetector and I–V response curves with (top) and without (bottom) UV excitation*
(Reproduced with permission from Ref. 144)

technological opportunities. Unfortunately, most complex oxides of the genre, exemplified by $YBa_2Cu_3O_{6.6}$ (YBCO), $La_{0.67}Ca_{0.33}MnO_3$ (LCMO), $PbZr_{0.58}Ti_{0.42}O_3$ (PZT) and Fe_3O_4 cannot be made as nanowires by VLS, chemical synthesis, solvothermal or template directed methods mainly because of complexities associated with precise control of elemental composition, avoidance of phase separation and suitable catalysts to facilitate growth.

A creative way to surmount this problem is to use vertically aligned arrays of single crystal MgO nanowires as templates to direct the growth of a uniform layer of the complex metal oxide on the surface of the nanowires, as depicted in Figure 5.51.[148] The coating method of choice was pulsed laser deposition, because of its well-documented ability to form high quality and controlled stoichiometry films, considerations that are crucial for defining materials properties, fabricating devices and optimizing performance.

The synthesis begins by flowing O_2 over Mg_3N_2 at 900 °C to create the precursors serving as feedstock for VLS growth of $\langle 100 \rangle$ oriented MgO nanowires. Growth occurs downstream, out of gold catalyst nanoclusters supported on {100} MgO as substrate. A Nd:YAG pulsed laser operating at $\lambda = 532$ nm was focused on a target of the complex metal oxide, generating a plume of material which deposits onto the surface of the MgO nanowire array. Single crystal core–shell MgO-complex metal oxide nanowires emerged from this synthesis, in which the coating grew conformally, stoichiometrically and epitaxially on the surface of the MgO nanowires.

The collective diagnostic information from PXRD, EDS, SAED and HREM studies of the coated nanowires established the dimensions of the core and shell elemental components as well as their crystal structure and orientation, epitaxial lattice match and elemental composition. In the case of LCMO coated MgO nanowires, temperature and magnetic field dependent transport studies revealed that the metal–insulator transition and giant magnetoresistance, well known for the bulk material, was retained at the nanometer scale.

The universality of this approach for making multilayer complex oxide nanowires with tailored surface layers will provide a cornucopia of opportunities for basic research aimed at understanding their properties as a function of composition and scale, and may suggest device opportunities that were not possible

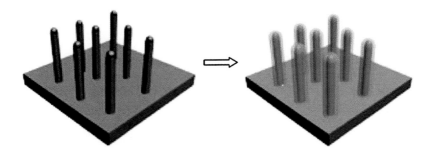

Figure 5.51 *Scheme depicting the coating of nanowires by pulsed laser deposition to form complex oxide wires*
(Reproduced with permission from Ref. 148)

with the materials in bulk and film form. There is no obvious reason why this approach to nanowires cannot be extended to other composition fields like metal borides, carbides, silicides, chalcogenides, and pnictides.

5.37 Nanowire Casting of Single-Crystal Nanotubes

In view of the fact that VLS growth can be used to make arrays of oriented single crystal nanowires on sapphire wafers, it is interesting to contemplate their use as templates for directing the growth of arrays of nanotubes of other materials. This idea has been realized in the case of GaN using oriented single crystal wurtzite nanowire arrays,[149] as illustrated in Figure 5.52. The process begins by placing ZnO nanowire arrays inside a reaction tube for GaN CVD. Trimethylgallium and ammonia were used as precursors and were fed into the system with argon or nitrogen carrier gas. The deposition temperature was set at 600–700 °C. After the GaN deposition, the samples were treated at 600 °C with 10% H_2 in argon, removing the ZnO nanowire templates by reducing them to Zn which is volatile at this temperature. The outcome of this nanotube casting synthesis is seen in the SEM images of Figure 5.53 with the GaN/ZnO core–shell nanowire array in the top panel giving the GaN nanotube array in the bottom panel after sacrificial etching of the ZnO nanowires. High-resolution TEM images establish that the process can be described as epitaxial casting because the GaN nanotubes so formed on the ZnO nanowire templates are found to be single crystal wurtzite with their axis orthogonal to the sapphire substrate, which mirrors the structure and orientation of the single crystal wurtzite ZnO nanowires that templated their growth. A similar strategy has been utilized for making silica nanotube arrays templated by silicon nanowires grown by VLS on a silicon substrate.[150] The twist in this synthesis is that thermal oxidation is used to create a sheath of silica on the silicon nanowire array

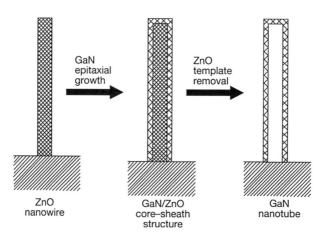

Figure 5.52 *Epitaxial casting of wurtzite GaN nanotubes using wurtzite ZnO nanowire templates*
(Reproduced with permission from Ref. 149)

Core-Shell Nanowire Array

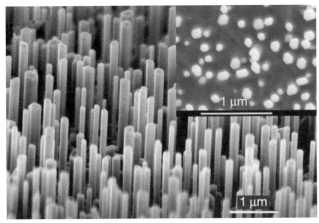

Nanotube Array

Figure 5.53 *GaN/ZnO core–shell nanowire array on sapphire substrate gives a GaN nanotube array upon removing the ZnO core* (Reproduced with permission from Ref. 149)

and aqueous KOH is used to etch away the remaining silicon core to create the desired silica nanotube array.

5.38 Solution-Phase Routes to Nanowires

Essentially all the examples we have seen thus far have been templating strategies. One-dimensional nanostructures were templated against porous membranes or their replicas, or were grown out of a template catalyst droplet in the VLS growth mechanism. However, it is possible to devise purely chemical routes to the synthesis of nanowires in solution.

A general way to modify the growth habit of nanocrystals is by using an additive which binds to particular crystal faces, or conditions that change the kinetics of crystal growth. If certain crystal planes are allowed to grow while the growth of others is inhibited, anisotropic structures can result. This process can often be as easy as changing the temperature in a semiconductor nanocrystal preparation, and has been used to create CdS and CdSe quantum rods (see Chapter 6).[151–153] Highly uniform silver nanowires can also be grown by performing the reduction of silver salts in the presence of a complexing surfactant such as polyvinylpyrrolidone (PVP).[154,155] In a typical synthesis, silver seeds are first generated by heating $AgNO_3$ at ~160 °C in ethylene glycol, which serves both as the reducing agent and the solvent. If a solution of $AgNO_3$ and PVP in ethylene glycol is then added dropwise to this seed solution, highly elongated silver structures result. It has been shown that the sides of these nanowires consist of {100} crystal facets to which PVP binds strongly, and the ends consist of {111} facets to which PVP binds less well, and surprisingly each nanowire has a five-fold twinned crystal structure.[156] Such uniform silver wires can be used to template nanotubes quite easily, simply by a galvanic displacement with a more noble metal.[157–159] For instance, when treated with a strong oxidant such as $(AuCl_4)^-$, silver will be oxidized and dissolved while gold deposited in elemental form. This leads to the formation of a gold shell around the silver nanostructure that grows at the expense of silver which is gradually dissolved away. SEM images of silver nanowires before and after galvanic displacement with gold are shown in Figure 5.54.

Another interesting solution route to nanowires utilizes a compound with a highly anisotropic crystal structure. Both selenium and tellurium are good candidates in this respect, since their trigonal crystal forms (t-Se, t-Te) are highly anisotropic, consisting of extended one-dimensional helical chains on a two-dimensional lattice.[160,161] Bonds are much stronger within the helical chains, making crystal growth in one dimension much more favorable than in the other two. The process of t-Se nanowire growth is illustrated in Figure 5.55. First, large colloids of amorphous selenium (a-Se) are generated by the reaction of selenous acid and hydrazine, which generates elemental selenium. The solution is then aged, during which the colloids aggregate and break apart randomly, and then quenched in an ice bath, which causes the nucleation of t-Se seeds. If the mixture is again refluxed, the more thermodynamically stable t-Se grows by depletion of the a-Se colloids, and grows as long whiskers because of its crystal anisotropy. A schematic of the crystal structure of the nanowires is shown in the figure. Modifying the concentration of reactants in such a synthesis, as well as the temperature, allows one to control the length and width of the nanowires, and can cause some unforeseen results such as hollow *t*-Te nanotubes.[162] Te nanowires are particularly interesting since they can be photoconductive, photovoltaic, and exhibit rectifying behavior when illuminated with white light, making them good prospects for many future devices.

Interestingly, exposure of these t-Se nanowires to an aqueous silver nitrate solution induces the topotactic transformation of oriented single crystal t-Se to oriented single crystal Ag_2Se nanowires[163] *via* the reaction

$$3Se(s) + 6Ag^+(aq) + 3H_2O \rightarrow 2Ag_2Se(s) + Ag_2SeO_3(aq) + 6H^+(aq)$$

A similar feat can be accomplished using Zn^{2+} and Cd^{2+} to convert t-Se to ZnSe and CdSe nanowires.

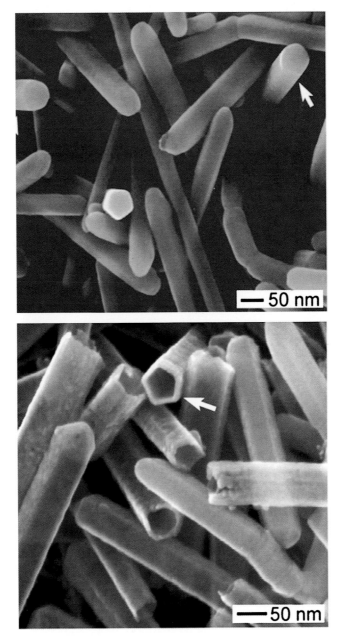

Figure 5.54 *Top, SEM of silver nanowires clearly showing their pentagonal cross-sections due to a fivefold twinned crystal structure. Below, uniform gold nanotubes synthesized from the silver nanowires through a galvanic displacement mechanism*
(Reproduced with permission from Ref. 156)

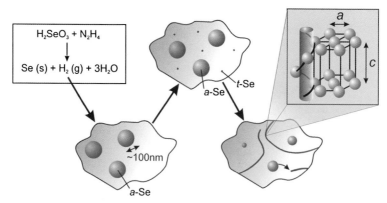

Figure 5.55 *Synthesis of selenium nanowires starting from selenous acid and hydrazine. First, large spheres of amorphous selenium (a-Se) are formed, then small seeds of trigonal selenium (t-Se) upon quenching at 0 °C. Refluxing causes the t-Se crystals to grow at the expense of the a-Se particles, which serve as Se reservoir*

5.39 Spinning Nanowire Devices

Spiders spin silk fibers into ornate web designs, so why not transfer this technology from nature and spin nanofibers and organize them into functional devices? Although this seems obvious in the field of polymers, it is only recently that the idea has been reduced to practice in the materials chemistry laboratory.

The strategy is based on nanofiber extrusion in a capillary jet, or electrospinning.[164] When a solution of a polymer such as polyvinylpyridine or polymer–sol–gel mixture like polyvinylpyridine-tetraalkoxytitanium passes through a high-voltage metal capillary, a thin charged stream undergoing stretching and bending motions as well as solvent evaporation emerges from the orifice. This continuous process generates a population of charged nanofibers that are driven to the ground electrodes on a substrate, as illustrated in Figure 5.56.

The dimensions of the electrospun nanowires depend on the solvent viscosity, conductivity and surface tension as well as the precursor concentration. By patterning the collector electrodes in various configurations it has been found that the nanowires can be coerced to undergo template directed ESA into organized arrays between the electrodes, (*e.g.*, parallel aligned and crossbars) compared to the random arrangements observed on non-patterned grounded substrates. An example of a parallel arrangement of gold collector electrodes on a quartz substrate and the resulting nanowire array is shown in the top of Figure 5.57. During nanowire electrospinning pairs of electrodes in the collector can be grounded either sequentially or simultaneously. The result of grounding six electrodes in three opposing pairs can be seen in the bottom of this figure, where the observed pattern of deposited nanowires can be either a mesh or aligned bundles depending mainly on the deposition time.

Using this approach single layer or multilayer nanowire architectures can be simply and reproducibly created with a wide range of dimensions, compositions

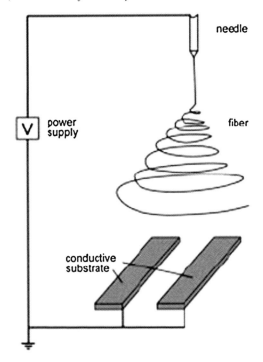

Figure 5.56 *Electrospinning nanowires with preferred orientation using patterned electrodes*
(Reproduced with permission from Ref. 164)

and designs, thereby enabling the properties of the system to guide function and ultimately utility.

5.40 Hollow Nanofibers by Electrospinning

Through a creative modification of the capillary spinneret for making long uniform diameter nanofibers of inorganic and polymeric materials that involved constructing the spinneret in the form of a co-axial double capillary, it has proven possible to prepare hollow nanofibers comprised of inorganic or inorganic-polymer hybrid materials.[165] An illustration of the modified spinneret and the means of introducing the reagents are shown in Figure 5.58. The technique involves ejection from the double-capillary spinneret of a continuous coaxial jet comprised of a heavy mineral oil core surrounded by a sheath of ethanol–acetic acid–PVP–Ti(OiPr)$_4$. The composite core–shell nanofibers are collected on an aluminum or silicon substrate and allowed to hydrolyze at room temperature in air, after which the oily core could be extracted with octane to leave behind a collection of nanotubes with the walls made of amorphous TiO$_2$ and PVP. Calcining in air at 500 °C causes oxidation of the organic polymer and crystallization of the amorphous titania into the anatase polymorph.

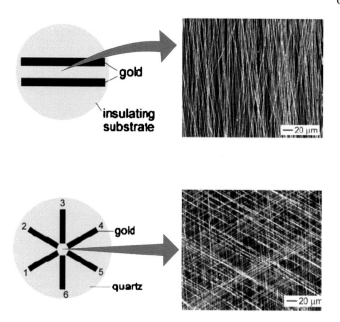

Figure 5.57 *Electrospun nanowires can be successively patterned in overlapping layers by switching the connectivity of the electrodes on the substrate. SEM images show nanowires patterned in one direction, and in three overlapping directions*
(Reproduced with permission from Ref. 164)

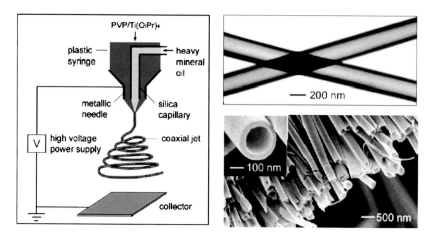

Figure 5.58 *Electrospinning hollow nanotubes. A coaxial stream with an oil core and solid precursor sheath is extruded from a capillary towards a counter-electrode substrate, and the oil subsequently dissolved away. TEM and SEM images confirm the morphology of the hollow tubes*
(Reproduced with permission from Ref. 165)

The dual capillary electrospinning technique appears quite general for making a wide variety of inorganic and polymer compositions. Through judicious variations of the precursor solution concentration and viscosity, choice of the oil phase and parameters selected for electrospinning, hollow nanofibers with uniform circular cross-sections over centimeter lengths can be obtained in large quantities. They could prove useful for catalysis, sensing, chemical storage and release, fluidic devices and optical waveguides.

5.41 Carbon Nanotubes

Carbon has been known since times immemorial, in gems, fuel, writing tools and dyes. Since the discovery of the perfectly symmetric Buckyball (C_{60}), the celestial star that fell to earth, chemical contortions with carbon have taken us from the familiar world of diamond (ccp), Lonsdaleite (hcp, meteorites) and graphite to exciting new forms, the most well known of which are mesoporous carbon, carbon opals and carbon nanotubes.

The addition of single and multi-walled carbon nanotubes (SWNT/MWNT) to the carbon zoo is currently having the greatest impact in nanoscience and nanotechnology.[166–169] The reason for this is because carbon nanotubes have unparalleled strength, high electrical and thermal conductivities,[170] making them ideal building blocks for a wide range of applications.

To put this in perspective, one of the most exciting scientific events of the last decade was the discovery by Iijima in 1991 of nanometer-diameter tubes seemingly made up of a rolled-up, crystalline graphite sheet.[171–173] These were formed using an electrical arc-discharge from a piece of graphite. The electrical energy in this process, as well as the energy associated with dangling bonds in a graphite sheet, seals rolled-up sheets as flawless nanotubes. It has since been proven that these tubes are indeed made of a single curved sheet of graphite connected at its edge. They can be synthesized as single-walled nanotubes,[174] or concentrically layered multi-walled nanotubes,[175] with drastically different physical and chemical properties.[176] Many methods for nanotube synthesis are available, the most prevalent being high-energy electrical, and CVD processes.[177,178] Not long after their discovery, the production of carbon nanotubes at the rate of 300 kg/day was reported, enabling their evaluation for a wide range of potential applications. Commercially available sources for these nano-building blocks have steadily increased, and their price has steadily decreased, making them widely available for proof-of-concept experimental demonstrations.

5.42 Carbon Nanotube Structure and Electrical Properties

These tubes have been found to have outstanding electrical properties,[179] surpassing in some ways standard conductors and semiconductors.[180] Depending on the angle at which the graphite sheet is rolled up, called the roll-up vector, electrical transport in carbon nanotubes changes dramatically.[181] Those tubes having a helical twist in

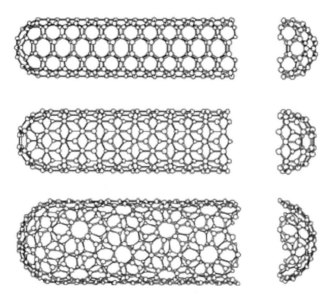

Figure 5.59 *Nanotube models: metallic zig-zag, metallic armchair structures, and semiconducting chiral nanotubes (top to bottom)*
(Reproduced with permission from physicsweb.org/box/world/11/1/9/world-11-1-9-2)

their structure have semiconducting properties, while achiral tubes are metallic.[182] The structure of metallic "zig-zag" and "armchair" tubes, as well as a chiral semiconducting tube, are shown in Figure 5.59. Semiconducting nanotubes (2 out of every 3 random tubes) have a band gap between their conduction and valence bands, while in metallic tubes two mini-bands occupy the mid-gap state and quantum mechanical tunneling effectively leads to electrical conductivity. Ballistic electron transport, electrical conductivity without phonon and surface scattering, has been observed in metallic carbon nanotubes[183,184] implying that the ultimate in mobility along a defect-free tube could be achieved with the minimum of resistive scattering losses. This is analogous to photons moving along a glass fiber with minimal absorption and scattering losses. Carbon nanotubes can also have structural and electronic defects, which can be successfully probed using electron scattering,[185] and can lead to non-ideal electrical properties. CNs are truly electronically unique, demonstrating many properties observed in classical semiconductors, and many more they lay claim to all their own.[186]

As a result of their semiconducting properties, helical carbon nanotubes have enabled the fabrication of the first transistor based on a single carbon nanotube[187,188] – a step towards molecular electronics. This prototype device called a TUBEFET was built of a SWNT with a length of 100 nm, and is shown schematically in Figure 5.60. It operated at RT with a switching speed of 10 THz. Voltage gain for a 300 nm oxide layer was found to be 0.35 and is expected to be <1 for a 5-nm oxide layer. Ballistic hole-transport is observed with $V_F = 8 \times 10^5$ m s^{-1}. The ultra-small TUBEFET based on a SWNT was described

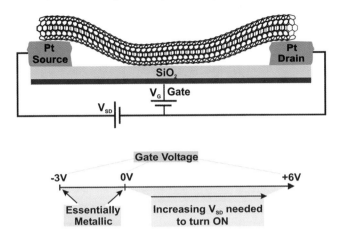

Figure 5.60 *A TUBEFET and its electrical properties*

electronically by the same semi-classical band bending, space charge, Fermi level pinning and Schottky–Ohmic junction model for devices used in today's computer industry. Since this first demonstration of a carbon nanotube transistor, a host of other devices and circuits have been assembled by directed self-assembly and their performance evaluated[189–192] – the results bode well for a future nanoelectronics founded on these tubular carbon constructs.[193]

There is considerable experimental effort now being applied to the synthesis of single length, diameter and chirality carbon nanotubes,[194] as well as for convenient ways of removing impurities and separating nanotubes by type following their synthesis.[195] Residual carbonaceous impurities and metal catalyst can be removed by high-temperature treatment in oxygen (~400 °C), and boiling in diluted mineral acid (*i.e.* 2.6 M nitric acid), respectively. Precipitation from a non-solvent followed by centrifugation, as well as elution chromatography, can further increase the material purity and can roughly sort nanotubes by length. Precipitation of carbon nanotubes by evaporation of an octadecylamine solution can separate metallic from semiconducting nanotubes, based on different propensities for charge transfer with the electron donating solvent.[196] The highly selective separation of semiconducting and metallic carbon nanotubes is also possible using electrophoresis.[197] When a suspension of carbon nanotubes is exposed to an alternating electric field, it has a different effect on semiconducting and conducting tubes. Because of their different dielectric constants, the two types of nanotubes migrate towards opposite electrodes when in the alternating field. The spatially segregated nanotubes can then be collected and used individually, albeit in small quantities (picograms).

5.43 Gone Ballistic

Advances in SWNT nanoelectronics has been progressing at a breathtaking pace worldwide. In this exciting field one of the most challenging technological hurdles that needed to be surmounted was the demonstration that a high electrical quality interface, essentially devoid of carrier scattering effects, could be achieved between a semiconducting SWNT and a high capacity, low current leakage and low power loss insulating material. These requirements would achieve desirable ballistic conductance in a practical SWNT device near the scaling limit (~50 nm length)

A soft-chemistry approach came to the rescue to solve this problem whereby ALD of the Hf(NEt)$_4$ precursor at 90 °C was utilized to create an 8-nm thick layer of the high dielectric constant (K~15) HfO$_2$ gate insulator on the top of the SWNT.[198] Device construction was completed by employing a highly directional e-beam deposition technique to make the top Al gate contact and the Pd source and drain electrodes, the latter having a low contact resistance to the SWNT. The overall SWNT–FET device architecture is illustrated in Figure 5.61 together with a SEM image showing the top view of the actual device.

The electrical characteristics measured for the device were found to be the best achieved to date for an SWNT–FET and a detailed temperature-dependent transport study with theoretical modeling confirmed that ballistic transport was indeed achieved. Significantly, the ALD method was successful for making a defect-free interface between the high dielectric constant hafnium dioxide gate insulator and the carbon nanotube semiconductor.

An extension of the aforementioned fabrication method enabled the construction of an array of SWNT–FETs. This was accomplished using an interdigitated symmetric double-gate electrode in which a single nanotube was strung across eight gate lines to create eight SWNT–FETs connected in parallel (Figure 5.61). A key demonstration was that an array of this genre is able to handle macroscopic currents, which speaks well for the utilization of carbon nanotubes in nanocircuitry.

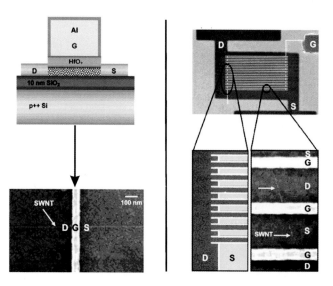

Figure 5.61 *Structure of a ballistic carbon nanotube transistor with high-quality interfaces, left, shown schematically and in a SEM image. The right part of the figure shows an array of transistors connected in series* (Reproduced with permission from Ref. 198)

5.44 Carbon Nanotube Nanomechanics

Carbon nanotubes have absolutely remarkable mechanical properties,[199] with a strength per weight ratio greater than any engineering material. Even as impure materials, they can be used as reinforcing agents, which makes them interesting as lightweight fibers for strengthening plastic, ceramic and metal composites. They have an incredible tensile strength, and have even been suggested for possible future use as a satellite tether (Figure 5.62) when assembled into microscopic or macroscopic bundles.[200] When these tubes are bent they deform reversibly, springing back into place once the strain is removed. The ability of carbon nanotubes to buckle without breaking makes them the ideally sharp scanning probe microscopy tip for exploring deep trenches in surfaces and shaped materials, aided by considerable development in the mounting of tubes onto tips.[201]

Carbon nanotube mechanical properties make them leading contenders for next generation, nanoelectro-mechanical systems. A convincing proof-of-concept demonstration of their mechanical potential used them as the active element of a high-precision rotational actuator.[202] To start, a dilute dispersion of carbon nanotubes was coated on a SiO_2/Si substrate. Once an isolated nanotube was identified by AFM, metal contacts around the nanotube and a metal plate in the nanotube center were defined by e-beam lithography. Etching of the oxide layer on the silicon wafer formed the final construct, shown schematically in Figure 5.63. This nanotube actuator can rotate in a highly controlled fashion, by tuning the voltage and frequency of the electric field applied to the S1, S2 and S3 electrodes shown in the figure. Real-time SEM images of the actuator in operation are shown in Figure 5.64, demonstrating incredible control over motion at a previously unthinkable length scale.

5.45 Carbon Nanotube Chemistry

One of the main problems with using nanotubes in basic research as well as industrial settings is their tendency to aggregate. Nanotubes can be dispersed in a solvent[203] using ultrasound, but strong van der Waals interactions between tube walls usually cause them to quickly precipitate. To eliminate this problem, a number

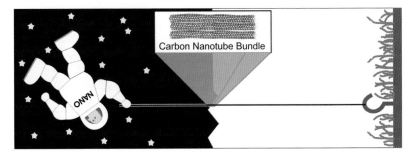

Figure 5.62 *Carbon nanotubes are so strong, it has been hypothesized they could one day function as satellite tethers and space elevators*

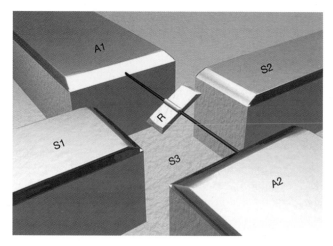

Figure 5.63 *Structure of a carbon nanotube electromechanical actuator. A1 and A2 are*
support beams, while S1, S2 and S3 are electrodes, which move the actuator
using electric fields
(Reproduced with permission from Ref. 202)

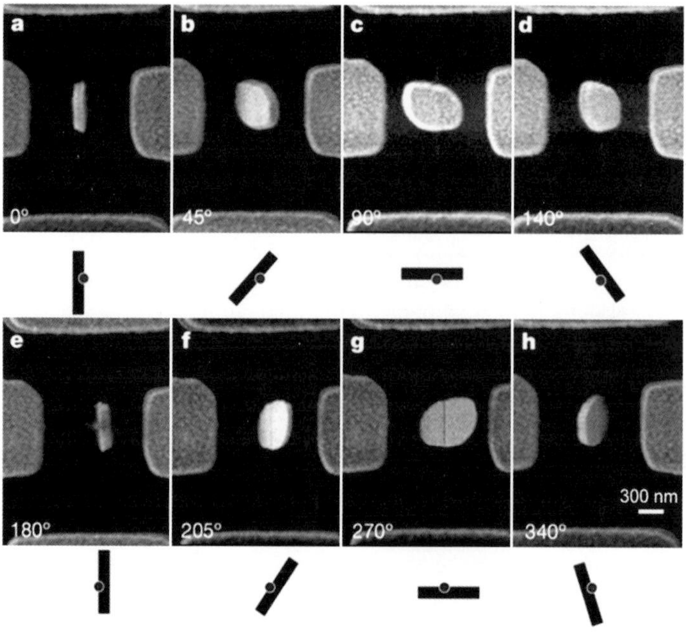

Figure 5.64 *SEM images of the carbon nanotube nanoactuator at different degrees of*
angular rotation, illustrated schematically below each image
(Reproduced with permission from Ref. 202)

of different methods have been used to chemically alter carbon nanotubes,[204,205] and make them soluble in a given matrix.[206,207] Oxidation using nitric acid or other strong acids opens the ends of carbon nanotubes and introduces carboxylate functionalities, imparting a modest solubility in polar solvents. Several chemical reactions can be employed to functionalize the walls of carbon nanotubes, either with molecular or polymeric species.[208] However, many chemical reactions with nanotube walls cause atomic defects in their structure, which most often compromises their desirable electrical and mechanical properties. To functionalize CNs with a minimal disruption of their properties, non-covalent modification methods may be more advantageous.[209] A simple and inexpensive method for doing so involves treating CNs with starch, an abundant and renewable natural resource.[210] Remarkably, when nanotubes are exposed to a starch solution this biomacromolecule wraps around the CN through cooperative van der Waals interactions, maintaining them in a solubilized state for extended periods of time. The re-precipitation of the solubilized CNs is also very easy, one simply needs to add the enzyme amylase (present in human saliva), which breaks down the starch and robs the nanotubes of their protective sheath.

The inside and outside of carbon nanotubes enables a range of materials to be replicated in the form of nanoscale wires and tubes for electronic and optical applications. Open-ended and end-functionalized carbon nanotubes with, for example, carboxylate or biotin facilitate AFM imaging with site-selective discrimination and molecular resolution.[211] The versatility in chemical species which can be anchored to the nanotube tip is promising for ultrahigh precision chemical force microscopy.[212] With chemically functionalized carbon nanotube AFM tips a nanoscale pH meter was demonstrated, surface anchored streptavidin was recognized, and in principle it becomes possible to interrogate and modify specific base pairs one-at-a-time down the double helix of DNA.

Carbon nanotubes may also find utility as the world's smallest sensors, and many successful demonstrations have appeared in the scientific literature.[213,214] For instance, due to their conjugated pi-electron system, carbon nanotubes can display selective recognition of pi-conjugated benzene derivatives.[215] The relative electron donating ability of various substituents on the benzene ring, the Hammett constant (*i.e.* $NH_2 < OH < OCH_3 < Cl < NO_2$), will influence the electron density of the aromatic ring. This can be directly measured by a FET carbon nanotube device illustrated in Figure 5.65. At the top left we see a carbon nanotube stretched between two patterned electrodes acting as source and drain, with the silica-on-silicon substrate acting as the gate. This configuration functions as a FET and when the voltage on the gate is sufficiently negative, positive charge carriers in the nanotube reach a critical concentration in the nanotube and enable conductance (transistor ON). If benzene derivatives are present around the tube, they can pi-stack with the curved graphene sheet and change its electron density. Electron withdrawing groups will deplete the nanotube of electron density and it will be easier to promote the flow of positive charge carriers, whereas electron-donating groups have the opposite effect. It was correspondingly found that the turn-on voltage for such a nanotube FET scaled linearly with the Hammett parameter for the benzene substituent (Figure 5.65, top right). The inherent and engineered

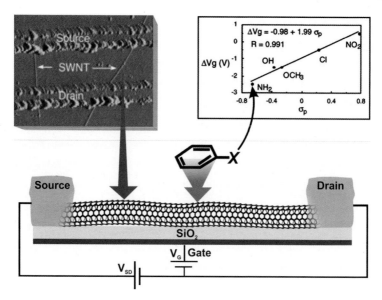

Figure 5.65 *Aromatic hydrocarbons can pi-stack on the carbon nanotube of a TUBEFET device. The electron donating ability of the substituent X, its Hammett constant, will determine the gate voltage necessary to induce current flow* (Reproduced with permission from Ref. 215)

selectivities in such nanotube sensors could make them ubiquitous and indispensable in future sensing technologies.

5.46 Carbon Nanotubes All in a Row

Since their discovery, several studies have been aimed at uncovering the properties of carbon nanotubes and a wealth of data has been compiled. Due to their anisotropic and structure-dependent properties, it has become ever more pressing to develop ways to align and organize carbon nanotubes onto supports to make functional constructs.[216]

The first strategy to alignment consists of starting with a collection of uniform carbon nanotubes and inducing these to organize. The nanotubes are first produced as dispersion in a solvent and can be treated as anisotropic colloids. One of the most effective ways of obtaining ordered colloid arrays is the technique of EISA.[217] As a colloidal dispersion dries there is a net flux towards the liquid interface from which evaporation is taking place, resulting in the transport of colloids along this flux. Thus, a solid surface immersed perpendicularly in a liquid will be wet by a liquid meniscus, whose increased surface area accelerates evaporation. As the meniscus evaporates, it leaves behind colloids bridged by a thin film of liquid. To reduce the capillary energy in this thin wetting film, colloidal particles are drawn together thus reducing the area of the liquid–air interface. This method of colloid assembly is illustrated in Figure 5.66. Consequently, it was found that a collection of carbon

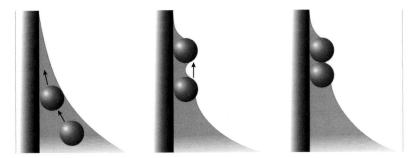

Figure 5.66 *EISA method for organizing colloids. The colloids are drawn to the meniscus through convective transport, then packed together by the minimization of capillary forces*

nanotubes could also be aligned using EISA.[218] A collection of CNs is first dispersed in water after an acid treatment to increase their uniformity and solubility. Simply letting this dispersion dry onto a hydrophilic substrate results in the growth of a CN film where the nanotubes are aligned parallel to the water interface. The alignment is not perfect and resembles a nematic liquid crystalline phase, where rod-shaped units loosely follow a common alignment axis, rather than a precisely ordered array. An approach has been recently developed which carried nanotube alignment to the next level of precision.[219] The method begins by making patterns of a hydrophilic SAM on a substrate, followed by hydrophobization of the remaining substrate area. When such a substrate, made either by dip-pen nanolithography or microcontact printing, was exposed to an aqueous suspension of CNs, it was found that the nanotubes were rapidly attracted and confined to the hydrophilic areas. If the hydrophilic regions were small enough, each could trap a single CN resulting in the predetermined organization of a collection of uniform nanotubes. Although the mechanism behind this alignment is not clear, it likely involves a local electrostatic field generated by the polar SAM patch, which attracts to it the readily polarized nanotubes. This method is very attractive in terms of its simplicity and effectiveness, and a portion of such an array is shown in Figure 5.67. If CNs are to be used successfully in electronic devices, they must be organized with a density approaching that of lithographically fabricated microchips, which currently can be produced with feature sizes of 100 nm. There is clearly a lot of work to be done to reduce the spacing, or pitch, between nanotubes in order to truly take advantage of their fascinating electrical properties. As mentioned earlier, promising steps in this direction have been realized.

The second strategy for alignment consists of using a pre-existing pattern to direct the aligned growth of an array of carbon nanotubes.[220] The disadvantage of this method is that there can be significant dispersity in the structure and purity of the formed nanotubes. The advantage, however, is that a very large number of nanotubes can be grown and used directly in applications. For example, intense electron field emission from carbon nanotubes has been observed and is of considerable interest in flat panel displays and TV screens based on carbon nanotube electron guns.[221] The field emission properties of nanotubes are quite robust to variations in nanotube

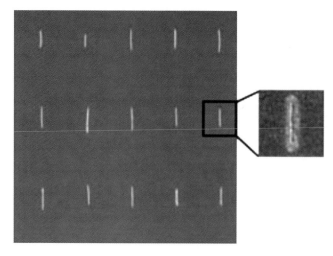

Figure 5.67 *An array of aligned single carbon nanotubes templated on nanopatterned hydrophilic patches*
(Reproduced with permission from Ref. 219)

size and occur for both single-walled and multi-walled CNs. In this regard, CVD of ethylene at 700 °C onto a regular array of catalytic iron nanoclusters anchored within the void spaces of an anodically etched p-doped Si<100> wafer has been used for the controlled nucleation, growth, self-orientation and patterning of massive arrays of monodispersed multi-walled carbon nanotubes.[222] This is shown schematically in Figure 5.68. Self-organized growth originates from a combination of (i) diffusion-controlled reaction of ethylene on fixed iron nanoclusters affixed in the porous silicon, (ii) flow of ethylene through the interior of the growing MWNT array and (iii) van der Waals interactions between nearest neighbor MWNT. Patterns of well-ordered MWNT, shown in Figure 5.69, can function as excellent field emission arrays. The CVD self-assembly synthesis can be easily integrated into silicon technology to enable the development of MWNT devices.

5.47 Carbon Nanotube Photonic Crystal

As mentioned, carbon nanotubes can be produced by metal cluster catalyzed growth from hydrocarbon precursors. By patterning metal clusters on a substrate and exposing them to the hydrocarbon source it is possible to grow periodic arrays of aligned nanotubes. This approach has been employed to make a two-dimensional carbon nanotube photonic crystal,[223] an intriguing development because of the opportunities to tune optical Bragg diffraction through the choice of materials parameters of the constituent carbon nanotubes. This depend on whether they are metallic or semiconducting, contain adsorbed hydrogen or oxygen, surface functional groups or dopants like lithium.

To amplify, nanosphere lithography was used to pre-define a hexagonal array of nickel clusters on a silicon substrate by sputtering through a sphere monolayer

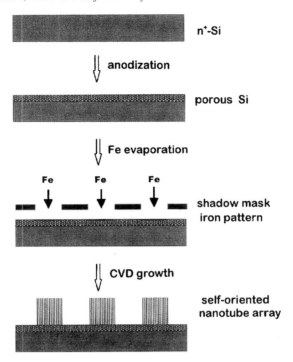

Figure 5.68 *Strategy for the growth of aligned carbon nanotube patterns*
(Reproduced with permission from Ref. 222)

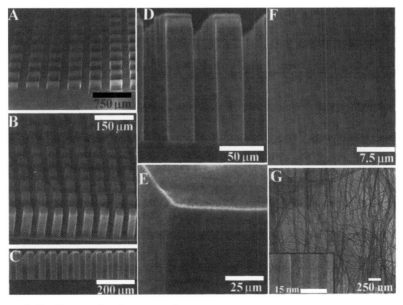

Figure 5.69 *Patterned carbon nanotube arrays*
(Reproduced with permission from Ref. 222)

mask, and acetylene was used as the carbon source in a plasma-enhanced CVD reaction chamber to catalyze nanotube growth. An aligned carbon nanotube array was observed to grow from the pattern of nickel clusters where the spacing between the nanotubes at the lightscale, controlled by the diameter of the spheres used in the nanosphere template, enables scaling of optical Bragg diffraction (Figure 5.70). Intense diffraction of light from the two-dimensional nanotube lattice in the visible wavelength range attests to its high degree of structural order. It is interesting to note that because the real part of the dielectric constant in the visible wavelength range for metallic nanotubes is negative, the photonic band gap for a two-dimensional carbon nanotube photonic crystal, which depends on refractive index contrast between the tubes and environment, could be very large. Maybe it will prove possible to electrically tune the photonic crystal properties of metallic carbon nanotube arrays. Finally, it is worth mentioning that carbon nanotube arrays may prove valuable as structure-directing templates to make two-dimensional inverted photonic crystals with wide ranging compositions using straightforward precursor deposition followed by sacrificial oxidation of the nanotubes.

5.48 Putting Carbon Nanotubes Exactly Where You Want Them

A noteworthy step towards the large-scale integration of carbon nanotubes into microelectronic circuitry has been described that involves the combination of alternating and direct current electric fields applied to a microelectrode array with 100 gaps to capture and align single nanotubes with about 90% reliability.[224] By varying the ratio of the applied E_{AC}/E_{DC} electric fields the conditions required for single nanotube deposition between electrodes could be established. A detailed theoretical analysis of the assembly mechanism implied that it is only the combined action of dielectrophoretic (polarization-induced movement), electro-osmotic (potential gradient induced flow) and electrophoretic (electrostatic-induced attraction) forces acting on the carbon nanotubes that serves to orient, move and trap them in the gap of the electrodes. For the optimum ratio of AC and DC electric fields, a single nanotube can be selectively placed in the gap of an individual electrode pair with around 90% certainty, as seen in Figure 5.71.

This ESA approach is a physical way of moving, organizing and connecting carbon nanotubes between electrodes for a range of nanoscale device applications rather than a chemical one that requires the nanotube or substrate to be chemically functionalized. The key to success of the ESA method is to establish, through theoretical modeling coupled with experimentation, just the right ratio of AC and DC fields to move and arrange a single nanotube of the right length between the electrodes. If a method can be devised that is able to create either semiconductive or metallic single-wall carbon nanotubes in the absence of any other tubes or particles then this ESA method could prove extremely useful as a rational way of assembling nanotube electronic circuitry.

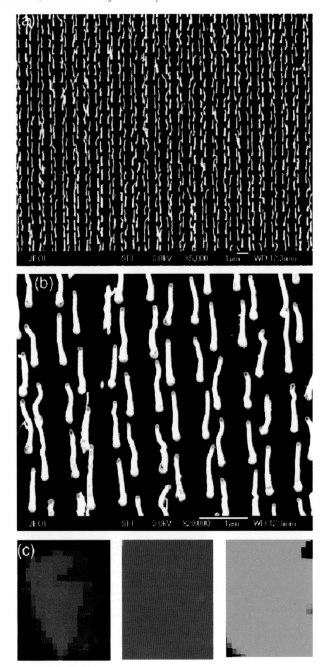

Figure 5.70 *Carbon nanotube photonic crystals, SEM, made by growth onto catalyst spots deposited by microsphere lithography. The bottom image shows optical pictures of samples with different lattice constants, resulting in different reflected colors*
(Reproduced with permission from Ref. 223)

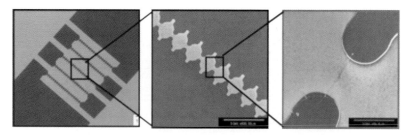

Figure 5.71 *An electrode array, at increasingly higher magnifications, shows the assembly of a carbon nanotube between bridging electrodes when a field is applied* (Reproduced with permission from Ref. 224)

5.49 The Nanowire Pitch Challenge

From the examples presented in this chapter it should be abundantly clear that nanowire diameter and length, crystalline and amorphous form, composition and doping, crystal orientation and surface structure, charge and functionality are under control. Complex multi-component architectures, like longitudinal hetero-structured superlattices and co-axial heterostructures, can be custom made and organized into different kinds of nanowire devices. Getting the nanowires to line up in made-to-measure geometries on substrates where the distance between nanowires is comparable to their diameter remains one of the foremost challenges in order to take advantage of their minute size in future nanoelectronic, nanophotonic and nanobiology applications. Such command over nanowire separation is known as the "pitch problem" and one creative approach to solving the problem involves the use of Langmuir–Blodgett techniques and surfactant-coated nanowires to create an organized array of parallel, densely packed nanowires located at the air–water interface. This layer can be transferred with the preservation of order and separation, as delineated schematically in Figure 5.72 and mentioned earlier. By making use of (i) core–corona co-axial architecture nanowires, for example, silicon–silica with a controlled diameter silicon core and thickness of the silica corona, (ii) by dispersing these nanowires coated with 1-octadecylamine in a solution of isooctane/2-propanol, (iii) by controlling the compression process of these nanowires in a Langmuir–Blodgett trough to form uniaxial close-packed arrays, (iv) by transferring the nanowire array to a planar substrate, for example, hydrophobic silicon and (v) by selectively removing the silica corona using, for example, CF_3H reactive ion etching, it is feasible to transfer, align and control the separation between the nanowires on the substrate (Figure 5.73). Moreover, by using these kinds of pitch-controlled nanowire arrays as a mask for nanolithography,[225] it is possible to deposit other materials, for example, Cr metal, between the nanowires, followed by removal of the nanowire mask by sonification in aqueous NH_4F to create an array of Cr nanowires on the silicon substrate.

Using this approach for delineating nanometer dimension lines with controlled separation between them one can envision a myriad of self-assembled nanowire devices and circuits for use in next generation integrated nanosystems.

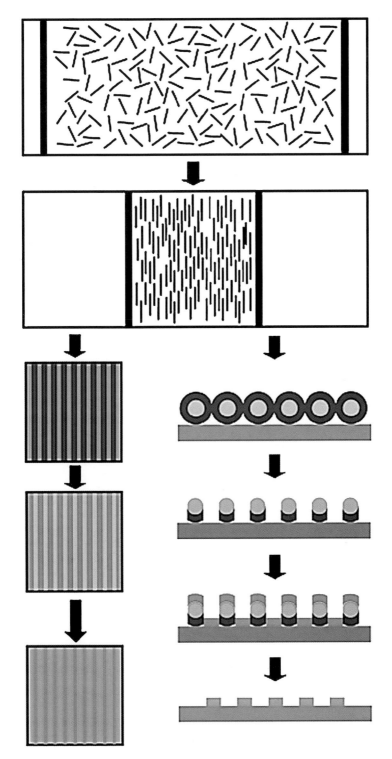

Figure 5.72 *Langmuir–Blodgett alignment of core–corona nanowires, transfer onto a substrate maintaining order and separation, and utilization as nanowire masks* (Reproduced with permission from Ref. 225)

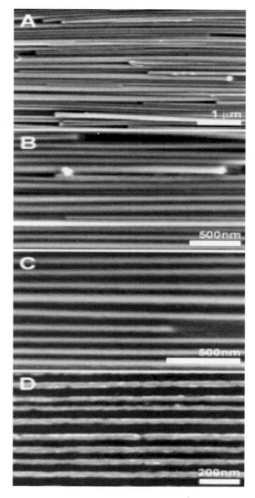

Figure 5.73 *SEM images showing the sequence of steps (A,B) Si–SiO₂ core–corona nanowire array on a Si substrate (C) removal of SiO₂ corona (D) Cr deposition and removal of Si nanowires*
(Reproduced with permission from Ref. 225)

5.50 Integrated Nanowire Nanoelectronics

If there really is to be bottom–up nanocircuitry based on nanowire components then strategies have to be developed having the ability to organize large numbers of nanowires in a massively parallel manner into electrically addressable nanoscale devices. While fluidic and electric field self-assembly methods can direct nanowires into limited scale structures, they are not suitable for maneuvering them in huge numbers into an integrated nanowire circuit. By contrast, the LB method mentioned earlier can be used to align nanowires over large areas with a lateral separation (pitch) between nanowires controlled by the compression force exerted on them in the LB trough.[226]

The process to obtain large area arrays of p-doped silicon NW-FETs on a substrate begins by first patterning a LB-aligned and fixed pitch nanowire film on a substrate using conventional photolithography and etching. This step is followed by depositing metal electrodes through a mask that is registered with the nanowire features patterned in the first step. A summary of the procedure is shown in Figure 5.74.[227] Because the pitch between aligned nanowires and nanoelectrodes on the chip is scalable it has proven feasible to make ~80% of thousands of possible wire–electrode connections. Interestingly, defective devices do not exert an adverse effect on the operation of correctly connected ones. An example of an integrated nanowire circuit made by this process is shown in Figure 5.75. The electrical characteristics of the devices showed that they behave as p-type NW-FETs and the reproducibility amongst devices indicated that they could serve as functional components of integrated nanowire systems.

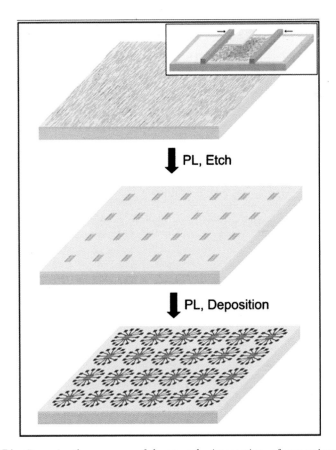

Figure 5.74 *Steps in the process of large-scale integration of nanowires including nanowire deposition, etching, and electrode deposition*
(Reproduced with permission from Ref. 227)

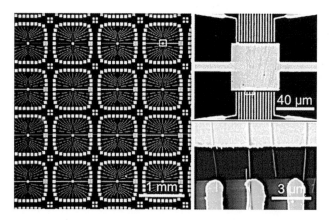

Figure 5.75 *A lower resolution optical image and two higher resolution SEM images of electrical arrays made with aligned arrays of nanowires. Electrodes are randomly connected, and can be used in devices after testing and calibration* (Reproduced with permission from Ref. 227)

5.51 A Small Thought at the End of a Large Chapter

The reader may have noticed on arriving, likely exhausted, at this point in the nanorod, nanowire, nanotube chapter that it is rather lengthy – in fact, it is by far the longest one in this nanochemistry textbook. While not intending to understate the significance of one-dimensional building blocks in the nanochemistry toolbox, it is worth noting that the chapter could have been divided into two, based on the fact that the diameter of nanorods and nanotubes derived from them are significantly larger than semiconductor nanowires and carbon nanotubes. This dimension has a profound effect on whether or not their properties are influenced by QSEs. Nevertheless, it seemed best to lump them altogether in one large chapter.

5.52 References

1. J. Hu, T.W. Odom and C.M. Lieber, Chemistry and physics in one dimension: synthesis and properties of nanowires and nanotubes, *Acc. Chem. Res.*, 1999, **32**, 435.
2. Y. Xia, P. Yang, Y. Sun, Y. Wu, B. Mayers, B. Gates, Y. Yin, F. Kim and H. Yan, One-dimensional nanostructures: synthesis, characterization, and applications, *Adv. Mater.*, 2003, **15**, 353.
3. Y. Wu, H. Yan, M. Huang, B. Messer, J.H. Song and P. Yang, Inorganic semiconductor nanowires: rational growth, assembly, and novel properties, *Chem. Eur. J.*, 2002, **8**, 1260.
4. A. Huczko, Template-based synthesis of nanomaterials, *Appl. Phys. A: Mater. Sci. Process*, 2000, **70**, 365.
5. P. Hoyer, Multistep replication processes, *Curr. Opin. Colloid Interface Sci.*, 1998, **3**, 160.

6. F. Schuth, Endo- and exotemplating to create high-surface-area inorganic materials, *Angew. Chem. Int. Ed.*, 2003, **42**, 3604.

7. C.T. Kresge, M.E. Leonowicz, W.J. Roth, J.C. Vartuli and J.S. Beck, Ordered mesoporous molecular sieves synthesized by a liquid-crystal template mechanism, *Nature*, 1992, **359**, 710.

8. J.S. Beck, J.C. Vartuli, W.J. Roth, M.E. Leonowicz, C.T. Kresge, K.D. Schmitt, C.T.W. Chu, D.H. Olson and E.W. Sheppard, A new family of mesoporous molecular sieves prepared with liquid crystal templates, *J. Am. Chem. Soc.*, 1992, **114**, 10834.

9. N.R.B. Coleman, N. O'Sullivan, K.M. Ryan, T.A. Crowley, M.A. Morris, T.R. Spalding, D.C. Steytler and J.D. Holmes, Synthesis and characterization of dimensionally ordered semiconductor nanowires within mesoporous silica, *J. Am. Chem. Soc.*, 2001, **123**, 7010.

10. T. Thurn-Albrecht, J. Schotter, G.A. Kästle, N. Emley, T. Shibauchi, L. Krusin-Elbaum, K. Guarini, C.T. Black, M.T. Tuominen and T.P. Russell, Ultrahigh-density nanowire arrays grown in self-assembled diblock copolymer templates, *Science*, 2000, **290**, 2126.

11. G.J. Liu, J.F. Ding, T. Hashimoto, K. Kimishima, F.M. Winnik and S. Nigam, Thin films with densely, regularly packed nanochannels: preparation, characterization, and applications, *Chem. Mater.*, 1999, **11**, 2233.

12. C.R. Martin, Nanomaterials: a membrane-based synthetic approach, *Science*, 1994, **266**, 1961.

13. H. Masuda and K. Fukuda, Ordered metal nanohole arrays made by a 2-step replication of honeycomb structures of anodic alumina, *Science*, 1995, **268**, 1466.

14. L.T. Canham ed., *Properties of Porous Silicon*, vol. 18, Short Run Press, London, 1997.

15. Y. Watanabe, Y. Arita, T. Yokoyama and Y. Igarashi, Formation and properties of porous silicon and its application, *J. Electrochem. Soc.*, 1975, **122**, 1351.

16. K.W. Kolasinski, The mechanism of Si etching in fluoride solutions, *Phys. Chem. Chem. Phys.*, 2003, **5**, 1270.

17. A.G. Cullis and L.T. Canham, Visible light emission due to quantum size effects in highly porous crystalline silicon, *Nature*, 1991, **353**, 335.

18. M.J. Sailor, J.L. Heinrich and J.M. Lauerhaas, Luminescent porous silicon: synthesis, chemistry, and applications, *Stud. Surf. Sci. Catal.*, 1997, **103**, 209.

19. J.R. Heath, Liquid-solution-phase synthesis of crystalline silicon, *Science*, 1992, **258**, 1131.

20. M.P. Stewart and J.M. Buriak, Chemical and biological applications of porous silicon technology, *Adv. Mater.*, 2000, **12**, 859.

21. R.L. Smith and S.D. Collins, Porous silicon formation mechanisms, *J. Appl. Phys.*, 1992, **71**, R1.

22. F. Muller, A. Birner, U. Gosele, V. Lehmann, S. Ottow and H. Foll, Structuring of macroporous silicon for applications as photonic crystals, *J. Porous Mater.*, 2000, **7**, 201.

23. H. Masuda, H. Yamada, M. Satoh, H. Asoh, M. Nakao and T. Tamamura, Highly ordered nanochannel-array architecture in anodic alumina, *Appl. Phys. Lett.*, 1997, **71**, 2770.

24. H. Masuda, H. Asoh, M. Watanabe, K. Nishio, M. Nakao and T. Tamamura, Square and triangular nanohole array architectures in anodic alumina, *Adv. Mater.*, 2001, **13**, 189.

25. G. Vincent, Optical properties of porous silicon superlattices, *Appl. Phys. Lett.*, 1994, **64**, 2367.

26. J. Schilling, F. Muller, S. Matthias, R.B. Wehrspohn, U. Gosele and K. Busch, Three-dimensional photonic crystals based on macroporous silicon with modulated pore diameter, *Appl. Phys. Lett.*, 2001, **78**, 1180.

27. S. Matthias, J. Schilling, K. Nielsch, F. Muller, R.B. Wehrspohn and U. Gosele, Monodisperse diameter-modulated gold microwires, *Adv. Mater.*, 2002, **14**, 1618.

28. S. Link and M.A. El-Sayed, Shape and size dependence of radiative, non-radiative and photothermal properties of gold nanocrystals, *Int. Rev. Phys. Chem.*, 2000, **19**, 409.

29. M.A. El-Sayed, Some interesting properties of metals confined in time and nanometer space of different shapes, *Acc. Chem. Res.*, 2001, **34**, 257.

30. S.R. Nicewarner-Pena, R.G. Freeman, B.D. Reiss, L. He, D.J. Pena, I.D. Walton, R. Cromer, C.D. Keating and M.J. Natan, Submicrometer metallic barcodes, *Science*, 2001, **294**, 137.

31. G.E. Thompson, Porous anodic alumina: fabrication, characterization and applications, *Thin Solid Films*, 1997, **297**, 192.

32. D.S. Xu, D.P. Chen, Y.J. Xu, X.S. Shi, G.L. Guo, L.L. Gui and Y.Q. Tang, Preparation of II–VI group semiconductor nanowire arrays by dc electro-chemical deposition in porous aluminum oxide templates, *Pure Appl. Chem.*, 2000, **72**, 127.

33. H. Masuda, H. Yamada, M. Satoh, H. Asoh, M. Nakao and T. Tamamura, Highly ordered nanochannel-array architecture in anodic alumina, *Appl. Phys. Lett.*, 1997, **71**, 2770.

34. O. Jessensky, F. Muller and U. Gosele, Self-organized formation of hexagonal pore arrays in anodic alumina, *Appl. Phys. Lett.*, 1998, **72**, 1173.

35. A.P. Li, F. Müller, A. Birner, K. Nielsch and U. Gösele, Hexagonal pore arrays with a 50–420 nm interpore distance formed by self-organization in anodic alumina, *J. Appl. Phys.*, 1998, **84**, 6023.

36. S. Ono, H. Asoh, M. Saito and M. Ishiguro, Relationship between pore diameter to cell diameter ratio and self-ordering of anodic porous alumina, *Electrochemistry*, 2003, **71**, 105.

37. O. Jessensky, F. Muller and U. Gosele, Self-organized formation of hexagonal pore arrays in anodic alumina, *Appl. Phys. Lett.*, 1998, **72**, 1173.

38. A. Huczko, Template-based synthesis of nanomaterials, *Appl. Phys. A-Mater. Sci. Proc.*, 2000, **70**, 365.

39. B.R. Martin, D.J. Dermody, B.D. Reiss, M. Fang, L.A. Lyon, M.J. Natan and T.E. Mallouk, Orthogonal self-assembly on colloidal gold–platinum nano-rods, *Adv. Mater.*, 1999, **11**, 1021.

40. J.J. Hickman, P.E. Laibinis, D.I. Auerbach, C.F. Zou, T.J. Gardner, G.M. Whitesides and M.S. Wrighton, Toward orthogonal self-assembly of redox active molecules on Pt and Au – selective reaction of disulfide with Au and isocyanide with Pt, *Langmuir*, 1992, **8**, 357.

41. J.L. Guesdon, Immunoenzymatic techniques applied to the specific detection of nucleic-acids – a review, *J. Immun. Methods*, 1992, **150**, 33.

42. J.S. Yu, J.Y. Kim, S. Lee, J.K.N. Mbindyo, B.R. Martin and T.E. Mallouk, Template synthesis of polymer-insulated colloidal gold nanowires with reactive ends, *Chem. Commun.*, 2000, 2445.

43. N.I. Kovtyukhova and T.E. Mallouk, Nanowires as building blocks for self-assembling logic and memory circuits, *Chem. Eur. J.*, 2002, **8**, 4354.

44. J.K.N. Mbindyo, B.D. Reiss, B.R. Martin, C.D. Keating and M.J. Natan, DNA-directed assembly of gold nanowires on complementary surfaces, *Adv. Mater.*, 2001, **13**, 249.

45. B.R. Martin, S.K. St. Angelo and T.E. Mallouk, Interactions between suspended nanowires and patterned surfaces, *Adv. Funct. Mater.*, 2002, **12**, 759.

46. T.M. Whitney, J.S. Jiang, P.C. Searson and C.L. Chien, Fabrication and magnetic-properties of arrays of metallic nanowires, *Science*, 1993, **261**, 1316.

47. J. Meier, A. Blondel, B. Doudin and J.P. Ansermet, Magnetic-properties of nanosized wires, *Helv. Phys. Acta*, 1994, **67**, 761.

48. M. Chen, L. Sun, J.E. Bonevich, D.H. Reich, C.L. Chien and P.C. Searson, Tuning the response of magnetic suspensions, *Appl. Phys. Lett.*, 2003, **82**, 3310.

49. M. Chen, P.C. Searson and C.L. Chien, Micromagnetic behavior of electrodeposited Ni/Cu multilayer nanowires, *J. Appl. Phys.*, 2003, **93**, 8253.

50. J.C. Love, A.R. Urbach, M.G. Prentiss and G.M. Whitesides, Three-dimensional self-assembly of metallic rods with submicron diameters using magnetic interactions, *J. Am. Chem. Soc.*, 2003, **125**, 12696.

51. R. Ferre, K. Ounadjela, J.M. George, L. Piraux and S. Dubois, Magnetization processes in nickel and cobalt electrodeposited nanowires, *Phys. Rev. B*, 1997, **56**, 14066.

52. A. Fert and L. Piraux, Magnetic nanowires, *J. Magn. Magn. Mater.*, 1999, **200**, 338.

53. J. Meier, B. Doudin and J.P. Ansermet, Magnetic properties of nanosized wires, *J. Appl. Phys.*, 1996, **79**, 6010.

54. B. Doudin, A. Blondel and J.P. Ansermet, Arrays of multilayered nanowires, *J. Appl. Phys.*, 1996, **79**, 6090.

55. K. Nielsch, R. Hertel, R.B. Wehrspohn, J. Barthel, J. Kirschner, U. Gosele, S.F. Fischer and H. Kronmuller, Switching behavior of single nanowires inside dense nickel nanowire arrays, *IEEE Trans. Magn.*, 2002, **38**, 2571.

56. J.M. Garcia, A. Thiaville and J. Miltat, MFM imaging of nanowires and elongated patterned elements, *J. Magn. Magn. Mater.*, 2002, **249**, 163.

57. G.P. Heydon, S.R. Hoon, A.N. Farley, S.L. Tomlinson, M.S. Valera, K. Attenborough and W. Schwarzacher, Magnetic properties of electro-deposited nanowires, *J. Phys. D: Appl. Phys.*, 1997, **30**, 1083.

58. U. Hartmann, Magnetic force microscopy, *Ann. Rev. Mater. Sci.*, 1999, **29**, 53.

59. R. Proksch, Recent advances in magnetic force microscopy, *Curr. Opin. Solid State Mater. Sci.*, 1999, **4**, 231.

60. V.V. Volkov, M.A. Schofield and Y. Zhu, Phase retrieval and induction mapping of artificially structured micromagnetic arrays, *Mod. Phys. Lett. B*, 2003, **17**, 791.

61. M. Lehmann and H. Lichte, Tutorial on off-axis electron holography, *Microsc. Microanal.*, 2002, **8**, 447.

62. R.E. Dunin-Borkowski, M.R. McCartney, R.B. Frankel, D.A. Bazylinski, M. Posfai and P. Buseck, Magnetic microstructure of magnetotactic bacteria by electron holography, *Science*, 1998, **282**, 1868.

63. M. Tanase, L.A. Bauer, A. Hultgren, D.M. Silevitch, L. Sun, D.H. Reich, P.C. Searson and G.J. Meyer, Magnetic alignment of fluorescent nanowires, *Nano Lett.*, 2001, **1**, 155.

64. S.P. Li, M. Natali, A. Lebib, A. Pepin, Y. Chen and Y.B. Xu, Magnetic nanostructure fabrication by soft lithography and vortex-single domain transition in Co dots, *J. Magn. Magn. Mater.*, 2002, **241**, 447.

65. A.R. Urbach, J.C. Love, M.G. Prentiss and G.M. Whitesides, Sub-100 nm confinement of magnetic nanoparticles using localized magnetic field gradients, *J. Am. Chem. Soc.*, 2003, **125**, 12704.

66. K.B. Lee, S. Park and C.A. Mirkin, Multicomponent magnetic nanorods for biomolecular separations, *Angew. Chem. Int. Ed.*, 2004, **43**, 3048.

67. M.J. Rosen, *Surfactants and Interfacial Phenomena*, Wiley Interscience, NY, 1989.

68. R. Nagarajan and E. Ruckenstein, Theory of surfactant self-assembly – a predictive molecular thermodynamic approach, *Langmuir*, 1991, **7**, 2934.

69. F. Tiberg, J. Brinck and L. Grant, Adsorption and surface-induced self-assembly of surfactants at the solid-aqueous interface, *Curr. Opin. Colloid Interface Sci.*, 1999, **4**, 411.

70. R. Rajagopalan, Simulations of self-assembling systems, *Curr. Opin. Colloid Interface Sci.*, 2001, **6**, 357.

71. V.T. John, B. Simmons, G.L. McPherson and A. Bose, Recent developments in materials synthesis in surfactant systems, *Curr. Opin. Colloid Interface Sci.*, 2002, **7**, 288.

72. S. Park, J.H. Lim, S.W. Chung and C.A. Mirkin, Self-assembly of mesoscopic metal–polymer amphiphiles, *Science*, 2004, **303**, 348.

73. B. Messer, J.H. Song and P.D. Yang, Microchannel networks for nanowire patterning, *J. Am. Chem. Soc.*, 2000, **122**, 10232.

74. P.A. Smith, C.D. Nordquist, T.N. Jackson, T.S. Mayer, B.R. Martin, J. Mbindyo and T.E. Mallouk, Electric-field assisted assembly and alignment of metallic nanowires, *Appl. Phys. Lett.*, 2000, **77**, 1399.

75. J.K.N. Mbindyo, T.E. Mallouk, J.B. Mattzela, I. Kratochvilova, B. Razavi, T.N. Jackson and T.S. Mayer, Template synthesis of metal nanowires containing monolayer molecular junctions, *J. Am. Chem. Soc.*, 2002, **124**, 4020.

76. I. Kratochvilova, M. Kocirik, A. Zamboya, J. Mbindyo, T.E. Mallouk and T.S. Mayer, Room temperature negative differential resistance in molecular nanowires, *J. Mater. Chem.*, 2002, **12**, 2927.

77. N.I. Kovtyukhova, B.R. Martin, J.K.N. Mbindyo, T.E. Mallouk, M. Cabassi and T.S. Mayer, Layer-by-layer self-assembly strategy for template synthesis of nanoscale devices, *Mater. Sci. Eng. C-Biomimetic Supramol. Syst.*, 2002, **19**, 255.

78. N.I. Koytyukhova, B.R. Martin, J.K.N. Mbindyo, P.A. Smith, B. Razavi, T.S. Mayer and T.E. Mallouk, Layer-by-layer assembly of rectifying junctions in and on metal nanowires, *J. Phys. Chem. B*, 2001, **105**, 8762.

79. N.I. Kovtyukhova, B.R. Martin, J.K.N. Mbindyo, T.E. Mallouk, M. Cabassi and T.S. Mayer, Layer-by-layer self-assembly strategy for template synthesis of nanoscale devices, *Mater. Sci. Eng. C: Biomimetic Supramol. Syst.*, 2002, **C19**, 255.

80. G. Schmid, Materials in nanoporous alumina, *J. Mater. Chem.*, 2002, **12**, 1231.

81. C.R. Martin, L.S. Vandyke, Z.H. Cai, W.B. Liang, Template synthesis of organic microtubules, *J. Am. Chem. Soc.*, 1990, **112**, 8976.

82. M. Steinhart, J.H. Wendorff, A. Greiner, R.B. Wehrspohn, K. Nielsch, J. Schilling, J. Choi and U. Gosele, Polymer nanotubes by wetting of ordered porous templates, *Science*, 2002, **296**, 1997.

83. M. Steinhart, R.B. Wehrspohn, U. Gosele and J.H. Wendorff, Nanotubes by template wetting: a modular assembly system, *Angew. Chem. Int. Ed.*, 2004, **43**, 1334.

84. M. Zhang, Y. Bando and K. Wada, Silicon dioxide nanotubes prepared by anodic alumina as templates, *J. Mater. Chem.*, 2000, **15**, 387.

85. G.L. Che, B.B. Lakshmi, E.R. Fisher and C.R. Martin, Carbon nanotubule membranes for electrochemical energy storage and production, *Nature*, 1998, **393**, 346.

86. G. Che, B.B. Lakshmi, C.R. Martin, E.R. Fisher and R.S. Ruoff, Chemical vapor deposition based synthesis of carbon nanotubes and nanofibers using a template method, *Chem. Mater.*, 1998, **10**, 260.

87. J. Li, C. Papadopoulos, J.M. Xu and M. Moskovits, Highly-ordered carbon nanotube arrays for electronics applications, *Appl. Phys. Lett.*, 1999, **75**, 367.

88. Z.J. Liang, A.S. Susha, A.M. Yu and F. Caruso, Nanotubes prepared by layer-by-layer coating of porous membrane templates, *Adv. Mater.*, 2003, **15**, 1849.

89. M.J. MacLachlan, M. Ginzburg, N. Coombs, T.W. Coyle, N.P. Raju, J.E. Greedan, G.A. Ozin and I. Manners, Shaped ceramics with tunable magnetic properties from metal-containing polymers, *Science*, 2000, **287**, 1460.

90. N.I. Kovtyukhova, T.E. Mallouk and T.S. Mayer, Templated surface sol–gel synthesis of SiO_2 nanotubes and SiO_2-insulated metal nanowires, *Adv. Mater.*, 2003, **15**, 780.

91. K.S. Mayya, D.I. Gittins, A.M. Dibaj and F. Caruso, Nanotubles prepared by templating sacrificial nickel nanorods, *Nano Lett.*, 2001, **1**, 727.

92. Y.N. Xia, P.D. Yang, Y.G. Sun, Y.Y. Wu, B. Mayers, B. Gates, Y.D. Yin, F. Kim and Y.Q. Yan, One-dimensional nanostructures: synthesis, characterization, and applications, *Adv. Mater.*, 2003, **15**, 353.

93. A.M. Morales and C.M. Lieber, A laser ablation method for the synthesis of crystalline semiconductor nanowires, *Science*, 1998, **279**, 208.

94. X. Duan and C.M. Lieber, Laser-assisted catalytic growth of single crystal GaN nanowires, *J. Am. Chem. Soc.*, 2000, **122**, 188.

95. X. GaN and C.M. Lieber, General synthesis of compound semiconductor nanowires, *Adv. Mater.*, 2000, **12**, 298.

96. M.S. Gudiksen and C.M. Lieber, Diameter-selective synthesis of semiconductor nanowires, *J. Am. Chem. Soc.*, 2000, **122**, 8801.

97. J.T. Hu, T.W. Odom and C.M. Lieber, Chemistry and physics in one dimension: synthesis and properties of nanowires and nanotubes, *Acc. Chem. Res.*, 1999, **32**, 435.

98. Y.Y. Wu and P.D. Yang, Direct observation of vapor–liquid–solid nanowire growth, *J. Am. Chem. Soc.*, 2001, **123**, 3165.

99. Y. Cui, L.J. Lauhon, M.S. Gudiksen, J.F. Wang and C.M. Lieber, Diameter-controlled synthesis of single-crystal silicon nanowires, *Appl. Phys. Lett.*, 2001, **78**, 2214.

100. M.H. Huang, S. Mao, H. Feick, H.Q. Yan, Y.Y. Wu, H. Kind, E. Weber, R. Russo and P.D. Yang, Room-temperature ultraviolet nanowire nanolasers, *Science*, 2001, **292**, 1897.

101. Y. Wu, Y. Cui, L. Huynh, C.J. Barrelet, D.C. Bell and C.M. Lieber, Controlled growth and structures of molecular-scale silicon nanowires, *Nano Lett.*, 2004, **4**, 433.

102. F.M. Davidson, A.D. Schricker, R.J. Wiacek and B.A. Korgel, Supercritical fluid–liquid–solid synthesis of gallium arsenide nanowires seeded by alkanethiol-stabilized gold nanocrystals, *Adv. Mater.*, 2004, **16**, 646.

103. D.D.D. Ma, C.S. Lee, F.C.K. Au, S.Y. Tong and S.T. Lee, Small-diameter silicon nanowire surfaces, *Science*, 2003, **299**, 1874.

104. J.T. Hu, L.S. Li, W.D. Yang, L. Manna, L.W. Wang and A.P. Alivisatos, Linearly polarized emission from colloidal semiconductor quantum rods, *Science,* 2001, **292**, 2060.

105. X.F. Duan and C.M. Lieber, Laser-assisted catalytic growth of single crystal GaN nanowires, *J. Am. Chem. Soc.*, 2000, **122**, 188.

106. X.F. Duan, J.F. Wang and C.M. Lieber, Synthesis and optical properties of gallium arsenide nanowires, *Appl. Phys. Lett.*, 2000, **76**, 1116.

107. X.F. Duan and C.M. Lieber, General synthesis of compound semiconductor nanowires, *Adv. Mater.*, 2000, **12**, 298.

108. Y.F. Zhang, Y.H. Tang, X.F. Duan, Y. Zhang, C.S. Lee, N. Wang, I. Bello and S.T. Lee, Yttrium–barium–copper–oxygen nanorods synthesized by laser ablation, *Chem. Phys. Lett.*, 2000, **323**, 180.

109. C. Li, D.H. Zhang, S. Han, X.L. Liu, T. Tang and C.W. Zhou, Diameter-controlled growth of single-crystalline In_2O_3 nanowires and their electronic properties, *Adv. Mater.*, 2003, **15**, 143.

110. H.Z. Zhang, D.P. Yu, Y. Ding, Z.G. Bai, Q.L. Hang and S.Q. Feng, Dependence of the silicon nanowire diameter on ambient pressure, *Appl. Phys. Lett.*, 1998, **73**, 3396.

111. C.M. Lieber, Nanowire superlattices, *Nano Lett.*, 2002, **2**, 81.

112. J.D. Holmes, K.P. Johnston, R.C. Doty and B.A. Korgel, Control of thickness and orientation of solution-grown silicon nanowires, *Science*, 2000, **287**, 1471.

113. C.M. Lieber, The incredible shrinking circuit – researchers have built nanotransistors and nanowires. Now they just need to find a way to put them all together, *Sci. Am.*, 2001, **285**, 58.

114. C.J. Barrelet, Y. Wu, D.C. Bell and C.M. Lieber, Synthesis of CdS and ZnS nanowires using single-source molecular precursors, *J. Am. Chem. Soc.*, 2003, **125**, 11498.

115. P.A. Smith, C.D. Nordquist, T.N. Jackson, T.S. Mayer, B.R. Martin, J. Mbindyo and T.E. Mallouk, Electric-field assisted assembly and alignment of metallic nanowires, *Appl. Phys. Lett.*, 2000, **77**, 1399.

116. B. Messer, J.H. Song and P.D. Yang, Microchannel networks for nanowire patterning, *J. Am. Chem. Soc.*, 2000, **122**, 10232.

117. Y. Huang, X.F. Duan, Q.Q. Wei and C.M. Lieber, Directed assembly of one-dimensional nanostructures into functional networks, *Science*, 2001, **291**, 630.

118. F. Kim, S. Kwan, J. Akana and P. Yang, Langmuir–Blodgett nanorod assembly, *J. Am. Chem. Soc.*, 2001, **123**, 4360.

119. P. Yang, Wires on water, *Nature*, 2003, **425**, 243.

120. A. Tao, F. Kim, C. Hess, J. Goldberger, R.R. He, Y.G. Sun, Y.N. Xia and P.D. Yang, Langmuir–Blodgett silver nanowire monolayers for molecular sensing using surface-enhanced Raman spectroscopy, *Nano Lett.*, 2003, **3**, 1229.

121. D. Whang, S. Jin, Y. Wu. and C.M. Lieber, Large-scale hierarchical organization of nanowire arrays for integrated nanosystems, *Nano Lett.*, 2003, **3**, 1255.

122. X.F. Duan, Y. Huang, Y. Cui, J.F. Wang and C.M. Lieber, Indium phosphide nanowires as building blocks for nanoscale electronic and optoelectronic devices, *Nature*, 2001, **409**, 66.

123. D.H. Cobden, Molecular electronics – nanowires begin to shine, *Nature*, 2001, **409**, 32.

124. Y. Huang, X. Duan, Q. Wei and C.M. Lieber, Directed assembly of one-dimensional nanostructures into functional networks, *Science*, 2001, **291**, 630.

125. F. Qian, Y. Li, S. Gradecak, D. Wang, C.J. Barrelet and C.M. Lieber, Gallium nitride-based nanowire radial heterostructures for nanophotonics, *Nano Lett.*, 2004, ASAP.

126. Y. Cui and C.M. Lieber, Functional nanoscale electronic devices assembled using silicon nanowire building blocks, *Science*, 2001, **291**, 851.

127. R.K. Iler, *The Chemistry of Silica: Solubility, Polymerization, Colloid and Surface Properties and Biochemistry of Silica*, Wiley, NY, 1979.

128. Y. Cui, Q.Q. Wei, H.K. Park and C.M. Lieber, Nanowire nanosensors for highly sensitive and selective detection of biological and chemical species, *Science*, 2001, **293**, 1289.

129. Y. Zhang, A. Kolmakov, S. Chretien, H. Metiu and M. Moskovits, Control of catalytic reactions at the surface of a metal oxide nanowire by manipulating electron density inside it, *Nano Lett.*, 2004, **4**, 403.

130. M.T. Bjork, B.J. Ohlsson, T. Sass, A.I. Persson, C. Thelander, M.H. Magnusson, K. Deppert, L.R. Wallenberg and L. Samuelson, One-dimensional steeplechase for electrons realized, *Nano Lett.*, 2002, **2**, 87.

131. Y.Y. Wu, R. Fan and P.D. Yang, Block-by-block growth of single-crystalline Si/SiGe superlattice nanowires, *Nano Lett.*, 2002, **2**, 83.

132. M.S. Gudiksen, L.J. Lauhon, J. Wang, D.C. Smith and C.M. Lieber, Growth of nanowire superlattice structures for nanoscale photonics and electronics, *Nature*, 2002, **415**, 617.

133. L.J. Lauhon, M.S. Gudiksen, D. Wang and C.M. Lieber, Epitaxial core–shell and core–multishell nanowire heterostructures, *Nature*, 2002, **420**, 57.

134. D. Wang, F. Qian, C. Yang, Z. Zhong and C.M. Lieber Rational growth of branched and hyperbranched nanowire structures, *Nano Lett.*, 2004, **4**, 871.

135. H.T. Ng, J. Han, T. Yamada, P. Nguyen, Y.P. Chen and M. Meyyappan, Single crystal nanowire vertical surround-gate field-effect transistor, *Nano Lett.*, 2004, **4**, 1247.

136. Y. Wu, J. Xiang, C. Yang, W. Lu and C.M. Lieber, Single-crystal metallic nanowires and metal/semiconductor nanowire heterostructures, *Nature*, 2004, **430**, 61.

137. M.H. Huang, Y. Wu, H. Feick, N. Tran, E. Weber and P. Yang, Catalytic growth of zinc oxide nanowires by vapor transport, *Adv. Mater.*, 2001, **13**, 113.

138. P. Yang, H. Yan, S. Mao, R. Russo, J. Johnson, R. Saykally, N. Morris, J. Pham, R. He and H.-J. Choi, Controlled growth of ZnO nanowires and their optical properties, *Adv. Funct. Mater.*, 2002, **12**, 323.

139. J.C. Johnson, H. Yan, R.D. Schaller, L.H. Haber, R.J. Saykally and P. Yang, Single nanowire lasers, *J. Phys. Chem. B*, 2001, **105**, 11387.

140. J.C. Johnson, H.-J. Choi, K.P. Knutsen, R.D. Schaller, P. Yang and R.J. Saykally, Single gallium nitride nanowire lasers, *Nature Mater.*, 2002, **1**, 106.

141. H. Yan, R. He, J. Johnson, M. Law, R.J. Saykally and P. Yang, Dendritic nanowire ultraviolet laser array, *J. Am. Chem. Soc.*, 2003, **125**, 4728.

142. X. Duan, Y. Huang, R. Agarwal and C.M. Lieber, Single-nanowire electrically driven lasers, *Nature*, 2003, **421**, 241.

143. J.C. Johnson, H. Yan, R.D. Schaller, P.B. Petersen, P. Yang and R.J. Saykally, Near-field imaging of nonlinear optical mixing in single zinc oxide nano-wires, *Nano Lett.*, 2002, **2**, 279.

144. H. Kind, H. Yan, B. Messer, M. Law and P. Yang, Nanowire ultraviolet photodetectors and optical switches, *Adv. Mater.*, 2002, **14**, 158.

145. C.N.R. Rao and J. Gopalakrishnan, Synthesis of complex metal-oxides by novel routes, *Acc. Chem. Res.*, 1987, **20**, 228.

146. S. Yamamoto and S. Oda, Atomic layer-by-layer MOCVD of complex metal oxides and in situ process monitoring, *Chem. Vap. Deposition*, 2001, **7**, 7.

147. M.A. Keane, Ceramics for catalysis, *J. Mater. Sci.*, 2003, **38**, 4661.

148. S. Han, C. Li, Z. Liu, B. Lei, D. Zhang, W. Jin, X. Liu, T. Tang and C. Zhou, Transition metal oxide core–shell nanowires: generic synthesis and transport studies, *Nano Lett.*, 2004, **4**, 1241.

149. J. Goldberger, R. He, Y. Zhang, S. Lee, H. Yan, H.-J. Choi and P. Yang, Single-crystal gallium nitride nanotubes, *Nature*, 2003, **422**, 599.

150. R. Fan, Y. Wu, D. Li, M. Yue, A. Majumdar and P. Yang, Fabrication of silica nanotube arrays from vertical silicon nanowire templates, *J. Am. Chem. Soc.*, 2003, **125**, 5254.

151. Z.A. Peng and X.G. Peng, Mechanisms of the shape evolution of CdSe nanocrystals, *J. Am. Chem. Soc.*, 2001, **123**, 1389.

152. S.M. Lee, S.N. Cho and J. Cheon. Anisotropic shape control of colloidal inorganic nanocrystals, *Adv. Mater.*, 2003, **15**, 441.

153. S. Kan, T. Mokari, E. Rothenberg and U. Banin, Synthesis and size-dependent properties of zinc-blende semiconductor quantum rods, *Nat. Mater.*, 2003, **2**, 155.

154. Y.G. Sun and Y.N. Xia, Large-scale synthesis of uniform silver nanowires through a soft, self-seeding, polyol process, *Adv. Mater.*, 2002, **14**, 833.

155. Y.G. Sun, Y.D. Yin, B.T. Mayers, T. Herricks and Y.N. Xia, Uniform silver nanowires synthesis by reducing $AgNO_3$ with ethylene glycol in the presence of seeds and poly(vinyl pyrrolidone), *Chem. Mater.*, 2002, **14**, 4736.

156. Y.G. Sun, B. Mayers, T. Herricks and Y.N. Xia, Polyol synthesis of uniform silver nanowires: a plausible growth mechanism and the supporting evidence, *Nano Lett.*, 2003, **3**, 955.

157. Y.G. Sun, B.T. Mayers and Y.N. Xia, Template-engaged replacement reaction: a one-step approach to the large-scale synthesis of metal nanostructures with hollow interiors, *Nano Lett.*, 2002, **2**, 481.

158. Y.G. Sun and Y.N. Xia, Mechanistic study on the replacement reaction between silver nanostructures and chloroauric acid in aqueous medium, *J. Am. Chem. Soc.*, 2004, **126**, 3892.

159. Y.G. Sun and Y.N. Xia, Multiple-walled nanotubes made of metals, *Adv. Mater.*, 2004, **16**, 264.

160. B. Mayers, B. Gates, Y.D. Yin and Y.N. Xia, Large-scale synthesis of monodisperse nanorods of Se/Te alloys through a homogeneous nucleation and solution growth process, *Adv. Mater.*, 2001, **13**, 1380.

161. B. Gates, B. Mayers, B. Cattle and Y.N. Xia, Synthesis and characterization of uniform nanowires of trigonal selenium, *Adv. Funct. Mater.*, 2002, **12**, 219.

162. B. Mayers and Y.N. Xia, Formation of tellurium nanotubes through concentration depletion at the surfaces of seeds, *Adv. Mater.*, 2002, **14**, 279.

163. B. Gates, B. Mayers, Y.Y. Wu, Y.G. Sun, B. Cattle, P.D. Yang and Y.N. Xia, Synthesis and characterization of crystalline Ag_2Se nanowires through a template-engaged reaction at room temperature, *Adv. Funct. Mater.*, 2002, **12**, 679.

164. D. Li, Y. Wang and Y. Xia, Electrospinning nanofibers as uniaxially aligned arrays and layer-by-layer stacked films, *Adv. Mater.*, 2004, **16**, 361.

165. D. Li and Y. Xia, Direct fabrication of composite and ceramic hollow nanofibers by electrospinning, *Nano Lett.*, 2004, **4**, 933.

166. P.J. Harris, *Carbon Nanotubes and Related Structures*, Cambridge University Press, Cambridge, UK, 1999.

167. R.H. Baughman, A.A. Zakhidov and W.A. de Heer, Carbon nanotubes – the route toward applications, *Science*, 2002, **297**, 787.

168. H.J. Dai, Carbon nanotubes: synthesis, integration, and properties, *Acc. Chem. Res.*, 2002, **35**, 1035.

169. P.J.F. Harris, *Carbon Nanotubes and Related Structures*, Cambridge University Press, Cambridge, UK, 2002.

170. J. Bernholc, D. Brenner, M.B. Nardelli, V. Meunier and C. Roland, Mechanical and electrical properties of nanotubes, *Ann. Rev. Mater. Res.*, 2002, **32**, 347.

171. S. Iijima, Helical microtubules of graphitic carbon, *Nature*, 1991, **354**, 56.

172. M. Dresselhaus, G. Dresselhaus and P. Eklund, *Science of Fullerenes and Carbon Nanotubes: Their Properties and Applications,* Academic Press, London, 1996.

173. P.J. Harris, *Carbon Nanotubes and Related Structures*, Cambridge University Press, Cambridge, UK, 1999.

174. S. Iijima and T. Ichihashi, Single-shell carbon nanotubes of 1-nm diameter, *Nature*, 1993, **363**, 603.

175. L. Forro and C. Schonenberger, Physical properties of multi-wall nanotubes, *Top. Appl. Phys.*, 2001, **80**, 329.

176. T.W. Odom, J.L. Huang, P. Kim and C.M. Lieber, Atomic structure and electronic properties of single-walled carbon nanotubes, *Nature*, 1998, **391**, 62.

177. R.B. Little, Mechanistic aspects of carbon nanotube nucleation and growth, *J. Clust. Sci.*, 2003, **14**, 135.

178. C.N.R. Rao and A. Govindaraj, Carbon nanotubes from organometallic precursors, *Acc. Chem. Res.*, 2002, **35**, 998.

179. S.J. Tans, M.H. Devoret, H.J. Dai, A. Thess, R.E. Smalley, L.J. Geerligs and C. Dekker, Individual single-wall carbon nanotubes as quantum wires, *Nature*, 1997, **386**, 474.

180. N. Hamada, S. Sawada and A. Oshiyama, New one-dimensional conductors – graphitic microtubules, *Phys. Rev. Lett.*, 1992, **68**, 1579.

181. P.J. Harris, *Carbon Nanotubes and Related Structures*, Cambridge University Press, Cambridge, UK, 1999.

182. J.W. Mintmire, B.I. Dunlap and C.T. White, Are fullerene tubules metallic? *Phys. Rev. Lett.*, 1992, **68**, 631.

183. A. Bachtold, M.S. Fuhrer, S. Plyasunov, M. Forero, E.H. Anderson, A. Zettl and P.L. McEuen, Scanned probe microscopy of electronic transport in carbon nanotubes, *Phys. Rev. Lett.*, 2000, **84**, 6082.

184. M. Bockrath, D.H. Cobden, P.L. McEuen, N.G. Chopra, A. Zettl, A. Thess and R.E. Smalley, Single-electron transport in ropes of carbon nanotubes, *Science*, 1997, **275**, 1922.

185. M. Bockrath, W.J. Liang, D. Bozovic, J.H. Hafner, C.M. Lieber, M. Tinkham and H.K. Park, Resonant electron scattering by defects in single-walled carbon nanotubes, *Science*, 2001, **291**, 283.

186. Z. Yao, H.W.C. Postma, L. Balents and C. Dekker, Carbon nanotube intramolecular junctions, *Nature*, 1999, **402**, 273.

187. S.J. Tans, M.H. Devoret, H.J. Dai, A. Thess, R.E. Smalley, L.J. Geerligs and C. Dekker, Individual single-wall carbon nanotubes as quantum wires, *Nature*, 1997, **386**, 474.

188. S.J. Tans, A.R.M. Verschueren and C. Dekker, Room-temperature transistor based on a single carbon nanotube, *Nature*, 1998, **393**, 49.

189. M.S. Fuhrer, J. Nygard, L. Shih, M. Forero, Y.G. Yoon, M.S.C. Mazzoni, H.J. Choi, J. Ihm, S.G. Louie, A. Zettl and P.L. McEuen, Crossed nanotube junctions, *Science*, 2000, **288**, 494.

190. A. Bachtold, P. Hadley, T. Nakanishi and C. Dekker, Logic circuits with carbon nanotube transistors, *Science*, 2001, **294**, 1317–1320.

191. H.W.C. Postma, T. Teepen, Z. Yao, M. Grifoni and C. Dekker, Carbon nanotube single-electron transistors at room temperature, *Science*, 2001, **293**, 76.

192. S.J. Tans and C. Dekker, Molecular transistors – potential modulations along carbon nanotubes, *Nature*, 2000, **404**, 834.

193. P.C. Collins, M.S. Arnold and P. Avouris, Engineering carbon nanotubes and nanotube circuits using electrical breakdown, *Science*, 2001, **292**, 706.

194. R.B. Little, Mechanistic aspects of carbon nanotube nucleation and growth, *J. Clust. Sci.*, 2003, **14**, 135.

195. Special Issue: Advances in Carbon Nanotubes, *MRS Bull.* April 2004.

196. D. Chattopadhyay, L. Galeska and F. Papadimitrakopoulos, A route for bulk separation of semiconducting from metallic single-wall carbon nanotubes, *J. Am. Chem. Soc.*, 2003, **125**, 3370.

197. R. Krupke, F. Hennrich, H. von Lohneysen and M.M. Kappes, Separation of metallic from semiconducting single-walled carbon nanotubes, *Science*, 2003, **301**, 344.

198. A. Javey, J. Guo, D.B. Farmer, Q. Wang, E. Yenilmez, R.G. Gordon, M. Lundstrom and H. Dai, Self-aligned ballistic molecular transistors and electrically parallel nanotube arrays, *Nano Lett.*, 2004, **4**, 1319.

199. Z.L. Wang, R.P. Gao, Z.W. Pan and Z.R. Dai, Nano-scale mechanics of nanotubes, nanowires, and nanobelts, *Adv. Eng. Mater.*, 2001, **3**, 657.

200. A. Thess, R. Lee, P. Nikolaev, H.J. Dai, P. Petit, J. Robert, C.H. Xu, Y.H. Lee, S.G. Kim, A.G. Rinzler, D.T. Colbert, G.E. Scuseria, D. Tomanek, J.E. Fischer and R.E. Smalley, Crystalline ropes of metallic carbon nanotubes, *Science*, 1996, **273**, 483.

201. S.S. Wong, E. Joselevich, A.T. Woolley, C.L. Cheung and C.M. Lieber, Covalently functionalized nanotubes as nanometre-sized probes in chemistry and biology, *Nature*, 1998, **394**, 52.

202. A.M. Fennimore, T.D. Yuzvinsky, W.-Q. Han, M.S. Fuhrer, J. Cumings and A. Zettl, Rotational actuators based on carbon nanotubes, *Nature*, 2003, **424**, 408.

203. J. Hilding, E.A. Grulke, Z.G. Zhang and F. Lockwood, Dispersion of carbon nanotubes in liquids, *J. Disp. Sci. Tech.*, 2003, **24**, 1.

204. Y.P. Sun, K.F. Fu, Y. Lin and W.J. Huang, Functionalized carbon nanotubes: properties and applications, *Acc. Chem. Res.*, 2002, **35**, 1096.

205. S. Niyogi, M.A. Hamon, H. Hu, B. Zhao, P. Bhowmik, R. Sen, M.E. Itkis and R.C. Haddon, Chemistry of single-walled carbon nanotubes, *Acc. Chem. Res.*, 2002, **35**, 1105.

206. K.F. Fu and Y.P. Sun, Dispersion and solubilization of carbon nanotubes, *J. Nanosci. Nanotech.*, 2003, **3**, 351.

207. D. Tasis, N. Tagmatarchis, V. Georgakilas and M. Prato, Soluble carbon nanotubes, *Chem. Eur. J.*, 2003, **9**, 4001.

208. S.B. Sinnott, Chemical functionalization of carbon nanotubes, *J. Nanosci. Nanotech.*, 2002, **2**, 113.

209. A. Star, Y. Liu, K. Grant, L. Ridvan, J.F. Stoddart, D.W. Steuerman, M.R. Diehl, A. Boukai and J.R. Heath, Noncovalent side-wall functionalization of single-walled carbon nanotubes, *Macromolecules*, 2003, **36**, 553.

210. A. Star, D.W. Steuerman, J.R. Heath and J.F. Stoddart, Starched carbon nanotubes, *Angew. Chem. Int. Ed.*, 2002, **41**, 2508.

211. S.S. Wong, E. Joselevich, A.T. Woolley, C.L. Cheung and C.M. Lieber, Covalently functionalized nanotubes as nanometre-sized probes in chemistry and biology, *Nature*, 1998, **394**, 52.

212. A. Noy, D.V. Vezenov and C.M. Lieber, Chemical force microscopy, *Ann. Rev. Mater. Sci.*, 1997, **27**, 381.

213. F.S. Sheu and J.S. Ye, Advances in electrochemical sensors using multi-walled carbon nanotubes, *Mater. Tech.*, 2004, **19**, 11.

214. Q. Zhao, Z.H. Gan and Q.K. Zhuang, Electrochemical sensors based on carbon nanotubes, *Electroanalysis*, 2002, **14**, 1609.

215. A. Star, T.R. Han, J.C.P. Gabriel, K. Bradley and G. Gruner, Interaction of aromatic compounds with carbon nanotubes: correlation to the Hammett parameter of the substituent and measured carbon nanotube FET response, *Nano Lett.*, 2003, **3**, 1421.

216. A. Huczko, Synthesis of aligned carbon nanotubes, *Appl. Phys. A- Mater. Sci. Proc.*, 2002, **74**, 617.

217. P. Jiang, J.F. Bertone, K.S. Hwang and V.L. Colvin, Single-crystal colloidal multilayers of controlled thickness, *Chem. Mater.*, 1999, **11**, 2132.

218. H. Shimoda, S.J. Oh, H.Z. Geng, R.J. Walker, X.B. Zhang, L.E. McNeil and O. Zhou, Self-assembly of carbon nanotubes, *Adv. Mater.*, 2002, **14**, 899.

219. S.G. Rao, L. Huang, W. Setyawan and S.H. Hong, Large-scale assembly of carbon nanotubes, *Nature*, 2003, **425**, 36.

220. J. Kong, H.T. Soh, A.M. Cassell, C.F. Quate and H.J. Dai, Synthesis of individual single-walled carbon nanotubes on patterned silicon wafers, *Nature*, 1998, **395**, 878.

221. W.I. Milne, K.B.K. Teo, G.A.J. Amaratunga, P. Legagneux, L. Gangloff, J.-P. Schnell, V. Semet, V. Thien Binh and O. Groening, Carbon nanotubes as field emission sources, *J. Mater. Chem.*, 2004, **14**, 933.

222. S.S. Fan, M.G. Chapline, N.R. Franklin, T.W. Tombler, A.M. Cassell and H.J. Dai, Self-oriented regular arrays of carbon nanotubes and their field emission properties, *Science*, 1999, **283**, 512.

223. K. Kempa, B. Kimball, J. Rybczynski, Z.P. Huang, P.F. Wu, D. Steeves, M. Sennett, M. Giersig, D.V.G.L.N. Rao, D.L. Carnahan, D.Z. Wang, J.Y. Lao, W.Z. Li and Z.F. Ren, Photonic crystals based on periodic arrays of aligned carbon nanotubes, *Nano Lett.*, 2003, **3**, 13.

224. J.Y. Chung, K.H. Lee, J.H. Lee and R.S. Ruoff, Toward large-scale integration of carbon nanotubes, *Langmuir*, 2004, **20**, 3011.

225. D. Whang, S. Jin and C.M. Lieber, Nanolithography using hierarchically assembled nanowire masks, *Nano Lett.*, 2003, **3**, 951.

226. D. Whang, S. Jin, Y. Wu and C.M. Lieber, Large-scale hierarchical organization of nanowire arrays for integrated nanosystems, *Nano Lett.*, 2003, **3**, 1255.

227. S. Jin, D. Whang, M.C. McAlpine, R.S. Friedman, Y. Wu and C.M. Lieber, Scalable interconnection and integration of nanowire devices without registration, *Nano Lett.*, 2004, **4**, 915.

228. S.K. St. Angelo, C.C. Waraksa and T.E. Mallouk, Diffusion of gold nanorods on chemically functionalized surfaces, *Adv. Mater.*, 2003, **15**, 400.

229. D.J. Sellmyer, M. Zheng and R. Skomski, Magnetism of Fe, Co and Ni nanowires in self-assembled arrays, *J. Phys.-Condens. Matter*, 2001, **13**, R433.

Nanofood for Thought – Wires, Rods, Tubes, Low Dimensionality

1. Propose a synthetic route to a membrane of monodispersed carbon nanotubes and explain how it may be used to advantage as an electrode in a hydrogen–oxygen fuel cell.

2. Explain how you would go about synthesizing single-size molybdenum disulfide nanotubes and evaluating whether they are a superior lubricant than conventional molybdenum disulfide.

3. Provide a synthetic route to soluble carbon nanotubes. If you could find a way to make pure single-wall carbon nanotubes of a particular diameter how might you expect their electrical properties to change with the arrangement of benzene rings in the graphene tube?

4. Devise a way of performing LbL assembly of single-wall carbon nanotubes.

5. How could you orchestrate semiconductor nanowires to form the world's smallest (a) LED (b) laser (c) diode (d) FET (e) pH sensor (f) biological sensor (g) electrical circuit?

6. Why does nitric acid preferentially open the end of a closed carbon nanotube and how would you make use of this knowledge to make (a) a carbon nanotube pH meter (b) a vanadium oxide nanowire?

7. How many ways can you think of to chemically coerce through self-assembly monodispersed nanorods to organize (a) end-to-end (b) side-to-side, and why would you want to do this?

8. What approach would you use to make a nanorod crossbar and a nanorod photonic crystal? Think of a use for these nanorod architectures.

9. Think of a rational way of synthesizing silica nanotubes and silicon nanorings and imagine where they could find utility as a consequence of their form.

10. How could you make bimetallic nanorods made of Au–Pt where the gold end is selectively functionalized with a fluorescent rhodamine B dye? In what areas could such labeled nanorods find applications?

11. Figure out a synthetic route to a longitudinal heterostructured nGaAs-pGaAs nanowire of around 10 nm diameter connected to gold microelectrodes on a silicon substrate.

12. In the context of nanomaterials self-assembly and chemical forms of lithography, how you would synthesize an oriented single crystal wurtzite ZnO nanowire array parallel aligned and orthogonally configured on a sapphire substrate? Briefly outline how knowledge of the relation between structure and properties of the nanomaterial could point to a particular function and possible utility in a future nanotechnology.

13. What possible ways could be used to synthesize and assemble a nanorod actuator?

14. Could you integrate SAM and DPN methodologies to write nanoscale chemical patterns on the surface of a gold nanorod and what would be the first experiment you would do with this newfound capability?

15. Devise a materials self-assembly synthetic pathway to a proton conducting inorganic channel membrane and imagine how such a construct could prove useful in a hydrogen–oxygen fuel cell.

16. How would you synthesize a square grid of TiO_2 nanowires with nanoscale pitch between the nanowires?

17. How would you make a parallel array of boron nanowires orthogonally oriented on an MgO substrate, and then convert these nanowires to superconducting MgB_2 nanowires?

18. How would you synthesize a thin freestanding membrane of nanoporous luminescent silicon with either randomly arranged or well-ordered pore architectures? Explain the pore formation process and the origin of the light emission, keeping in mind that silicon in its bulk form does not emit light.

19. Why would you expect a single crystal silicon nanowire field effect transistor to outperform one made from either amorphous silicon or the organic semiconductor sexithiophene?

20. Can you imagine a means of making a nanoroller out of nanorods and what would be a use for this dynamic nanoscale machine?

21. Think of a materials self-assembly approach for preparing a low-friction nanobearing comprised of a nanorod inside a nanotube and dream about where it might find utility in the future world of nanomachines.

22. If you could synthesize to order a collection of barcoded helical nanorods with control of composition, dimensions and handedness what would you do with them that was previously impossible in the materials world?

23. How would you grow calcite crystals at predetermined locations on a nanorod, and what would this newfound ability permit you to do never previously deemed possible?

24. Is it possible to conceive of a way of synthesizing axial and toroidal barcoded nanorings? If you could, what opportunities would such novel constructs enable in materials self-assembly?

25. Of all the new forms of carbon recently discovered, which would you expect to have the best performance as the anode in a solid-state lithium battery compared to conventional carbon-based anodes?

26. What would you expect to observe through the eyepiece of an optical microscope on immersing Au–Ag barcoded nanorods in a droplet of aqueous nitric acid supported on a glass slide? Can you think of some creative research projects that you could undertake based on your observations?

27. Why are single crystal nanowires proving to be viable candidates for replacing amorphous silicon and sexithiophene semiconductors in thin film transistor flexible electronics?

28. How could you make a barcoded nanorod with alternating Ti and Se segments of controlled length and explain how you would use this

construct as a platform to study the factors which control the thermally driven solid-state reaction of Ti with Se to form the layered titanium diselenide $TiSe_2$? What interesting properties might you expect for a nanorod made of $TiSe_2$ and how might you make use of this knowledge to make something potentially useful?

29. Describe strategies that would allow you to place VLS active catalytic nanoclusters at specific locations down the length of a semiconductor nanowire to direct site-selective VLS growth of nanowire branches at the points that you designated on the body of the nanowire?

30. How could you use an as-synthesized random mixture of metallic and semiconducting single-wall carbon nanotubes to make transistors and diodes on a flexible substrate? If you could make this work, why would this be technologically significant?

31. Create a synthetic scenario that enables nanorods to roll up a slope either with or without power.

32. What would happen to a Pt nanorod in aqueous hydrogen peroxide?

33. Can you think of a way of making a spring-back nanorod?

34. How and why would you make a barcoded swellable–shrinkable nanorod?

35. Think of a way of making a nanorod windmill and why would that be rather interesting?

36. Do you think nanorods could be coerced to form thermotropic and lyotropic liquid crystals? How might this be accomplished and why would this be an interesting thing to do?

37. Can you think of ways of connecting one or both ends of a single 2-nm diameter nanowire to (a) the end or side of a 20-nm diameter nanorod and (b) to the surface of a 200-nm diameter microsphere? Why would this be an interesting accomplishment for particular compositions? Can you think how to do this for a collection of nanowires on the shapes in (a) and (b) and how this construct suggests adventurous new research ideas?

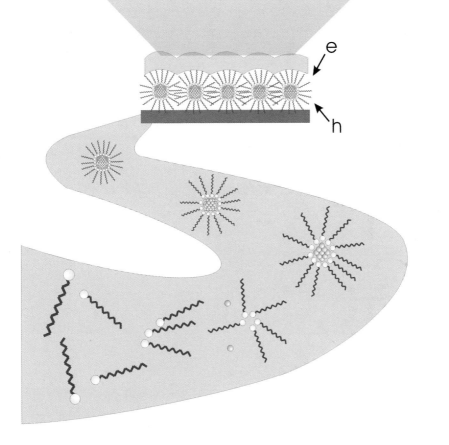

Nanoclusters

e

h

Nanocluster Self-Assembly

"A small rock holds back a great wave."
Homer (800–700 BC)

6.1 Building-Block Assembly

The ultimate state of finely dividing bulk matter to nanoscale dimensions is the quantum dot, a zero-dimensional construct in which charge carriers are confined in all three spatial dimensions. Confinement of electrons and holes transforms the size-invariant continuum of electronic states of bulk matter into size dependent discrete electronic states in a quantum dot.[1,2] This is the size regime, between the molecule and infinity, where the properties of materials can be tuned through their physical dimensions rather than by the traditional way of varying structure and composition.[3] From luminescence to catalytic activity, almost all properties seem to be subject to some form of size dependence.[4–6]

Many years of intensive research have been aimed at making monodispersed nanoclusters in order to study the size evolution of their chemical and physical properties and to utilize this information for a range of applications.[7] The pre-eminent synthetic method that has emerged for accomplishing this goal, for both semiconductors and metals, involves a ligand capping technique referred to as arrested nucleation and growth. In this process, small nanoclusters capped by organic monolayers are rapidly nucleated, and then slowly accrete precursors in solution to increase in size. Capped metal and semiconductor nanoclusters currently dominate most of the research and development in nanocluster self-assembly.[8–11]

In this chapter, we will examine different ways of synthesizing nanoscale dimension capped semiconductor and metal clusters that have strict chemical control over their size, shape and surfaces, describe how they can be used as seeds to grow nanorods and nanotetrapods as well as how they can be self-assembled into different architectures and enable such diverse applications as light emitting diodes (LEDs), full color displays, electronic switches, biological labels and recording media.[12]

6.2 When is a Nanocluster a Nanocrystal or Nanoparticle?

A question worth asking at the outset of this chapter is: when is a nanocluster called a nanocrystal or nanoparticle? These terms are often used interchangeably in the literature and some confusion might arise as to the differences between the notation, if any. In this chapter, a nanocluster will refer to a nanometer scale particle with well-defined positions of the constituent atoms. They nucleate from atoms and enter a size range where they behave electronically as molecular nanoclusters. As the number of atoms increases further, they cross over into the nanoscale size domain where quantum size effects (QSEs) dominate their electronic properties; they become quantum dots or boxes. Above this size regime they become the familiar bulk material with conventional behavior that is not scale dependent. A nanocrystal in this text will refer to a single-crystalline nanoparticle that may be faceted in a way to create a morphology or habit that depends on the growth conditions. It is a term that may be used interchangeably with nanocluster, although the latter seems to denote a better understanding of how the atoms are positioned in the crystal. A nanoparticle denotes simply a particle in the nanometer range, and encompasses the previous two definitions of nanoclusters and nanocrystals as well as amorphous nano-objects or those with a crystal structure which is undetermined or ill defined.

6.3 Synthesis of Capped Semiconductor Nanoclusters

The synthetic method of choice is known as arrested nucleation and growth, where capping groups serve to coat and stabilize the nanocluster at a particular size. The capping groups have numerous other attributes; for instance, they endow the nanoclusters with solubility properties enabling their size-selective crystallization, solubility in water, control over surface functionality and assembly into a range of morphologies. The synthetic strategy for making capped semiconductor nanoclusters can be illustrated for CdSe,[13] although a similar approach can be used to make a range of metal chalcogenide, oxide and other semiconductor compositions.[14–19] A schematic of the synthesis and manipulation of such formed nanocrystals is shown in Figure 6.1. The desired synthesis can be achieved using the following type of reaction:

$$n\text{Me}_2\text{Cd} + n^n\text{Bu}_3\text{PSe} + m^n\text{Oct}_3\text{PO} \longrightarrow (^n\text{Oct}_3\text{PO})_m(\text{CdSe})_n + n/2\text{C}_2\text{H}_6 + n^n\text{Bu}_3\text{P}$$

In practice, tributylphosphine selenide in a syringe is rapidly injected into a 300 °C solution of dimethyl cadmium in a trioctylphosphine oxide surfactant–ligand–solvent, known as TOPO. The capped CdSe nanoclusters (nc-CdSe) rapidly nucleate and then evolve in size to a diameter determined by the amount of precursors in the solution. One ends up with $(^n\text{Oct}_3\text{PO})_m(\text{CdSe})_n$, that is TOPO capped CdSe nanoclusters. They can be precipitated, filtered and washed with acetone to remove impurities. Re-dissolving them in toluene and gradually adding a non-solvent like acetone causes size-selected nanoclusters to precipitate from

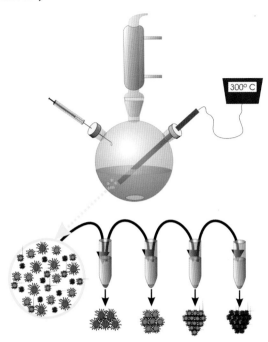

300° C

Figure 6.1 *Arrested nucleation and growth synthetic method for making semiconductor nanoclusters in a high-boiling solvent. Adding a non-solvent causes the larger nanocrystals to precipitate first, allowing size-selective precipitation*

solution with the larger clusters precipitated first. Every time a bit of acetone is added the solution is centrifuged and the precipitate collected, and more acetone is added to the clear supernatant. The result is highly monodisperse nanoclusters collected in each fraction, with decreasing diameter nanoclusters precipitated upon increasing precipitation cycles.

The basic idea behind this synthesis is to rapidly create a large number of viable CdSe seeds, or embryonic nanoclusters, from suitable precursors at supersaturation, and then allow all these thermodynamically viable seeds to grow at the same rate below supersaturation condition. This strategy for monodisperse nanocluster synthesis remains essentially unchanged in the many available synthetic procedures. It enables growth of nanoclusters with a very narrow size distribution, and this process can be monitored with *in situ* absorbance or fluorescence spectroscopy.[20] To facilitate crystallization of the nanoclusters to the thermodynamically stable form, the nucleation and growth is performed in a high-temperature solvent. At the same time, the solvent coordinates to the nanocluster surface as it grows facilitating control of its size, chemical and thermal stability. The solvent-stabilizing molecule of choice is most often the surfactant trioctylphosphineoxide, which ends up capping the CdSe nanocluster surface through coordination of the oxygen of the phosphine oxide head group to the Cd^{2+} ions on the nanocluster surface. The hydrophobic sheath of alkane chains of the surfactant make the nanoclusters soluble in non-polar solvents like toluene. This solubility is a crucial property for achieving successful purification and size selective crystallization of the nanoclusters, as well as the processing of these materials for use in applications.

Fractionated samples have been shown by electron microscopy and laser-induced fluorescence spectroscopy to contain monodisperse capped CdSe nanoclusters with diameters tunable with Angstrom precision in the size range of 1–5 nm. Growth of CdSe in this manner forms faceted nanocrystals of the wurtzite polymorph with well-defined morphologies. Alkane chains of the capping TOPO surfactants are found to interdigitate in a bunching fashion when they are organized as a capped nanocluster crystal, film or pattern. Low angle powder X-ray diffraction (PXRD) provides the intercluster separation, while high angle PXRD can be used to obtain the intracluster structure, and finally mapping the carbon distribution using X-ray photoelectron spectroscopy (XPS) in high resolution TEM images can reveal the bunching interdigitation between the surfactant tails of adjacent nanoclusters.

6.4 Electrons and Holes in Nanocluster Boxes

The impressive control that has been achieved over cluster size and dispersity at the nanoscale enables tuning of nanocluster properties with a degree of precision that is unprecedented in cluster science.[21] The tiny 1–5 nm length scale attainable for capped CdSe nanoclusters is a size range where the continuous electron bands of the bulk semiconductor parent CdSe become discrete.[22] These nanoclusters are called quantum dots or boxes and their dimension resides in the so-called quantum size regime, where the properties of this semiconductor begin to scale in a predictable fashion with size.[23] Almost every material has a unique length scale, comparable to the size of a characteristic electron-hole pair (Bohr exciton), at which such QSEs start to appear.[24]

The size dependence of the electronic band gap of semiconductor nanoclusters has been studied by optical absorption and fluorescence spectroscopy,[25] photoelectron spectroscopy[26] and electronic structure theory.[27] The approximate solution to the Schrödinger wave equation for an electron and hole in a sphere has been used to model the electronic properties of semiconductor nanoclusters. These calculations provide an analytical expression for how the electronic band gap (first excitonic transition) of the semiconductor nanocluster (E_g^{NC}) is modified relative to that of the bulk semiconductor (E_g^B):

$$E_g^{NC} = E_g^B + (h^2\pi^2/2R^2)(1/m_e^* + 1/m_h^*) - 1.8e^2/\varepsilon R$$

where R is the radius of the nanocluster, $m_{e,h}^*$ the reduced masses of the electron and hole and ε the dielectric constant of the bulk form of the semiconductor. The third term in the expression is a small coulomb correction factor. It can be appreciated that there is a blue QSE shift in the energy of the first exciton band or band gap that scales with the reciprocal of the square of the radius R of the nanocluster. Another characteristic of these monodispersed nanoclusters is their intense narrow photoluminescence and electroluminescence, the energy of which scales in a similar way to the first excitonic band gap absorption.[28,29] Due to discrete electronic orbitals, nanoclusters can also display quantized electron charging[30,31] and electronic doping,[32] which can lead to single-electron nanoelectronics.[33–35]

The size dependence of the optical properties observed for capped CdSe nanoclusters is quite dramatic, and increasing the nanocluster size from about 10 to over 100 Å red shifts the optical absorption in a controlled fashion. The colors are concomitantly altered from yellow to red determined solely by the physical size of the nanoclusters and not their composition, although the latter provides additional control over optical properties. Broadly tunable optical emission spectra are observed for capped nanoclusters of InAs, InP, CdS, CdSe, ZnS, ZnSe, PbS, PbSe and core–shell versions thereof. Examples of this tunable photoluminescence for different semiconductor compositions are shown in Figure 6.2.[36]

By encasing a narrow band semiconductor nanocluster by a sheath of a wide band gap semiconductor like ZnO, ZnS or others,[37–39] excited electron hole pairs in the CdSe core are confined by the ZnO shell, which enhances the probability of radiative recombination and enhances the quantum efficiency for fluorescence. The ZnO shell also protects the core from photochemical reactions at a core–solvent interface that can inhibit or completely quench its luminescence.[40] Playing with the energy gap of the shell material affords an additional influence on electronic properties, such as allowing the electron hole pair to be more easily separated which may be useful in photovoltaic applications.[41] Analysis of absorption and emission spectra indicate that the nanoclusters are very nearly

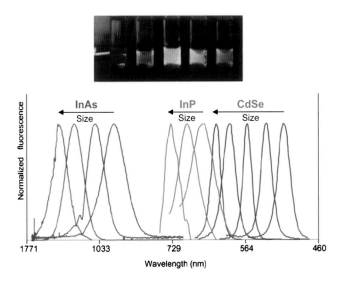

Figure 6.2 *A fluorescence image showing the bright colors achievable with quantum dots, and the size and composition dependence of the optical emission spectra of capped InAs, InP and CdSe nanoclusters*
(Reproduced with permission from Ref. 36)

monodispersed, TEM images show size distributions below one lattice plane, and the calculated and observed single nanocluster absorption spectra are very similar.

6.5 Watching Nanoclusters Grow

An *in situ* methodology has been developed using a microfluidic reactor that enables the synthesis and optical spectroscopy monitoring of the growth of size-controlled capped semiconductor nanoclusters.[42,43] A microliter volume continuous flow chip-based microfluidic reactor (Figure 6.3) constructed out of glass with built-in heaters, thermocouples, syringe pumps and interfaced to a spectrometer is used for a one-pot parallel synthesis of capped CdSe nanoclusters. The source reagents are Se, Me_2Cd and Bu_3P with various capping ligands/solvents. This continuous, on-chip synthesis is found to yield capped nc-CdSe with size distributions similar to those found in conventional serial macroscale syntheses like that described above.

Experimentally, the nc-CdSe size is probed by fluorescence spectroscopy. The size is controlled by independently varying the temperature, flow rate and

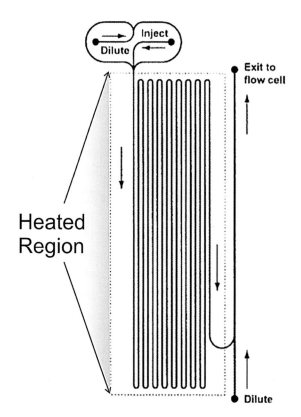

Figure 6.3 *Microfluidic reactor for* in situ *synthesis and optical probing of capped nanoclusters*
(Reproduced with permission from Ref. 42)

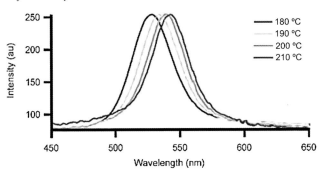

Figure 6.4 *Optical fluorescence probe of* in situ *microfluidic chip synthesis of capped CdSe nanoclusters performed at four different temperatures* (Reproduced with permission from Ref. 42)

concentration of precursor solution flowing through heated microchannels. An illustration of the effect of temperature is seen in the expected red spectral shift of the nanocluster fluorescence with increasing reaction temperature, diagnostic of increasing the nanocluster diameter (Figure 6.4). Results of this kind demonstrate the ability to continuously and easily fine-tune nanocluster physical properties and test wide ranges of synthesis conditions precisely and rapidly inside chip-based microfluidic reactors.[44] The chip-based methodology provides a facile, rational, cost-effective and environmental-friendly (green) way of making capped nanoclusters with a wide range of sizes, compositions and capping groups. In the materials-intensive near future, microchip microreactors of this genre will likely occupy increasingly prevalent roles. Combinatorial materials chemistry techniques based on these may be developed to create large libraries of nanoclusters in order to optimize composition, size and dispersity relations. Their nucleation and growth in outer space, maybe on the International space station under microgravity sedimentation and convection-free conditions, might provide a useful way of making them even more to measure with the added bonus of fewer defects!

6.6 Nanocrystals in Nanobeakers

A common thread running throughout nanocrystal synthetic methods is the development of a strategy for minimizing size dispersity in a sample of nanocrystals. The goal here is to capture all possible benefits of QSEs for fundamental studies and practical applications. Many examples achieving this objective have been documented including arrested or confined nucleation and growth of nanocrystals using surfactants, micelles, dendrimers, zeolites, mesoporous materials, micropatterned SAMs and PDMS microwells. All of these approaches for controlling nanocrystal size distribution in some way or another depend on controlling mass-limited growth in a confined one-dimensional, two-dimensional or three-dimensional nanoreactor. In this context, any method that utilizes nanoreactors made of temperature-sensitive organic molecules or polymers will not be well suited for nanocrystal syntheses from inorganic precursors using high temperature, direct solid-state reaction techniques.

In order to perform this type of solid-state chemistry in the spatial confines of a nanoreactor, it would need to be made of thermally robust materials like silica, alumina, zirconia, carbon, silicon, silicon carbide, niobium or iridium. Examples of high-temperature nanoreactor architectures for synthesis and growth of nanocrystals could be inverted carbon colloidal crystals or cubic mesoporous silicas and carbons. However, nanocrystals produced in this way will end up being trapped within the nanoreactor.

If one requires arrays of isolated and readily accessible nanocrystals then one could make use of nanosphere-imprinted nanobeaker arrays of the kind illustrated in Figure 6.5.[45] Such periodic hemispherical nanowells can be obtained by embossing a silicon wafer with a layer of well-ordered microspheres. By pressing an ordered microsphere mold against a silicon wafer and transiently melting just the surface of the wafer with a pulsed laser, molten silicon flows into the intersphere voids of the mold. On cooling and removing the mold one obtains a hexagonal close packed hemispherical surface imprinted in silicon, a replica of the original microsphere array and suitable for high-temperature nanocrystal syntheses. The size of these silicon nanobeakers is predicated upon the diameter of the microspheres, which can be as little as 100 nm. The nanobeakers can be wet with a solution of precursors and thermally post-treated to grow, under mass-limited conditions, "nanocrystals in nanobeakers". Ultrasound can be used to shake the nanocrystals from their nanobeakers, which can then be collected for subsequent studies. The only downside of this approach is that the amount of nanocrystals obtained will be quite small, and will depend on the two-dimensional area of the embossed wafer.

This simple approach would enable access to a wide compositional range of isolated and addressable nanocrystals, like perovskite-based high T_c superconducting, giant magnetoresistant, fast-ion conducting, ferroelectric and non-linear

Figure 6.5 *Nanocrystals produced in "beakers" templated by a monolayer of silica spheres*
(Reproduced with permission from Ref. 45)

optical materials, which would not be easily accessible in controlled size nano-crystal form by other methods.

6.7 Nanocluster Semiconductor Alloys and Beyond

As mentioned previously, a wide range of semiconductors have successfully been grown as size-selected nanoclusters. Two examples will be used here to highlight the extent of this range.

The first concerns the synthesis of nanoclusters made of homogenous semiconductor alloys.[46–48] When a high-temperature semiconductor nanocluster synthesis is performed using precursors for two different types of semiconductors, we could expect the system simply to phase-separate and form two types of nanoclusters. However, it was found that highly homogenous particles with a composition of $Zn_xCd_{1-x}S$ could be formed where x could be up to 0.5. The synthesis of these particles was accomplished using oleic acid complexes of ZnO and CdO with elemental sulfur as S source, in a weakly coordinating octadecene solvent. It was found by PXRD that increasing the content of Zn lead to a gradually decreased lattice constant, expected for a homogenous alloy. This synthesis also allows one to tune the luminescence of the alloy nanoclusters. In these nanoclusters one has the expected size-dependent luminescence due to tunable confinement of the electron–hole pair (exciton) inside the nanocrystal "box". As well, by increasing the Zn content and keeping the size similar, the luminescence of these nanoclusters shifts to lower energy. This is rationalized by considering that ZnS has a slightly smaller excitonic Bohr radius than CdS (2.2 *vs.* 3.0 nm), leading to weaker confinement for the same size nanocluster.

Another example illustrates the diversity of semiconductors accessible by the high-temperature growth technique. Indium oxide is a semiconductor highly relevant to commercial products. As an alloy with tin oxide (indium–tin oxide, ITO) this material is widely used as optically transparent electrically conductive electrodes for devices such as liquid-crystal displays (LCDs). It is interesting, therefore, that this important semiconductor has been prepared as monodisperse nanoclusters with tunable size and photoluminescence.[49,50] The chemical synthesis begins with an indium complex bearing three bidentate acetylacetonato (acac) ligands. This compound, $In(acac)_3$, is mixed with oleylamine as a ligand-solvent and heated to 250 °C for different times. Smaller or larger nanoclusters could be obtained by changing the In/amine ratios, or adding more In precursor and re-growing formed nanoclusters. An In_2O_3 nanocluster superlattice is shown in Figure 6.6, highlighting the monodispersity available in this synthesis. Clearly, it is now feasible to include $Sn(acac)_4$ into the synthesis to make capped nc-ITO nanoclusters and extend the idea to make important related cousins, nc-ATO (antimony tin oxide) and nc-FTO (fluorine tin oxide), thereby capturing the n-doped and p-doped versions of this class of materials for future use as building blocks in the nanocluster tool box!

These are just a few examples showing the diversity available in nanocluster synthesis. The take-home message is that with experimental effort, monodisperse nanoclusters of any given semiconductor should be available. Careful choice of

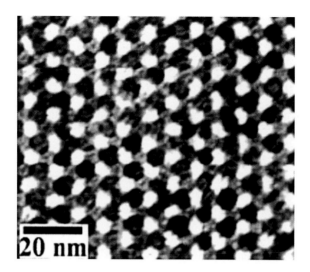

Figure 6.6   *TEM image of a superlattice composed of monodisperse In_2O_3 nanocrystals* (Reproduced with permission from Ref. 50)

solvent, capping ligand and the appropriate precursors are essential to the successful realization of this goal.

6.8 Nanocluster Phase Transformations

One of the defining features of a crystalline solid-state material is its crystal structure, and many analytical techniques are devoted to the study of the arrangement of atoms on a characteristic lattice. Problems can occur, however, when trying to probe the crystal structure of a nanostructure. First of all, a small crystal size can be a source of experimental noise and line broadening (so-called Scherrer broadening can be used to determine the size of nanocrystals). Secondly, and perhaps more importantly, the crystal structure of nanocrystalline materials may be totally different from the bulk parent due to the need to minimize both the bulk and surface-free energies. An excellent example of dimension-defined crystal structures is seen in ultra-small semiconductor nanocrystals, isolated from a synthesis by mass-spectrometry.[51,52] These small nanocrystals shown in Figure 6.7 clearly do not adopt the bulk wurtzite form of CdSe, resembling more closely the carbon fullerenes, yet are remarkably stable.

If we consider nanoclusters larger than these ultra-small ones, they are found to have crystal structures of the bulk form of the material. For instance, bulk CdSe typically has the wurtzite crystal structure, but a metastable rock-salt structure exists at higher pressures. Typically, a transition between these two phases has a prohibitive activation energy and consequently does not occur or occurs only very slowly. Phase transformations of nanoclusters, however, can occur much more readily.[53] Because the energy of a nanocluster is a combination of its bulk and surface-free energies, phase transformations also show a distinct dependence on nanocrystal size.[54,55] A sensitivity to the external environment can also be

(CdSe)$_{13}$ (CdSe)$_{34}$

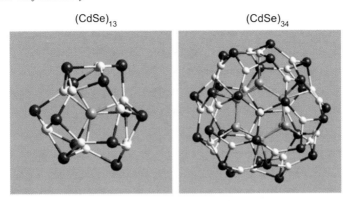

Figure 6.7 *The unexpected structure of (CdSe)$_{13}$ and (CdSe)$_{34}$ nanocrystallites isolated by mass spectrometry*
(Reproduced with permission from Ref. 52)

observed, and it has been shown that simply exposing nanocrystals to a different chemical environment can effect a phase transformation at room temperature.[56]

6.9 Capped Gold Nanoclusters – Nanonugget Rush

The aurophilicity (love of gold) of alkanethiols provides an excellent way of arresting the growth of gold nanoclusters in a synthesis that bears some resemblances to that used for preparing monodisperse semiconductor nanoclusters. Alkanethiols similarly have a high affinity for silver, and this metal forms nanoclusters with very similar properties to gold. For the sake of simplicity, we will describe the properties of gold clusters. Alkanethiolate capped gold nanoclusters, $Au_n(SR)_m$, represent a novel class of materials, which reside structurally and thermodynamically between linear alkanethiolate gold polymers at one extreme $(AuSR)_n$, and self-assembled alkanethiolate monolayers on planar gold surfaces at the other[57] (see Chapter 2). This idea is represented in Figure 6.8.

The synthesis of alkanethiolate capped gold nanoclusters is achieved using the following reactions:

$$HAuCl_4(aq) + Oct_4NBr(Et_2O) \longrightarrow Oct_4NAuCl_4(Et_2O)$$

$$nOct_4NAuCl_4(Et_2O) + mRSH + 3nNaBH_4 \longrightarrow Au_n(SR)_m$$

In practice, one mixes an aqueous solution of $HAuCl_4$ with a solution of Oct_4NBr in diethyl ether containing the RSH capping agent and $NaBH_4$ reducing agent. The hydrophobically protected tetrachloroaurate(III) anion in the form of $nOct_4NAuCl_4$ is transferred to the non-aqueous ether phase where the Au(III) is reduced to Au(0). This reduced gold rapidly nucleates and evolves to a diameter determined by the amount of precursor in solution. One ends up with $Au_n(SR)_m$, that is alkanethiolate capped Au nanoclusters that can be precipitated, filtered, washed with acetone,

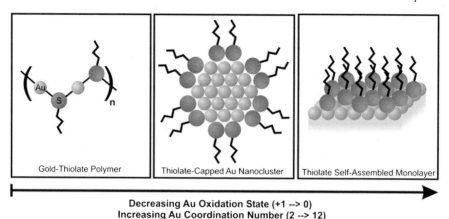

Decreasing Au Oxidation State (+1 --> 0)
Increasing Au Coordination Number (2 --> 12)

Figure 6.8 *Relationship between an alkanethiolate polymer, nanocluster and self-assembled monolayer*

dissolved in toluene and then crystallized in a size-selective fashion. This is realized in the same manner as for semiconductor nanocrystals. Gradually adding a non-solvent such as acetone to a toluene solution of capped gold nanoclusters first causes larger crystals to precipitate, then smaller and smaller crystals as the non-solvent concentration increases. When a non-solvent is added, nanocrystal–nanocrystal contacts start to become more favorable than nanocrystal–solvent interactions. The larger diameter capped gold nanoclusters interact *via* the chains of the alkanethiolate capping ligands more strongly than the smaller ones due to the smaller curvature of their surface and the resulting greater interaction area. As a result they are caused to flocculate, that is aggregate and crystallize first. The process can be repeated to obtain the next lower size nanoclusters and the procedure repeated and refined to obtain essentially monodispersed alkanethiolate capped gold nanoclusters. In this way, one can size select the desired diameter nanocluster of gold for a range of investigations. Note that the same idea applies to the size-selective precipitation of semiconductor nanoclusters by addition of a non-solvent. Improvements in nanocluster synthesis are appearing in droves, and nanoclusters of almost any metal can now be obtained in relatively large quantities.[58–60]

Being at the nanometer length scale, $Au_n(SR)_m$ nanoclusters are in the quantum size regime where the continuous electron bands of bulk Au become discrete. In this size regime, certain properties of the noble metal nanoclusters, such as the conduction electron plasmon resonance, begin to scale in a predictable fashion with size and state of aggregation of the nanoclusters. The plasmon resonance of gold and silver are also sensitive to the surrounding environment, which enables a plasmon resonance spectroscopy of the nanoclusters that is sensitive to the nature of the capping group and dispersing medium.[61–63] This size dependence can cause these capped metal nanoclusters to be conducting, semiconducting and even insulating as the nanocluster size is decreased.[64] QSEs in alkanethiolate capped nanoclusters have been observed in optical spectra and scanning tunneling conductance measurements and utilized in single nanocluster single electron transistor devices.[65]

6.10 Alkanethiolate Capped Nanocluster Diagnostics

One of the most powerful techniques for characterizing these very high-molecular weight alkanethiolate gold and silver nanoclusters is matrix assisted laser desorption ionization time-of-flight mass spectroscopy (MALDI TOF MS).[66,67] The method allows the nanoclusters to be identified by their characteristic molecular weights with and without capping groups. This provides the nuclearity (n) of the nanocluster core, from which it has been discovered that stable "magic" nanocluster sizes predominate containing 225, 314 and 459 atoms in the case of gold nanoclusters.[68] The information that emerges from MALDI TOF MS of the gold core and the capping alkanethiolates, combined with low and high angle PXRD, electron diffraction and HR-TEM, provides incredibly detailed information about the structure of the capped nanoclusters and nanocluster superlattices for a broad range of nuclearities and capping alkanethiolates.

When detailed mass spectroscopy and structural information were combined, it allowed the development of a very appealing core–corona structural model of capped nanoclusters and nanocluster superlattices.[69,70] The model is based upon a dimensionless parameter X:

$$X = 2L/D_{core}$$

where L is the thickness of the corona equated to the fully extended all-trans alkane chain length and D_{core} the equivalent core diameter of the nanocrystal core, D_{nn} the center-to-center distance between nearest neighbors in the solid-state and is a measure of the effective size of the assembly and $D_{nn}-D_{core}$ is the measure of the distance of closest approach of neighboring nanocrystals, as defined in Figure 6.9.

Another very useful diagnostic is UV-visible spectroscopy.[71] Because of their small size, free electrons in the nanoclusters become confined. These confined electrons can be excited to oscillate within the cluster, resulting in intense absorptions in the visible range called plasmon resonances, which are very sensitive

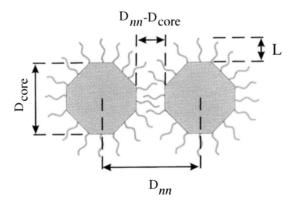

Figure 6.9 *Structural model of interdigitated alkanethiolate capped gold and silver nanoclusters*
(Reproduced with permission from Ref. 70)

to the size of the cluster.[72] A UV-visible spectrum in solution can therefore tell us a lot about the size of nanoclusters, as well as their monodispersity an increase in which decreases the plasmon absorption width, provided the size-dependent optical properties have already been probed using supporting techniques mentioned above.

6.11 Periodic Table of Capped Nanoclusters

The collected information that emerged from these structural studies for large and small diameter gold and silver nanoclusters with capping groups having short and long alkane chains led to the appealing idea that there exists a new kind of "periodic table" of capped nanoclusters that classifies the geometry of packing of core–corona spheres in much the same way that the elements of the periodic table pack in the solid state.[73] This concept is illustrated in Figure 6.10.

6.12 There's Gold in them Thar Hills!

Monodispersed capped metal nanoclusters and nanocluster superlattices are a novel and exciting class of self-assembling materials. While most of the work has focused on gold and silver, other capped metal clusters based on iron, alloys of silver–gold and iron–platinum, with capping groups other than alkanethiolates like amines and carboxylic acids have appeared in the literature. Syntheses can be performed in a one-pot fashion, or conversely the seed nucleation and growth steps can be separated to obtain greater control over monodispersity.[74,75] The ability to stabilize and solublize the metal nanoclusters in organic solvents by capping the

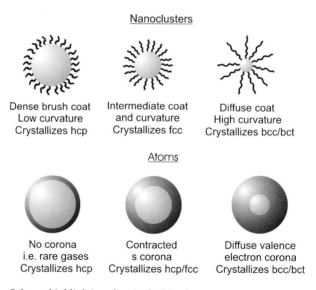

Figure 6.10 *Scheme highlighting the similarities between core–corona nanoclusters and atoms. A diffuse corona imparts similar behavior as diffuse valence electrons*

nanoclusters with surfactant ligands makes it possible to selectively precipitate a particular size nanocluster from solution. The size dispersity of the nanoclusters is so narrowly defined that they are able to spontaneously assemble and crystallize into a nanocluster superlattice in which the alkane chains of the capping ligands experience bunching-interdigitation. The ability to finely tune the size of the nanocluster core plus the length and end-functional groups of the caps allows the distance and electronic coupling between the nanoclusters to be controlled. This provides tremendous opportunities for tailoring the chemical and physical properties of the individual nanoclusters as well as the collective behavior of nanoclusters self-assembled in the form of dimers and trimers, strings, networks, crystals, films, patterns and colloidal photonic crystals.

6.13 Water-Soluble Nanoclusters

The development of nanoclusters of varied composition bearing well defined capping ligands has enabled the field to reach previously unforeseen heights. The capping group is critical to most of the envisioned applications for these materials. They protect the nanoclusters from luminescence or charge-carrier quenching, and impart solubility and processability necessary for any practical use. So far, we have seen many examples of nanoclusters with organic solvent solubility, confered by hydrophobic capping groups. However, a very attractive goal is the synthesis of various types of nanocrystalline materials which are soluble in water. Water is cheap, widely available and in a commercial setting avoids the disposal of organic solvents, which can be quite harmful for the environment. This drive towards aqueous systems, one facet of so-called green chemistry, is fueled in part by an ever-increasing environmental awareness amongst chemists and the general public. Due to these considerations a great deal of effort has been expended to produce water-soluble nanoclusters, with great success.

We saw in Chapter 2 that the greatest feature of self-assembled monolayers (SAMs) is the ability to change surface properties through tuning of the terminal group on the layer. Therefore, it is easy to make a hydrophilic SAM by choosing a hydrophilic terminal group such as a carboxylate, ammonium or oligo-ethylene glycol.[76] Knowing the principles of SAM surface chemistry allows us to simply apply them to capped nanoclusters. In the case of gold, a thiol with a terminal carboxyl group gives an ionized, water-loving carboxylate when in aqueous solution. In certain cases, the desired capping group can be installed directly during synthesis, or afterwards by a ligand-displacement step. Capping group control has allowed a wide range of water-soluble metal, semiconductor and insulator nanoclusters.[77–79] Using this approach, each type of nanocluster requires a chemically specific α,ω-bi-functional ligand in order to achieve solubility in water.

A particularly appealing method for achieving water solubility that appears quite universal in its applicability involves wrapping the hydrophobic nanoclusters with an amphiphilic polymer, poly(maleic anhydride alt-1-tetradecane), in which the alkane chain of the polymer inserts between the alkane chains on the surface of the nanoclusters while the anhydride groups are oriented towards the aqueous environment to impart water solubility.[80] The polymer coating is stabilized by partially cross linking the anhydride groups with bis(6-aminohexyl)amine. The entire polymer coating cross linking scheme is illustrated in Figure 6.11. A wide range of compositions were investigated using this coating technique and were found to maintain key physical properties in an aqueous environment, such as fluorescence for CdSe/ZnS and magnetism for $CoPt_3$ nanoclusters. Further, they remained in a highly dispersed state with little evidence of aggregation and showed remarkably narrow bands in gel electrophoresis, diagnostic of a narrow distribution of sizes and charges. Significantly, the polymer coating method for making hydrophobic nanoclusters water soluble could be applicable to other polymers like poly(ethyleneoxide), useful for aqueous electrolyte solutions, and other nano-object shapes like sheets, rings, wires and tetrapods and may even be suitable for more complex constructs like DNA hybrids.

Another general approach that can be utilized is the coating of nanoclusters with a shell of a water-compatible material such as silica.[81,82] Many types of materials can be coated with silica, provided that surface property requirements are met. For instance, silver and gold colloids can be treated with aminopropyltrimethoxysilane

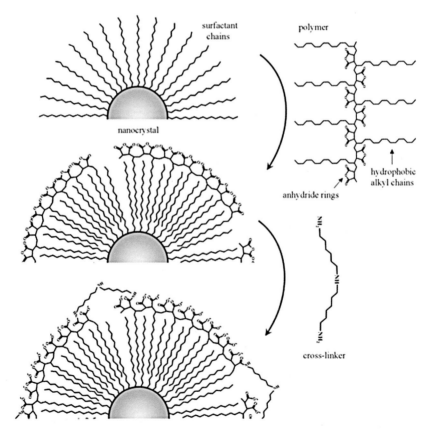

Figure 6.11   *Making nanocrystals water soluble with an amphiphilic polymer*
(Reproduced with permission from Ref. 80)

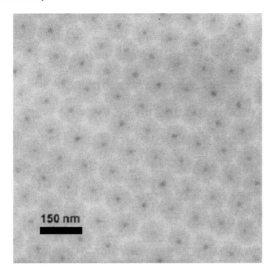

Figure 6.12 *Superlattice of gold nanocrystals coated with silica shells*
(Reproduced with permission from Ref. 81)

(APS) to form a capping shell terminated by methoxysilyl groups. These primed
nanoclusters can then be used as seeds in the growth of a silica shell by alkaline
hydrolysis of tetraethoxysilane. The APS treatment provides a silica-like surface
onto which nanosize silicate clusters can accrete to produce a silica shell. A TEM
image of such coated nanoclusters is shown in Figure 6.12. As well as providing
water solubility, a silica shell can protect nanoclusters from photodegradation by
preventing photochemistry at the solid–liquid interface.[83]

A wide range of capping groups, SAMs, as well as hydrophilic polymers and
inorganic shells can be used to impart water-solubility. Often, the resulting crystals
bear a surface charge, which allows their use in electrostatic layer-by-layer
deposition[84] (see Chapter 3). We are likely to see a large increase in water-soluble
systems in the nanochemistry and bionanochemistry of the near future.

6.14 Capped Nanocluster Architectures and Morphologies

Following a synthesis of capped nanoclusters one normally ends up with a
distribution of nanocluster sizes. As mentioned earlier, this can be narrowed
considerably by size-selective crystallization strategies to yield monodisperse,
capped nanoclusters that differ in size by less than a lattice plane of material. In
this way, one can size select the desired diameter nanocluster for self-assembly
into a given architecture and morphology like a single cluster between two
microelectrodes, linked dimers and trimers,[85] chains, wires[86] and networks, single
crystals, monolayer and multilayer films and shapes, and lithographically defined
patterns.[87,88] The architecture of the assembly will ultimately determine its
properties. More than ever before, form determines function!

6.15 Alkanethiolate Capped Silver Nanocluster Superlattice

The morphology and structure of capped silver nanoclusters have been elegantly probed by transmission electron microscopy and electron diffraction.[70] Faceted polyhedral nanoclusters have been observed to self-assemble in crystals and films consisting of cluster superlattices with long-range order. This effect is quite general, and has been observed for gold and silver[89] clusters, as well as a host of other compositions.[90,91] Alkanethiolate capped gold nanoclusters are mainly decahedral and cubo-octahedral, and consequently pack in bcc and fcc arrangements, respectively, with the former shown in Figure 6.13.

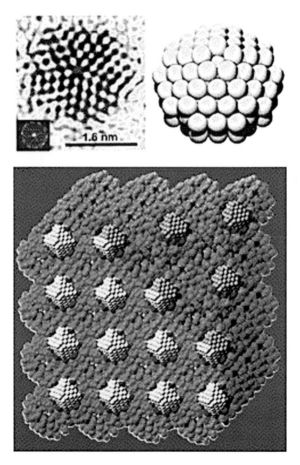

Figure 6.13   *TEM image and model of a single gold nanocluster, top, and computational model of an alkanethiolate capped gold nanocluster superlattice showing interdigitation of the alkane chains, bottom*
(Reproduced with permission from Ref. 70)

Energy filtered Carbon K-edge (core electron orbitals) TEM has provided exquisite detail concerning the geometry of the capping alkanethiolates and their bunching and interdigitation in a capped silver nanocluster superlattice. A carbon map of the alkanethiols clearly shows the distribution of capping molecules between the silver nanoclusters.[69]

Figure 6.14 shows the cubo-octahedral morphology of the nanoclusters, with the {111} and {100} facet structure and the ⟨011⟩ direction designated. A [110] projection of the truncated octahedral nanocluster is also shown with the capping alkanethiolates on the {111} facets. A [110] projection of the nanocluster superlattice model shows the distribution of capping ligands between the nanoclusters that is consistent with the Carbon K-edge TEM image shown in Figure 6.15. A projected unit cell is represented by the rectangle in the C map as well as for the bunching-interdigitation in the model of the nanocluster superlattice, Figure 6.14. The agreement with the nanocluster and capping group packing arrangements shown in this figure is reasonable.

Some very unexpected properties can arise from metal nanocrystals when they are organized on a lattice. Silver, of course, is a metal in its bulk form. If a dilute monolayer of capped silver nanoclusters is spread on an air–water interface, however, the ensemble behaves as an insulator. Using a Langmuir–Blodgett (LB) trough, this monolayer can gradually be compressed until a compact monolayer is formed. At some point in the process, the monolayer is found to undergo an insulator-to-metal transition, with properties such as a metallic reflection appearing

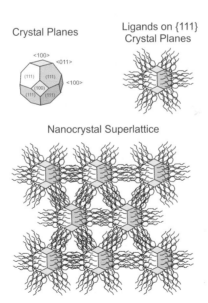

Crystal Planes

Ligands on {111} Crystal Planes

Nanocrystal Superlattice

Figure 6.14 *Faceted cubo-octahedral silver nanoclusters top left, capping alkanethiolates on the nanocluster facets top right, and nanocluster superlattice packing and bunching interdigitation of the alkane chains of the capping ligands bottom*

Figure 6.15   *Energy filtered C K-edge TEM image of the alkanethiolate capped silver nanocluster superlattice showing the distribution of carbon arising from the capping ligands*
(Reproduced with permission from Ref. 69)

at a certain surface pressure. As they are compressed, the nanoclusters become sufficiently close and electrons can tunnel through the inter-cluster gaps leading to metallic behavior.[92]

6.16   Crystals of Nanocrystals

Nanocrystal superlattices or "crystals of nanocrystals" (ordered assemblies of crystalline nanoparticles), have attracted increasing attention since their discovery in 1989.[93] Since then, there have been many reports of ordered two-dimensional and three-dimensional assemblies of nanocrystals of various materials.[94–96] One attractive method for achieving this goal involves the use of controlled diffusion crystallization for growing single crystals with well-defined habits, that is crystal morphologies.[97] This is achieved using three immiscible solvents that are arranged to form separate layers in a test tube as illustrated in Figure 6.16. The bottom layer contains capped CdSe nanoclusters dissolved in toluene, the middle layer is a buffer layer of 2-propanol and the top layer is a non-solvent for the nanoclusters, such as methanol. The buffer layer is selected for its poor solvent properties with regard to the capped nanoclusters. The tri-layer crystallization process involves slow diffusion of the CdSe nanoclusters from the toluene bottom layer and the methanol from the top layer into the buffer layer, where the change in solvent

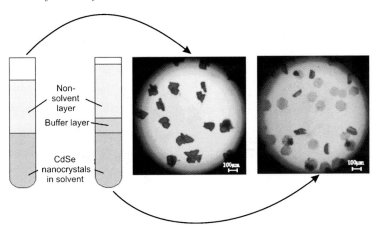

Figure 6.16 *Tri-layer solvent diffusion crystallization of capped nanocluster single crystals. Omitting the buffer layer creates ill-defined crystals* (Reproduced with permission from Ref. 97)

properties causes a slow and controlled nucleation and growth of capped CdSe nanocluster single crystals, that is a crystal built of nanocrystals, with a well-defined hexagonal crystal morphology, as seen in Figure 6.16. This method seems to be generally applicable to a wide range of nanocluster compositions.

6.17 Beyond Crystals of Nanocrystals – Binary Nanocrystal Superlattices

A very exciting development in nanocluster research is the discovery of synthetic routes to heterocrystals, ordered assemblies of more than one component. Close-packed structures of spheres of two different sizes were first observed in natural opal, but mimicking these lattices by synthetic methods proved much more difficult. The first true synthetic nanocrystal heterosuperlattice consisted of a bimodal assembly of gold nanoparticles, which assembled into a variety of crystal structures related to that of binary metal alloys[98–100] and observed in the studies of larger colloids.[101] A recent report shows that nanocrystals with two different sizes and made of different materials can be organized into long-range, well-ordered lattices of AB_{13} or AB_2 crystal structures.[102] Here, the materials used are monodisperse nanocrystals of semiconducting PbSe and magnetic Fe_2O_3, and the resulting so-called supercrystals display remarkable long-range order and structural regularity, as shown in Figure 6.17. The properties of such composites are virtually unknown. The ability to assemble different nanoclusters with size-tunable optical, electronic and magnetic properties into well-defined structures gives us the opportunity to examine new effects due to electronic and magnetic coupling between constituent units. This is another case in nanotechnology where the whole may well be greater than the sum of its parts.

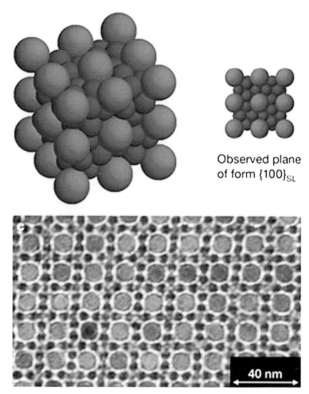

Observed plane
of form {100}$_{SL}$

40 nm

Figure 6.17   *Computational model, and TEM cross-section observed along a {100}*
crystal plane, of binary nanocrystal superlattices
(Reproduced with permission from Ref. 102)

6.18 Capped Magnetic Nanocluster Superlattice – High Density Data Storage Materials

In addition to the widespread research in the electronic and optical properties of nanoclusters, another important avenue in this field is the synthesis and assembly of magnetic nanoparticles,[103–107] for obvious uses in applications such as magnetic memory storage. Other uses include bioconjugable magnetic "dyes" for MRI imaging,[108] targeted killing of cells through magneto-resistive heating, and composites with tailored magnetic properties.

Amongst other routes, organoplatinum and iron pentacarbonyl precursors and surfactant capping groups have been used to synthesize monodispersed iron and iron–platinum nanoclusters.[109] They form with a close-packed metal core, an oxidized surface and a monolayer coating of stabilizing surfactant capping ligands. They can be self-assembled into nanocluster superlattice films and soft lithographic patterns, with a structure of the kind shown in Figure 6.18. They can also be assembled into superlattice single crystals by using the tri-layer diffusion technique described previously.[110] Their uniform size and well ordered packing make these

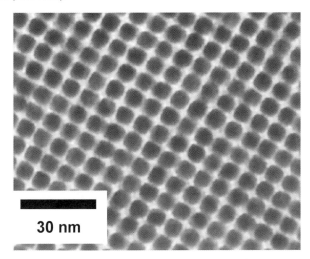

30 nm

Figure 6.18 *TEM image of a superlattice of highly monodisperse magnetic nanocrystals* (Reproduced with permission from Ref. 109)

magnetic nanoclusters useful for very high-density data storage. In the case of the iron–platinum nanoclusters, XPS and TEM together show that capped (oleic acid)$_n$(PtFe)$_m$ nanoclusters have a core–corona structure and are monodisperse with the oleic acid capping ligands binding to oxidized surface Fe sites.

Temperature-dependent ^{57}Fe Mössbauer and magnetization measurements define the nanoclusters to be superparamagnetic single domains and undergo a super-paramagnetic–ferromagnetic transition at 10 K. This is the so-called blocking temperature, below which the magnetic dipole of the cluster is no longer under-going thermally driven dynamic fluctuations and is statically locked in position. Using cobalt nanocrystals it has been demonstrated that ordered arrays of these exhibit spin-dependent conductivity behavior, making them perfect for amperome-trically detecting magnetic fields (termed magnetoresistance).[111]

Making perfect building blocks and organizing them into arrays is only one-half of the challenge. The other is to interface this array with other nanocomponents in order to make use of their properties. This may well be the more challenging of the two tasks.

6.19 Alloying Core–Shell Magnetic Nanoclusters

The thermally driven inter-diffusion of the elemental constituents of core–shell semiconductor nanocrystals to form solid-solution semiconductor nanocrystals can be used to tune their optical absorption and luminescence properties. If this process is harnessed for core–shell metal nanocrystals, a similarly precise command over their magnetic properties may be possible. The materials challenge is to find a reliable synthetic strategy for creating a homogeneous metal sheath of one metal over the core of another at the nanoscale.[112]

A creative redox transmetallation reaction provided a nice solution to this problem. In this approach, 6.3 nm diameter 2-ethylhexylsulfosuccinate stabilized Co nanocrystals in nonane were refluxed for 6 h with $Pt(hfac)_2$ in the presence of the capping ligand dodecyl isonitrile. In this process, surface Co(0) becomes oxidized to Co(II) with concurrent reduction of Pt(II) to Pt(0). As expected, $Pt(hfac)_2$ is converted to $Co(hfac)_2$. In the course of this reaction the cobalt core is found to uniformly diminish in diameter with the concomitant growth of a uniform thickness platinum shell capped by the isonitrile ligand. The original 6.3 nm fcc Co nanocrystals are transformed to a 6.4 nm fcc $Co_{core}Pt_{shell}$ nanocrystals, which can be imaged by high-resolution TEM. The measured lattice fringes, higher electron contrast of the shell, EDX elemental composition of $Co_{0.45}Pt_{0.55}$ and EXAFS Co K-edge and Pt L_{III}-edge analysis of interatomic distances together provide strong support for the proposed core–shell nanocluster model.

When these core–shell nanocrystals were thermally annealed at 600–700 °C for 12 h it was found that the Co and Pt atoms had inter-diffused from their core and shell regions to form solid–solution alloy fct nanocrystals, evidenced by structural characterization using a similar set of analytical measurements. Concomitantly, a change in the magnetic behavior was observed from room temperature super-paramagnetism to ferromagnetism, as seen by the appearance of hysteresis in the fct alloy nanocrystals when there was none in the fcc core–shell nanocrystals.

Nanocrystals made of magnetic materials, as mentioned above, are often found to lose their magnetization in the absence of a magnetic field due to their size being smaller than the Weiss domain, the minimum size required for permanent magne-tization. Overcoming these so-called superparamagnetic limitations in magnetic nanocrystals is important for the future development of ultra-high density magnetic memory, bioanalytical magnetic separations and magnetic sensing.

6.20 Soft Lithography of Capped Nanoclusters

One way of achieving this goal is to begin with a Langmuir–Blodgett (LB) monolayer film of compressed capped nanoclusters.[113–115] The capped clusters in a solvent like chloroform are syringed onto the water surface in an LB trough. Evaporation of the solvent leaves the clusters at the interface of the air and water. Organically modified nanoclusters will very often float on water because of its surface tension, and these floating "rafts" can be compressed into close-packed monolayers. This is shown schematically in Figure 6.19 for the printing of magnetic nanocluster monolayers,[116] left panel. A patterned polydimethylsiloxane (PDMS) stamp is dipped into the LB nanocluster monolayer. The nanocluster monolayer can then be transferred to the surface of a silicon wafer by contacting the wafer with the nanocluster-inked stamp (Figure 6.19, right panel). In this way, a monolayer of capped nanoclusters can be patterned on silicon, and has been further elaborated to printing multilayers.[117] This method can be used to form patterns of capped metal, metal chalcogenide, metal oxide and core–shell metal oxide-metal nanoclusters in a parallel fashion. The transfer of monolayers of a variety of colloids can be affected by the LB technique, which was discussed in some detail in Chapter 5 with regard to nanowires.

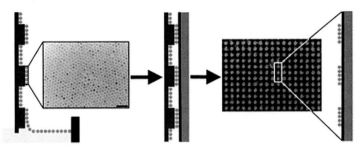

Figure 6.19 *Strategy for soft lithographic patterning of capped nanoclusters* (Reproduced with permission from Ref. 116)

6.21 Organizing Nanoclusters by Evaporation

Evaporation-induced self-assembly (EISA) is a great way of organizing nanoscale building blocks. The strong capillary forces in an evaporating water meniscus drives the nanocomponents into close-packing, and boundary conditions such as nanotrenches can direct and organize units into highly ordered patterns. Patterns made by electron beam lithography can be fine enough to direct very small semiconductor nanoclusters and form a huge variety of periodic and aperiodic surface patterns.[118] This concept is further explored in the synthesis of in-chip photonic crystals (Chapter 7) in an anisotropically etched silicon wafer. In Figure 6.20, we see the strategy for organizing nanoclusters into small trenches. The surface tension of the evaporating liquid attracts the nanoclusters to the trench bottoms where they accumulate and close-pack. A regular array of pits can give a pattern like the one shown at the left of the figure. A diversity of patterns is shown in the bottom of the figure, formed simply by varying the size and shape of lithographically defined trenches.

6.22 Electroluminescent Semiconductor Nanoclusters

The size and composition, wavelength tunable, narrow, intense and stable photoluminescence behavior of capped semiconductor nanoclusters provides interesting opportunities for making electroluminescent devices such as LEDs and lasers using the nanoclusters as the active layer.[119,120] The first example that showed the feasibility of this idea was a device based on an ITO supported hole transporting polyphenylenevinylidene (PPV) film interfaced to a capped CdSe nanocluster film with a Mg top electrode, that is Mg/5 layers 3.5 nm CdSe/PPV/ITO.[121] The current voltage profile obtained for this kind of thin film device is shown in the left-hand side of Figure 6.21 and best fits a mechanism of charge diffusion across the Mg to CdSe interfacial barrier according to $I \sim \exp[eV/\ln kT]$.

An interesting observation concerns the voltage dependent emission of a nanocluster assembly in an electro-luminescent device, of the type shown on the right of the figure. At lower voltages yellow-red light is emitted from the CdSe nanoclusters, whereas at higher voltages green emission instead emerges from the

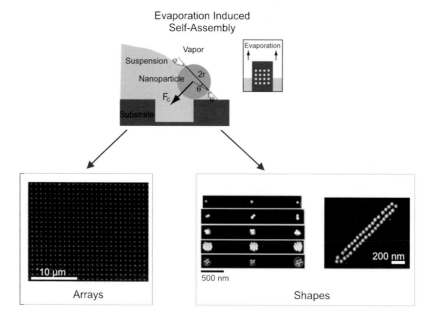

Figure 6.20 *Evaporation-induced self-assembly can be used to make ordered regular arrays and arbitrary patterns of semiconductor nanocrystals*
(Reproduced with permission from Ref. 118)

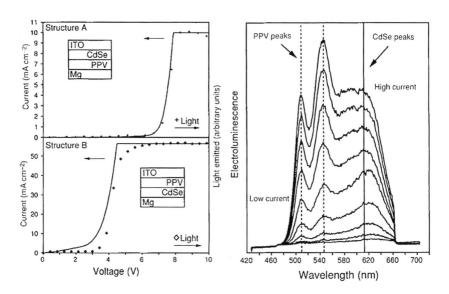

Figure 6.21 *Right, I–V profile for Mg/5 layer 3.5 nm CdSe/PPV/ITO electroluminescent device. Left, voltage-dependent electroluminescence from Mg/5 layer 3.5 nm CdSe/PPV/ITO thin film device*
(Reproduced with permission from Ref. 121)

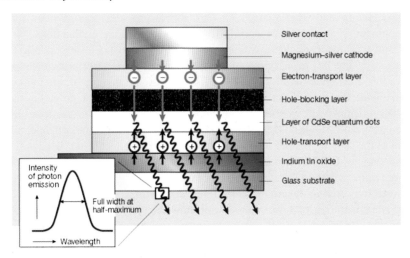

Silver contact

Magnesium–silver cathode

Electron-transport layer

Hole-blocking layer

Layer of CdSe quantum dots

Hole-transport layer

Indium tin oxide

Glass substrate

Intensity of photon emission

Full width at half-maximum

Wavelength

Figure 6.22 *The structure and light-emission properties of a device based on an active monolayer of semiconductor quantum dots*
(Reproduced with permission from Ref. 123)

PPV. Although the mechanism of this phenomenon is not completely understood, it does raise the intriguing idea of building a multicolor pixel from a single electroluminescent device through tunable charge-transfer effects.

To increase the efficiency of this kind of device it is important to optimize charge transport in the capped semiconductor nanocluster active layer to maximize the recombination efficiency of electrons and holes therein. The key to making a very high quantum efficiency light emitting diode, other than correctly selecting the Fermi energies (chemical potential) of the electrodes and the electron affinities and ionization potentials of the active electron and hole charge transporting layers, is to devise a way of making a nanocluster monolayer as the charge transporting emissive region of the device. This may be done by spin coating or dip coating the nanoclusters in the right choice of liquid matrix or possibly by microtransfer printing the nanoclusters as mentioned above. An example of this kind of LED is displayed in Figure 6.22.[122,123] This device is constructed such that both holes and electrons get trapped in the active nanocluster layer, leading to quantum efficiencies for luminescent recombination higher than 50%. The electroluminescence wavelength of quantum-confined nanoclusters can be tuned by their size, or simply by the applied voltage.[124] In addition, these nanoclusters can exhibit tunable electrochromism,[125] a property currently used in smart windows and next generation flat-screen display prototypes.[126] All in all, nanoclusters have a bright and colorful future in optoelectronic devices!

6.23 Full Color Nanocluster-Polymer Composites

It is possible to tune the colors of capped semiconductor nanoclusters through all the colors within the chromaticity scale, a term used to describe the purity of colors in displays, paints and light sources.[127] This ability provides opportunities for the development of composite materials for full color display applications.[128] One interesting approach to achieve this goal is based on nanoclusters stabilized in polymers.[129] The key to success is to find a polymer that prevents nanocluster phase separation and agglomeration, and does not cause quenching of luminescence. To maintain the capped nanocluster in a dispersed state in a polymer matrix, polylaurylmethacrylate is used as the stabilizing medium where the long alkane chain lauryl side groups on the polymer backbone interact with similarly long alkane chains of the capping ligands to effectively solubilize the nanocluster in the polymer.

The choice of II–VI semiconductor nanoclusters and the ability to control their size provides access to a range of nearly pure colors. Bright mixed colors can be obtained by creating layered architectures containing nanoclusters of different sizes. An illustration of achievable colors shown against the chromaticity scale is shown in Figure 6.23. High color efficiency can be achieved since core–shell nanoclusters enhance photon emission due to confinement of e–h pairs of the small band gap core by the large band gap shell. Photoluminescent quantum yields for a range of II–VI nanocluster compositions and sizes show that the quantum yields for emission are only slightly lower (20–25%) in the polymer matrix compared to hexane. This is most likely caused by more efficient radiative relaxation of the capped nanoclusters because of strong hydrophobic interactions of the alkane chains of the capping ligands of the nanocluster with the lauryl groups of the polymer. These kinds of nanocluster–polymer composite materials have a wide range of potential applications. One appealing example is a down conversion LED made by encapsulating a GaN LED in a sheath of capped semiconductor nanoclusters in polylaurylmethacrylate, as illustrated in Figure 6.24. Here, a 425 nm wavelength emitting GaN LED evokes 590 nm light emission from the nanocluster-polymer sheath, thereby achieving the desired effect of light energy down conversion.

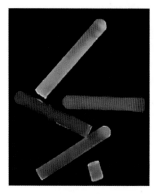 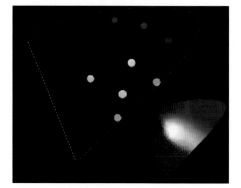

Figure 6.23   *Color fluorescence image of UV excited TOPO capped core–shell (CdSe)ZnS nanocluster – poly(laurylmethacrylate) composites rods, for 1, 1.3, 2.3, 2.8 nm size clusters (left) and end-on image of the rods excited from below and placed on the chromaticity diagram (right)*
(Reproduced with permission from Ref. 129)

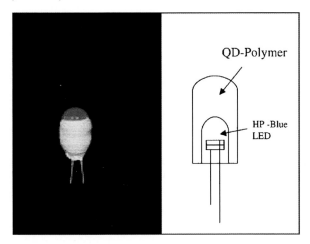

Figure 6.24 *Down conversion quantum dot-polymer LED, optical image and schematic diagram*
(Reproduced with permission from Ref. 129)

The strategy of ligand–matrix interaction stabilization can be carried further. By replacing the trioctylphosphine capping ligands by phosphines bearing terminal hydroxyl groups, the nanocrystals become polar without affecting their optical properties. This modification allows the fluorescent nanoclusters to be incorporated into a sol–gel titania matrix, one of the promising composites which could be used as gain media for nanocluster-based lasers or robust matrices for nanocluster LEDs.[130–133]

6.24 Capped Semiconductor Nanocluster Meets Biomolecule

One might have expected based on the above discussion of capped semiconductor nanoclusters that the first application for these materials would likely have come from optoelectronics. As it turns out, in biology is where these nanoclusters really shine.[36,134–136] The reason for this surprising happening relates to the fact that organic fluorescent dyes, typically used in biology for identifying specific sites on DNA, antibodies or cells and in materials chemistry for probing building block self-assembly, are not very photo-stable and show broad, low quantum efficiency emissions. In sharp distinction, inorganic dye labels based on capped semiconductor nanoclusters produce narrower and brighter fluorescence than conventional organic dyes, and keep on shining for about $1000\times$ longer. The bright and narrow fluorescence means that several colors can be used in parallel to label different cell organelles or macromolecules, giving one a chance to observe biological interactions in real-time. The well-defined surface chemistry of these nanoclusters in turn allows them to be coated with a wide range of active biological agents so they can interact with various biological systems, and be targeted to a certain part of an organism.[137,138]

The first requirement for biological labeling is water-solubility, and a coating providing biocompatibility. In one way to make a CdSe@ZnS core–shell nanocluster water-soluble, a mercaptoacetic acid capping group is affixed to the nanocluster surface.[134] The carboxylic acid group allows the nanocluster to bind selectively to an amine site on a protein or peptide molecule *via* an amide link. Another method involves growing a silica shell over the nanocluster surface, which allows one to take advantage of the very well understood chemistry of the silica surface.[36] A proof-of-concept experiment that showed the feasibility of this kind of brightly emissive, capped semiconductor nanocluster functioning as a new class of biological stain is seen in the selective labeling of mouse cells with large and small, red and green fluorescing nanoclusters having different surface functionalization, as depicted in Figure 6.25.

Another recent report on the use of capped nanoclusters in biology involves phospholipid bilayer-coated CdSe nanoclusters.[139] The preparation of such surface protected nanoclusters is shown in the top of Figure 6.26. Because they are surrounded by a membrane, which is present in all living cells, these highly luminescent tags are extremely biocompatible and can be used to monitor cellular events and organism development *in vivo*. In a striking example, a collection of coated nanoclusters were injected into one of the cells of a *Xenopus* frog embryo, which had recently started mitosis (cell division). The progress of this cell could then be monitored because at every cell division the nanoclusters were evenly divided between the two daughter cells. The embryos were grown into mature tadpoles without any observable toxicity (see Appendix B), and surprisingly the nanoclusters were still highly luminescent. Every single luminescent cell in the

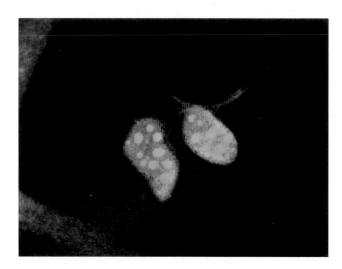

Figure 6.25 *Selective staining of mouse cells with large and small red and green fluorescing capped ZnS(CdSe) core–corona nanoclusters*
(Reproduced with permission from Ref. 36)

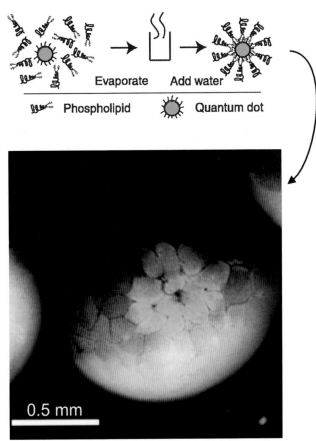

Figure 6.26 *Preparation of biocompatible bilayer coated luminescent quantum dots, top. On the bottom is shown a fluorescent cluster of cells in a growing frog embryo, derived from a single cell injected with biocompatible quantum dots* (Reproduced with permission from Ref. 139)

mature tadpole came from that *one* cell that was labeled early in embryogenesis, allowing an unprece-dented experiment in organism development. A result of one of these experiments is shown in the bottom of Figure 6.26. Different color nanoclusters could be used in conjunction to monitor different parts of the organism, and in theory this could be extended to hundreds of nanoclusters of different colors all being used in parallel. Finally, by observing that green nanoclusters permeate into the nucleus (where the cell stores its genetic material) and that red ones, which are larger (the larger the nanoclusters the more red-shifted the luminescence maximum), cannot penetrate the nucleus, the authors were able to deduce the size of the pores in the nuclear membrane of a living and developing organism. These are just a few of the examples showing how QSE capped nanoclusters are poised to lead to quantum leaps in our understanding of biological systems by allowing us to perform previously inconceivable experiments.

6.25 Nanocluster DNA Sensors – Besting the Best

It was the unique size dependent physical and chemical properties of capped semiconductor and metal nanoclusters that inspired research aimed at the development of nanoscale electronic, photonic and magnetic devices, but it seems that biological applications of these nanoclusters are likely closer to the marketplace. As discussed earlier, this is certainly the case for the use of capped semiconductor nanoclusters as biological labels because of their superior luminescence properties compared to organic dye labels (intensity, color tunability and stability) and ability to selectively target specific biomolecular sites in cells and tissues through target-specific capping ligands. The biological promise of nanoclusters may be extended further, using metal nanoclusters for DNA amplification and detection with polymerase chain reaction (PCR) selectivity and sensitivity.[140]

PCR amplifies DNA through heating and cooling cycles using a primer-mediated enzymatic method, with n cycles giving 2^n copies of the target DNA. While it is true that medical diagnostic systems for detecting and quantifying DNA targets have been revolutionized by PCR, in practice they are quite complicated, costly, lengthy and labor intensive. The challenge is to find a new strategy for detecting 5–10 DNA strands, and at the same time overcoming the aforementioned difficulties with PCR. In this context, a two-component system comprised of capped gold nanoclusters and capped magnetic iron oxide nanoclusters has been creatively orchestrated to detect DNA with PCR-like sensitivity.

The strategy is illustrated in Figure 6.27. In brief, the method begins with the preparation of gold and magnetic iron oxide nanoclusters capped with DNA strands. First, a citrate stabilized gold nanocluster is capped with a mixed thiol-terminated DNA monolayer consisting of a 1:100 ratio of target strand and "bar code" strand. The bar code oligonucleotide is then hybridized with the complementary oligonucleotide strand. Next a magnetic iron oxide nanocluster is coated with an amino-modified silane and coupled to a thiol-terminated oligonucleotide strand complementary to one portion of the target sequence. The next step involves hybridizing the target DNA, the one being detected, with a mixture of the DNA-capped magnetic iron oxide and gold nanocluster probes, followed by magnetic separation from solution phase components. The bar code DNA is then de-hybridized from the gold and iron oxide nanoclusters by warming at 55 °C for 3 min, and the released DNA is detected using a chip-based DNA detection method that relies on the binding of DNA-modified gold nanocluster probes and amplification with a coating of electroless deposited silver.[141] This architecture provides the required signal amplification, by enhanced light scattering, to be able to detect as little as 500 zeptomolar ($zM = 10^{-21}$ M) of target DNA. (Interesting, the example used was the anthrax lethal factor, which is important to detect quickly and with very high sensitivity in bioterrorism and biowarfare situations). This represents the detection of only 10 DNA strands in the whole solution, thereby providing proof-of-concept that the bio-bar code-based DNA strategy described above achieves PCR-like detection sensitivity without the need for enzymatic amplification, in a shorter time, and with less cost. Noteworthy, the bio-barcode method was also demonstrated to have the

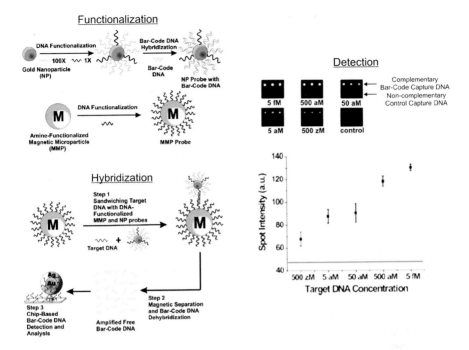

Figure 6.27 *Schematic showing the steps involved in the derivatization of metallic and magnetic nanoparticles and using these for ultra-sensitive DNA detection. The right part of the image shows the results of the experiment, with sensitivity down into the zeptomolar range*
(Reproduced with permission from Ref. 140)

capability of differentiating DNA targets with a single base mismatch at atomolar $(aM = 10^{-18} M)$ sensitivity.

6.26 Semiconductor Nanoclusters Extend and Branch Out

Soluble capped nanocrystals are spreading like wildfire across all the scientific disciplines. With the advent of high yielding and reproducible nanocluster syntheses, these nanomaterials have been used to effect functions ranging from biological labeling to solar cells. It may then seem at first glance that the field of nanocrystal synthesis has reached its peak, and further developments are likely to be evolutionary rather than revolutionary. This is far from being the case, and a good example is illustrated with the synthesis of complex architecture semiconductor nanocrystals.[142–152] Highly monodisperse nanoarchitectures, from elongated nanocrystal rods which display polarized emission,[153,154] to tetrapod CdTe nanocrystals with four tetrahedrally disposed arms[155] can now be routinely accessed.

The synthesis of CdTe tetrapods is achieved through an understanding of crystal stability, and clever manipulation of the temperature and capping agent. CdTe has two common crystal polymorphs, wurtzite (hexagonal) and zinc blende (cubic), whose energy difference is substantial but not exceedingly great. The synthesis temperature was chosen such that the growth of the wurtzite structure is preferred, but nucleation sites only form in the zinc blende structure. These seeds serve as templates for the growth of wurtzite "arms", the wurtzite {0001} facets growing outwards from the four {111} crystal faces of the zinc blende seeds. Use of a long chain alkanephosphonic acid, *n*-octadecylphosphonic acid (ODPA), which selectively binds to the lateral facets of hexagonal CdTe, serves to confine wurtzite CdTe growth along only one spatial dimension. A schematic showing crystal plane orientation is seen in Figure 6.28.

Just as remarkable as the control of crystal nucleation and growth was the observation that both the length and width of the wurtzite arms could be independently tuned by changing the Cd:Te and Cd:ODPA ratios, respectively. Anisotropy of the obtained structure is achieved through fast growth, and since the growth is limited by the availability of Cd a higher Cd:Te ratio ensures faster growth and longer arms. On the other hand, complexation of Cd by ODPA decreases its availability and thus the driving force for addition of Cd to the growing crystal, thereby decreasing the growth rate for a given Cd concentration. This results in less anisotropic rods with a large diameter for a given length.

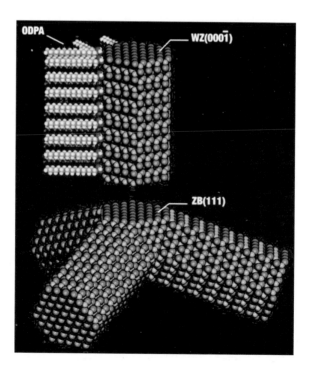

Figure 6.28   *A model of a synthetically produced CdSe tetrapod architecture, showing the lattice plane orientations*
(Reproduced with permission from Ref. 155)

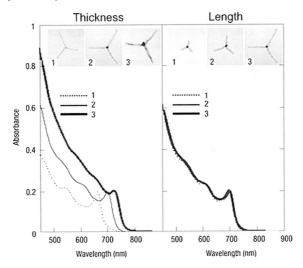

Figure 6.29 *The dependence of the fluorescent emission of semiconductor tetrapods on the thickness and length of the arms*
(Reproduced with permission from Ref. 155)

Being in the nanoscale, these tetrapods correspondingly display QSEs, with the lowest energy optical absorption peak blue-shifting as the arm diameter decreases. Surprisingly enough, the confinement energy leading to QSEs is only dependent on the arm diameter, and tetrapods with an equal arm diameter but different arm length show virtually identical optical properties (Figure 6.29). This effectively demonstrates the nature of quantum confinement, that is, only variations in the dimensions which are in the quantum size regime have a noticeable effect on nanoparticle QSEs. This is only one of the impressive examples of the control over crystal shape achievable through chemical techniques.[156]

6.27 Branched Nanocluster Solar Cells

Nanoclusters figure prominently in candidates for next-generation solar collecting devices.[157] As we have seen, these nanomaterials can be very efficient at producing light through electron–hole recombination. If we want to operate the other way around, that is, if we want to separate the electron from the hole, things can get more difficult. If we try to use a spherical nanocluster for this purpose, charge separation can be inefficient since the exciton is bound in an electronic box surrounded by an energetic barrier only surmounted by quantum tunneling. Things can get easier when using quantum rods, since we are providing a transport pathway for charge carriers,[158] and nanorod-based solar cells have been convincingly demonstrated.[159] The problem in this case is achieving nanorod alignment, since maximum efficiency is achieved when the nanorods are aligned such that they connect the cathode and the anode.

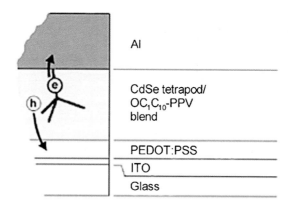

Figure 6.30 *Schematic showing the architecture of a semiconductor nano-tetrapod-based solar cell*
(Reproduced with permission from Ref. 160)

One way to circumvent the alignment problem is by using semiconductor tetrapods described in the previous section. If a tetrapod is deposited onto a substrate, there is only one way it can lay down, that is, three of the arms will form a tripod, and the fourth arm will stick up upwards perpendicular to the substrate. This principle has been used to reproducibly synthesize tetrapod-based solar cells with improved collection efficiency.[160] A schematic illustrating the architecture of such a solar cell is shown in Figure 6.30. Tailor-made nanoparticles provide an attractive means to increase the efficiency of many types of devices, and allow one to use less material to make better technology.

6.28 Tetrapod of Tetrapods – Towards Inorganic Dendrimers

We have seen earlier in this chapter how the synthesis, growth conditions and selection of capping agents and solvents used for making semiconductor nanocrystals can be manipulated to favor the production of either nanorods or nanotetrapods for the II–VI metal–chalogenide compositions. What would happen if at some stage in the creation of these anisotropic nanocrystals one of the elements were changed? If epitaxial growth of the added reagent materials continues on the end of the parent nanostructure, novel heterostructures could be obtained with crystallographically well-defined junctions. The resulting well-defined interfacial energy barriers could, for instance, promote long-range charge separation, and enable the development of novel photovoltaic devices. This concept and its reduction to practice is illustrated in Figure 6.31 for growing nanorod extensions and branches on nanorods and tetrapods for the CdS–CdSe and CdSe–CdTe system.[161] Epitaxial growth is possible at the nanoscale, even for nanocrystal components with large lattice dimension mismatches.

In a typical synthesis of these nanoscale heterostructures, elemental chalcogenide dissolved in a tri-(*n*-alkane) phosphine was injected at a temperature between

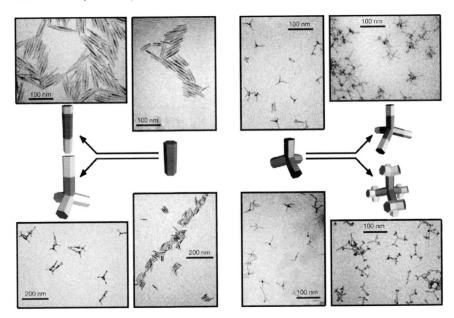

Figure 6.31   *Different architecture linear and branched nanorods shown schematically and in TEM images*
(Reproduced with permission from Ref. 161)

280 and 320 °C into a mixture of CdO complexed to alkylphosphonic acids. Growth of nanorods and nanotetrapods under excess cation concentrations was allowed to proceed to a predetermined size at which point the anion was changed to promote end-selective extension or branching of the pre-grown nanostrucutures. The process also works well with nanorods and nanotetrapods that have been removed from a synthesis mixture and re-dispersed and re-grown as described above, indicating that the ends of the nanostructures can be extended with the same or different metal and chalcogenide, thereby increasing the compositional domain of the heterostructures.

A detailed microscopy analysis with element mapping confirms the growth proceeds more-or-less in the way illustrated, where the diameter of the rods in the nanostructures changes at most by an atomic layer while the length increases by an amount dependent on synthesis conditions and growth time. This approach enables impressive control over the composition, dimensions and architecture of these novel nanoscale heterostructures, which in turn allows command over the separation of photogenerated electrons and holes across their designer interfaces. These may find uses in a range of perceived applications, ranging from photosynthesis to photovoltaics.

6.29 Golden Tips – Making Contact with Nanorods

While gold nanoclusters have been shown to catalyze VLS growth of semiconductor nanowires in which the nanocluster decorates only one end of the nanowire it is not so easy to conceive a bottom–up synthesis that directs nanoclusters to both ends.

Such golden tips, if reduced to practice, would provide a range of fascinating opportunities for modifying semiconductor nanowire optical behavior through plasmon–exciton coupling and enhancing nanowire electrical transport through synergistic metal-semiconductor binding. In addition, these tips could facilitate directed self-assembly of end-to-end nanowire chains *via* bifunctional capping ligands, and nanowire contact to electrodes in nanowire circuitry. If selective metal end tipping can be achieved one could envisage applying the know-how to other anisotropically shaped nanostructures like tetrapods, prisms and cubes.[162]

A first step in this direction has been demonstrated by adding a toluene solution of AuCl$_3$, dodecyldimethylammonium bromide and dodecylamine to a toluene dispersion of nanoscale CdSe rods. The resulting nanorods are then precipitated using methanol and are collected by centrifugation. Electron microscopy images coupled with X-ray diffraction and energy dispersive X-ray spectroscopy show that gold nanoclusters grow on both ends of the nanorods (Figure 6.32).

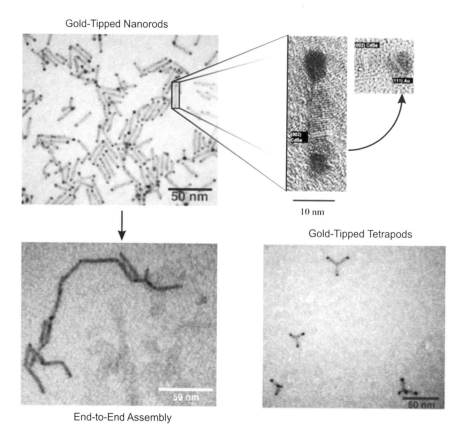

Figure 6.32 *Gold tips can be grown on the ends of semiconductor nanorods, shown in wide-area and high-resolution TEM images. These gold tips can facilitate assembly of nanorods into linear chains, and the tips of other structures such as tetrapods can also be coated*
(Reproduced with permission from Ref. 162)

It was surmised that because of the high surface free-energy and/or incomplete ligand passivation of the tips of the nanorods together with the presence of the base dodecylamine, the Au(III) precursor is reduced to Au(0) which selectively nucleates and grows to form gold nanoclusters on each end of the CdSe nanorods. An expanded scale TEM image clearly depicts the characteristic lattice planes of the CdSe nanorod and Au nanocluster tips. The size of the gold tips was found to scale with the gold concentration, thereby providing size-tunability. A similar set of observations was observed for CdSe tetrapods in which a gold nanocluster is exclusively formed on the end of each arm of the tetrahedrally shaped object. Further, the method works equally well for core–shell nanorods exemplified by CdSe–ZnS.

The electronic outcome of putting gold tips at the ends of the nanorods is seen as a strong perturbation of the nanorod optical absorption, photoluminescence and electrical transport properties. In brief, the absorption properties were not simply a convolution of excitonic CdSe nanorod, and plasmonic Au nanocluster spectra. This implies that the entire density-of-states of the nanocluster–nanorod hybrid has been modified by the coupling between semiconductor and metal electronic states. Similar clues emerge from the observed Au nanocluster tip size dependence of the quenching of the CdSe nanorods photoluminescence. An increased quenching with larger size gold tips implies enhanced non-radiative relaxation pathways *via* electron transfer from nanorod to nanocluster, facilitated by strong interfacial bonding between the gold and selenium. AFM tunneling spectroscopy shows a distinct spatial dependence of the tunneling current, where electrical transport from AFM tip to conducting graphite substrate is favored at the gold nanocluster ends of the nanorods. By functionalizing the gold ends with α,ω-alkanedithols, nanorods were found to self-assemble into chains of nanorods connected in an end-to-end to end fashion (Figure 6.32). All of the above developments bode well for using bottom–up, directed self-assembly approaches for making electrical contacts between nanorods, as well as connecting them to electrodes in nanocircuitry.

6.30 Flipping a Nanocluster Switch

The idea of this demonstration device is to exploit the electrochemically tunable state of charge and charge transport conductivity of a bipyridinium molecule anchored between two gold electrodes by alkanethiol groups attached to the quaternary ammoniums,[163,164] (Figure 6.33). One of the bipyridinium alkanethiols is anchored to a planar gold electrode to form a SAM while the other alkanethiol displaces a capping alkanethiolate ligand on a capped gold nanocluster. The tip of a STM is used to record the current flowing through the molecular switch by changing the substrate bias V. In the left panel of Figure 6.33, the bipyridinium molecule is in the oxidized bipy(2+) state and no current flows. Electrons can be injected into the bipyridinium molecule by applying a suitable V through the STM tip. In the right panel of Figure 6.33, when an electron is added the bipyridinium molecule is reduced to bipy(1+) and a large current flows. This device can therefore be seen to function as a molecular electronic switch by changing the chemical state of a molecule.

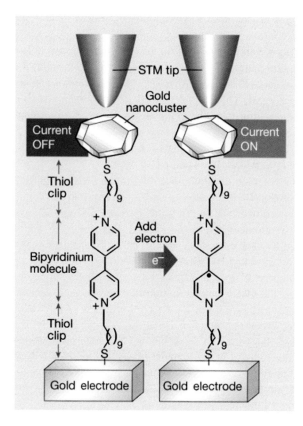

Figure 6.33 *Molecular electronic switch based on a gold nanocluster and bipyridine molecules*
(Reproduced with permission from Ref. 164)

The size of this molecular switch is less than 10 nm and uses less than 60 molecules and 30 electrons to operate. The caveat is that these nanoswitches can only be operated one at a time, making addressable arrays the biggest challenge at the moment.

6.31 Photochromic Metal Nanoclusters

One of the most exciting applications of nanochemistry is paradoxically one of the oldest sciences, that of tunable color. The study of nanoscale building blocks and their assemblies is uncovering a multitude of novel pathways to colored materials, from layer-by-layer electrochromic assemblies (Chapter 3), to multicolor quantum dots, to wavelength tunable reflection from photonic crystals (Chapter 7). One of the main goals of nano-based color technology is to achieve tunable and reversible pigmentation. That is, make a material having a certain color, and be able to switch it reversibly between two or more color states. A very interesting way to achieve this has recently been demonstrated using silver nanoclusters embedded in a titania matrix.[165] To prepare a film with a low silver loading, a sol–gel titania film is spin-coated on a substrate and then dipped in silver nitrate followed by a UV treatment to photo reduce the metal salt. To load more silver into the film the titania sol–gel precursors are mixed with $AgNO_3$, exposed to UV then spin-coated onto a substrate and again UV-exposed. The films produced initially are blackish-brown, with the pigmentation originating from the plasmon resonance of silver nanoclusters with a broad distribution of sizes.

A blackish-brown material is not that interesting, but something quite interesting happens when this material is exposed to white light without UV. Visible light promotes the silver nanoclusters to their excited states, and these then react with oxygen *via* titania, which is a photocatalyst. The reaction with oxygen effectively quenches the plasmon absorbances, and the material becomes white and colorless. Simply exposing the material to UV light once again regenerates the silver nanoclusters *via* photoreduction. Now, if the white light exposing the sample is first passed through a narrow band-pass filter, which absorbs a small range of wavelengths, the bleaching process still occurs. However, since the filter wavelength range could not excite the sample, silver nanoclusters having this color were never excited and thus were never bleached through photo-oxidation. The result is a material having the same color as the filter, as shown in Figure 6.34. This type of multicolor nanocluster-based photochromism is a very attractive property since the same material can be switched at will between any colors of the rainbow. While the achievable colors in this example are not especially brilliant, it is an important step towards the achievement of multi-functional active´ pigments.

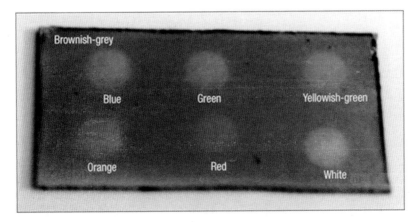

Figure 6.34 *Optical image of spots irradiated with visible light used in conjunction with color filters. Thus, the blue, green, yellow, orange and red spots were irradiated using filters at 450, 530, 560, 600 and 650 nm, respectively, while on the white spot no color filter was used*
(Reproduced with permission from Ref. 165)

6.32 Carbon Nanoclusters – Buckyballs

For centuries it was assumed we knew everything there was to know about carbon. In its elemental form it existed as three varieties – graphite, diamond and amorphous carbon. While these elements continued to gain increased industrial and commercial use such as fuel, filler, electrodes, jewellery and cutting tools, fundamental research had almost reached the point of diminishing returns. One of the major scientific breakthroughs of the 20th century was the discovery of a new form of carbon. This carbon consisted of isolated molecules made up of 60 carbon atoms, C_{60}, connected together as hexagons and pentagons as in a soccer ball.[166–168] This molecule was named buckminsterfullerene, after the curved geometrical designs of the architect Buckminster Fuller, and its discovery was cause for a 1996 Nobel Prize in Chemistry. An image of such architectural designs compared to a C_{60} molecule is shown in Figure 6.35, highlighting their highly symmetrical and esthetically pleasing structures. At about 1 nm in diameter, this carbon construction narrowly enters the realm of nano-chemistry. This molecule was first isolated using mass spectrometry on a plume of vapor generated by laser ablation of a graphite target. Nowadays, it is commercially available from several chemical suppliers with its price steadily decreasing as preparation methods improve.[169]

As one of the most studied molecules in chemistry, C_{60} deserves a book all its own, in fact several of these have been published; there are even scientific journals entirely dedicated to the so-called fullerene science. Amongst its unique characteristics are such things as superconductivity,[170,171] reversible redox behavior,[172] polymerization,[173] non-linear optical properties,[174] magnetism,[175] catalysis,[176] a rich chemistry[177,178] and the ability to trap atomic guests in their hollow interiors.[179,180] Much effort has therefore been expended both in the study of fullerene fundamental properties as well as their integration into materials and devices.[181]

Figure 6.35 *Porcelain model of buckminsterfullerene, C_{60} left, and a geodesic dome designed by Buckminster Fuller for the 1967 World Exposition in Montreal, right*
(Left image reproduced with permission from www.porcelainia.com/bet.html, Right image reproduced with permission from "Wikipedia, The free encyclopedia, http://en.wikipedia.org")

Out of the plethora of exciting examples, only a couple will be covered here, the rest being left for the readers to discover themselves!

6.33 Building Nanodevices with Buckyballs

One of the main thrusts in nanochemistry is the miniaturization of microelectronic devices, eventually aiming to self-assemble devices and interconnects in the 1–10 nm size range. With this huge increase in area density, potentially levied by a much lower cost than current lithographic techniques, we may soon be seeing a quantum leap in computing power in the near future. For such a task, C_{60} may well be the smallest building block we could use, and has been fashioned into the smallest ever transistor.[182] First, a pair of electrodes were patterned by electron lithography onto an insulator-on-semiconductor gate electrode, such that they touched at a very thin point. After depositing a coating of C_{60} from toluene solution the two electrodes were separated by breaking the electrode junction using electromigration,[183] leaving gaps as small as 1 nm in size. A single C_{60} molecule can find itself bridging this gap, and its electrical characteristics then probed. Not only do these spherical molecules act as single-electron transistors (Figure 6.36) but also it was found that the hopping of the electron was coupled to a nanomechanical oscillation. The carbon cage was deforming in response to the quantized transfer of electrons, and this type of mechanical–electrical coupling has no precedent. This is only one way in which buckyball may disrupt established standards, such as the silicon transistor in this case.

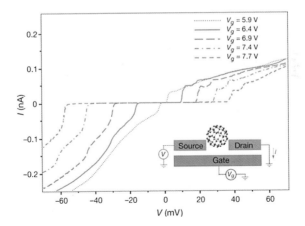

Figure 6.36 *The current passing through a single C_{60} molecule as a function of source-drain voltage for a given gate voltage. The schematic architecture of this transistor is shown on the bottom right*
(Reproduced with permission from Ref. 182)

6.34 Carbon Catalysis with Buckyball

The perfectly symmetrical structure of buckminsterfullerene has more than just an esthetic appeal. All the carbons are conjugated through a huge circular pi-cloud and this conjugated system can easily be reduced, and loaded with up to four electrons. The resulting anionic species can be good reducing agents, and can perform some unexpected chemistry such as reducing nitrogen to ammonia.[184] The industrial route to producing ammonia, called the Haber process, uses temperatures in excess of 300 °C, very large pressures, and an expensive rhodium catalyst which only functions at low conversions, requiring complex re-circulating and distilling. Nature on the other hand is quite good at making ammonia at room temperature, for instance by the nitrogenase enzyme found in anaerobic microbes. Chemistry has approached this milestone using organometallic catalysts,[185,186] but none even come close to the efficiency and longevity of natural nitrogen reducers. The key to a gentle chemistry route to ammonia may be C_{60}, chemically protected and activated by encapsulating organic groups. In this particular example the C_{60} molecule is capped by two cyclodextrin molecules as shown in Figure 6.37. Upon exposure of this complex to 1 atm nitrogen pressure for 1 h, with white light irradiation and bisulfite reducing agent, ammonia could be produced in a 33% yield. Nitrogen is thought to bind and be reduced on the exposed C_{60} graphitic surfaces. Light in this case promotes the formation of a charge-transfer complex, which speeds up the reaction by lowering corresponding activation energies.

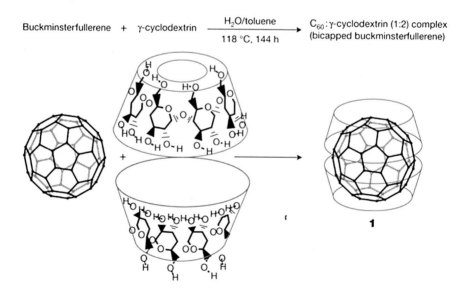

Buckminsterfullerene + γ-cyclodextrin $\xrightarrow[\text{118 °C, 144 h}]{\text{H}_2\text{O/toluene}}$ C_{60}:γ-cyclodextrin (1:2) complex (bicapped buckminsterfullerene)

1

Figure 6.37 *Self-assembly of a 2:1 cyclodextrin-C_{60} complex. A reducing agent feeds electrons to the fullerene, which then transfers them to coordinated nitrogen* (Reproduced with permission from Ref. 184)

6.35 References

1. A.P. Alivisatos, Semiconductor clusters, nanocrystals, and quantum dots, *Science*, 1996, **271**, 933.
2. A.P. Alivisatos, Semiconductor nanocrystals, *MRS Bull.*, 1995, **20**, 23.
3. D.M. Wood and N.W. Ashcroft, Quantum size effects in the optical properties of small metallic particles, *Phys. Rev. B.*, 1982, **25**, 6255.
4. L.N. Lewis, Chemical catalysis by colloids and clusters, *Chem. Rev.*, 1993, **93**, 2693.
5. C.J. Murphy and J.L. Coffer, Quantum dots: a primer, *Appl. Spectrosc.*, 2002, **56**, 16A.
6. A.D. Yoffe, Semiconductor quantum dots and related systems: electronic, optical, luminescence and related properties of low dimensional systems, *Adv. Phys.*, 2001, **50**, 1.
7. M.P. Pileni, Nanocrystals: fabrication, organization and collective properties, *C. R. Chim.*, 2003, **6**, 965.
8. J.R. Heath, The chemistry of size and order on the nanometer scale, *Science*, 1995, **270**, 1315.
9. R.M. Crooks, B.I. Lemon, L. Sun, L.K. Yeung and M.Q. Zhao, Dendrimer-encapsulated metals and semiconductors: synthesis, characterization and applications, *Top. Curr. Chem.*, 2001, **212**, 81.
10. C.B. Murray, C.R. Kagan and M.G. Bawendi, Synthesis and characterization of monodisperse nanocrystals and close-packed nanocrystal assemblies, *Ann. Rev. Mater. Sci.*, 2000, **30**, 546.
11. C.N.R. Rao, G.U. Kulkarni, P.J. Thomas, V.V. Agrawal, U.K. Gautam and M. Ghosh, Nanocrystals of metals, semiconductors and oxides: novel synthesis and applications, *Curr. Sci.*, 2003, **85**, 1041.
12. P. Alivisatos, Colloidal quantum dots. From scaling laws to biological applications, *Pure Appl. Chem.*, 2000, **72**, 3.
13. C.B. Murray, D.J. Norris and M.G. Bawendi, Synthesis and characterization of nearly monodisperse CdE (E = S, Se, Te) semiconductor nanocrystallites, *J. Am. Chem. Soc.*, 1993, **115**, 8706.
14. J. Joo, H.B. Na, T. Yu, J.H. Yu, Y.W. Kim, F.X. Wu, J.Z. Zhang and T. Hyeon, Generalized and facile synthesis of semiconducting metal sulfide nanocrystals, *J. Am. Chem. Soc.*, 2003, **125**, 11100.
15. J.D. Holmes, D.M. Lyons and K.J. Ziegler, Supercritical fluid synthesis of metal and semiconductor nanomaterials, *Chem. Eur. J.*, 2003, **9**, 2144.
16. A.C.C. Esteves and T. Trindade, Synthetic studies on II/VI semiconductor quantum dots, *Curr. Opin. Solid State Mater. Sci.*, 2002, **6**, 347.
17. M. Green, Solution routes to III–V semiconductor quantum dots, *Curr. Opin. Solid State Mater. Sci.*, 2002, **6**, 355.
18. D. Battaglia and X.G. Peng, Formation of high quality InP and InAs nanocrystals in a noncoordinating solvent, *Nano Lett.*, 2002, **2**, 1027.
19. T. Trindade, P. O'Brien and N.L. Pickett, Nanocrystalline semiconductors: synthesis, properties, and perspectives, *Chem. Mater.*, 2001, **13**, 3843.

20. L.H. Qu, W.W. Yu and X.P. Peng, In situ observation of the nucleation and growth of CdSe nanocrystals, *Nano Lett.*, 2004, **4**, 465.

21. C.B. Murray, C.R. Kagan and M.G. Bawendi, Synthesis and characterization of monodisperse nanocrystals and close-packed nanocrystal assemblies, *Ann. Rev. Mater. Sci.*, 2000, **30**, 545.

22. V.N. Soloviev, A. Eichhofer, D. Fenske and U. Banin, Molecular limit of a bulk semiconductor: size dependence of the "band gap" in CdSe cluster molecules, *J. Am. Chem. Soc.*, 2000, **122**, 2673.

23. T. Kippeny, L.A. Swafford and S.J. Rosenthal, Semiconductor nanocrystals: a powerful visual aid for introducing the particle in a box, *J. Chem. Educ.*, 2002, **79**, 1094.

24. W.L. Wilson, P.F. Szajowski and L.E. Brus, Quantum confinement in size-selected, surface-oxidized silicon nanocrystals, *Science*, 1993, **262**, 1242.

25. U. Banin and O. Millo, Tunneling and optical spectroscopy of semiconductor nanocrystals, *Ann. Rev. Phys. Chem.*, 2003, **54**, 465.

26. J.E.B. Katari, V.L. Colvin and A.P. Alivisatos, X-ray photoelectron spectroscopy of CdSe nanocrystals with applications to studies of the nanocrystal surface, *J. Phys. Chem.*, 1994, **98**, 4109; J.R. Heath, J.J. Shiang and A.P. Alivisatos, Germanium quantum dots: optical properties and synthesis, *J. Chem. Phys.*, 1994, **101**, 1607

27. A.L. Efros and M. Rosen, The electronic structure of semiconductor nanocrystals, *Ann. Rev. Mater. Sci.*, 2000, **30**, 475.

28. M. Nirmal and L. Brus, Luminescence photophysics in semiconductor nanocrystals, *Acc. Chem. Res.*, 1999, **32**, 407.

29. S. Empedocles and M. Bawendi, Spectroscopy of single CdSe nanocrystallites, *Acc. Chem. Res.*, 1999, **32**, 389.

30. W.K. Woo, K.T. Shimizu, M.V. Jarosz, R.G. Neuhauser, C.A. Leatherdale, M.A. Rubner and M.G. Bawendi, Reversible charging of CdSe nanocrystals in a simple solid-state device, *Adv. Mater.*, 2002, **14**, 1068.

31. M. Shim, C.J. Wang, D.J. Norris and P. Guyot-Sionnest, Doping and charging in colloidal semiconductor nanocrystals, *MRS Bull.*, 2001, **26**, 1005.

32. M. Shim and P. Guyot-Sionnest, n-type colloidal semiconductor nanocrystals, *Nature*, 2000, **407**, 981.

33. M. Kastner, Technology and the single electron, *Nature*, 1997, **389**, 667.

34. D.L. Klein, R. Roth, A.K.L. Lim, A.P. Alivisatos and P.L. McEuen, A single-electron transistor made from a cadmium selenide nanocrystal, *Nature*, 1997, **389**, 699.

35. D.L. Feldheim and C.D. Keating, Self-assembly of single electron transistors and related devices, *Chem. Soc. Rev.*, 1998, **27**, 1.

36. M. Bruchez, Jr., M. Moronne, P. Gin, S. Weiss and A.P. Alivisatos, Semiconductor nanocrystals as fluorescent biological labels, *Science*, 1998, **281**, 2013.

37. A.R. Kortan, R. Hull, R.L. Opila, M.G. Bawendi, M.L. Steigerwald, P.J. Carroll and L.E. Brus, Nucleation and growth of CdSe on ZnS quantum crystallite seeds, and vice versa, in inverse micelle media, *J. Am. Chem. Soc.*, 1990, **112**, 1327.

38. X.G. Peng, M.C. Schlamp, A.V. Kadavanich and A.P. Alivisatos, Epitaxial growth of highly luminescent CdSe/CdS core/shell nanocrystals with photo-stability and electronic accessibility, *J. Am. Chem. Soc.*, 1997, **119**, 7019.

39. M. Danek, K.F. Jensen, C.B. Murray and M.G. Bawendi, Synthesis of lumine-scent thin-film CdSe/ZnSe quantum dot composites using CdSe quantum dots passivated with an overlayer of ZnSe, *Chem. Mater.*, 1996, **8**, 173.

40. A. Hasselbarth, A. Eychmuller and H. Weller, Detection of shallow electron traps in quantum sized CdS by fluorescence quenching experiments, *Chem. Phys. Lett.*, 1993, **203**, 271.

41. S. Kim, B. Fisher, H.J. Eisler and M. Bawendi. Type-II quantum dots: CdTe/CdSe(core/shell) and CdSe/ZnTe(core/shell) heterostructures, *J. Am. Chem. Soc.*, 2003, **125**, 11466.

42. E.M. Chan, R.A. Mathies and A.P. Alivisatos, Size-controlled growth of CdSe nanocrystals in microfluidic reactors, *Nano Lett.*, 2003, **3**, 199.

43. B.K.H. Yen, N.E. Stott, K.F. Jensen and M.G. Bawendi, A continuous-flow microcapillary reactor for the preparation of a size series of CdSe nanocrystals, *Adv. Mater.*, 2003, **15**, 1858.

44. J. DeMello and A. DeMello, Microscale reactors: nanoscale products, *Lab on a Chip.*, 2004, **4**, 11N.

45. J.E. Barton and T.W. Odom, Mass-limited growth in zeptoliter beakers: a general approach for the synthesis of nanocrystals, *Nano Lett.*, 2004, **11**, 1525.

46. X.H. Zhong, M.Y. Han, Z.L. Dong, T.J. White and W. Knoll, Composition-tunable $Zn_xCd_{1-x}Se$ nanocrystals with high luminescence and stability, *J. Am. Chem. Soc.*, 2003, **125**, 8589.

47. X.H. Zhong, Y.Y. Feng, W. Knoll and M.Y. Han, Alloyed $Zn_xCd_{1-x}S$ nanocrystals with highly narrow luminescence spectral width, *J. Am. Chem. Soc.*, 2003, **125**, 13559.

48. R.E. Bailey and S.M. Nie, Alloyed semiconductor quantum dots: tuning the optical properties without changing the particle size, *J. Am. Chem. Soc.*, 2003, **125**, 7100.

49. A. Murali, A. Barve, V.J. Leppert, S.H. Risbud, I.M. Kennedy and H.W.H. Lee, Synthesis and characterization of indium oxide nanoparticles, *Nano Lett.*, 2001, **1**, 287.

50. W.S. Seo, H.H. Jo, K. Lee and J.T. Park, Preparation and optical properties of highly crystalline, colloidal, and size-controlled indium oxide nanoparticles, *Adv. Mater.*, 2003, **15**, 795.

51. G. Seifert. Nanomaterials – Nanocluster magic, *Nat. Mater.*, 2004, **3**, 77.

52. A. Kasuya, R. Sivamohan, Y.A. Barnakov, I.M. Dmitruk, T. Nirasawa, V.R. Romanyuk, V. Kumar, S.V. Mamykin, K. Tohji, B. Jeyadevan, K. Shinoda, T. Kudo, O. Terasaki, Z. Liu, R.V. Belosludov, V. Sundararajan and Y. Kawazoe, Ultra-stable nanoparticles of CdSe revealed from mass spectrometry, *Nat. Mater.*, 2004, **3**, 99.

53. K. Jacobs, D. Zaziski, E.C. Scher, A.B. Herhold and A.P. Alivisatos, Activation volumes for solid–solid transformations in nanocrystals, *Science*, 2001, **293**, 1803.

54. S.H. Tolbert and A.P. Alivisatos, Size dependence of a first-order solid–solid phase-transition – the wurtzite to rock-salt transformation in CdSe nanocrystals, *Science*, 1994, **265**, 373.

55. C.C. Chen, A.B. Herhold, C.S. Johnson and A.P. Alivisatos, Size dependence of structural metastability in semiconductor nanocrystals, *Science*, 1997, **276**, 398.

56. H.Z. Zhang, B. Gilbert, F. Huang and J.F. Banfield, Water-driven structure transformation in nanoparticles at room temperature, *Nature*, 2003, **424**, 1025.

57. R.H. Terrill, T.A. Postlethwaite, C.-H. Chen, C.-D. Poon, A. Terzis, A. Chen, J.E. Hutchison, M.R. Clark, G. Wignall, J.D. Londono, R. Superfine, M. Falvo, C.S. Johnson, E.T. Samulski, R.W. and Murray, Monolayers in three dimensions: NMR, SAXS, thermal, and electron hopping studies of alkanethiol stabilized gold clusters, *J. Am. Chem. Soc.*, 1995, **117**, 12537.

58. N.R. Jana and X.G. Peng, Single-phase and gram-scale routes toward nearly monodisperse Au and other noble metal nanocrystals, *J. Am. Chem. Soc.*, 2003, **125**, 14280.

59. S.W. Kim, J. Park, Y. Jang, Y. Chung, S. Hwang, T. Hyeon and Y.W. Kim, Synthesis of monodisperse palladium nanoparticles, *Nano Lett.*, 2003, **3**, 1289.

60. C.N.R. Rao, G.U. Kulkarni, P.J. Thomas and P.P. Edwards, Metal nanoparticles and their assemblies, *Chem. Soc. Rev.*, 2000, **29**, 27.

61. K. Kneipp, H. Kneipp, I. Itzkan, R.R. Dasari, M.S. Feld and M.S. Dresselhaus, Nonlinear Raman probe of single molecules attached to colloidal silver and gold clusters, *Top. Appl. Phys.*, 2002, **82**, 227.

62. Y.Y. Yang and S.W. Chen, Surface manipulation of the electronic energy of subnanometer-sized gold clusters: an electrochemical and spectroscopic investigation, *Nano Lett.*, 2003, **3**, 75.

63. P. Mulvaney, Surface plasmon spectroscopy of nanosized metal particles, *Langmuir*, 1996, **12**, 788.

64. S.W. Chen, R.S. Ingram, M.J. Hostetler, J.J. Pietron, R.W. Murray, T.G. Schaaff, J.T. Khoury, M.M. Alvarez and R.L. Whetten, Gold nanoelectrodes of varied size: transition to molecule-like charging, *Science*, 1998, **280**, 2098.

65. D. Roy and J. Fendler, Reflection and absorption techniques for optical characterization of chemically assembled nanomaterials, *Adv. Mater.*, 2004, **16**, 479.

66. T.G. Schaaff, M.N. Shafigullin, J.T. Khoury, I. Vezmar, R.L. Whetten, W.G. Cullen, P.N. First, C. GutierrezWing, J. Ascensio and M.J. JoseYacaman, Isolation of smaller nanocrystal Au molecules: robust quantum effects in optical spectra, *J. Phys. Chem. B*, 1997, **101**, 7885.

67. R. Serna, R.W. Dreyfus, J. Solis, C.N. Afonso, D.A. Allwood, P.E. Dyer and A.K. Petford-Long, Matrix assisted laser desorption/ionisation studies of metallic nanoclusters produced by pulsed laser deposition, *Appl. Surf. Sci.*, 1998, **129**, 383.

68. M.M. Alvarez, J.T. Khoury, T.G. Schaaff, M. Shafigullin, I. Vezmar and R.L. Whetten. Critical sizes in the growth of Au clusters, *Chem. Phys. Lett.*, 1997, **266**, 91.

69. Z.L. Wang, S.A. Harfenist, R.L. Whetten, J. Bentley and N.D. Evans, Bundling and interdigitation of adsorbed thiolate groups in self-assembled nanocrystal superlattices, *J. Phys. Chem. B*, 1998, **102**, 3068.

70. R.L. Whetten, M.N. Shafigullin, J.T. Khoury, T.G. Schaaff, I. Vezmar, M.M. Alvarez and A. Wilkinson, Crystal structures of molecular gold nanocrystal arrays, *Acc. Chem. Res.*, 1999, **32**, 397.

71. M.M. Alvarez, J.T. Khoury, T.G. Schaaff, M.N. Shafigullin, I. Vezmar and R.L. Whetten, Optical absorption spectra of nanocrystal gold molecules, *J. Phys. Chem. B*, 1997, **101**, 3706.

72. M.A. El-Sayed, Some interesting properties of metals confined in time and nanometer space of different shapes, *Acc. Chem. Res.*, 2001, **34**, 257.

73. M.J. Hostetler, J.E. Wingate, C.J. Zhong, J.E. Harris, R.W. Vachet, M.R. Clark, J.D. Londono, S.J. Green, J.J. Stokes, G.D. Wignall, G.L. Glish, M.D. Porter, N.D. Evans and R.W. Murray, Alkanethiolate gold cluster molecules with core diameters from 1.5 to 5.2 nm: core and monolayer properties as a function of core size, *Langmuir*, 1998, **14**, 17.

74. H. Yu, P.C. Gibbons, K.F. Kelton and W.E. Buhro, Heterogeneous seeded growth: a potentially general synthesis of monodisperse metallic nanoparticles, *J. Am. Chem. Soc.*, 2001, **123**, 9198.

75. N.R. Jana, L. Gearheart and C.J. Murphy, Seed-mediated growth approach for shape-controlled synthesis of spheroidal and rod-like gold nanoparticles using a surfactant template, *Adv. Mater.*, 2001, **13**, 1389.

76. W.P. Wuelfing, S.M. Gross, D.T. Miles and R.W. Murray, Nanometer gold clusters protected by surface-bound monolayers of thiolated poly(ethylene glycol) polymer electrolyte, *J. Am. Chem. Soc.*, 1998, **120**, 12696.

77. P. Pengo, S. Polizzi, M. Battagliarin, L. Pasquato and P. Scrimin, Synthesis, characterization and properties of water-soluble gold nanoparticles with tunable core size, *J. Mater. Chem.*, 2003, **13**, 2471.

78. S.H. Wang, S. Sato and K. Kimura, Preparation of hexagonal-close-packed colloidal crystals of hydrophilic monodisperse gold nanoparticles in bulk aqueous solution, *Chem. Mater.*, 2003, **15**, 2445.

79. S.F. Wuister, I. Swart, F. van Driel, S.G. Hickey and C.D. Donega, Highly luminescent water-soluble CdTe quantum dots, *Nano Lett.*, 2003, **3**, 503.

80. T. Pellegrino, L. Manna, S. Kudera, T. Liedl, D. Koktysh, A.L. Rogach, S. Keller, J. Radler, G. Natile and W.J. Parak, Hydrophobic nanocrystals coated with an amphiphilic polymer shell: a general route to water soluble nanocrystals, *Nano Lett.*, 2004, **4**, 703.

81. M. Alejandro-Arellano, T. Ung, A. Blanco, P. Mulvaney and L.M. Liz-Marzan, Silica-coated metals and semiconductors. Stabilization and nanostructuring, *Pure Appl. Chem.*, 2000, **72**, 257.

82. P. Mulvaney, L.M. Liz-Marzan, M. Giersig and T. Ung, Silica encapsulation of quantum dots and metal clusters, *J. Mater. Chem.*, 2000, **10**, 1259.

83. M.A. Correa-Duarte, M. Giersig and L.M. Liz-Marzan, Stabilization of CdS semiconductor nanoparticles against photodegradation by a silica coating procedure, *Chem. Phys. Lett.*, 1998, **286**, 497.

84. N.A. Kotov, I. Dekany and J.H. Fendler, Layer-by-layer self-assembly of polyelectrolyte–semiconductor nanoparticle composite films, *J. Phys. Chem.*, 1995, **99**, 13065.

85. X.G. Peng, T.E. Wilson, A.P. Alivisatos and P.G. Schultz, Synthesis and isolation of a homodimer of cadmium selenide nanocrystals, *Angew. Chem. Int. Ed.*, 1997, **36**, 145.

86. B.A. Korgel and D. Fitzmaurice, Self-assembly of silver nanocrystals into two-dimensional nanowire arrays, *Adv. Mater.*, 1998, **10**, 661.

87. C.B. Mao, J.F. Qi and A.M. Belcher, Building quantum dots into solids with well-defined shapes, *Adv. Funct. Mater.*, 2003, **13**, 648.

88. A.L. Rogach, D.V. Talapin, E.V. Shevchenko, A. Kornowski, M. Haase and H. Weller, Organization of matter on different size scales: monodisperse nanocrystals and their superstructures, *Adv. Funct. Mater.*, 2002, **12**, 653.

89. S.A. Harfenist, Z.L. Wang, R.L. Whetten, I. Vezmar and M.M. Alvarez, Three-dimensional hexagonal close-packed superlattice of passivated Ag nanocrystals, *Adv. Mater.*, 1997, **9**, 817.

90. C.B. Murray, C.R. Kagan and M.G. Bawendi, Synthesis and characterization of monodisperse nanocrystals and close-packed nanocrystal assemblies, *Ann. Rev. Mater. Sci.*, 2000, **30**, 545.

91. Z.L. Wang, Structural analysis of self-assembling nanocrystal superlattices, *Adv. Mater.*, 1998, **10**, 13.

92. C.P. Collier, R.J. Saykally, J.J. Shiang, S.E. Henrichs and J.R. Heath, Reversible tuning of silver quantum dot monolayers through the metal-insulator transition, *Science*, 1997, **277**, 1978.

93. M.D. Bentzon, J.V. Wonterghem, S. Morup, A. Thoelen and C.J.W. Koch, Ordered aggregates of ultrafine iron oxide particles: "super crystals", *Philos. Mag. B*, 1989, **60**, 169.

94. C.B. Murray, C.R. Kagan and M.G. Bawendi, Self-organization of CdSe nanocrystallites into 3-dimensional quantum dot superlattices, *Science*, 1995, **270**, 1335.

95. L.M. Liz-Marzan and P. Mulvaney, The assembly of coated nanocrystals, *J. Phys. Chem. B*, 2003, **107**, 7312.

96. C.P. Collier, T. Vossmeyer and J.R. Heath. Nanocrystal superlattices, *Ann. Rev. Phys. Chem.*, 1998, **49**, 371.

97. D.V. Talapin, E.V. Shevchenko, A. Kornowski, N. Gaponik, M. Haase, A.L. Rogach and H. Weller, A new approach to crystallization of CdSe nanoparticles into ordered three-dimensional superlattices, *Adv. Mater.*, 2001, **13**, 1868.

98. C.J. Kiely, J. Fink, M. Brust, D. Bethell and D.J. Schiffrin, Spontaneous ordering of bimodal ensembles of nanoscopic gold clusters, *Nature*, 1998, **396**, 444.

99. W. Hume-Rothery, R.E. Smallman and C.W. Haworth, *The Structure of Metals and Alloys*, Metals and Metallurgy Trust, London, 1969.

100. S. Haschisu and S. Yoshimura, Optical demonstration of crystalline super-structures in binary mixtures of latex globules, *Nature*, 1980, **283**, 188.

101. P. Bartlett, R.H. Ottewill and P.N. Pusey, Superlattice formation in binary-mixtures of hard sphere colloids, *Phys. Rev. Lett.*, 1992, **68**, 3801.

102. F.X. Redl, K.-S. Cho, C.B. Murray and S. O'Brien, Three-dimensional binary superlattices of magnetic nanocrystals and semiconductor quantum dots, *Nature*, 2003, **423**, 968.

103. E.V. Shevchenko, D.V. Talapin, A.L. Rogach, A. Kornowski, M. Haase and H. Weller, Colloidal synthesis and self-assembly of CoPt$_3$ nanocrystals, *J. Am. Chem. Soc.*, 2002, **124**, 11480.

104. T. Hyeon, S.S. Lee, J. Park, Y. Chung and H.B. Na, Synthesis of highly crystalline and monodisperse maghemite nanocrystallites without a size-selection process, *J. Am. Chem. Soc.*, 2001, **123**, 12798.

105. S. Sun and C.B. Murray, Synthesis of monodisperse cobalt nanocrystals and their assembly into magnetic superlattices, *J. Appl. Phys.*, 1999, **85**, 4325.

106. T. Hyeon, Chemical synthesis of magnetic nanoparticles, *Chem. Commun.*, 2003, **8**, 927.

107. S.H. Sun and H. Zeng, Size-controlled synthesis of magnetite nanoparticies, *J. Am. Chem. Soc.*, 2002, **124**, 8204.

108. C.J. Xu, K.M. Xu, H.W. Gu, X.F. Zhong, Z.H. Guo, R.K. Zheng, X.X. Zhang and B. Xu, Nitrilotriacetic acid-modified magnetic nanoparticles as a general agent to bind histidine-tagged proteins, *J. Am. Chem. Soc.*, 2004, **126**, 3392.

109. S.H. Sun, C.B. Murray, D. Weller, L. Folks and A. Moser, Monodisperse FePt nanoparticles and ferromagnetic FePt nanocrystal superlattices, *Science*, 2000, **287**, 1989.

110. E. Shevchenko, D. Talapin, A. Kornowski, F. Wiekhorst, J. Kotzler, M. Haase, A. Rogach and H. Weller, Colloidal crystals of monodisperse FePt nano-particles grown by a three-layer technique of controlled oversaturation, *Adv. Mater.*, 2002, **14**, 287.

111. C.T. Black, C.B. Murray, R.L. Sandstrom and S.H. Sun, Spin-dependent tunneling in self-assembled cobalt-nanocrystal superlattices, *Science*, 2000, **290**, 1131.

112. J.-I. Park, M.G. Kim, Y.-W. Jun, J.S. Lee, W.-R. Lee and J. Cheon, Characterization of superparamagnetic "core-shell" nanoparticles and moni-toring their anisotropic phase transition to ferromagnetic "solid solution" nanoalloys, *J. Am. Chem. Soc.*, 2004, **126**, 9072.

113. B.O. Dabbousi, C.B. Murray, M.F. Rubner and M.G. Bawendi, Langmuir–Blodgett manipulation of size-selected CdSe nanocrystallites, *Chem. Mater.*, 1994, **6**, 216.

114. J.R. Heath, C.M. Knobler and D.V. Leff, Pressure/temperature phase diagrams and superlattices of organically functionalized metal nanocrystal monolayers: the influence of particle size, size distribution, and surface passivant, *J. Phys. Chem. B*, 1997, **101**, 189.

115. C.P. Collier, R.J. Saykally, J.J. Shiang, S.E. Henrichs and J.R. Heath, Reversible tuning of silver quantum dot monolayers through the metal-insulator transition, *Science*, 1997, **277**, 1978.

116. Q.J. Guo, X.W. Teng, S. Rahman and H. Yang, Patterned Langmuir–Blodgett films of mondisperse nanoparticles of iron oxide using soft lithography, *J. Am. Chem. Soc.*, 2003, **125**, 630.

117. V. Santhanam and R.P. Andres, Microcontact printing of uniform nanoparticle arrays, *Nano Lett.*, 2004, **4**, 41.

118. Y. Cui, M.T. Björk, J.A. Liddle, C. Sönnichsen, B. Boussert and A.P. Alivisatos, Integration of colloidal nanocrystals into lithographically patterned devices, *Nano Lett.*, 2004, **4**, 1093.

119. V.I. Klimov, A.A. Mikhailovsky, S. Xu, A. Malko, J.A. Hollingsworth, C.A. Leatherdale, H.-J. Eisler and M.G. Bawendi, Optical gain and stimulated emission in nanocrystal quantum dots, *Science*, 2000, **290**, 314.

120. H.J. Eisler, V.C. Sundar, M.G. Bawendi, M. Walsh, H.I. Smith and V. Klimov, Color-selective semiconductor nanocrystal laser, *Appl. Phys. Lett.*, 2002, **80**, 4614.

121. V.L. Colvin, M.C. Schlamp and A.P. Alivisatos, Light-emitting diodes made from cadmium selenide nanocrystals and a semiconducting polymer, *Nature*, 1994, **370**, 354.

122. S. Coe, W.K. Woo, M. Bawendi and V. Bulovic, Electroluminescence from single monolayers of nanocrystals in molecular organic devices, *Nature*, 2002, **420**, 800.

123. T. Tsutsui. A light-emitting sandwich filling, *Nature*, 2002, **420**, 752.

124. S.A. Empedocles and M.G. Bawendi, Quantum-confined stark effect in single CdSe nanocrystallite quantum dots, *Science*, 1997, **278**, 2114.

125. C.J. Wang, M. Shim and P. Guyot-Sionnest, Electrochromic nanocrystal quantum dots, *Science*, 2001, **291**, 2390.

126. www.ntera.com

127. www.webster-dictionary.org/definition/chromaticity

128. K. Rajeshwar, N.R. de Tacconi and C.R. Chenthamarakshan, Semiconductor-based composite materials: preparation, properties, and performance, *Chem. Mater.*, 2001, **13**, 2765.

129. J. Lee, V.C. Sundar, J.R. Heine, M.G. Bawendi and K.F. Jensen, Full color emission from II–VI semiconductor quantum dot-polymer composites, *Adv. Mater.*, 2000, **12**, 1102.

130. H.-J. Eisler, V.C. Sundar, M.G. Bawendi, M. Walsh, H.I. Smith and V. Klimov, Color-selective semiconductor nanocrystal laser, *Appl. Phys. Lett.*, 2002, **80**, 4614.

131. V.C. Sundar, H.-J. Eisler and M.G. Bawendi, Room-temperature, tunable gain media from novel II–VI nanocrystal-titania composite matrices, *Adv. Mater.*, 2002, **14**, 739.

132. M.A. Petruska, A.V. Malko, P.M. Voyles and V.I. Klimov, High-performance, quantum dot nanocomposites for nonlinear optical and optical gain applications, *Adv. Mater.*, 2003, **15**, 610.

133. S.T. Selvan, C. Bullen, M. Ashokkumar and P. Mulvaney, Synthesis of tunable, highly luminescent QD-glasses through sol–gel processing, *Adv. Mater.*, 2001, **13**, 985.

134. W.C.W. Chan and S. Nile, Quantum dot bioconjugates for ultrasensitive nonisotopic detection, *Science*, 1998, **281**, 2016.

135. D.R. Larson, W.R. Zipfel, R.M. Williams, S.W. Clark, M.P. Bruchez, F.W. Wise and W.W. Webb, Water-soluble quantum dots for multiphoton fluorescence imaging in vivo, *Science*, 2003, **300**, 1434.

136. W.J. Parak, D. Gerion, T. Pellegrino, D. Zanchet, C. Micheel, S.C. Williams, R. Boudreau, M.A. Le Gros, C.A. Larabell and A.P. Alivisatos, Biological applications of colloidal nanocrystals, *Nanotechnology*, 2003, **14**, R15.

137. E.R. Goldman, E.D. Balighian, H. Mattoussi, M.K. Kuno, J.M. Mauro, P.T. Tran and G.P. Anderson, Avidin: a natural bridge for quantum dot-antibody conjugates, *J. Am. Chem. Soc.*, 2002, **124**, 6378.

138. S.J. Rosenthal, A. Tomlinson, E.M. Adkins, S. Schroeter, S. Adams, L. Swafford, J. McBride, Y.Q. Wang, L.J. De Felice and R.D. Blakely, Targeting cell surface receptors with ligand-conjugated nanocrystals, *J. Am. Chem. Soc.*, 2002, **124**, 4586.

139. B. Dubertret, P. Skourides, D.J. Norris, V. Noireaux, A.H. Brivanlou and A. Libchaber, In vivo imaging of quantum dots encapsulated in phospholipid micelles, *Science*, 2002, **298**, 1759.

140. J.M. Nam, S.I. Stoeva and C.A. Mirkin, Bio-bar-code-based DNA detection with PCR-like sensitivity, *J. Am. Chem. Soc.*, 2004, **126**, 5932.

141. S.J. Park, T.A. Taton and C.A. Mirkin, Array-based electrical detection of DNA with nanoparticle probes, *Science*, 2002, **295**, 1503.

142. M.P. Pileni, The role of soft colloidal templates in controlling the size and shape of inorganic nanocrystals, *Nat. Mater.*, 2003, **2**, 145.

143. S. Kan, T. Mokari, E. Rothenberg and U. Banin, Synthesis and size-dependent properties of zinc-blende semiconductor quantum rods, *Nat. Mater.*, 2003, **2**, 155.

144. S.M. Lee, S.N. Cho and J. Cheon, Anisotropic shape control of colloidal inorganic nanocrystals, *Adv. Mater.*, 2003, **15**, 441.

145. X.G. Peng, Mechanisms for the shape-control and shape-evolution of colloidal semiconductor nanocrystals, *Adv. Mater.*, 2003, **15**, 459.

146. J. Yang, C. Xue, S.H. Yu, J.H. Zeng and Y.T. Qian, General synthesis of semiconductor chalcogenide nanorods by using the monodentate ligand *n*-butylamine as a shape controller, *Angew. Chem. Int. Ed.*, 2002, **41**, 4697.

147. L. Manna, E.C. Scher and A.P. Alivisatos, Shape control of colloidal semiconductor nanocrystals, *J. Cluster. Sci.*, 2002, **13**, 521.

148. L. Manna, E.C. Scher and A.P. Alivisatos, Synthesis of soluble and processable rod-, arrow-, teardrop-, and tetrapod-shaped CdSe nanocrystals, *J. Am. Chem. Soc.*, 2000, **122**, 12700.

149. Y.W. Jun, S.M. Lee, N.J. Kang and J. Cheon, Controlled synthesis of multi-armed CdS nanorod architectures using monosurfactant system, *J. Am. Chem. Soc.*, 2001, **123**, 5150.

150. W.W. Yu, Y.A. Wang and X.G. Peng, Formation and stability of size-, shape-, and structure-controlled CdTe nanocrystals: ligand effects on monomers and nanocrystals, *Chem. Mater.*, 2003, **15**, 4300.

151. N. Pinna, K. Weiss, J. Urban and M.P. Pileni, Triangular CdS nanocrystals: structural and optical studies, *Adv. Mater.*, 2001, **13**, 261.

152. Z.A. Peng and X.G. Peng, Mechanisms of the shape evolution of CdSe nanocrystals, *J. Am. Chem. Soc.*, 2001, **123**, 1389.

153. J.T. Hu, L.S. Li, W.D. Yang, L. Manna, L.W. Wang and A.P. Alivisatos, Linearly polarized emission from colloidal semiconductor quantum rods, *Science*, 2001, **292**, 2060.

154. M. Kazes, D.Y. Lewis, Y. Ebenstein, T. Mokari and U. Banin, Lasing from semiconductor quantum rods in a cylindrical microcavity, *Adv. Mater.*, 2002, **14**, 317.

155. L. Manna, D.J. Milliron, A. Meisel, E.C. Scher and A.P. Alivisatos, Controlled growth of tetrapod-branched inorganic nanocrystals, *Nat. Mater.*, 2003, **2**, 382.

156. X.G. Peng, L. Manna, W.D. Yang, J. Wickham, E. Scher, A. Kadavanich and A.P. Alivisatos, Shape control of CdSe nanocrystals, *Nature*, 2000, **404**, 59.

157. R.P. Raffaelle, S.L. Castro, A.F. Hepp and S.G. Bailey, Quantum dot solar cells, *Prog. Photovol.*, 2002, **10**, 433.

158. H. Yu, J.B. Li, R.A. Loomis, P.C. Gibbons, L.W. Wang and W.E. Buhro, Cadmium selenide quantum wires and the transition from 3D to 2D confinement, *J. Am. Chem. Soc.*, 2003, **125**, 16168.

159. W.U. Huynh, J.J. Dittmer and A.P. Alivisatos, Hybrid nanorod-polymer solar cells, *Science*, 2002, **295**, 2425.

160. B.Q. Sun, E. Marx and N.C. Greenham, Photovoltaic devices using blends of branched CdSe nanoparticles and conjugated polymers, *Nano Lett.*, 2003, **3**, 961.

161. D.J. Milliron, S.M. Hughes, Y. Cui, L. Manna, J.B. Li, L.W. Wang and A.P. Alivisatos, Colloidal nanocrystal heterostructures with linear and branched topology, *Nature*, 2004, **430**, 190.

162. T. Mokari, E. Rothenberg, I. Popov, R. Costi and U. Banin, Selective growth of metal tips onto semiconductor quantum rods and tetrapods, *Science*, 2004, **304**, 1787.

163. D.I. Gittins, D. Bethell, D.J. Schiffrin and R.J. Nichols, A nanometer-scale electronic switch consisting of a metal cluster and redox-addressable groups, *Nature*, 2000, **408**, 67.

164. D. Feldheim, Flipping a molecular switch, *Nature*, 2000, **408**, 45.

165. Y. Ohko, T. Tatsuma, T. Fujii, K. Naoi, C. Niwa, Y. Kubota and A. Fujishima, Multicolour photochromism of TiO_2 films loaded with silver nanoparticles, *Nat. Mater.*, 2003, **2**, 29.

166. H.W. Kroto, J.R. Heath, S.C. O'Brien, R.F. Curl and R.E. Smalley, C(60) Buckminsterfullerene, *Nature*, 1985, **318**, 14.

167. M.S. Dresselhaus, G. Dresselhaus and P. Eklund, *Science of Fullerenes and Carbon Nanotubes*, Academic Press, New York, 1996.

168. K.M. Kadish and R.S. Ruoff, *Fullerenes: Chemistry, Physics, and Technology*, Wiley-Interscience, New York, 2000.

169. A.A. Bogdanov, D. Deininger and G.A. Dyuzhev, Development prospects of the commercial production of fullerenes, *Tech. Phys.*, 2000, **45**, 521.

170. S. Suzuki, T. Chida and K. Nakao, Electron correlation and Jahn–Teller effect in alkali-metal-doped C-60, *Adv. Quantum. Chem.*, 2003, **44**, 535.

171. Y. Iwasa, Current status of doped C-60 solids: superconductors and related materials, *New Diamond Front. Carbon Technol.*, 2001, **11**, 415.

172. L. Echegoyen and L.E. Echegoyen, Electrochemistry of fullerenes and their derivatives, *Acc. Chem. Res.*, 1998, **31**, 593.

173. T.L. Makarova, Electrical and optical properties of pristine and polymerized fullerenes, *Semiconductor*, 2001, **35**, 243.

174. G. Brusatin and R. Signorini, Linear and nonlinear optical properties of fullerenes in solid state materials, *J. Mater. Chem.*, 2002, **12**, 1964.

175. D. Arcon and K. Prassides, Magnetism in fullerene derivatives, *Struct. Bond.*, 2001, **100**, 129.

176. N.F. Goldshleger, Fullerenes and fullerene-based materials in catalysis, *Fullerene Sci. Technol.*, 2001, **9**, 255.

177. A. Hirsch, Principles of fullerene reactivity, *Fullerenes Relat. Struct.; Top. Curr. Chem.*, 1999, **199**, 1.

178. E. Nakamura and H. Isobe, Functionalized fullerenes in water. The first 10 years of their chemistry, biology, and nanoscience, *Acc. Chem. Res.*, 2003, **36**, 807.

179. H. Shinohara, Endohedral metallofullerenes, *Rep. Prog. Phys.*, 2000, **63**, 843.

180. S.Y. Liu and S.Q. Sun, Recent progress in the studies of endohedral metallofullerenes, *J. Organomet. Chem.*, 2000, **599**, 74.

181. A. Penicaud, Building solids with Buckminsterfullerene (C-60), *Fullerene Sci. Technol.* 1998, **6**, 731.

182. H. Park, J. Park, A.K.L. Lim, E.H. Anderson, A.P. Alivisatos and P.L. McEuen, Nanomechanical oscillations in a single C-60 transistor, *Nature*, 2000, **407**, 57.

183. H. Park, A.K.L. Lim, J. Park, A.P. Alivisatos and P.L. McEuen, Fabrication of metallic electrodes with nanometer separation by electromigration, *Appl. Phys. Lett.*, 1999, **75**, 301.

184. Y. Nishibayashi, M. Saito, S. Uemura, S.-I. Takekuma, H. Takekuma and Z.-I. Yoshida, A non-metal system for nitrogen fixation, *Nature*, 2004, **428**, 279.

185. C.E. Laplaza and C.C. Cummins, Dinitrogen cleavage by a 3-coordinate molybdenum(III) complex, *Science*, 1995, **268**, 861.

186. D.V. Yandulov and R.R. Schrock, Catalytic reduction of dinitrogen to ammonia at a single molybdenum center, *Science*, 2003, **301**, 76.

Nanofood for Thought –
Nanoclusters, Nanocrystals, Quantum Dots, Quantum Size Effects

1. Explain why a non-solvent like acetone when added drop-wise to a toluene solution of polydispersed alkanethiol capped gold nanoclusters causes size-selective precipitation of nanoclusters with the larger diameter ones emerging first followed sequentially by progressively smaller ones?

2. Would you expect the optical properties of an isolated alkanethiol capped gold nanocluster to alter with (a) diameter of the gold core (b) number of methylene groups in the alkane chain of the capping thiol and (c) what would you expect to happen to the optical properties of these classes of nanoclusters in (a,b) when they are self-assembled to form a nanocluster superlattice?

3. What does one mean by the statement that a capped aurothiol nanocluster superlattice undergoes a melting transition?

4. How would you go about proving whether or not a monodispersed capped semiconductor nanocluster is actually a semiconductor and similarly whether a monodispersed capped gold cluster behaves as a metal?

5. What methods would you apply to determine the nature of surface gold atoms in an aurothiol nanocluster and whether or not the surface gold atoms are distinct to those of the ones in the bulk of the nanocluster?

6. It has been generally accepted in the literature although contested at times that an alkanethiol binds to the surface of an alkanethiol capped gold nanocluster as a thiolate, but how would you prove it is not actually a disulfide or a thiol form of bonding instead?

7. Devise self-assembly methods for organizing alkanethiolate capped gold clusters into a (a) duplex (b) linear triplet (c) bent triplet (d) network as well as micron dimension (e) chain (f) toroid (g) shape of the word "CLUSTER".

8. Formulate synthetic strategies for stabilizing a film of capped gold clusters.

9. What kinds of defects are expected to occur in capped metal cluster crystals?

10. Explain what is meant by bunching interdigitation in alkanethiolate capped gold cluster crystals, and how could you probe the effect with molecule scale resolution?

11. Give synthesis, structure and bonding details of how you would size-tune the electrical and optical properties of a nanocrystalline II–VI semiconductor.

12. How can a single electron transistor (SET) be made from a single 5 nm diameter capped CdSe nanocluster?

13. Why do you think an ensemble of monodisperse 5 nm capped CdS nanoclusters, excited with UV light, displays continuous bright green-blue luminescence, whereas a single nanocluster shows flashing green?

14. Give a purely synthetic method for making a metal–insulator–gold nanocluster–insulator–metal MINIM device. What function could you evoke from such a device?

15. Provide a synthetic method for preparing an aurothiol nanocluster super-lattice. How would you expect the properties of the cluster crystal to differ from that of a single nanocluster?

16. How would you make a Langmuir–Blodgett monolayer of monodispersed alkanethiolate capped silver clusters and use it to study electronic coupling between individual clusters? What evidence would you seek to show the existence of a metal–nonmetal transition in the film?

17. Can you think of a way of using a LB film of ferromagnetic, capped PtFe nanoclusters to create a patterned monolayer on a silicon substrate?

18. Briefly explain which techniques you would employ to reveal quantization of energy levels in monodispersed alkanethiolate capped gold nanoclusters.

19. Describe a method for synthesizing compositionally tunable, monodisperse and capped $Zn_xCd_{1-x}Se$ alloy nanocrystals and explain what experiments they might enable.

20. What could one learn by self-assembling pre-determined mixtures of capped metal nanoclusters and capped semiconductor nanoclusters in the form of a film, and what would you do with this newfound knowledge?

21. Describe how you would probe and what you would expect to happen when an aurothiol nanocluster superlattice film is gradually heated from 25 to 200 °C and imagine where you could usefully exploit this knowledge.

22. Why might it prove interesting to investigate how capped luminescent semiconductor nanoclusters self-assemble with various kinds of thermotropic liquid crystals?

23. Give synthetic details of how you would make and characterize the electrochromic performance (color contrast, color switching time, cycle stability) of a single nanocrystal of anatase?

24. How and why would you synthesize quantum dot barcoded nanorods with three segments comprised of capped gold nanoclusters at the ends and capped ZnS/CdSe nanoclusters in the middle of the nanorod?

25. Devise a robust self-assembly strategy whereby you could make electrical contact between a capped nanocluster and two nanoelectrodes.

26. How could you convert CdSe nanoclusters to CdSe nanorods and then induce them to self-assemble in an end-to-end fashion to form chains of nanorods?

27. Can you think of a way of self-assembling different types of open-framework lattices from semiconductor nanotetrapod building blocks? If this proved feasible would you expect these kinds of nanotetrapod networks to display novel properties, provide unexpected functions and offer opportunities for developing new products, processes or devices?

28. Can you come up with a way of self-assembling a nano-traffic light made of a single red, orange and green emitting semiconductor quantum dot? Could there ever be a use for something so far fetched?

29. Invent a way of making capped mercury nanoclusters and devise diagnostics that would differentiate between a liquid or solid mercury nanocluster core.

Shaping Color

CHAPTER 7

Microspheres – Colors from the Beaker

"The hues of the opal, the light of the diamond, are not to be seen if the eye is too near."

Ralph Waldo Emerson (1803–1882)

7.1 Nature's Photonic Crystals

Nature is the archetype sculptor and painter, but not all of nature's shapes are Platonic, and not all of nature's colors originate from the absorption of light by a pigment.[1–3] Bio-organic matrices in Nature mold organics and inorganics into biomaterials with complex form, and color can emerge from these materials if they are cast at the scale of light into bio-optical diffraction gratings.[4,5]

Natural gratings that diffract light are based on dielectric lattices with periodicity at optical wavelengths. They can have lattices that are 1D as in the coating of nacre in the shell of mollusks such as abalone.[6] They can be 2D, like whiskers of the sea mouse,[7] and 3D in species of blue iridescent butterfly from Peru[8–11] (Figure 7.1). While the shimmering quality of nature's iridescent biomaterials has been around for eons, coloration of this type has also been revered since antiquity in the sparkling mineral opal (Figure 7.2). The iridescence of opal originates from optical diffraction of light from a 3D grating made of close-packed silica microspheres, while in the butterfly wing it comes from a 3D grating of close-packed air holes in a chitin matrix.

This is an important lesson from Nature, unquestionably the world's most experienced materials chemist: she teaches that 3D optical diffraction gratings can be synthesized with dielectric lattices that are geometrically complementary, *i.e.*, one is the inverse of the other. Such a capability must be within our grasp, since, in the words of Jean-Marie Lehn: "If it exists, it can be synthesized".

7.2 Photonic Crystals

The long time fascination with geological and biological diffractive optics has recently found a new lease on life with the theoretical discovery of Yablonovitch[12]

325

Figure 7.1 Morpho didius *iridescent butterfly, with an SEM micrograph showing a representative example of such a diffracting biological microstructure*
(Left image reproduced with permission from encyclopedia.thefreedictionary. com/Morpho%20butterfly; Right image reproduced with permission from Ref. 11)

and John.[13] The combination of high dielectric contrast and periodicity at the light scale creates a unique class of materials, called photonic crystals. When made with the right materials and structure they can be considered to be the optical analogue of the semiconductor[14,15] enabling them to be used as active materials in all-optical transistors, diodes, and other devices.[16,17] Both top-down engineering and bottom-up self-assembly strategies can be used advantageously to fabricate photonic crystals.

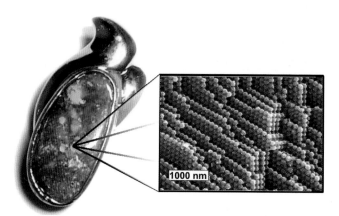

Figure 7.2 *A gemstone quality iridescent natural opal and an SEM image showing its microstructure comprised of close-packed silica spheres*
(Left image reproduced with permission from www.opalmine.com/graphics/ 4071.jpg; Right image reproduced with permission from www.adelaide.edu.au/ microscopy/services/instrumentation/gallery.html)

7.3 Photonic Semiconductors

The solution of Maxwell's equations for the propagation of light in a dielectric lattice shows that discontinuities appear in frequency $\omega(k)$ *versus* wavevector k photon dispersion due to optical Bragg diffraction.[18,19] This concept is expressed in Figure 7.3, where the analogy is brought forth between an electron propagating in a 1D crystal lattice and a photon in a 1D photonic lattice. The continuous electronic and photonic density of states for an electron and photon in a vacuum breaks up in a crystal and photonic lattice into what are called stopgaps at wavevector k values (momentum) that correspond to standing waves, that is the condition for electron and photon Bragg diffraction.

The above idea can be extended to 2D and 3D photonic lattices, however, it is only for certain lattice structure 3D photonic crystals made from high dielectric constant materials like silicon, that an omni-directional or full photonic band gap (PBG) can develop in the photon density of states (Figure 7.4). The PBG is a frequency range $\omega(k)$ in reciprocal wavevector space k, where the propagation of light is forbidden in all directions in the photonic crystal. In this energy range, light incident on the photonic crystal is completely reflected and light created within the crystal is completely trapped. The gap can be incomplete, called a stopgap, when there are certain directions in which the range of frequencies is allowed to propagate. For instance, in Figure 7.4 we can see that a 2D photonic crystal can inhibit the propagation of light in a range of angles around the direction

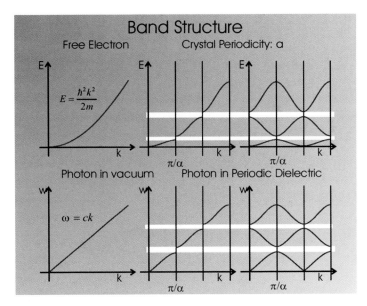

Figure 7.3 *Bragg diffraction of electrons in 1D atomic crystal, and photons in 1D photonic lattices. The lighter bands are ranges of electron or photon frequencies, which cannot propagate through the crystal because of coherent diffraction (Figure courtesy of H. Miguez)*

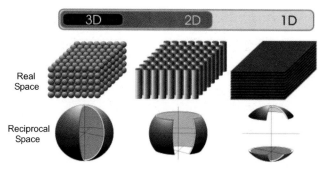

Figure 7.4   *Real space photonic lattice and reciprocal space photon density of states for 1D, 2D, and 3D photonic crystals. The areas delineated in gray represent the directions over which photon propagation is inhibited because of a PBG* (Figure courtesy of H. Miguez)

of periodicity, but light can "leak" perpendicular to it. The PBG in a 3D silicon photonic crystal is analogous to the electronic band gap (EBG) in a silicon electronic crystal in that it extends in all directions. Because of this analogy, the PBG can control the flow of light in a silicon photonic crystal in ways akin to EBG control of the transport of charge carriers in silicon. In both cases it is standing waves, of photons or electrons, which are the condition for band gap formation.

7.4 Defects, Defects, Defects

A material free of defects is almost non-existent, and a good thing too. It is the imperfection of solid-state materials rather than their perfection, which provides them with interesting properties and ultimately their function and utility. The same is true for photonic crystals where theory shows that by introducing micron scale point, line or bend defects into a perfect photonic lattice, electromagnetic modes emerge within the PBG which can localize light at a vacancy and confine and guide light along and around tiny waveguides. Point and line defects are illustrated in Figure 7.5. Micro-optical components made from photonic crystals on chips could form the basis of all-optical devices and circuits for next generation optical tele-communication technologies.

7.5 Computing with Light

Joannopoulos's[20] "photonic crystal micropolis" (Figure 7.6) is believed to represent the all-optical chip of the future. It inspired a generation of materials researchers to dream about the scientific and technological challenges of light computing with chips constructed from new miniature optical components. The all-optical chip is comprised of integrated microphotonic crystals with 1D, 2D, or 3D periodicity. Coupling of light from waveguide to photonic crystal may be accomplished *via* extrinsic defects built into the photonic lattice. Integrated

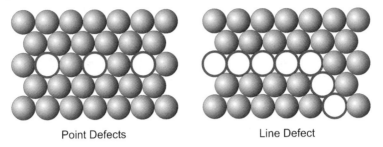

Point Defects   Line Defect

Figure 7.5 *Point and line defects, two possible types of defects in photonic crystals*

photonic crystal components on all-optical chips are envisioned to function as low threshold lasers, wavelength division multiplexers, wavelength dispersion compensators, and switches amongst others.

While photonic crystals exist on the road map of photonics technology, the science and applications of photonic crystals are still in the earliest phases of development, and there are many crucial questions that still need to be addressed. These questions include: is computing with light in microphotonic crystal chips science fiction or reality? What are the materials challenges that have to be surmounted to reduce such an integrated microphotonic crystal chip to practice? Will top-down microfabrication or bottom-up self-assembly or a creative fusion of these

Figure 7.6 *Vision of a futuristic all-optical chip comprised of different types of photonic crystals and other optical elements*
(Reproduced with permission from Ref. 20)

two methodologies, directed self-assembly, enable microphotonic crystal chips to be manufactured? Can the degree of microstructural perfection, level of intrinsic defects and quality of extrinsic defects, all demanded for effective photon control in miniature optical circuits with minimal light losses, be engineered to photonic crystal specifications?

7.6 Color Tunability

Maxwell's equations for photonic crystals are scalable, in contrast to the Schrödinger equation for electronic crystals. This is because the electron wavelengths of atomic crystals are more or less fixed at the Angstrom scale whereas the length scale for photonic crystals can traverse ultraviolet to microwave wavelengths simply by changing the dimensions of the photonic lattice components. Therefore, one way to tune the photonic crystal properties is by manipulating the lattice structure and dimensions.

Maxwell teaches that another way to obtain tunability involves manipulating the dielectric or refractive index contrast of the materials that comprise the photonic crystal. Altering refractive index contrast can be achieved in a "passive" way by changing the composition of the photonic crystal from say silicon to germanium. On the other hand this can be obtained in an "active" mode by exploiting photonic crystal materials properties, such as optical non-linearity, metal–non-metal transition, liquid crystal, and ferroelectric birefringence, polymer gel swelling and shrinking, and so forth, to effect the refractive index change.

7.7 Transferring Nature's Photonic Crystal Technology to the Chemistry Laboratory

The blue iridescent butterfly is an impressive example of Nature's skill at creating complex protein forms. Its wing's inverse opal microstructure has obvious practical consequences – the creation of a successful diffractive optic for signaling, communication, defense, and temperature control. The survival of this biological photonic crystal technology over evolutionary time scales depended upon the morphogenesis of a protein into an optimized cellular structure with a modulated refractive index contrast variation across the entire surface of the butterfly wings.

Just as Nature perfected the shimmering diffractive optic wing-scale of the photonic crystal butterfly, materials chemists are using microspheres as building blocks to self-assemble synthetic opals, known as colloidal crystals, as templates to micro-sculpt diverse kinds of inorganic, organic, polymeric, and composite materials into a periodic table of micro-optical diffractive components. This diffraction toolbox does not always require the perfection demanded of optical telecommunications or optical computing devices. A panacea of interesting and useful effects arise from the ability to engineer the properties of materials comprising the periodic dielectric. Luminescence, optical non-linearity, magnetism, ferroelectricity, electrochromicity, liquid crystallinity, dimensional and phase changes, intercalation, chemical sensing, and optical amplification can all be used in synergy

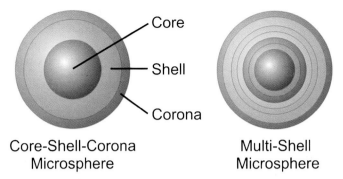

Core

Shell

Corona

Core-Shell-Corona
Microsphere

Multi-Shell
Microsphere

Figure 7.7 *A composite microsphere consisting of a central core, an intermediate shell, and an outer corona. Such layers can be repeated to give a complex multi-shelled architecture*

with the unique optical properties of photonic crystals that derive their color from the interaction of light with a microstructure.

7.8 Microsphere Building Blocks

A great deal of knowledge exists in colloid science concerning the chemistry of making and crystallizing silica and polystyrene microspheres with a very narrow size distribution. Monodispersed colloidal microspheres can be most easily made with diameters from tens of nanometers to several microns. They can also have multi-shell architectures like the core–shell–corona microsphere with functionality built into either of the components of the composite shown in Figure 7.7. The synthetic approaches to microspheres are generally a one-stage or a re-growth process. In the former, reagents are mixed and the microsphere is obtained, while the latter involves production of a seed in the first reagent-mixing step and then the seed is grown to the desired microsphere by the controlled addition of more reagent. In each case minimization of surface tension controls the spherical shape, and growth is most rapid on high-surface smaller particles providing an inherent size "focusing" mechanism. These methods provide access to microsphere building blocks with an extended range of sizes and excellent control over dispersity. The dispersity of a collection of microspheres is typically quantified as a percentage equal to the standard deviation in size over the mean diameter. It follows that to accurately quote sphere dispersity, at least several hundred spheres must be measured to obtain statistical significance.

7.9 Silica Microspheres

Monodisperse silica microspheres are made by the Stöber[21] or modified Stöber process involving the base catalyzed hydrolytic polycondensation of tetraethoxyorthosilicate:

$$Si(OEt)_4 + 2H_2O \longrightarrow SiO_2 + 4EtOH$$

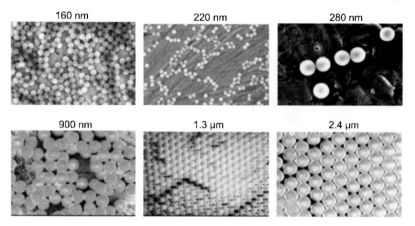

Figure 7.8   *Polystyrene microspheres of different diameters made by heterogeneous radical polymerization*
(Figure courtesy of V. Kitaev)

The single stage process is limited to microsphere sizes below 700–800 nm while the re-growth process extends to the range from about 20 nm to 3–4 μm.

7.10 Latex Microspheres

Polymer microspheres are primarily made by heterogeneous radical polymerization.[22,23] Different size ranges require different types of polymerization processes, like microemulsion, emulsion, emulsifier free, re-growth, and dispersion. Different kinds of initiators, surfactants, and co-monomers are used to establish size, charge, and surface functionality. Some examples of latex microspheres made in these ways are shown in Figure 7.8.

7.11 Multi-Shell Microspheres

Microspheres with composite layer structures offer a number of interesting opportunities for the study and utilization of colloidal photonic crystals.[24,25] They provide a means of introducing several, sometimes contrasting functionalities into the photonic lattice. Shells can be fluorescent, magnetic, photoactive, semiconductive, or sacrificial, to name only a few.

Core–corona and core–shell–corona microspheres are usually synthesized using the one stage or re-growth Stöber methodology for silica and some kind of heterogeneous radical polymerization for latex components, as described above. The approach to these multi-shell microspheres can be illustrated with a silica architecture based on a 500 nm non-fluorescent core, a 250 nm fluorescent shell and a 250 nm non-fluorescent corona (Figure 7.9). First the silica core is grown, and then used as a seed for the re-growth of a modified silica shell in which is incorporated a trimethoxysilane derivatized fluorescein molecule. This is achieved by coupling

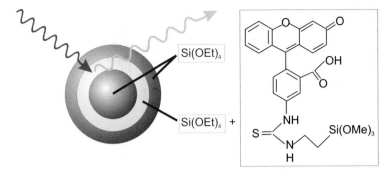

Figure 7.9 *Core-shell-corona silica microspheres with a fluorescent shell synthesized by incorporating a silane-derivatized fluorescein dye molecule during particle re-growth*

fluorescein isothiocyanate with the amine group of an aminoethyltrimethoxysilane. Finally, an outer corona of silica is grown on the core–shell to give the desired core–shell–corona fluorescent microspheres.

Fluorescent microspheres of this type have been used for probing colloidal crystals grown in surface relief patterns in silicon chips using depth imaging confocal optical microscopy.[26,27] An example of a core–shell–corona silica colloidal crystal confined within a square pyramidal shaped microwell in a silicon wafer reveals the <100> orientation and the individual layers containing 3×3, 2×2, and 1 microspheres (Figure 7.10).

7.12 Basics of Microsphere Self-Assembly

Microspheres are driven to assemble into colloidal crystals as a result of the balance of repulsing electrostatic forces and attracting van der Waals forces between them. These colloidal interactions are amenable to a wide range of tuning because of the ability to control the surface properties of the microspheres by tuning surface charge and surface functional groups.[28] To appreciate the process of microsphere self-assembly and crystal growth it is useful to take a cursory look at the under-pinnings of the colloidal interactions responsible for crystallization.

Electrostatic colloidal interactions between charged microspheres results from a hierarchy of many body interactions. The main model used to study such inter-actions is the DLVO theory. It takes into account mutual interactions amongst the microspheres as well as with a sea of surrounding ions. The local electric potentials will be dependent on the concentration, type, and charge of ions, as well as the dielectric constant of the fluid. By assuming the microspheres are fixed in space relative to the simple ions in the surrounding fluid it is possible to calculate the screened Coulomb repulsion for the case of pairs of microspheres given a radius and charge number. The electrostatic repulsion is balanced against the van der Waals attraction arising from sympathetic fluctuations in the particles electron distributions, where a Hamaker constant accounts for the scale of the attraction and

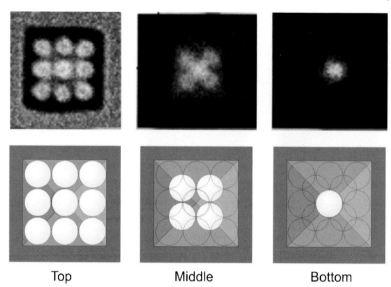

Top   Middle   Bottom

Figure 7.10   *Confocal fluorescent depth images of an (100) oriented core-fluorescent shell-corona silica colloidal crystal confined in a square pyramidal-shaped microwell ansiotropically etched in a (111) silicon wafer. The top images are obtained through confocal fluorescence microscopy, while the bottom images show schematically how the particles are packed into a pyramidal pit* (Reproduced with permission from Ref. 26)

depends on the dielectric constant mismatch between the microspheres and the surrounding fluid.

The lessons learned from this cursory glance at colloidal interactions is that surface electrical double layer and van der Waals forces between microspheres determine their assembly. Therefore, to grow high structural quality colloidal crystals it is crucial to properly clean microsphere surfaces of ions or molecules adventitiously adsorbed in the synthesis process. In this way, one can tailor microsphere surface charge either positive or negative, which can be measured as a zeta potential, or anchor neutral surface layers. Control of double layer repulsions is affected by added electrolytes, which screen surface charge. Armed with this toolbox of microspheres and building rules, it is now possible to plan a rational approach to self-assemble them into different kinds of colloidal crystal morphologies.

7.13 Microsphere Self-Assembly – Crystals and Films

Investigations of the underlying chemistry and physics for the organization of microspheres into colloidal crystals has a long history, however, it is only recently that methods for creating high structural and optical quality samples have been developed. The approaches for growing colloidal crystals and films include (i) sedimentation, (ii) electrophoresis, (iii) hydrodynamic shear, (iv) spin coating,

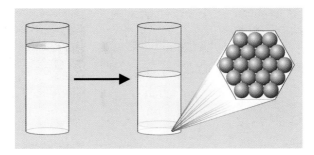

Figure 7.11 *Sedimentation growth of a colloidal photonic crystal*

(v) Langmuir–Blodgett layer-by-layer, (vi) parallel plate confinement, and (vii) evaporation induced self-assembly (EISA).

To illustrate just two of these methods let us begin with the sedimentation strategy, which involves an aqueous dispersion of microspheres in a container that are allowed to settle to the bottom of the container under gravity (Figure 7.11). The velocity of the moving front can be monitored, and is given by the Stokes equation: $v = d^2(\rho_s - \rho_w)g/18\eta$, where d is the microsphere diameter; ρ_s, ρ_w the density of the microsphere and water, respectively; and η the viscosity of water. This is a way for determining the diameter of the microspheres as well as growing reasonable quality colloidal crystals. Usually when growing crystals by sedimentation it is necessary to increase the density and viscosity of the solvent so that settling occurs very slowly. One of the foremost ways to make colloidal crystal films is known as EISA, shown in Figure 7.12, in which a substrate is placed vertically in a dispersion

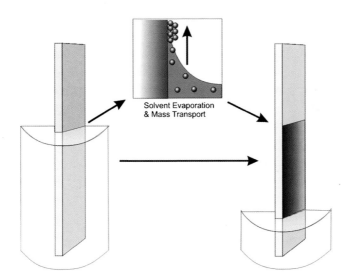

Figure 7.12 *EISA of a colloidal crystal film. Spheres are driven to the meniscus by convective forces, and capillary forces between them drive their assembly into an ordered array*

of microspheres in ethanol.[29] The formation of microsphere multilayers using this technique is based on a series of in-depth studies on the formation of 2D arrays.[30–34] As the ethanol evaporates, the microspheres are driven by convective forces to the meniscus, whereby they are coerced by capillary forces to self-assemble into a colloidal crystal film, with a layer thickness determined by the microsphere diameter, volume concentration, and the wetting properties of the ethanol on the substrate of interest. To grow films of silica spheres with diameter larger than about 450 nm, an accelerated evaporation must be employed to counter rapid sedimentation. Both heat[35,36] and vacuum[37] can be used for this purpose.

7.14 Colloidal Crystalline Fluids

As we have seen, it is fairly straightforward to assemble microspheres into a close-packed crystal. Another useful way of assembling microspheres is by electrostatic repulsion in the fluid phase. In a typical aqueous solution the charges on the surface of a microsphere will be more or less screened by solvated ions, causing sphere–sphere repulsion only when spheres get very close together. To reduce charge screening it is necessary to reduce the level of ionic impurities in the medium to negligible levels. This is done first by transferring the sphere suspension into distilled deionized water, then removing trace ions with the use of ion exchange resin. When such a process is diligently carried out, charge screening is reduced to the point that sphere–sphere interactions become long ranged and extremely repulsive. At a certain volume fraction of spheres, the entire suspension will crystallize into a non-close-packed assembly having a cubic (bcc, fcc) or hexagonal (hcp) crystal structure.[38,39] Colloidal crystals consisting of polystyrene spheres in deionized water are shown in Figure 7.13.

While such non-close-packed colloidal crystals in solution have been known for some time, they are of limited utility due to their fluid and fluxional nature. A major

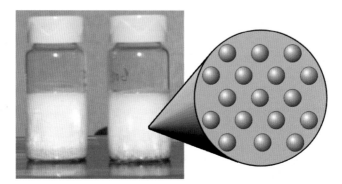

Figure 7.13   *Iridescence of colloidal crystals formed in a de-ionized dispersion of polystyrene particles of diameter 150 nm, left, and 200 nm, right. Their diffracting properties are due to a non-close-packed crystal of spheres assembled through repulsive electrostatic interactions*

breakthrough in this area occurred when charged spheres were instead assembled in a deionized aqueous solution of polymerizable monomers. Following the polymerization, the sphere array is trapped inside a hydrogel, which preserves the order against external disturbances.[40] In addition, the hydrogel matrix can be swollen or shrunk by modifying environmental conditions such as pH and temperature[41,42] or the concentration of a predetermined analyte. Dimensional changes in the hydrogel are manifest as a change in the crystal lattice dimensions, which concomitly modifies the diffracted wavelength. A wide range of functional monomers have been evaluated leading to a range of highly sensitive detectors.[43] The nature of the spheres can be modified to allow tuning by light[44–46] or magnetic fields.[47] The water can be removed from such a system and replaced by a polymerizable monomer, giving solid and robust polymeric replicas.[48]

7.15 Beyond Face Centered Cubic Packing of Microspheres

Materials chemistry has risen to the challenge of making photonic crystals with 3D microstructures comprised of well-ordered lattices of microspheres, as well as inverted replicas thereof. These are known as colloidal and inverse colloidal photonic crystals, respectively.[49] Because the most stable packing arrangement of microspheres is face centered cubic,[50] research on colloidal photonic crystals has been dominated by this structure. That is not to say that other microsphere lattices and their colloidal photonic crystal properties have gone unnoticed, quite the opposite. In fact, the diamond lattice arrangement of silicon microspheres has represented a long sought after colloidal crystal because of a very large predicted PBG. The tetrahedral lattice of microspheres is metastable, and all colloidal assembly techniques have so far failed to stabilize this structure. Nano-robotic maipulation can position microspheres one-at-a-time to make the diamond colloidal crystal – a monumental scientific accomplishment but not necessarily a practical technology.[51] Besides diamond constructs, attempts to self-assemble binary colloidal crystals with lattices built from microspheres of two distinct sizes also have a long and interesting history,[52–54] but not many examples are known in which binary structures and optical properties have been reported to be under control. It is known from theory that certain binary colloidal lattices with a sufficiently large refractive index contrast can develop a complete PBG, but efforts to demonstrate this experimentally have not been reported. Making structurally well-defined binary colloidal crystals of different kinds can be achieved by a layer-by-layer assembly technique whereby alternately large and small microspheres are deposited on a substrate to build-up a binary lattice. It has proven difficult, however, to grow well-ordered binary microsphere layers much beyond two using this approach.[55] The synchronous assembly of two different sphere sizes is also possible by employing accelerated evaporation to offset disparate sedimentation rates.[37] This method has thus far only been developed to form well-ordered surface patterns, samples of which are shown in Figure 7.14. These patterns are produced over large areas, and in some cases were observed to form 3D structures.

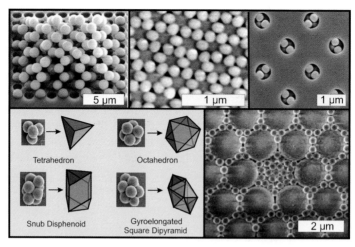

Figure 7.14 *Examples of different non-fcc sphere assemblies. Top-left, a diamond symmetry colloidal lattice formed through nano-robotic manipulation. Top-middle, a two-layer binary colloidal crystal formed through layer-by-layer deposition. Top-right, tetrahedral sphere aggregates made by templating in cylindrical pits. Bottom-left, a series of sphere aggregates made by sphere segregation into emulsion droplets. Bottom-right, representative example of a binary sphere surface layer made by tandem accelerated evaporative deposition*
(Reproduced with permission from Refs. 37,51,55–57)

An impressive library of colloidal aggregates have been formed by templating in surface patterns[56] as well as by aggregation in a homogenous emulsion.[57–59] In the latter method, a dispersion of cross-linked polystyrene microspheres with sulfonated surfaces is produced in toluene. This dispersion is then added to a solution of Pluronics surfactant in water, and an emulsion produced by vigorous agitation. As the toluene is allowed to evaporate, the toluene droplets shrink and the microspheres in the droplet are pulled together into a stable cluster through capillary forces. For a given number of microspheres a single geometry cluster is observed, and collections of identical clusters can be produced by centrifugation in a density gradient. Asymmetric spheres can also be made by directional deposition of a coating onto a sphere monolayer,[60] and these spheres subsequently coupled into dimers.[61] The self-assembly of such exotic aggregates, however, has yet to be explored.

A cross-section of the structural diversity achievable when producing non-fcc assemblies is shown in Figure 7.14. The diversity of sphere arrangements most likely extends beyond the few types of 2D lattices and finite 3D clusters here mentioned, and the discovery of new ways to make 3D colloidal crystal structures analogous to the ones described remains a daunting but worthy challenge.

7.16 Templates – Confinement and Epitaxy

The self-assembly of microspheres is governed by surface forces interacting between them. It is not surprising then, that the assembly process be quite sensitive to boundary conditions in the form of template surfaces. Wafer processing techniques allow access to a great variety of surface features. For instance, optical or electron lithography can create pits in a substrate matching particular colloidal crystal faces. This colloidal epitaxy can orient the colloidal crystal with its templated face parallel to the substrate, as well as increase the order in the system by nucleating a perfect crystal.[62-64] Other useful types of surface features are angular walls anisotropically etched into silicon. When a single crystal silicon (100) wafer is exposed to alcoholic base, such as KOH in isopropanol, it will be etched preferentially exposing Si {111} crystal planes. By patterning a silicon wafer with an etch resist, structures such as V-grooves, rectangular channels, and pyramidal pits can be generated. Micromolding in capillaries (see Chapter 2) can also generate surface features with flat walls, which can be ideal for confining and organizing spheres. A flat surface will most easily template a close-packed plane, thus the colloidal crystal close-packed {111} planes grow parallel to flat templating walls. This in-silicon chip confined colloidal crystal growth can result in single-crystal assemblies with a preset crystallographic orientation and rigorously understood optical properties.[65-67]

Self-assembly of a stirred ethanol dispersion of silica microspheres onto a vertically held wafer results in the growth of silica colloidal crystals exclusively within the surface relief patterns. Agitation of the dispersion maintains the spheres in a levitated state while ethanol evaporation and capillary interactions drive the spheres into the surface patterned regions where they self-assemble to form in-silicon chip colloidal crystal patterns. The assembly of colloids into a V-shaped microchannel, and the orientation of the resulting crystal planes, is shown in Figure 7.15.

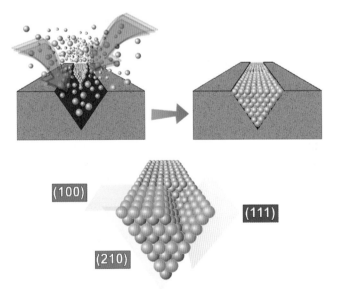

Figure 7.15 *Evaporative assembly of spheres into V-grooves anisotropically etched into a single crystal silicon (100) substrate results in a single crystalline fiber with well-defined crystallographic orientations. The orientation of highlighted crystal planes is shown below*

7.17 Photonic Crystal Fibers

Shape is everything in materials chemistry and form is at the heart of function. Nowhere can this be truer than in glass optical fibers for guiding and transmitting light by total internal reflection over huge distances. The optical fiber can be considered as a new optical component that revolutionized optical tele-communications. It came as quite a surprise to the materials community when a new generation of optical fibers emerged in the nineties known as photonic crystal fibers.[68] These microstructured fibers are based on 1D and 2D constructs, the former being described as a periodic dielectric based on a coaxial, micro-laminate architecture while the latter comprises a micropattern of air holes. Both of these microstructures traverse the entire length of the fiber. The 1D photonic crystal fibers are usually structured in the form of a polymer–inorganic multilayer and display a 1D PBG in a direction orthogonal to the axis of the fiber with a corres-ponding high-reflection efficiency in that direction.[69] These PBG fibers are being touted for soldier battle uniforms, designed to reflect a "friendly" IR wavelength, protect from radiation and alter their color for camouflage.

In the 2D photonic crystal fibers, there are two main categories of micro-structures. One type has a high index solid core and the other a low index air core, both types being surrounded by a regular micropattern of air holes. The former guides light by total internal reflection in the core whereas the latter guides light by core confinement due to the existence of a 2D PBG.[70] These air core microfibers can be designed to be single mode over an unlimited wavelength range and are rather insensitive to bend light losses, and may find applications in high-capacity transmission of light and switching and shaping of light pulses.

The utilization of controlled size and shape microchannel surface relief patterns lithographically defined in silicon substrates leads to a straightforward route to 3D photonic crystal fibers. Highly ordered and oriented colloidal photonic crystal microchannel templates are used for creating a silicon–silica photonic crystal through silicon infiltration. Upon removal of the silica template by etching in a fluoride-based medium the fibers are released from the template to give oriented free standing 3D inverse silicon colloidal photonic crystal microfibers.[71] Repre-sentative examples of microfibers with rectangular and V-shaped cross-sections, as well as 1D and 2D photonic crystal fibers, are displayed in Figure 7.16. These novel 3D photonic crystal fiber constructs provide a new class of optical components with a complete PBG along transverse and longitudinal directions of the microfiber axis that can be tailored to lie in the optical telecommunication wavelength range. One can foresee a myriad of opportunities where 3D colloidal photonic crystal micro-fibers are self-assembled into optically functional microphotonic crystal devices and circuits.

7.18 Photonic Crystal Marbles

EISA of a dispersion of colloidal dimension particles (nm to μm) in aqueous or non-aqueous solvents is a general way to organize colloids of different size, shape, and composition into regular arrays and with particular morphologies. Examples

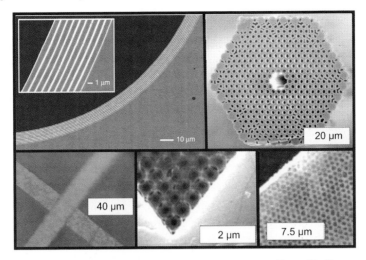

Figure 7.16 *Representative examples of photonic crystal microfibers. 1D fiber, top-left, 2D fiber, top-left, and 3D fibers on the bottom*
(Reproduced with permission from Refs. 69–71)

include surfactants and inorganics to make periodic mesoporous metal oxide films, capped semiconductor nanoclusters to make nanocluster superlattice films, and silica or latex microspheres to make colloidal crystal films. When EISA is performed in a confined geometry, directed evaporation induced self-assembly (DEISA), colloidal particles can still organize into a periodic array but with a form that is directed by the size and shape of the space in which the process occurs. Thus, DEISA performed with surfactants and inorganics in aerosol droplets yields periodic mesoporous silica nanospheres, and DEISA conducted with capped metal nanoclusters or microspheres in surface relief patterns that have been lithographically defined in a substrate yields nanocluster designs and colloidal crystal patterns.

When an aqueous solution is forced under pressure through a small aperture in the presence of an AC electric field, dubbed the electro spray process, the surface charge on the resulting thin liquid jet can make it break into homogenously sized spherical droplets. When this is performed with an aqueous dispersion of microspheres, each uniform droplet contains a preset quantity of microspheres. As the water evaporates from the microsphere suspension droplets, the capillary force in conjunction with colloidal forces cause the microspheres to self-assemble within the spherical droplets to create a close-packed colloidal crystal with a sphere shape.[72] These so-formed photonic crystal marbles have been formed with diameters in the range 60–100 μm, and most interestingly are found to display angle independent optical Bragg diffraction of light according to the Bragg–Snell equation. The explanation of the angle independence is related to formation of the thermodynamically stable {111} plane of the fcc colloidal lattice at the air–water interface of the spherical droplet during the DEISA process. Therefore, whenever light is reflected from one of these spheres it is occurring at normal incidence on the

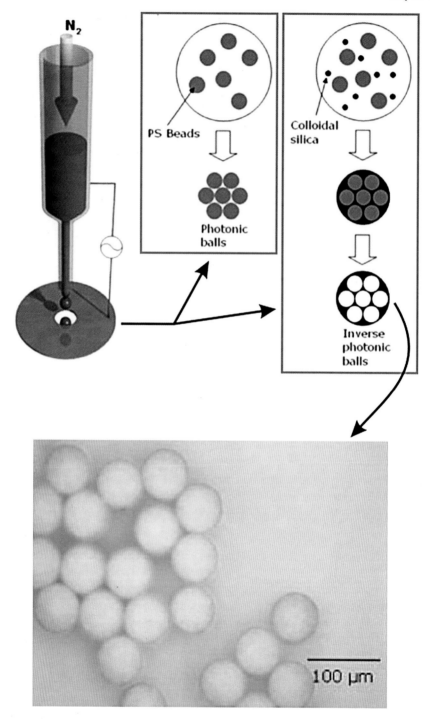

Figure 7.17 *Scheme depicting the assembly of regular and inverse photonic crystal marbles, the latter of which is shown in an optical image on the bottom* (Reproduced with permission from Ref. 72)

surface {111} planes. When the droplet confined DEISA process is performed using an electro-sprayed mixture of latex microspheres and colloidal silica, composite photonic balls of latex and silica form, from which the latex can be removed by calcination to create inverse colloidal crystal replicas of the colloidal crystal balls of the type displayed in Figure 7.17. The angle-independent hues of photonic balls speak well for their utility as color pigments in a range of applications. A related way of making colloidal crystal spheres as well as other shapes like donuts, involves droplet confined DEISA on the surface of liquid perfluorodecalin held in a Petri dish at a temperature where the microsphere dispersion solvent is volatile.[73]

7.19 Optical Properties of Colloidal Crystals – Combined Bragg–Snell Laws

One of the simplest ways to probe the structural and optical quality of colloidal crystals is by observation of their optical reflectance or transmission spectra. Most colloidal crystals form with the {111} lattice planes parallel to the substrate surface, as shown in Figure 7.18. One then normally observes, in the optical spectrum at lowest frequencies, the first stop band described to a first approximation by the combination of the diffraction Bragg law:

$$\lambda = 2n_{\text{eff}}d_{(hkl)}\cos\theta_{\text{in}}$$

and refraction Snell law:

$$n_{\text{out}}\sin\theta_{\text{out}} = n_{\text{in}}\sin\theta_{\text{in}}$$

to give the master equation for the wavelength of the first stop band:

$$\lambda_{c(hkl)} = 2d_{(hkl)}\sqrt{\langle\varepsilon\rangle - \sin^2\theta}$$

where the effective dielectric constant $\langle\varepsilon\rangle$ of the colloidal crystal (dielectric constant corresponds to the refractive index squared):

$$\langle\varepsilon\rangle = \sum_i f_i\varepsilon_i$$

is the volume fraction weighted average of air and spheres which is 26 and 74%, respectively, for a face centered cubic colloidal crystal.

7.20 Basic Optical Properties of Colloidal Crystals

The Bragg–Snell law has been often applied to analyze the optical spectra of colloidal crystals, like that shown in Figure 7.19 for a silica colloidal crystal film. The band observed around 650 nm is the first stop band and corresponds to optical Bragg diffraction from the {111} set of lattice planes as indicated in Figure 7.18.

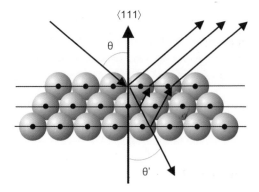

Figure 7.18   *The diffraction of light from close-packed {111} crystal planes in a colloidal photonic crystal*

The optical spectra respond to changes in microsphere diameter, refractive index, crystal orientation, and inversion of the structure to the so-called inverse colloidal crystal, according to that predicted by the Bragg–Snell equation. Furthermore, by angle tuning the colloidal crystal film and measuring the optical reflectance off-normal, one observes that the Bragg peak shifts according to that expected by the Bragg–Snell equation. The fringes to the high and low wavelength sides of the first stop band arise from a Fabry–Perot interference effect due to the finite thickness of the film. The high-energy bands correspond to higher order diffraction of the light in the colloidal crystal film. All of these optical bands are well predicted by a full vectorial solution of Maxwell's equations for the silica fcc colloidal crystal lattice.

Armed with the basics of silica and latex microsphere synthesis, colloidal-assembly and diagnostic optical characterization of colloidal crystals and films, one

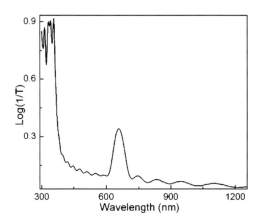

Figure 7.19   *Normal incidence spectrum of a colloidal crystal film made of ~270 nm diameter silica spheres. A high quality film will display a characteristic Bragg stop band (~700 nm), periodic fringes due to Fabry–Perot oscillations, and well-defined high-energy peaks due to higher order diffractions*

is in a position to be able to utilize them as templates for creating composite and inverted colloidal crystals out of a broad range of materials as well as using the inverted colloidal crystal as a mold for making colloidal crystals of compositions well beyond just silica and latex.[74] In what follows some building rules and case histories will be described that exemplify the power of this materials self-assembly approach for making colloidal photonic crystals out of materials that provide them with some interesting opportunities in the fields of digital imaging, optical switching, chemical sensing, lighting, and solar cells.

7.21 How Perfect is Perfect?

The greatest challenge that remains to be overcome before colloidal crystals can be engineered as photonic crystal devices is their experimentally achievable degree of structural perfection and optical quality. It is still not clear what type and level of flaws can be tolerated in a lattice of microspheres before it is rendered unacceptable as a colloidal photonic crystal device in optical telecommunication applications, or in future optical computers. Much discussion has centered on various kinds of intrinsic defects that can form in microsphere crystals self-assembled by sedimentation in gravitational and electric fields[75] and films formed by convection driven evaporative self-assembly.[76] It has been of great interest to comprehend how size and positional disorder of microspheres, point, line and planar defects, surface roughness, grain boundaries, and fractures in colloidal lattices affect photonic crystal properties like optical band structure, PBGs, photon group velocity, and scattering losses.[77–79] While much of these discussions falls in the realm of photonic crystal theoretical and experimental physics and engineering, it is ultimately the responsibility of materials chemistry to devise new ways to eliminate troublesome imperfections in colloidal crystals in order to enhance desired optical properties to levels deemed acceptable for photonic crystal device applications.[80]

It is clear that monodispersity of the spherical building blocks is essential. Monodisperse, however, is a confusing term because absolutely single size silica microspheres are impossible to obtain. There is always some degree of spread in size, which depends on the synthetic method used, post-synthetic techniques for further narrowing the size variation, and the size range of interest. Larger micro-sphere diameters, around 0.7–1.5 μm, are required for colloidal and inverse photonic crystal applications in the near infrared optical telecommunication wavelength region around 1.5 μm. However, several difficulties arise when working in this size range. One originates from the relatively large mass of large silica microspheres coupled to a rapid settling, which renders conventional tech-niques[75,76] unsuitable for the growth of high quality silica colloidal crystals, films or patterned samples. Either strategies of accelerated evaporation, through heat[35,36] for instance, or agitation,[67] to keep the spheres in suspension, must be employed to circumvent this. Another less recognized complication relates to microsphere doublets, or fused particles, which always exist in a population of as-synthesized silica microspheres. Even at the level of less than 1 in 10^4, doublets are

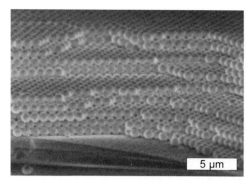

Figure 7.20 *860* nm *diameter silica sphere film made by high-temperature deposition (IHEISA)*

responsible for the formation of significant numbers of intrinsic defects, particularly dislocations and grain boundaries, which cause deterioration of the structural and optical quality of colloidal crystal films. An example of a large diameter silica microsphere colloidal crystal film produced by the high-temperature deposition and in which doublet concentrations have been reduced to undetectable levels is shown in Figure 7.20. Quite close to the highest achievable perfection, these types of samples clearly demonstrate the level of self-assembly success when all relevant structural influences are controlled.

The minimum criteria for acceptable structural quality of a colloidal crystal film is expressed in terms of measured optical quality, which embraces observation of the low energy $<111>$ optical stop band as well as resolved Fabry–Perot fringes arising from finite thickness effects of the colloidal crystal film, together with the detection of resolved high energy bands in the so-called flat band region of the transmittance spectra of the film (Figure 7.19).

7.22 Cracking Controversy

Recently, much attention has been paid to microsphere self-assembly aimed at growing colloidal crystal films and patterned samples of high structural and optical quality. Yet, one discovers that not much is documented about the pronounced tendency of the samples to crack after film formation due to capillary stresses and further fracture on drying and sintering.[81] The reality is that colloidal crystals in any form are found to crack after crystallization, thermal and chemical post-treatment, in ways that render them essentially useless as photonic crystal optical devices for telecommunication applications. A striking example of thermally induced cracking of a colloidal crystal micro-channel sample during silica microsphere necking (annealing to join adjacent spheres) is displayed in Figure 7.21 (left panel). This image highlights a major difficulty that needs to be addressed and overcome if colloidal photonic crystals are to be usefully engineered as miniature optical components for all-optical chips.

Figure 7.21 *Comparison between a templated silica colloidal crystal, which has been thermally necked (left), and the same type of sample necked by a room-temperature CVD process (right). Cracking can be a major problem, but is tractable if the right process is employed*

Cracking during crystallization is very hard to prevent. The initial crystal consists of spheres separated by a nanometer thick hydration layer, and this layer subsequently evaporates and causes cracking through lattice contraction. Slow drying may reduce this, or conversely, if the spheres are hydrophobic the hydration layer may be eliminated though these spheres are not repulsive and may not order as well. The use of templates can also reduce cracking by localizing the cracks at the crystal edges.

Assuming that a well ordered, relatively crack-free crystal can be obtained, the spheres must then be connected to stabilize the structure, and to allow sphere etching after infiltration of the voids. This process has traditionally been accomplished by heating at a temperature high enough to cause mass flow at sphere contact points ($\sim$600 °C for silica). Such temperatures can cause catastrophic cracking, see Figure 7.21, and is incompatible with many materials such as polymers one would want to use as substrates. One solution is to perform the necking process at room temperature, using vapor phase alternating chemical reactions. A straightforward method that has proven rather efficacious for attaining this objective is based on a layer-by-layer chemical vapor deposition (CVD) of silica by alternating treatment in silicon tetrachloride and water vapors onto the colloidal crystal.[82] A diagram depicting this process, leading to a buildup of silicon dioxide over the crystal surfaces, is shown in Figure 7.22. Using this method, structural parameters are strictly under control by varying the amount of silica layers deposited. Control over the microsphere connectivity and pore size using the CVD process is also found to be a valuable way of enhancing the mechanical properties of colloidal crystals. This can be appreciated from the measurements of the Young's modulus, or stiffness, and hardness of a silica colloidal crystal film sample after different number of silicon tetrachloride CVD cycles using depth-sensing nanoindentation. Typical results show that the stiffness increases from 1.5 GPa in an as-grown crystal film up to 15 GPa after four CVD treatments, while the hardness changes from 0.01 to 0.8 GPa. Typical values of these parameters for dense silica glass are 70 and 5.5 GPa, respectively. CVD necking can be seen to have a dramatic effect on the mechanical properties of the colloidal crystals.

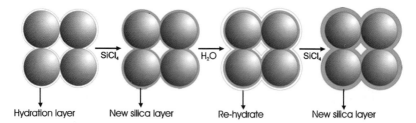

Figure 7.22   *A room temperature CVD method for coating a colloidal crystal with a*
stabilizing coating of silica or other oxide. Alternating gas-phase treatments
of SiCl₄ and water vapors are applied to the crystal, leading to an increase in
connectivity of the structure without thermal treatment
(Reproduced with permission from Ref. 82)

Optical properties can be monitored between deposition cycles, the red shift of the
Bragg peak signaling the amount of silica that has been deposited.

Another strategy is to treat the spheres at high temperature before self-assembly.[83] Once re-dispersed and crystallized they can be annealed at high temperatures
without worrying about concomitant volume contraction. Preventing agglomeration
at these temperatures may be quite difficult and if used should include rigorous
post-purification.

7.23 Synthesizing a Full Photonic Band Gap

The silicon woodpile was the first 3D photonic crystal made by top-down microfabrication techniques with a complete PBG at 1.5 μm.[84] It consists of a stacking of
layers of silicon rods in which adjacent layers are arranged perpendicularly. The
silicon woodpile is made by a complex layer-by-layer procedure that involves
lithographic patterning of silicon, infilling the surface relief pattern with a deposited layer of silica, planar machining of the patterned silica–silicon composite layer,
and deposition of the next silicon layer. These steps are repeated to give the desired
number of layers and selective etching of silica provides the desired silicon
woodpile photonic crystal.

The silicon inverse opal was the first 3D silicon photonic crystal made by bottom-up self-assembly with a complete PBG at 1.5 μm.[85] It presents a face centered
cubic photonic lattice of interconnected air spheres in a silicon matrix. The silicon
inverse opal is made by the CVD of silicon by thermal decomposition of disilane
within the interstitial voids of a silica colloidal crystal template. The silica is then
selectively etched from the composite silicon–silica colloidal crystal to create the
desired inverse silicon opal. First optical characterization results showed good
agreement between the optical spectra of the inverse silicon opal measured along
the $\Gamma - L$ <111> direction of the photonic crystal and the calculated band structure.
The presence of a complete PBG between the eighth and ninth photonic bands was
further confirmed by optical reflectance microspectroscopy measurements along the
symmetry directions $\Gamma - X$ <100>, $\Gamma - K$ <110> and $\Gamma - W$ <210> of cleaved
inverse silicon opals.[86] The results obtained for these different symmetry points in

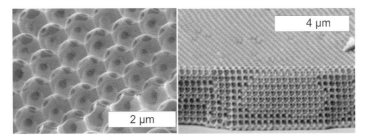

Figure 7.23 *Full PBG silicon microstructures by self-assembly. (Left) The first silicon inverse colloidal crystal, confirmed theoretical predictions of a full PBG in these materials. (Right) A shift from bulk to thin film crystallization allows subsequent integration of silicon inverse colloidal crystals in optical devices* (Reproduced with permission from Refs. 35,85)

the Brillouin zone, especially along $\Gamma - W$ $<210>$ where the width of the PBG is narrowest, demonstrate that there exists a range of frequencies around 1.5 μm that produces a reflectance peak invariant to orientation, an observation that is consistent with the existence of a complete PBG for the inverse silicon opal. Further confirmation of the PBG in these macroporous silicon materials has been performed using ultra-low defect density colloidal crystal films made by evaporative deposition. Images of both bulk and planar silicon inverse colloidal crystals are shown in Figure 7.23.

Significant efforts have also been expended into the development of alternative colloidal based semiconductor photonic crystals. One new approach centered on the principle of Micromolding in Inverse Silica Opals (MISO) has led to the creation of a new Si backbone topology with a theoretical 3D PBG of 12%.[87] In short, a polymeric latex opal template is first infiltrated with a sol–gel silica preparation,[88] and the template removed using O_2 plasma etching, solvent removal or thermal evaporation. The inverted SiO_2 opal is then infiltrated with silicon by CVD, followed by silica etching. Higher silicon filling and tubular connectivity are the two main structural differences with the original Si inverted opal. In addition to silicon, germanium also has a sufficiently high refractive index to display a full PBG, although this material is absorbing at frequently used telecommunications wavelengths.[89] The synthesis of a layered Si/Ge/air inverted opal structure has also been reported.[90] This approach makes use of ultra-fine control for the CVD deposition of Si and Ge as well as selective oxidation and etching of the Ge layer in order to microengineer the structures in terms of individual layer thickness and refractive index. It is clear that many great breakthroughs are awaiting to be uncovered in this exciting field.

7.24 Writing Defects

To exploit the unique optical properties of colloidal photonic crystals it is necessary, as mentioned earlier, to be able to control intrinsic defects. Naturally arising defects limit the quality of the optical properties of colloidal photonic crystals.

At the same time, it is also of paramount importance to be able to introduce in a predetermined and predictable manner different kinds of extrinsic defects into the photonic lattice. Defects such as point or line vacancies can be envisioned to make micron-scale lasers, waveguides, while larger scale structures may couple and switch light in a specific region of the photonic lattice.

Several strategies have been employed to introduce micron-scale defects into photonic lattices. In one, multiphoton laser writing using a confocal optical microscope induced polymerization of an organic monomer imbibed within the colloidal crystal and created micron-scale lines inside the photonic lattice.[91] Another employed colloidal crystal growth in and around lithographically defined micron-scale polymer microchannels, followed by removal of the polymer.[92] Modulated heterostructures with different colloidal crystal lattice dimensions and thickness have also been reported as an approach towards band gap engineering of photonic crystals. For instance, different colloidal crystals can form an alternating superlattice in a plane, or a superlattice of opal colloidal crystal films can be built by the consecutive layer-by-layer deposition.[93]

If the PBG of silicon inverse opals is to be exploited in optical devices, it is of paramount importance to be able to introduce extrinsic defects into a full PBG material. A step in this direction, using laser microwriting, has recently been reported.[94] By using a laser attached to a scanning optical microscope, micro-annealing of an amorphous hydrogenated silicon inverse opal causes a permanent local phase change to the nanocrystalline phase. A representative example that demonstrates laser writing of a line defect at the micron-length scale is shown in Figure 7.24, where the microstructure is probed by micro-Raman spectroscopy to confirm the refractive index modulation. This phase transition reduces the refractive index of the material, from 3.95 to 3.45, and a demonstrated decrease of about 12% is enough to introduce defect states in the complete PBG. Phase and refractive index properties were probed by microscope Raman and near infrared spectroscopy. These studies show that phase change based refractive index patterns can be rapidly and straightforwardly written with a laser in silicon inverse opals to create designed defects at different length scales, a step towards optical functionality.

7.25 Getting Smart with Planar Defects

There are a great variety of methods available for covering surfaces with 2D coatings of almost any imaginable material (see Chapter 3). Considering this, it is not surprising that a straightforward and general synthetic method has been recently described that enables a silica planar defect to be sandwiched between two silica opals.[95,96] It begins with EISA to deposit a uniform colloidal template thin film, composed of polystyrene spheres. The growth of a silica coating is then performed by CVD, which fills up the pores and is over-infiltrated to create a homogeneous silica over-layer of controlled uniform thickness. A second colloidal crystal is grown on top and fully infiltrated as well. Microscopy and optical characterization results of these new structures indicate they have high structural and optical quality due to the top and bottom layer near-perfect registry with the trapped silica layer

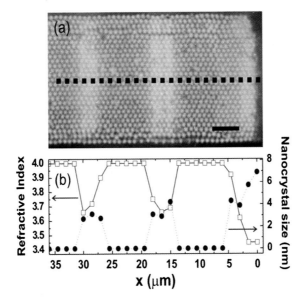

Figure 7.24 *Laser micro-writing and micro-annealing of a micron scale refractive index gradient and modulated extrinsic defects in an inverse amorphous silicon colloidal crystal, top. The refractive index is obtained by an evaluation of nanocrystal size and fraction in the amorphous silicon matrix by micro-Raman spectroscopy*
(Reproduced with permission from Ref. 94)

behaving as a planar defect. This is seen by the appearance of a transmission dip within the stopgap in good agreement with the allowed states predicted by a model based on a scalar wave approximation (SWA).[97] Changing the thickness of the defect changes its energetic position and modulates the position of the transmission dip relative to the band edges.

Structural defects can act analogously to electronic defects within the forbidden electronic gap in a silicon semiconductor. While it is relatively straightforward to tune the energetic position of an electronic defect by controlling the type of dopant and electric field, it is conceptually much harder to imagine the same process occurring in a photonic crystal. To reduce this idea to practice the defect must be actuated in some fashion either by changing its dimensions or refractive index. Conversely, changing these parameters in the surrounding photonic crystal would also change the defect state position relative to the photonic band edges. In this context, the first smart defect in a photonic crystal has recently been reported.[98] Such a construction consists of a polyelectrolyte multilayer, made by layer-by-layer electrostatic deposition, sandwiched between two silica colloidal crystal films. This architecture can be made by growing the multilayer directly onto the first crystal, or printing the multilayer from a PDMS stamp, and then growing a second crystal on top as illustrated in Figure 7.25. It is known that polyelectrolyte multilayers can have a great number of functionalities by virtue of the infinite choice of charged

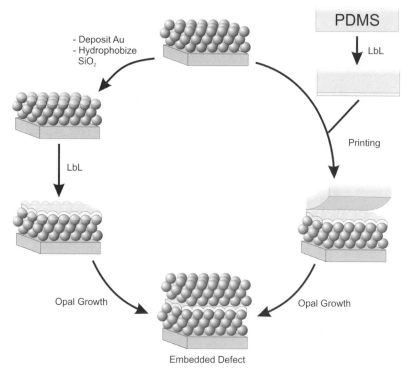

PDMS

- Deposit Au
- Hydrophobize
  SiO₂

LbL

Printing

LbL

Opal Growth

Opal Growth

Embedded Defect

Figure 7.25   *Scheme for the fabrication of a polyelectrolyte-based smart defect in a colloidal photonic crystal*

building units. One property that is common to most multilayer combinations is an intrinsic porosity, which confers to them the ability to absorb vapors. By exposing the photonic crystal with the smart polyelectrolyte defect to a controlled pressure of solvent vapor, this solvent condenses inside the defect pores and changes its refractive index. At higher pressures, the solvent vapor swells the defect, changing its dimensions. A combination of these effects has a dramatic influence on the optical properties, as seen in Figure 7.26. The defect state can be seen as a very sharp dip (width ~1 nm) in the middle of what would usually be a continuous stop-band peak. Upon increasing vapor pressure the position of this dip gradually red shifts over a range of 10 nm, mirroring the changes predicted by theoretical calculations. One of the most direct applications of this phenomenon could be in tunable laser sources. If a laser dye is incorporated into the polyelectrolyte mul-tilayer between colloidal photonic crystal mirrors in this type of sandwich het-erostructure, and has an emission peak within the photonic stop band, the excitation energy will be funneled to the position of the transmission dip of the defect. By changing the vapor pressure while exciting the dye, the laser wavelength can be easily and rapidly tuned over an arbitrary range of energies. Such smart defects in photonic crystals may form the basis of optical elements in future opto-electronic devices, networks and computing systems.

Defect State   Defect State Transmission   Defect State Tuning

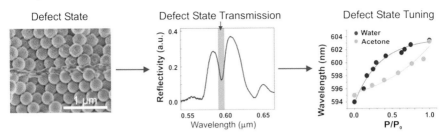

Figure 7.26   *Shown on the left is an SEM of an embedded polyelectrolyte multilayer defect between two colloidal crystals. Such a sample displays a sharp transmission window in the middle of the PBG. This dip can be shifted by controlled pressure vapor sorption through the swelling of the defect (water) or condensation inside the colloidal crystals (acetone)*

7.26 Switching Light with Light

If inverse silicon opals are ever to function as miniature components in an all-optical telecommunication or computing system, the PBG will need to be manipulated in time and space by purely optical means. One way of achieving ultrafast switching in silicon is by exploiting its ultrafast non-linear optical properties. The idea is to increase the density of free-carriers in silicon and hence its refractive index by using an intense femtosecond laser pulse at the band-edge of silicon. Optically excited electrons and holes induce ultrafast refractive index alterations in silicon causing shifts in the optical Bragg diffraction of the photonic crystal from its original position. Experimentally, a femtosecond pump laser incident on the photonic crystal shifts the PBG and a probe laser records the change in reflectivity. Ultrafast optical switching has recently been achieved for a silicon–silica opal where 30 fs reflectivity changes of around 1% were measured near the stop band for relatively low energy excitation pulses.[99] The experimental setup was specifically designed to separate diffracted from specular light thereby allowing surface reflection effects to be distinguished from Bragg diffraction of the silicon–silica opal.

7.27 Internal Light Sources

One of the holy grails of photonic crystal research is to observe complete inhibition of spontaneous emission at the PBG. Localization of light would enable the development of zero threshold photonic crystal lasers. A step in this direction has recently been made with the implantation of Er^{3+} ions in an inverse silicon opal.[100] The idea of this experiment is to overlap the Er^{3+} emission at 1.5 μm with the PBG of the silicon inverse opal, where the photonic density of states is expected to be very low and the Er^{3+} emission decay time should therefore be increased. Er^{3+} was implanted by ion-bombardment in the surface regions of a silicon–silica opal, followed by annealing at either 400 or 750 °C. At these temperatures Er^{3+}

photoluminescence (PL) can be selectively excited and radiative relaxation dynamics observed in a-Si but not a-SiO$_2$ at 400 °C or c-Si and a-SiO$_2$ at 750 °C. Selective etching of silica from these Er^{3+}-silicon–silica opals creates Er^{3+} doped inverse silicon opals. The PL lifetimes determined for Er^{3+} in the a-Si, c-Si and -SiO$_2$ phases of these opals were 0.3, 0.9 and 7 ms, respectively. Results from these preliminary experiments provide the design criteria to probe the effect of the PBG of the inverse silicon opal on Er^{3+} spontaneous emission in samples where exact overlap is achieved.

It has recently been shown that the emission from quantum dots can also be modified by a PBG.[101] Both the spectral distribution and lifetime of visible light luminescence from semiconductor nanoclusters can be greatly modified by encasing them in a titania inverse opal, even if such a photonic crystal does not have a complete PBG. These examples bode well for photonic crystal light sources operating at visible and infrared wavelengths.

7.28 Photonic Inks

One of the most striking aspects of a colloidal crystal is the brilliant opalescence observed when it is illuminated with white light. It is immediately obvious that whatever high-tech uses they might have, these self-assembled photonic crystals are a very efficient means to achieve color, *i.e.*, structural color through the interaction of light with a periodic dielectric microstructure rather than from a pigment or chromophore. This is a simple and robust use for these materials, with an almost infinite choice of construction materials, and with the necessary requirements for perception by the human eye being much less demanding than for an optical transistor. For instance, a series of high refractive index contrast inverse opals can be made of titania and zirconia.[102] Changing the template sphere size can achieve bright colors spanning the entire visible spectrum, and grinding the materials provides powdered opal "dyes" ready for use in various pigment requiring industrial applications, shown in Figure 7.27. However, it is important to note that if any fluid or solid finds its way into the void spaces, it can change both the color and color intensity by its refractive index, degree of filling, and filling homogeneity. Using photonic crystal pigments in this way will therefore require a thorough understanding of all material processing steps.

If colloidal crystal materials are to become part of next generation digital displays, supplanting LCD screens for instance, a more stringent set of requirements has to be met. First, the material should be provided as a thin layer on a planar substrate, since if it is not integration costs might be prohibitive. Second, the material should be subject to *dynamic* tunability, that is the ability to change the color of the material by an applied stimulus after its fabrication. Here changing the microsphere size is not an option, and tuning must be done by either modulating the refractive index of the components, or by somehow changing the lattice constant of the colloidal crystal. Third, the material should be spatially addressable by a stimulus that is straightforward to apply, in order to achieve an array of pixels. All these requirements have been met by a recent material, dubbed

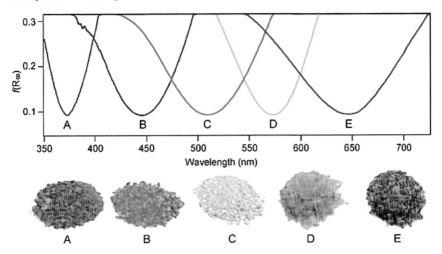

Figure 7.27 *Powdered opal "dyes" made of high refractive index inverse colloidal crystals, along with optical spectra above*
(Reproduced with permission from Ref. 102)

Photonic Ink (P-Ink).[103] P-Ink consists of a planar array of silica microspheres embedded in a redox-active partially cross-linked metallopolymer network made of polyferrocenylsilane.[104] The metallopolymer network can swell by absorbing solvent, leading to an increase in the lattice spacing of the colloidal crystal and a red shift in the back-reflected Bragg diffraction. Shrinking of the polymer network by ejection of non-solvent has the opposite effect, a decrease of the lattice spacing and a blue shift in reflected wavelength. This process can also be conducted in the gas phase, using a controlled pressure of a condensable gas to effect these same dimensional changes.[105] A plot showing the optical changes upon increasing then decreasing the vapor pressure of dichloromethane can be seen in Figure 7.28, with comparison to theoretically simulated curves, highlighting the optical quality and tunability of the material.

The most fascinating aspect of this material is that the amount of solvent retained inside the network can be controlled by changing the oxidation state of the iron atoms in the polymer backbone. The metal atoms in the backbone are coupled to each other, leading to a continuum of oxidation states, which is manifested as a continuous shift in the colors of these materials across a given wavelength range. These materials can be fabricated on conductive substrates, and subsequently actuated by a localized electrochemical stimulus.[106] The material is thus planar, can be conveniently tuned by an electric current applied to a spatially localized region, and display sub-second response times, portending an ideal platform for next-generation displays. A schematic view of how P-Ink works is shown in Figure 7.29. The absorption of solvent (first equilibrium arrows) modulates the lattice spacing, and the solvent absorption is in turn modulated by the state of charge of the metallopolymer network (second equilibrium arrows).

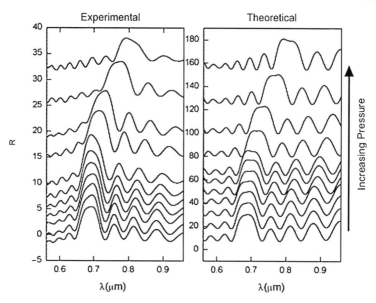

Figure 7.28   *Swelling of a polymer gel photonic crystal in dichloromethane vapor*
                  (Reproduced with permission from Ref. 105)

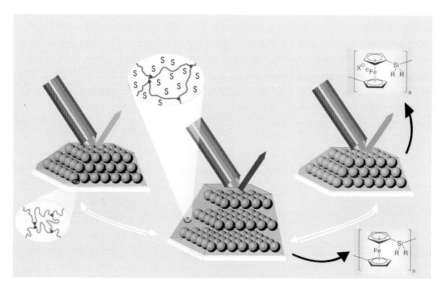

Figure 7.29   *How P-Ink works. The polyferrocenylsilane cross-linked polymer gel matrix*
                  can reversibly absorb solvent, increasing the lattice constant and red-shifting
                  the diffraction. Oxidation of the iron-based polymer gel can affect the
                  interaction with solvent, leading to a shrinking or expansion
                  (Reproduced with permission from Ref. 15)

7.29 Color Oscillator

Can rhythms and patterns in Nature be emulated in the laboratory? A well-known example of an abiological system that can generate spatiotemporal dynamic patterns is the activator–inhibitor chemical reaction, the most famous of which is that of Belousov and Zhabotinsky (BZ). This is a diffusion-controlled system where metal ions will periodically switch between two redox states, with a characteristic time period. This type of reaction forms the basis of the popular "chemical clock" science demonstration, where a solution will oscillate between two or more colors at constant time intervals depending on the initial concentrations of reactants. Imagine what you would get if you could integrate the self-sustained oscillations of a BZ chemical reaction with a swellable–shrinkable polymer-gel fashioned as a colloidal photonic crystal! This concept raises the possibility of a color oscillator, a material that displays structural color that swings between hues with a frequency and wavelength that is dependent on the chemical makeup and temperature of the polymer-gel as well as the dimensions and refractive index contrast of the colloidal lattice.

Proof-of-concept that such a photonic crystal based color oscillator could work in practice was first achieved by immersing an inverted colloidal crystal film made of a partially cross-linked co-polymer comprised of *N*-isopropylacrylamide and a redox active ruthenium bipyridyl complex into a typical BZ aqueous solution of nitric acid, malonic acid, and sodium bromate held at different temperatures.[107] As the ruthenium complex flips between Ru(II) and Ru(III) oxidation states, water is driven out of and into the colloidal lattice, induced by osmotic pressure and polarity changes in the polymer-gel. This causes reflected color from the film to reversibly alter between lower and higher wavelengths, corresponding to the shrunken and swollen dimensions of the photonic lattice. Observed oscillation times for this system in the range 4–19 °C were found to be around 7–2 min, respectively, and during the cycling, color rings were seen to spread from the center to extremities of the film with a wave velocity at 19 °C of about 0.05 mm s^{-1}. Optical images as well as recorded optical spectra are shown in Figure 7.30.

In this strategy, dynamic patterns in space and time created by a coupled chemical oscillator, can be transformed to an optical signal that can be temperature and chemically tuned. It seems that creative exploitation of this idea could lead to new kinds of dynamic micromachines and actuators and maybe pulsed chemical delivery systems.

7.30 Photonic Crystal Sensors

Since colors coming from photonic crystals are a function of the microstructure and its components, it follows that the coloration can be an easily detected signal which reports on the properties of the crystal. If the property change leading to changes in optical properties is coupled to an analyte of interest, we have the foundations of a sensing material with a color-based readout and almost limitless possibilities.

One of the most promising approaches for photonic crystal sensing makes use of non-close-packed polymerized colloidal crystalline arrays (mentioned earlier) that

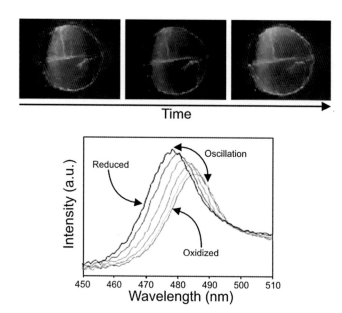

Figure 7.30 *An oscillating colloidal photonic crystal based on the Belousov–Zhabotinsky reaction. Optical images can be correlated with optical spectra below* (Reproduced with permission from Ref. 107)

can be made to swell and shrink reversibly by some external influence. As the diffraction colors of a colloidal crystal depend on the spacing between constituent microspheres then by placing a polyacrylamide hydrogel in the interstitial spaces of a colloidal crystal whose volume depends on how much water it contains, the colloidal crystal will swell or shrink in response to changes of its temperature, pH or ionic strength. Because the swelling of the colloidal crystal hydrogel increases the spacing between the microspheres this causes a shift in Bragg diffracted light of the colloidal crystal hydrogel and the detection of these color changes forms the basis of a new kind of chemical sensor.

An example is a crown ether functionalized polyacrylamide, which selectively binds Pb^{2+} ions.[108] Coordination of the divalent ion makes the backbone of the polymer positively charged, with some counter-anion from the solution to balance charge. The mobile counter anions can diffuse out into the bulk solution but the cations are fixed, and there is, therefore, a deficiency of anions inside the gel which accumulates a so-called Donnan potential. Charge screening between chains is thus decreased and they repel one another causing the gel network to swell and alters the color of the colloidal crystal hydrogel. The degree of swelling is proportional to the number of covalently attached charged groups. Removal of the lead causes the colloidal crystal hydrogel to shrink to its original volume. These Pb^{2+} induced color shifts can be detected for lead concentrations as small as 20 μmol. The observed order of cation complexation is $Pb^{2+} > K^+ > Li^+$ and an observed red shift of the Bragg reflection occurs with increased cation binding in support

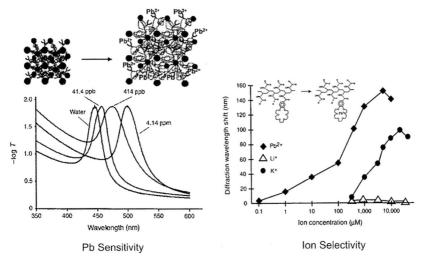

Pb Sensitivity   Ion Selectivity

Figure 7.31 *Crown ether functionalized colloidal crystal polyacrylamide hydrogel sensor selective for Pb^{2+} ions, showing the shift of the optical stop band at different Pb concentrations (left) and the sensitivity of the structure towards different ions (right)*
(Reproduced with permission from Ref. 108)

of the above charge-induced swelling model (Figure 7.31). Other transition metal ions can be sensed, simply by changing the chelating motif in the hydrogel.[109] A wide variety of biological species can also be sensed in this way by directly attaching enzymes to the hydrogel.[110] When the enzyme substrate is present it chemically acts upon it, changing the chemical makeup of the hydrogel and causing either an influx or outflow of solvent. This has been employed notably for sensing glucose in diabetes patients by incorporating the enzyme glucose oxidase inside the photonic crystal sensor. These materials can even operate in body fluids, and could be used for *in vivo* monitoring.[111] By attaching selective ion and molecule recognition sites to the polyacrylamide chains, there are almost limitless possibilities for sensing using colloidal photonic crystals.[112]

7.31 Colloidal Photonic Crystal Solar Cell

While PBGs and optical diffraction properties of photonic crystals have been well studied, less attention has been paid to another interesting property. The speed of light is simply a function of a material's refractive index, but the net group velocity of photons can be decreased if they experience back-and-forth reflections within the material. It turns out that photons propagating through a photonic crystal may present a very low group velocity for certain propagation directions and frequencies. If the photonic crystal is made of a material that has optical absorptions or interactions (stimulated emission in LASERs for instance), this slow motion of photons imply a higher probability of the photon being captured than in the disordered material. Therefore, a colloidal photonic crystal made of dye-sensitized nanocrystalline anatase TiO_2 could harvest more photons and display a higher photon to electron conversion efficiency than a conventional Grätzel type solar cell made of a random collection of dye-sensitized anatase nanocrystallites.[113–115]

This idea has been used recently with great success to modify the absorption properties of dye-sensitized titanium dioxide solar cells. A conventional sensitized solar cell material is first coupled to an inverse opal, also made of titania. The absorption enhancement at the photonic stop-band edges leads to an increase in optical absorption, and consequently to a more efficient solar cell.[116] The optical absorption enhancement arises from the slow group velocity of light at the low energy edge of the stop band. It is interesting that amplification is not observed on the blue side of the peak, given that this optical mode propagates mainly in the air-portion of the crystal and does not interact with the photoactive titania. This principle has been demonstrated to result in an increase in the short circuit pho-tocurrent efficiency across the visible spectrum (400–750 nm) by about 26%, relative to the same material with no photonic crystal lattice.

The optical amplification properties of photonic crystals could be used in many other areas, for instance, by building an opal analogue of a solid-state silicon solar cell. Nanostructured titania is also a great photo-oxidant, and microstructuring it in the form of an inverse opal could lead to devices which quickly and efficiently break down environmental pollutants when exposed to sunlight. Photonic crystals promise better solar cells, better photocatalysts, and better photoconductors. Essentially by taking any photoactive material and fashioning it in the form of an inverse opal, effects which transcend the sum of the parts can be observed and harnessed for practical purposes.

7.32 Thermochromic Colloidal Photonic Crystal Switch

This active colloidal crystal device is based on the thermally driven metal–non-metal transition of rutile type VO_2.[117] It can be synthesized by sol–gel infiltration or deposition[118] of precursors such as $V(O^{iso}Pr)_4$ or CVD[119] of various precursors including VCl_4 into a latex colloidal crystal template, followed by sacrificial removal of the latex template either with toluene, an oxygen plasma or calcination in air. A femtosecond laser pulse incident on an inverse colloidal crystal made of VO_2 induces a transition around 60 °C from the low temperature Peierls distorted non-metal form to the high temperature undistorted metallic form.[120,121] The large refractive index change between the non-metal and metallic forms of VO_2 cause the optical Bragg peak to shift within the femtosecond time frame of the laser pulse. This is an example of a synthetic thermochromic colloidal photonic crystal device that can manipulate light with light and could form the basis of an all-optical transistor. Clearly, there are uses for thermochromics besides color indicating thermometers and smart heat saving windows whose color and hence transmission varies with temperature!

7.33 Liquid Crystal Photonic Crystal

A liquid crystal is neither a liquid nor crystal but is an intermediate state of matter, known as a mesophase, that lacks the long range order of the crystalline state yet does not exhibit the randomness of the liquid state.[122] The most well-known class of liquid crystals is represented by the thermotropics, which are usually obtained by melting a solid comprised of molecules with anisotropic shapes like rods or discs into a mobile liquid in which they are partially aligned. In the liquid-state rods and discs have a tendency, through intermolecular interactions, to align in different ways along a major axis called the director field. This kind of molecular organization gives rise to different structural phases of liquid crystals, two of the most well known of which are the nematics or discotics that exhibit a degree of molecular order defined by an order parameter with a value between 1 and 0, defining the most ordered and most disordered states, respectively.

Liquid crystal phases can be identified by observing diagnostic optical birefringence patterns in a polarizing optical microscope, when the liquid crystal is placed between crossed polarizers. The birefringence has its origin in topological defects, which are director field patterns around defect lines that always exist in liquid crystals. On substrates, rod-shaped liquid crystals align themselves mainly in two ways, parallel and perpendicular to the surface, denoted as planar and homeotropic. Surface energy differences between these two states determine which is favored. Application of electric and magnetic fields to a liquid crystal can cause the director field to align along the E or H field direction, thereby providing a means of controlling refractive index and hence optical properties. A change in temperature can also switch the anisotropic state to the isotropic disordered one in which the properties of the liquid crystal are the same in any spatial direction.

Properties of the kind mentioned above enable the construction of liquid crystal light valves like those used in liquid crystal displays. Liquid crystals in confined geometries such as droplets in polymer film have intriguing director field patterns, like the bipolar one defined by the droplet sphere shape. The axis of the bipolar sphere can be controlled by an E field to be coherently aligned or random, which makes the film either transparent or reflective to incident light. This property forms the basis of polymer dispersed liquid crystals (PDLCs) used for instance in privacy windows.[123]

This brief introduction to liquid crystals allows one to imagine the consequences of introducing a liquid crystal into a colloidal photonic crystal. Busch and John[124,125] first proposed that such a hybrid construction could provide a means of electrically or thermally tuning the optical properties of a colloidal crystal by exploiting the birefringence properties of the liquid crystal. The idea here is that the refractive index differences between the anisotropic and isotropic states of the liquid crystal can be used to tune the optical stop band or PBG of a colloidal crystal. This can be achieved in principle by infiltrating a nematic liquid crystal into a colloidal or inverse colloidal crystal and using E field or temperature to control the director field with respect to the crystallographic axes of the colloidal crystal.

It turns out that the silica colloidal crystal provides a complex confinement geometry for a nematic liquid crystal, namely a periodic network of interconnected tetrahedral and octahedral interstitial sites. Because rod shaped molecules prefer to align parallel to the surface of the silica spheres, planar anchoring, the director field patterns and topological defects of a nematic in a silica colloidal crystal are quite complicated and difficult to define. Moreover, the anchoring energy of the nematic to the sphere surface makes switching times quite slow, a concern when trying to make a tunable photonic crystal device.

By contrast, the void spaces in an inverse silica colloidal crystal are essentially a periodic array of interconnected air spheres and the director field patterns of a nematic imbibed therein are simpler and in fact resemble the bipolar ones in liquid crystal droplets mentioned above.[126] Such an architecture is illustrated in Figure 7.32. The optical reflectance spectrum of the well-known rod-shaped nematic 4-pentyl-4$'$-cyanobiphenyl (5CB), infiltrated into an inverse silica colloidal crystal is shown in the figure, and displays two weak peaks at 25 °C around 640 and 600 nm. With increasing temperature these gradually shift and converge at 40 °C to a single more intense peak around 610 nm. The sample changes from white to red in the process, originating from the alteration in optical Bragg diffraction of the colloidal crystal induced by switching the liquid crystal from the nematic to the isotropic phase. The refractive indices obtained from fitting the optical data shown to the Bragg–Snell equation described earlier associates the two weak peaks at 600 and 640 nm to the ordinary n_o and extraordinary n_e refractive indices of the nematic in the void spaces of the inverse colloidal crystal. The former

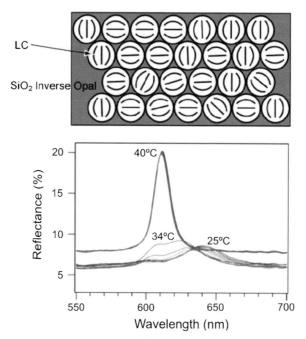

Figure 7.32 *A photonic crystal architecture comprising a nematic liquid crystal inside a silica inverse opal, top, can act as a tunable photonic crystal element, bottom* (Reproduced with permission from Ref. 126)

value is associated to domains with the bipolar sphere axis parallel, and the latter one to domains perpendicular to the direction of the incident light. This was proven by a creative experiment that involved placing a polyimide alignment film in contact with the liquid crystal colloidal crystal to induce either planar or homeotropic alignment and then observing the reflectance spectra of the two samples using linearly polarized light. The results defined the two-weak peaks in the optical reflectance spectrum in Figure 7.32 as being associated with n_o and n_e. This work could form the underpinnings of developing an optical switch tuned thermally, electrically, or even optically.

Reduction to practice of such a light-driven liquid crystal photonic crystal switch was achieved by mixing the liquid crystal 4-butyl-4′-methoxyazobenzene with 5CB, where the former can be reversibly photoisomerized from the *trans*- to the *cis*-form and *vice-versa* in a two color process at irradiation wavelengths 360 and 450 nm. The *trans*-form is sterically compatible with the rod-like nematic state of 5CB but the latter is not and forces 5CB to go isotropic. This interesting phenomenon provides the basis of tuning light with light in a liquid crystal–colloidal crystal switch.

7.34 Encrypted Colloidal Crystals

Imagine a colloidal crystal built from n close-packed microspheres, where every sphere encodes a bit of color-coded information that can be deciphered on demand. In a centimeter cube of such a recording medium one can therefore in principle address around 10^{12} bits of data. This form of 3D encryption can be seen to scale as m^n if one could create m individually readable colors in every one of n spheres. This is a gigantic storage density, which could if reduced to practice revolutionize information technologies like security systems and data storage. It turns out that colloidal crystals can rise to this challenge if constructed from smart microsphere building units.

The strategy involves synthesizing monodisperse microspheres with a core–corona or core–shell–corona architecture in which each component of the binary or ternary sphere comprises a different color dye anchored to a different polymer.[127] These microspheres are then crystallized in the form of a well-ordered colloidal lattice, by any means described previously. The glass transition temperatures of the polymer components of the microspheres are arranged, through judicious polymer design, to increase from the outside corona to the inside shell and core such that a gentle thermal treatment of the colloidal crystal causes the spheres to consolidate into a necked and mechanically stable colloidal lattice while maintaining the multilayer structure of the spheres. For the system to successfully function, dyes in spatially distinct regions of the multilayer spheres must be fluorescent and photo-bleachable at distinct wavelengths in a one or two photon-driven process, so that the color of a particular sphere in a specific plane of the colloidal crystal can be annihilated and read, on demand, with a laser coupled to a fluorescence confocal microscope.

Key requirements for success of this 3D recording system include the choice of polymer anchored dye combinations that ensure negligible energy transfer "cross talk" between dyes, as well as long-term stability of the dyes to the reading

Figure 7.33 *Two letters recorded at the same position but at different spectral positions in a dye-labeled colloidal crystal. This is made possible by selective photobleaching of two different dyes incorporated into the core or shells of composite microspheres*
(Reproduced with permission from Ref. 127)

laser and daylight. An example that fulfills these demands and which demonstrates color-coding in different sphere planes in the multi-dye colloidal crystal recording medium is illustrated in Figure 7.33, where two different letters have been written at different wavelengths on the same portion of the material.

7.35 Gazing in the Photonic Crystal Ball

Unquestionably, one of the most exciting recent discoveries to emerge from condensed matter physics is the photonic crystal. One of the most exciting developments that soon after followed in materials chemistry was the ability to synthesize these crystals by microsphere self-assembly into an impressive range of microstructures, compositions, feature sizes, and morphologies. These bottom-up colloidal photonic crystal constructs, through the phenomenon of optical Bragg diffraction in all three spatial dimensions, have demonstrated the ability to control the flow of photons by structuring at the light scale. At the same time, electronic elements of data transfer, data storage, and recently, super-fast processor chips[128] are today's reality. Undoubtedly, miniaturization will remain the driving force in photonics to mirror those remarkable achievements in electronics. There is a big question currently in the minds of university and industrial materials chemists, physicists, scientists, and engineers alike: can microsphere assembly compete with micromachining strategies in photonic crystal optical telecommunications technology? Can self-assembled photonic crystals actually be engineered with sufficiently high structural and optical quality to fulfill their promised potential? It is possible that bottom-up self-assembly strategies might provide some examples of 3D photonic crystal microstructures that are able to outshine their microfabricated analogues. Sustained materials chemistry and engineering collaborative research aimed at enhancing structural and optical quality of colloidal photonic crystals, controlling their morphology and integrating them with, and coupling them to waveguides, will be required to stay the course!

Whatever the answer to this question, it is already absolutely clear from the creative contributions appearing daily in the materials science literature that materials chemists are already having a heyday finding a myriad of unexpected uses for this new class of light-scale microstructures whose compositions seem limitless and whose colors originate from the interaction of light with the colloidal photonic lattice – "structural color" – rather than from the absorption of light by a chromophore or pigment.

7.36 References

1. C.W. Mason, Structural colors in insects I, *J. Phys. Chem.*, 1926, **30**, 383.
2. C.W. Mason, Structural colors in insects II, *J. Phys. Chem.*, 1927, **31**, 321.
3. C.W. Mason, Structural colors in insects III, *J. Phys. Chem.*, 1927, **31**, 1856.
4. M. Srinivasarao, Nano-optics in the biological world: beetles, butterflies, birds, and moths, *Chem. Rev.*, 1999, **99**, 1935.
5. R.O. Prum, R.H. Torres, S. Williamson and J. Dyck, Coherent light scattering by blue feather barbs, *Nature*, 1998, **396**, 28.
6. S.W. Wise, Microarchitecture and deposition of gastropod nacre, *Science*, 1970, 167, 1486; N. Watabe, Studies on shell formation. 11. Crystal–matrix relationships in inner layers of mollusk shells, *J. Ultrastruct. Res.*, 1965, **12**, 351; D.L. Kaplan, Mollusc shell structures: novel design strategies for synthetic materials, *Curr. Opin. Solid State Mater. Sci.*, 1998, **3**, 232.

7. R.C. McPhedran, N.A. Nicorovici, D.R. McKenzie, L.C. Botten, A.R. Parker and G.W. Rouse, The sea mouse and the photonic crystal, *Aust. J. Chem.*, 2001, **54**, 241.

8. H. Ghiradella, Light and color on the wing – structural colors in butterflies and moths, *Appl. Opt.*, 1991, **30**, 3492.

9. H. Tada, S.E. Mann, I.N. Miaoulis and P.Y. Wong, Effects of a butterfly scale microstructure on the iridescent color observed at different angles, *Appl. Opt.*, 1998, **37**, 1579.

10. D.J. Brink and M.E. Lee, Confined blue iridescence by a diffracting microstructure: an optical investigation of the *Cynandra Opis* butterfly, *Appl. Opt.*, 1999, **38**, 5282.

11. L.P. Biro, Z. Balint, K. Kertesz, Z. Vertesy, G.I. Mark, Z.E. Horvath, J. Balazs, D. Mehn, I. Kiricsi, V. Lousse and J.P. Vigneron, Role of photonic crystal-type structures in the thermal regulation of a *Lycaenid* butterfly sister species pair, *Phys. Rev. E*, 2003, **67**, 021907.

12. E. Yablonovitch, Inhibited spontaneous emission in solid-state physics and electronics, *Phys. Rev. Lett.*, 1987, **58**, 2059.

13. S. John, Strong localization of photons in certain disordered dielectric superlattices, *Phys. Rev. Lett.*, 1987, **58**, 2486.

14. C. Lopez, Materials aspects of photonic crystals, *Adv. Mater.*, 2003, **15**, 1679.

15. A. Arsenault, S. Fournier-Bidoz, B. Hatton, H. Míguez, N. Tétreault, E. Vekris, S. Wong, S.M. Yang, V. Kitaev and G.A. Ozin, Towards the synthetic all-optical computer: science fiction or reality? *J. Mater. Chem.*, 2004, **14**, 781.

16. S. John and K. Busch, Photonic bandgap formation and tunability in certain self-organizing systems, *J. Lightwave Technol.*, 1999, **17**, 1931.

17. D. Dragoman and M. Dragoman, Optical analogue structures to mesoscopic devices, *Prog. Quant. Electron.*, 1999, **23**, 131.

18. E. Yablonovitch, Photonic band-gap crystals, *J. Phys. Condens. Matter*, 1993, **5**, 2443.

19. S.L. Kuai, Y.Z. Zhang and X.F. Hu, Band structures, applications and preparations of photonic crystals, *J. Inorg. Mater.*, 2001, **16**, 193.

20. J.D. Joannopoulos, P.R. Villeneuve and S.H. Fan, Photonic crystals: putting a new twist on light, *Nature*, 1997, **386**, 143.

21. W. Stöber, A. Fink and E. Bohn, Controlled growth of monodisperse silica spheres in micron size range, *J. Colloid Interface Sci.*, 1968, **26**, 62.

22. P. Guiot and P. Couvreur, *Polymeric Nanoparticles and Microspheres*, CRC Press, Boca Raton, FL, 1986.

23. H. Kawaguchi, Functional polymer microspheres, *Prog. Polym. Sci.*, 2000, **25**, 1171.

24. F. Caruso, Nanoengineering of particle surfaces, *Adv. Mater.*, 2001, **13**, 11.

25. K. Ishizu, Synthesis and structural ordering of core–shell polymer microspheres, *Prog. Polym. Sci.*, 1998, **23**, 1383.

26. V. Kitaev, S. Fournier-Bidoz, S.M. Yang and G.A. Ozin, Internal photonic crystal lattice structures of planarized opal-patterned chips probed by laser scanning confocal fluorescence microscopy, *J. Mater. Chem.*, 2002, **12**, 966.

27. M.H. Chestnut, Confocal microscopy of colloids, *Curr. Opin. Colloid Interface Sci.*, 1997, **2**, 158.
28. P.C. Hiemenz and R. Rajagopalan, *Principles of Colloid and Surface Chemistry*, Marcel Dekker, New York, 1997.
29. P. Jiang, J.F. Bertone, K.S. Hwang and V.L. Colvin, Single-crystal colloidal multilayers of controlled thickness, *Chem. Mater.*, 1999, **11**, 2132.
30. P.A. Kralchevsky and K. Nagayama, Capillary interactions between particles bound to interfaces, liquid films and biomembranes, *Adv. Colloid Interface Sci.*, 2000, **85**, 145.
31. N.D. Denkov, O.D. Velev, P.A. Kralchevsky, I.B. Ivanov, H. Yoshimura and K. Nagayama, Mechanism of formation of 2-dimensional crystals from latex-particles on substrates, *Langmuir*, 1992, **8**, 3183.
32. A.S. Dimitrov and K. Nagayama, Continuous convective assembling of fine particles into two-dimensional arrays on solid surfaces, *Langmuir*, 1996, **12**, 1303.
33. K. Nagayama, Two-dimensional self-assembly of colloids in thin liquid films, *Colloid Surf. A – Physicochem. Eng. Aspects*, 1996, **109**, 363.
34. P.A. Kralchevsky and K. Nagayama, Capillary forces between colloidal particles, *Langmuir*, 1994, **10**, 23.
35. Y.A. Vlasov, X.Z. Bo, J.C. Sturm and D.J. Norris, On-chip natural assembly of silicon photonic bandgap crystals, *Nature*, 2001, **414**, 289.
36. S. Wong, V. Kitaev and G.A. Ozin, Colloidal crystal films: advances in universality and perfection, *J. Am. Chem. Soc.*, 2003, **125**, 15589.
37. V. Kitaev and G.A. Ozin, Self-assembled surface patterns of binary colloidal crystals, *Adv. Mater.*, 2003, **15**, 75.
38. R. Kesavamoorthy, S. Tandon, S. Xu, S. Jagannathan and S.A. Asher, Self-assembly and ordering of electrostatically stabilized silica suspensions, *J. Colloid Interface Sci.*, 1992, **153**, 188.
39. R.J. Carlson and S.A. Asher, Characterization of optical diffraction and crystal-structure in monodisperse polystyrene colloids, *Appl. Spectrosc.*, 1984, **38**, 297.
40. S.A. Asher, J. Holtz, L. Liu and Z.J. Wu, Self-assembly motif for creating submicron periodic materials – polymerized crystalline colloidal arrays, *J. Am. Chem. Soc.*, 1994, **116**, 4997.
41. K. Lee and S.A. Asher, Photonic crystal chemical sensors: pH and ionic strength, *J. Am. Chem. Soc.*, 2000, **122**, 9534.
42. J.M. Weissman, H.B. Sunkara, A.S. Tse and S.A. Asher, Thermally switchable periodicities and diffraction from mesoscopically ordered materials, *Science*, 1996, **274**, 959.
43. J.H. Holtz and S.A. Asher, Polymerized colloidal crystal hydrogel films as intelligent chemical sensing materials, *Nature*, 1997, **389**, 829.
44. C.E. Reese, A.V. Mikhonin, M. Kamenjicki, A. Tikhonov and S.A. Asher, Nanogel nanosecond photonic crystal optical switching, *J. Am. Chem. Soc.*, 2004, **126**, 1493.
45. M. Kamenjicki, I.K. Lednev, A. Mikhonin, R. Kesavamoorthy and S.A. Asher, Photochemically controlled photonic crystals, *Adv. Funct. Mater.*, 2003, **13**, 774.

46. G.S. Pan, R. Kesavamoorthy and S.A. Asher, Nanosecond switchable polymerized crystalline colloidal array Bragg diffracting materials, *J. Am. Chem. Soc.*, 1998, **120**, 6525.

47. X.L. Xu, G. Friedman, K.D. Humfeld, S.A. Majetich and S.A. Asher, Superparamagnetic photonic crystals, *Adv. Mater.*, 2001, **13**, 1681.

48. S.H. Foulger, P. Jiang, Y.R. Ying, A.C. Lattam, D.W. Smith and J. Ballato, Photonic bandgap composites, *Adv. Mater.*, 2001, **13**, 1898.

49. Y. Xia, B. Gates and Z.-Y. Li, Self-assembly approaches to three-dimensional photonic crystals, *Adv. Mater.*, 2001, **13**, 409.

50. D. Frenkel and A.J.C. Ladd, New Monte-Carlo method to compute the free-energy of arbitrary solids – application to the fcc and hcp phases of hard-spheres, *J. Chem. Phys.*, 1984, **81**, 3188.

51. F. Garcia-Santamaria, H.T. Miyazaki, A. Urquia, M. Ibisate, M. Belmonte, N. Shinya, F. Meseguer and C. Lopez, Nanorobotic manipulation of micro-spheres for on-chip diamond architectures, *Adv. Mater.*, 2002, **14**, 1144.

52. M.D. Eldridge, P.A. Madden and D. Frenkel, Entropy-driven formation of a superlattice in a hard-sphere binary mixture, *Nature*, 1993, **365**, 35.

53. P. Bartlett, R.H. Ottewill and P.N. Pusey, Superlattice formation in binary-mixtures of hard-sphere colloids, *Phys. Rev. Lett.*, 1992, **68**, 3801.

54. P. Bartlett, R.H. Ottewill and P.N. Pusey, Freezing of binary mixtures of colloidal hard-spheres, *J. Phys. Chem.*, 1990, **93**, 1299.

55. K.P. Velikov, C.G. Christova, R. Dullens and A. van Blaaderen, Layer-by-layer growth of binary colloidal crystals, *Science*, 2002, **296**, 106.

56. Y. Yin, Y. Li, B. Gates and Y. Xia, Template-assisted self-assembly: a practical route to complex aggregates of monodispersed colloids with well-defined sizes, shapes, and structures, *J. Am. Chem. Soc.*, 2001, **123**, 8718.

57. V.N. Manoharan, M.T. Elsesser and D.J. Pine, Dense packing and symmetry in small clusters of microspheres, *Science*, 2003, **301**, 483.

58. G.R. Yi, T. Thorsen, V.N. Manoharan, M.J. Hwang, S.J. Jeon, D.J. Pine, S.R. Quake and S.M. Yang, Generation of uniform colloidal assemblies in soft microfluidic devices, *Adv. Mater.*, 2003, **15**, 1300.

59. V.N. Manoharan and D.J. Pine, Building materials by packing spheres, *MRS Bull.*, 2004, **29**, 91.

60. Z.N. Bao, L. Chen, M. Weldon, E. Chandross, O. Cherniavskaya, Y. Dai and J.B.H. Tok, Toward controllable self-assembly of microstructures: selective functionalization and fabrication of patterned spheres, *Chem. Mater.*, 2002, **14**, 24.

61. Y. Lu, H. Xiong, X.C. Jiang, Y.N. Xia, M. Prentiss and G.M. Whitesides, Asymmetric dimers can be formed by dewetting half-shells of gold deposited on the surfaces of spherical oxide colloids, *J. Am. Chem. Soc.*, 2003, **125**, 12724.

62. A. van Blaaderen, R. Ruel and P. Wiltzius, Template-directed colloidal crystallization, *Nature*, 1997, **385**, 321.

63. J.P. Hoogenboom, A.K. van Langen-Suurling, J. Romijn and A. van Blaaderen, Hard-sphere crystals with hcp and non-close-packed structure grown by colloidal epitaxy, *Phys. Rev. Lett.*, 2003, **90**, 138301.

64. A. van Blaaderen, J.P. Hoogenboom, D.L.J. Vossen, A. Yethiraj, A. van der Horst, K. Visscher and M. Dogterom, Colloidal epitaxy: playing with the boundary conditions of colloidal crystallization, *Faraday Discuss.*, 2003, **123**, 107.

65. S.M. Yang and G.A. Ozin, Opal chips: vectorial growth of colloidal crystal patterns inside silicon wafers, *Chem. Commun.*, 2000, **24**, 2507.

66. G.A. Ozin and S.M. Yang, The race for the photonic chip: colloidal crystal assembly in silicon wafers, *Adv. Funct. Mater.*, 2001, **11**, 95.

67. S.M. Yang, H. Miguez and G.A. Ozin, Opal circuits of light – planarized microphotonic crystal chips, *Adv. Funct. Mater.*, 2002, **12**, 425.

68. P. Russel, Photonic crystal fibers, *Science*, 2003, **299**, 358.

69. B. Temelkuran, S.D. Hart, G. Benoit, J.D. Joannopoulos and Y. Fink, Wavelength-scalable hollow optical fibres with large photonic bandgaps for CO_2 laser transmission, *Nature*, 2002, **420**, 650.

70. R.F. Cregan, B.J. Mangan, J.C. Knight, T.A. Birks, P.S. Russell, P.J. Roberts and D.C. Allan, Single-mode photonic band gap guidance of light in air, *Science*, 1999, **285**, 1537.

71. H. Miguez, S.M. Yang, N. Tetreault and G.A. Ozin, Oriented free-standing three-dimensional silicon inverted colloidal photonic crystal microfibers, *Adv. Mater.*, 2002, **14**, 1805.

72. J.H. Moon, G.R. Yi, S.M. Yang, D.J. Pine and S. Bin Park, Electrospray-assisted fabrication of uniform photonic balls, *Adv. Mater.*, 2004, **16**, 605.

73. O.D. Velev, A.M. Lenhoff and E.W. Kaler, A class of microstructured particles through colloidal crystallization, *Science*, 2000, **287**, 2240.

74. A. Stein and R.C. Schroden, Colloidal crystal templating of three-dimensionally ordered macroporous solids: materials for photonics and beyond, *Curr. Opin. Solid State Mater. Sci.*, 2001, **5**, 553.

75. H. Miguez, F. Meseguer, C. Lopez, A. Misfud, J.S. Moya and L. Vazquez, Evidence of FCC crystallization of SiO_2 microspheres, *Langmuir*, 1997, **13**, 6009.

76. P. Jiang, J.F. Bertone, K.S. Hwang and V.L. Colvin, Single-crystal colloidal multilayers of controlled thickness, *Chem. Mater.*, 1999, **11**, 2132.

77. D. Cassagne, A. Barra and C. Jouanin, Defects and diffraction in photonic crystals, *Superlattices Microstruct.*, 1999, **25**, 343.

78. Z.-Y. Li, X. Zhang and Z.-Q. Zhang, Disordered photonic crystals understood by a perturbation formalism, *Phys. Rev. B*, 2000, **61**, 15738.

79. J.F. Lopez and W.L. Vos, Angle-resolved reflectivity of single-domain photonic crystals: effect of disorder, *Phys. Rev. E*, 2002, **66**, 036616.

80. C. Lopez, Materials aspects of photonic crystals, *Adv. Mater.*, 2003, **15**, 1679.

81. M.A. McLachlan, N.P. Johnson, R.M. De La Rue and D.W. McComb, Thin film photonic crystals: synthesis and characterization, *J. Mater. Chem.*, 2004, **14**, 144.

82. H. Miguez, N. Tetreault, B. Hatton, S.M. Yang, D. Perovic and G.A. Ozin, Mechanical stability enhancement by pore size and connectivity control in colloidal crystals by layer-by-layer growth of oxide, *Chem. Commun.*, 2002, **22**, 2736.

83. A.A. Chabanov, Y. Jun and D.J. Norris, Avoiding cracks in self-assembled photonic band-gap crystals, *Appl. Phys. Lett.*, 2004, **84**, 3573.

84. S.Y. Lin, J.G. Fleming, D.L. Hetherington, B.K. Smith, R. Biswas, K.M. Ho, M.M. Sigalas, W. Zubrzycki, S.R. Kurtz and J. Bur, A three-dimensional photonic crystal operating at infrared wavelengths, *Nature*, 1998, **394**, 251.

85. A. Blanco, E. Chomski, S. Grabtchak, M. Ibisate, S. John, S.W. Leonard, C. Lopez, F. Meseguer, H. Miguez, J.P. Mondia, G.A. Ozin, O. Toader and H.M. van Driel, Large-scale synthesis of a silicon photonic crystal with a complete three-dimensional bandgap near 1.5 micrometres, *Nature*, 2000, **405**, 437.

86. E. Palacios-Lidon, A. Blanco, M. Ibisate, F. Meseguer, C. Lopez and J. Sanchez-Dehesa, Optical study of the full photonic band gap in silicon inverse opals, *Appl. Phys. Lett.*, 2002, **81**, 4925.

87. H. Miguez, N. Tetreault, S.M. Yang, V. Kitaev and G.A. Ozin, A new synthetic approach to silicon colloidal photonic crystals with a novel topology and an omni-directional photonic bandgap: micromolding in inverse silica opal (MISO), *Adv. Mater.*, 2003, **15**, 597.

88. A. Stein, Sphere templating methods for periodic porous solids, *Microporous Mesoporous Mater.*, 2001, **44**, 227.

89. H. Miguez, E. Chomski, F. Garcia-Santamaria, M. Ibisate, S. John, C. Lopez, F. Meseguer, J.P. Mondia, G.A. Ozin, O. Toader and H.M. van Driel, Photonic bandgap engineering in germanium inverse opals by chemical vapor deposition, *Adv. Mater.*, 2001, **13**, 1634.

90. F. Garcia-Santamaria, M. Ibisate, I. Rodriguez, F. Meseguer and C. Lopez, Photonic band engineering in opals by growth of Si/Ge multilayer shells, *Adv. Mater.*, 2003, **15**, 788.

91. W. Lee, S.A. Pruzinsky and P.V. Braun, Multi-photon polymerization of waveguide structures within three-dimensional photonic crystals, *Adv. Mater.*, 2002, **14**, 271.

92. Y.H. Ye, T.S. Mayer, I.-C. Khoo, I.B. Divliansky, N. Abrams and T.E. Mallouk, Self-assembly of three-dimensional photonic-crystals with air-core line defects, *J. Mater. Chem.*, 2002, **12**, 3637.

93. R. Rengarajan, P. Jiang, D.C. Larrabee, V.L. Colvin and D.M. Mittleman, Colloidal photonic superlattices, *Phys. Rev. B: Condens. Matter Mater. Phys.*, 2001, **64**, 205103.

94. N. Tetreault, H. Miguez, S.M. Yang, V. Kitaev and G.A. Ozin, Refractive index patterns in silicon inverted colloidal photonic crystals, *Adv. Mater.*, **15**, 2003, 1167.

95. N. Tetreault, A. Mihi, H. Miguez, I. Rodriguez, G.A. Ozin, F. Meseguer and V. Kitaev, Dielectric planar defects in colloidal photonic crystal films, *Adv. Mater.*, 2004, **16**, 346.

96. E. Palacios-Lidon, J.F. Galisteo-Lopez, B.H. Juarez and C. Lopez, Engineered planar defects embedded in opals, *Adv. Mater.*, 2004, **16**, 341.

97. K.W.K. Shung and Y.C. Tsai, Surface effects and band measurements in photonic crystals, *Phys. Rev. B*, 1993, **48**, 11265.

98. N. Tetreault, A.C. Arsenault, A. Mihi, S. Wong, V. Kitaev, H. Miguez and G.A. Ozin, Building tunable planar defects into photonic crystals using polyelectrolyte multilayers, *Adv. Mater.*, 2005, in press.

99. D.A. Mazurenko, R. Kerst, J.I. Dijkhuis, A.V. Akimov, V.G. Golubev, D.A. Kurdyukov, A.B. Pevtsov and A.V. Sel'kin, Ultrafast optical switching in three-dimensional photonic crystals, *Phys. Rev. Lett.*, 2003, **91**, 213903.

100. J. Kalkman, E. de Bres, A. Polman, Y. Jun, D.J. Norris, D.C. 't Hart, J.P. Hoogenboom and A. van Blaaderen, Selective excitation of erbium in silicon-infiltrated silica colloidal photonic crystals, *J. Appl. Phys.*, 2004, **95**, 2297.

101. P. Lodahl, A.F. van Driel, I.S. Nikolaev, A. Irman, K. Overgaag, D.L. Vanmaekelbergh and W.L. Vos, Controlling the dynamics of spontaneous emission from quantum dots by photonic crystals, *Nature*, 2004, **430**, 654.

102. R.C. Schroden, M. Al-Daous, C.F. Blanford and A. Stein, Optical properties of inverse opal photonic crystals, *Chem. Mater.*, 2002, **14**, 3305.

103. A.C. Arsenault, H. Miguez, V. Kitaev, G.A. Ozin and I. Manners, A polychromic, fast response metallopolymer gel photonic crystal with solvent and redox tunability: a step towards photonic ink (P-ink), *Adv. Mater.*, 2003, **15**, 503.

104. I. Manners, Putting metals into polymers, *Science*, 2001, **294**, 1664.

105. A.C. Arsenault, V. Kitaev, I. Manners, G.A. Ozin, A. Mihi and H. Miguez, Vapor swellable colloidal photonic crystals with pressure tunability, *J. Mater. Chem.*, 2005, **15**, 133.

106. A.C. Arsenault, G.A. Ozin and I. Manners, Manuscript in preparation.

107. Y. Takeoka, M. Watanabe and R. Yoshida, Self-sustaining peristaltic motion on the surface of a porous gel, *J. Am. Chem. Soc.*, 2003, **125**, 13320.

108. J.H. Holtz and S.A. Asher, Polymerized colloidal crystal hydrogel films as intelligent chemical sensing materials, *Nature*, 1997, **389**, 829.

109. S.A. Asher, A.C. Sharma, A.V. Goponenko and M.M. Ward, Photonic crystal aqueous metal cation sensing materials, *Anal. Chem.*, 2003, **75**, 1676.

110. A.C. Sharma, T. Jana, R. Kesavamoorthy, L.J. Shi, M.A. Virji, D.N. Finegold and S.A. Asher, A general photonic crystal sensing motif: creatinine in bodily fluids, *J. Am. Chem. Soc.*, 2004, **126**, 2971.

111. V.L. Alexeev, A.C. Sharma, A.V. Goponenko, S. Das, I.K. Lednev, C.S. Wilcox, D.N. Finegold and S.A. Asher, High ionic strength glucose-sensing photonic crystal, *Anal. Chem.*, 2003, **75**, 2316.

112. S.A. Asher, V.L. Alexeev, A.V. Goponenko, A.C. Sharma, I.K. Lednev, C.S. Wilcox and D.N. Finegold, Photonic crystal carbohydrate sensors: low ionic strength sugar sensing, *J. Am. Chem. Soc.*, 2003, **125**, 3322.

113. M. Gratzel, Dye-sensitized solar cells, *J. Photochem. Photobiol. C – Photochem. Rev.*, 2003, **4**, 145.

114. M. Gratzel, Applied physics – solar cells to dye for, *Nature*, 2003, **421**, 586.

115. M. Gratzel, Photoelectrochemical cells, *Nature*, 2001, **414**, 338.

116. S. Nishimura, N. Abrams, B.A. Lewis, L.I. Halaoui, T.E. Mallouk, K.D. Benkstein, J. van de Lagemaat and A.J. Frank, Standing wave enhancement of red absorbance and photocurrent in dye-sensitized titanium dioxide photoelectrodes coupled to photonic crystals, *J. Am. Chem. Soc.*, 2003, **125**, 6306.

117. A. Zylbersztejn and N.F. Mott, Metal-insulator transition in vanadium dioxide, *Phys. Rev. B*, 1975, **11**, 4383; V.G. Golubev, V.Y. Davydov, N.F. Kartenko, D.A. Kurdyukov, A.V. Medvedev, A.B. Pevtsov, A.V. Scherbakov and E.B. Shadrin, Phase transition-governed opal–VO$_2$ photonic crystal, *Appl. Phys. Lett.*, 2001, **79**, 2127.

118. S. Deki, Y. Aoi and A. Kajinami, A novel wet process for the preparation of vanadium dioxide thin film, *J. Mater. Sci.*, 1997, **32**, 4269.

119. H.L.M. Chang, Y. Gao, J. Guo, C.M. Foster, H. You, T.J. Zhang and D.J. Lam, Heteroepitaxial growth of TiO$_2$, VO$_2$, and TiO$_2$/VO$_2$ multilayers by MOCVD, *J. Phys. II*, 1991, **1**(C2), 953.

120 M.F. Becker, A.B. Buckman, R.M. Walser, T. Lepine, P. Georges and A. Brun, Femtosecond laser excitation of the semiconductor–metal phase-transition in VO$_2$, *Appl. Phys. Lett.*, 1994, **65**, 1507.

121. A.V. Scherbakov, A.V. Akimov, V.G. Golubev, A.A. Kaplyanskii, D.A. Kurdyukov, A.A. Meluchev and A.B. Pevtsov, Optically induced Bragg switching in opal–VO$_2$ photonic crystals, *Physica E-Low-Dimens. Syst. Nanostruct.*, 2003, **17**, 429.

122. P.J. Collings, M. Hird and M. Hird, *An Introduction to Liquid Crystals: Chemistry and Physics* (Liquid Crystals Book Series), CRC Press, New York, 1997.

123. D.A. Higgins, Probing the mesoscopic chemical and physical properties of polymer-dispersed liquid crystals, *Adv. Mater.*, 2000, **12**, 251.

124. K. Busch and S. John, Liquid-crystal photonic-band-gap materials: the tunable electromagnetic vacuum, *Phys. Rev. Lett.*, 1999, **83**, 967.

125. S. John and K. Busch, Photonic bandgap formation and tunability in certain self-organizing systems, *J. Lightwave Technol.*, 1999, **17**, 1931.

126. S. Kubo, Z.-Z. Gu, K. Takahashi, A. Fujishima, H. Segawa and O. Sato, Tunable photonic band gap crystals based on a liquid crystal-infiltrated inverse opal structure, *J. Am. Chem. Soc.*, 2004, **126**, 8314.

127. H.H. Pham, I. Gourevich, J.K. Oh, J.E.N. Jonkman and E. Kumacheva, A multidye nanostructured material for optical data storage and security data encryption, *Adv. Mater.*, 2004, **16**, 516.

128. See http://www.lenslet.com/products.asp for more details.

Nanofood for Thought – Colloidal Assembly, Colloidal Crystals, Colloidal Crystal Devices, Structural Color

1. Mother of pearl, whiskers of the sea mouse, and wing tiles of *Morpho Didius* the blue butterfly from Peru are examples of Nature's 1D, 2D, and 3D photonic crystals, respectively, they all display impressive iridescence originating from the interference of light with a microstructure fashioned at the light scale. By examining the structure of these bio-photonic crystals at different length scales try to rationalize why Nature has created these optical architectures and what technology can be transferred from this knowledge to the world of materials science and technology.

2. Devise methods for (i) patterning a colloidal crystal (ii) patterning a silicon wafer using colloidal nanolithography (iii) growing a designer colloidal crystal using colloidal epitaxy.

3. Propose a purely synthetic route to a material with a structure based upon a silicon inverse colloidal photonic crystal. What is the significance of the ability to make such a structure? Briefly delineate the pros and cons of making it out of silicon rather than germanium or tin.

4. How many ways can you think of reversibly opening and closing the complete PBG at 1.5 μm of a silicon inverse colloidal photonic crystal? Why would this be a useful thing to achieve?

5. Conceive of an entirely synthetic approach to fcc silica colloidal crystal that contains a missing line of spheres running along the $<100>$ or the $<111>$ directions. What structure–property–function relations would be expected for a colloidal crystal containing such an extrinsic micron scale defect?

6. Explain why it would be scientifically and technologically interesting to be able to make a colloidal photonic crystal out of (a) $YBa_2Cu_3O_7$ (b) Li_xCoO_2 (c) Li_xSn (d) nc-Y_2O_3–ZrO_2 (d) CrO_2 (e) MoS_2 (f) $Pb_xZr_{1-x}TiO_3$ (f) periodic mesoporous benzenesilica (g) C_{60} (h) Teflon (i) SiC (j) Ag (j) polyaniline (k) DNA, (l) polylactic acid, and propose synthetic pathways for making these constructs to particular structural and optical specifications for achieving the goals that you have in mind.

7. How would you use self-assembly strategies to make (a) stacked (b) side-by-side and (c) binary colloidal crystals comprised of microspheres of two different diameters?

8. Invent a synthetic continuously tunable Bragg reflector for visible light.

9. Devise a synthetic method for storing and retrieving information in a colloidal crystal array.

10. Can you imagine how you could self-assemble a photonic crystal from monodispersed capped semiconductor or capped metal nanoclusters and if they could be made what would they be good for?

11. Can you think of a way of making a photonic crystal from carbon nanotubes, and why one would have this goal?

12. How could you make a 1D, 2D, and 3D photonic crystal fiber and for what might they prove to be useful?

13. How would you self-assemble a lattice of (a) large and small silica microspheres (b) large and small capped gold nanoclusters?

14. How could you use the classic metal–non-metal transition in VO_2 to assemble a light driven colloidal photonic crystal femtosecond switch?

15. What approaches would you utilize to build a polychrome display out of monodispersed microspheres?

16. Can you think of a chemical strategy for making a photonic crystal from anodically oxidized porous silicon? Why would you do this?

17. Devise a way to make bipolar microspheres. Why would you do this?

18. How would you orchestrate a colloidal crystal made of nanocrystalline titania to be a better solar cell than just a film of nanocrystalline titania?

19. Design a self-assembled p–n junction amorphous silicon colloidal photonic crystal solar cell that could outperform a conventional p–n junction amorphous silicon thin film device.

20. Think of a way of synthesizing a nanocluster inside a hollow microsphere that is free to rattle around and then devise a way to assemble them into a microsphere lattice and imagine why you might want to do this.

21. Develop a purely chemical strategy for making a pulsating colloidal photonic crystal and explain how such an oscillator might find utility.

22. How could you utilize anodic oxidation of a silicon wafer to synthesize a 1D photonic crystal and think of a use for such a construct?

23. Can you think of a way of making a colloidal crystal out of calcium hydroxyphosphate? Explain why would you want to do this?

24. Why would the invention of an "optical chromatography", founded upon a self-assembling colloidal or inverted colloidal crystal capillary column, likely have a transformative effect on current high pressure liquid phase chromatography HPLC technology especially with respect to the contributions that it could make to emerging fields like proteonomics and genomics?

25. How and why would it be interesting to synthesize a collection of red, green, and blue microspheres?

26. Think of a way of making microspheres self-assemble in a line and in a circle and propose something useful you could achieve with these unusual microsphere constructions.

27. How would you synthesize nanoscale hollow hemispherical capsules and what might such nanobowls be used for?

28. Think of a strategy by which you could decorate the outside surface of a 300 μm outside diameter glass tube having a 100 μm diameter inner capillary, with a periodic array of 10 μm wide rings separated by 10 μm, where each ring is comprised of a silica colloidal crystal having 10 microsphere

layers – can you think why such a colloidal crystal architecture might prove interesting?

29. What would you expect to observe if you brought two parallel-aligned silica colloidal crystal films, each supported on a glass slide, closer and closer together? How could you expand upon this concept to build a functional device?

30. Imagine a means of making charged SAM patterns on a mercury surface – quickly add an ethanol dispersion of silica microspheres and let the ethanol evaporate hopefully to organize the negative microspheres on the positive patches – this is an interesting project in its own right – then watch the system under the optical microscope evolve in time – what would one expect to see – a static or dynamic system?

31. What chemical methods can you think of to improve the mechanical stability of a silica microsphere based colloidal crystal film? Why is this a very important thing to do?

32. How many ways can you think of making controlled size and geometry silica microsphere clusters and having gained access to reasonable quantities dream up some innovative experiments that you could perform with them.

33. Think of a synthetic route for making an iridium inverse opal and why would this be a potentially technologically relevant development?

34. How and why would you make a biodegradable photonic crystal?

35. Suppose polymer microspheres could be extrusion molded into flexible large area polymer colloidal crystal films, how would you exploit this knowledge to create a product and raise venture capital for a spin-off company?

36. Formulate a rationale and develop a plan for making a lithium solid-state battery anode out of an inverse silicon opal.

37. Put together a proposal that expounds the potential benefits of sandwiching a thin film p-i-n junction silicon solar cell between two inverse silicon opals.

38. How would you synthesize monodispersed microspheres made of titania and lead?

Materials with Pores

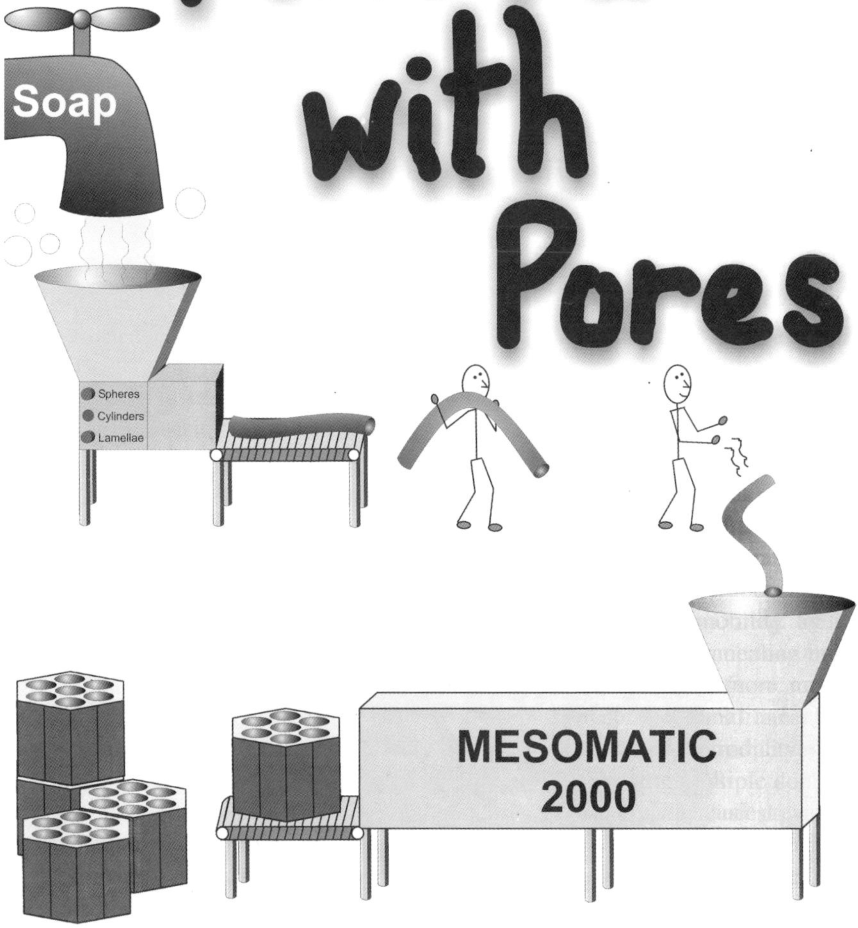

Microporous and Mesoporous Materials from Soft Building Blocks

"A hole is nothing at all, but you can break your neck in it."
Austin O'Malley (1858–1932)

8.1 Escape from the Zeolite Prison

It is often said that Nature abhors a vacuum. Likewise, Nature abhors open space in solid-state materials. The thermodynamic tendency is for a structure to minimize its energy by close packing and drive a structure filled with holes into a dense state. Atoms, ions, molecules, clusters, colloids, and polymers will pack in a solid as closely as minimum energy considerations will allow. However, given the opportunity kinetics will take the initiative and metastable structures permeated with voids will prevail.[1,2] Although thermodynamically unstable to collapse, depending on conditions open-framework materials can survive indefinitely. Nowhere is this seen more dramatically than in biology's siliceous radiolarian and diatom filigree micro-skeletons[3] (see Chapter 10) and geology's naturally occurring zeolites[4,5] aluminosilicates riddled with regular arrays of micropores. Silica in these biomineralized and geomineralized porous solids is, respectively, amorphous and crystalline, yet both are unstable with respect to quartz.

In the assembly of these siliceous materials, space is created and preserved through the mediation of structure-directing templates. These additives are often organics – they fill space, balance charge, and guide inorganics to form a particular structure.[6,7] It is a universal co-assembly strategy that forms the basis of template syntheses of porous solids. Lowenstam's organic matrix mediated assembly of biomineral shapes[8] and Barrer's organic molecule templated synthesis of zeolites and molecular sieves (silicate versions of zeolite aluminosilicates)[9] can be considered to have paved the way to the fields of biomineralization and biomimetic

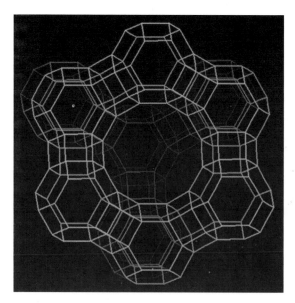

Figure 8.1 *Crystal structure of Faujasite, one of the many naturally occurring zeolites*
(Reproduced with permission from http://zeolites.ethz.ch/IZA-SC/Atlas/data/
pictures/FAU_mod.html)

materials chemistry.[10] A crystal structure of Faujasite, a naturally occurring zeolite, is shown in Figure 8.1, which highlights its intrinsic structural order, impressive topology, and open-framework architecture. Barrer's approach provided crystalline microporous materials having pore and channel sizes tunable from about 3 to 10 Å with 0.1 Å crystallographic precision, summarized in the classic text on zeolites by Breck.[11] Note that the International Union of Pure and Applied Chemistry (IUPAC) has established a convention that classifies porous solids in terms of the range of dimensions of the pore and channel spaces contained therein: microporous (2 nm, mesoporous 2–50 nm, macroporous >50 nm.

8.2 A Periodic Table of Materials Filled with Holes

Flanigen and Bedard[12,13] expanded Barrer's microporous aluminosilicates and silicates, with pore sizes below 2 nm, to encompass a periodic table of solid-state compositions, including aluminophosphates, metalloaluminophosphates, metal sulfides, and metal selenides (Figure 8.2).[14] These seminal breakthroughs greatly expanded the scientific and technological scope of the field of zeolites and established open-frameworks as a new class of materials in mainstream solid sate chemistry. From their traditional shape selective catalytic[15] and separation properties, zeolites and molecular sieves now embrace all of materials chemistry. *There are as many papers appearing these days on inorganic materials filled with holes as dense packed solids!*

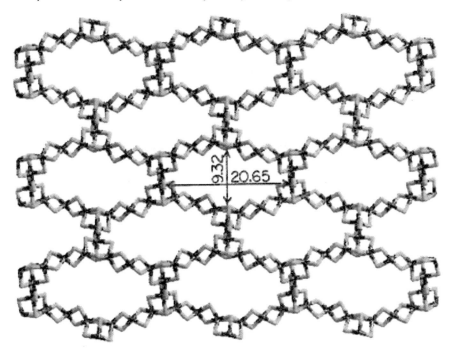

Figure 8.2　*Schematic showing the crystal structure of a microporous tin sulfide material* (Reproduced with permission from Ref. 14)

8.3 Modular Self-Assembly of Microporous Materials

A notable departure from the traditional hydrothermal synthesis of zeolites and molecular sieves, where building-blocks in the assembly process are typically not well defined and the chemistry is not particularly well understood, concerns a modular chemistry approach to constructing microporous solids.[16,17,18] The method exploits self-assembly of modular construction units guided by templates, where "modules are well defined," to form materials with open-frameworks within which the integrity of the module is maintained within the structure. Organic chemistry provides the knowledge necessary to synthesize an infinite number of rigid modules with ligands at predetermined positions and angles. Combining these units with metal ions having preferred coordination geometry can lead to extended solids with predictable structures. However, if the metal centers are fluxional, the structure may lack the required stiffness to sustain drying and guest removal without collapse. The use of chelating ligands with a fixed arrangement of two or more donor atoms can further enforce geometric constraints on the final structure, allowing a greater synthetic control. This type of materials self-assembly chemistry has led to a myriad of porous solids, comprised of coordination networks when built from interconnected metal–ligand modules,[19] (Figure 8.3), or cluster frameworks when made of interconnected cluster modules[20] (Figure 8.4). The structures of these

Molecular Complexes   Extended Solids

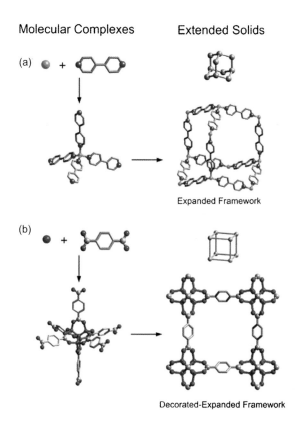

Expanded Framework

Decorated-Expanded Framework

Figure 8.3   *Modular self-assembly of coordination open-framework materials. Linkers can be flexible, like metal–pyridine links above, or rigid, like metal–carboxylate links shown below*
(Reproduced with permission from Ref. 16)

materials can extend indefinitely in one-, two-, or three-dimensions. *To illustrate the diversity of this class of materials, the Cambridge Structural database currently has well over 3000 entries for 3-D extended materials bearing metal ions or clusters connected by cyanide, pyridyl, phosphate, or carboxylate groups!*

The attractiveness of the modular self-assembly method is the almost unlimited choice of construction units and the mild conditions utilized for their assembly. A challenge in modular chemistry is to prevent self-interpenetration of one open-framework within another.[21] Interpenetration is quite common and the Guinness record appears to be nine times. This type of framework entanglement serves to fill space and tends to defeat the primary objective of building functional porosity into open-framework materials. Another goal is to be able to retain the porosity of the open-framework after removing imbibed template or solvent from the material.

Modular assembly has provided a new and rational approach to "designer" open-frameworks with a myriad of 1D, 2D, and 3D structural motifs. Frameworks may

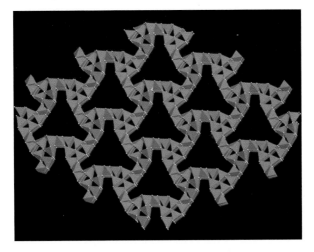

Figure 8.4 *3D representation of a self-assembled metal sulfide cluster framework found to display remarkable ion conductivity*
(Reproduced with permission from Ref. 20)

be cationic,[22] anionic, or neutral, centrosymmetric, or chiral,[23] and the modules can be intentionally tailored to provide porous solids with desirable adsorption,[24] ion exchange, chemical and catalytic, electronic, and optical properties.

Utility for cationic frameworks is anticipated in selective anion exchange, which would be beneficial in the recovery and removal of inorganic and organic anions from industrial wastewater. Selective inclusion of organic molecules by neutral frameworks is perceived to be extremely relevant to the removal and recovery of aromatics and halogenated hydrocarbons from industrial processes.

8.4 Hydrogen Storage Coordination Frameworks

A potentially important application of coordination frameworks is safe storage of hydrogen for fueling motor vehicles and driving portable electronics. Currently, the main materials contenders for this task are metal hydrides and different forms of carbon, but while promising they still have some problems. The goal of hydrogen storage for safe mobile fueling is to discover lightweight and compact materials that can rapidly adsorb and desorb large amounts of hydrogen under ambient temperature and pressure conditions. The US Department of Energy has set 6.5 wt% hydrogen as the target for automobile hydrogen storage capacity. For coordination frameworks to meet this challenge they will need to satisfy these criteria plus the as-synthesized material must be able to maintain its structural integrity on removing template or occluded water or solvent. A candidate for this role is the metal–organic open-framework compound $Zn_4O(BDC)_3$ where BDC = 1,4-benzene-dicarboxylate.[25,26] It has a sturdy cubic open-framework structure with a very large surface area (2500–3000 m^2 g^{-1}) and impressive thermal stability to 300–400 °C. From a manufacturing point of view, the synthesis

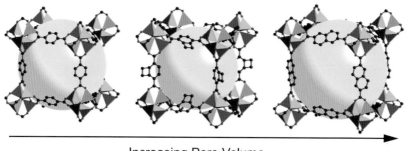

Increasing Pore Volume

Figure 8.5 *Hydrogen storage metal–organic open-framework compound Zn₄O(RDC)₃* (Reproduced with permission from Ref. 25)

is simple and utilizes readily available low cost precursors:

$$4Zn^{2+} + 3H_2BDC + 8OH^- \rightarrow Zn_4O(BDC)_3 + 7H_2O$$

The $[OZn_4]^{6+}$ inorganic cluster is interconnected by an octahedral array of organic dicarboxylate linkers. The choice of organic group R provides control over the size of the cubic unit cell and concomitantly the adsorption capacity of the material (Figure 8.5) where the empty volume is shown by the central spheres. Cubic frameworks of this type show Type I gas adsorption isotherms for hydrogen at 78 K and linear up-take of hydrogen at room temperature. (Note that the form of an isotherm is diagnostic of the dimension of pores in an open-framework material, where Type I signifies microporosity. Isotherms are obtained by graphing the amount of nitrogen adsorbed at 77 K as a function of the partial pressure of nitrogen.) In the case of BDC, the hydrogen up-take at 78 K is 4.5 wt% while at 20 bar and room temperature it is 1 wt%. Inelastic neutron scattering studies show two structurally well-defined sites for adsorbed hydrogen in these porous solids, one in the vicinity of zinc and the other around the organic linker. Adsorption capacities for the related RDC compounds (where R is the bridging organic) shown in Figure 8.5 increase left to right, and are superior to commercial carbons and comparable to the best carbon nanotubes.[27] These frameworks can also be adapted to storage of methane,[28] and likely a host of other gases.

8.5 Overview and Prospects of Microporous Materials

Crystalline microporous materials have made, and will continue to make, important contributions to catalytic[29,30] and separation science, ion exchange and detergency, chiral transformations or separations,[31] nuclear and toxic waste clean-up,[32,33] to name a few. These successes have occurred because the framework is crystalline offering structurally well-defined sites to molecules contained therein and because of the 2 nm or less pore size in these materials, which is comparable to most molecule dimensions, despite the fact that this size was a barrier imposed by single organic molecule templating. But larger pore sizes are interesting too, with nature's soft templates guiding the biomineralization of impressive architectures with complex form, inspiring us and showing us the way towards the synthesis of mesoporous materials with pore sizes in the 2–10 nm intermediate size range and a new world of fascinating morphologies.

8.6 Mesoscale Soft Building Blocks

An elegant and natural "escape from the <2 nm zeolite micropore size prison" emerged from the pioneering work of Charles Kresge and researchers at Mobil.[34,35] They discovered that a micellar assembly of surfactant molecules could function as a supramolecular template for organizing silicates into a silica replica material (Figure 8.6). This led to the first fully characterized examples of periodic hexagonal and cubic forms of mesoporous silica with pore diameters of first generation materials in the 2–10 nm range – a type of "medusa chemistry" in which a soft template is "turned to stone"!

Diagnostics of the new mesoscale materials came from a combination of transmission electron microscopy,[36] low angle X-ray diffraction, and gas adsorption isotherms. Together these analytical techniques provided visual images of channels in the hexagonal and bicontinuous labyrinths in the cubic mesoporous silica materials and a means of evaluating the surface area, pore diameter, pore size

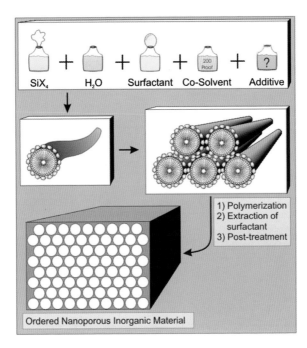

Figure 8.6 *Supramolecular templating of mesoporous inorganics*

distribution, and wall thickness of the mesostructure. Typically surface areas were around 1000 m²/g and wall thickness in the 1–2 nm range.

The microstructure of the siliceous material constituting the channel walls, by all diagnostic probes, such as ^{29}Si magic angle spinning nuclear magnetic resonance (MAS-NMR),[37] vibrational infrared (IR) and Raman spectroscopy, extended X-ray absorption fine structure analysis (EXAFS) and X-ray absorption near edge spectroscopy (XANES), was found to be glassy. Thermal gravimetric analysis and ^{13}C MAS-NMR provided a detailed insight into the filling fraction and structure, bonding and dynamics of the templating surfactant assembly imbibed within the mesostructure.

Collected information that emerged from numerous investigations of these archetype mesoporous silicas provided the physicochemical information needed to understand their mode of formation[38] and generalize the synthesis methodology[39] to enable control over structure type, pore size, composition, and morphology. Pore size was dependent upon micelle diameter, which is a function of surfactant alkane chain length. The diameter of micelles could also be controlled by hydrophobic molecules like mesitylene and cyclohexane, which cause the micelle to swell in proportion to the amount of additive. This provided a rational way of synthesizing mesoporous silicas with Angstrom precision control of pore size in the range 2–10 nm, as shown in Figure 8.7.

The genie was out of the mesobottle! Materials self-assembly, using supramolecular templates, represented a new way of thinking about the synthesis of inorganic materials over length scales not seen before in solid-state chemistry. As a more

Figure 8.7 *TEM image of hexagonal mesoporous silica along with X-ray diffraction data demonstrating the hexagonal order*
(Reproduced with permission from Ref. 36)

detailed understanding of supramolecular templating developed, it became apparent that it also provided a vital and inspirational link with nature's processing of biominerals.[40,41] Often materials appeared with morphologies resembling those found in the biological world having curved shapes and patterns on their surfaces, inspiration from which sprung a type of morphosynthesis. Complexity was emerging from simple chemistry.

8.7 Micelle Versus Liquid Crystal Templating Paradox

It was discovered early in the development of the field that mesoporous silica could be assembled from micelles under dilute aqueous surfactant conditions as well as lyotropic liquid crystals at much higher concentrations.[42] The paradox of how the same mesoporous silica material could be templated both by surfactant micelles and liquid crystals was resolved by studies of its mode of formation under dilute and concentrated surfactant synthesis conditions. These studies revealed that silicate micelles could self-assemble into a silicatropic liquid crystal just as a lyotropic liquid crystal could imbibe silicate species and form the same silicate mesophase.

In the context of colloid chemistry, the micelle can be viewed as a colloidal particle and the lyotropic liquid crystal as a colloidal crystal.[43] Therefore, the former process involves colloidal crystallization of a micelle-silicate co-assembly while the latter one involves a pre-formed colloidal crystal into which silicates infuse. In both cases mesoporous silica forms from the hydrolytic poly-condensation of silicates in the microphase separated domains of the colloidal crystal. The analogy with colloidal microsphere self-assembly of colloidal crystals discussed in Chapter 7 is an important one when planning the synthesis of periodic mesoporous and macroporous materials. Using either mesoscale (2–50 nm) micelle or macroscale (50–3000 nm) microsphere or other colloid building blocks, one can self-assemble periodic mesoporous and macroporous materials starting either from micelle or microsphere, liquid crystal or colloidal crystal templates – *a sort of bottom-up versus top-down colloidal-assembly paradigm!*

8.8 Designing Function into Mesoporous Materials

Armed with this basic understanding of the dual-templating behavior of surfactants as micelles and lyotropic liquid crystals, it is possible to develop a rational synthetic strategy to mesoporous materials with compositions and properties beyond those of mesoporous silica.[44] This is best expressed in terms of an illustration of a single channel of a mesoporous material and a classification of the various ways in which the composition of the channels can be altered, the contents of the mesopore changed, and the surface of the mesopore modified (Figure 8.8). In the remainder of the chapter, the various aspects of this illustration will be expanded upon, showing how the simple elegance of this concept can be extended to give rise to a plethora of multifunctional materials.

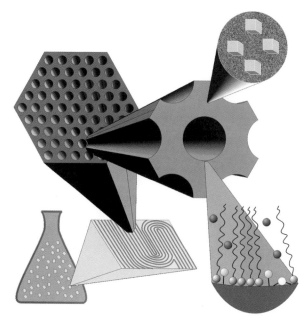

Figure 8.8 *Putting function into mesoporous materials at a hierarchy of length scales. Synthesized material displays hexagonal cylindrical order displaying defects like any liquid crystal. The composition and texture of the walls can be tuned, and the interior of the channels can be functionalized by a choice of surfactant, bridging or side-groups, and additives*

8.9 Tuning Length Scales

These synthesis variables have enabled pore diameters to be tuned from about 2 to 10 nm. The charge on the head group determines whether the co-assembly of surfactant (S) and inorganic (I) is described as S^+I^- or S^-I^+ ion pairs, $S^+X^-I^+$ or $S^-X^+I^-$ ion-clusters, or S^0I^0 hydrogen bonded pairs,[45] where X is a charge-balancing cation or anion.[46] Mesoporous materials have been formed with surfactant alkane chain lengths between $n = 8$ and 20. Below $n = 8$, micelle assembly is not favored and zeolites are templated instead. Microphase separated[47] amphiphilic tri-block copolymers $PEO_xPPO_yPEO_x$ have proven to be ideal templates for expanding pore dimensions of mesoporous materials beyond the surfactant templating limit of 10 nm.[48,49] Block copolymers are well studied (see Chapter 9) and many methods are available to control the structure and orientation of their mesophases, knowledge of which can then be transferred to an inorganic mesostructure. In aqueous solution at low concentrations, $PEO_xPPO_yPEO_x$ copolymers exist as isolated chains, which transform at increasing concentrations to flower (rosette) micelles and eventually to micelle aggregates.[50]

Like surfactant micelles and lyotropic liquid crystals, block copolymers[51] co-assemble with a wide variety of inorganic precursors to form mesoporous inorganic materials with structural features that can traverse 5–35 nm length scales[52,48]

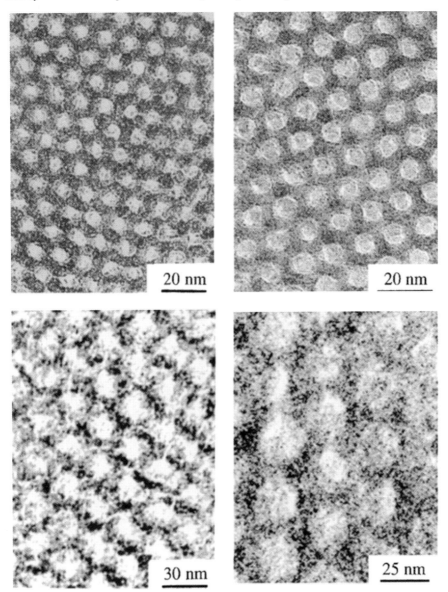

Figure 8.9 *Block copolymer templated mesoporous silicas with pore diameters in the 5–35 nm range*
(Reproduced with permission from Ref. 48)

(Figure 8.9). As-synthesized mesoporous materials contain the surfactant or block copolymer template. They may be removed and the mesoporosity retained using proton, ion, or surfactant exchange, thermal treatment in air, plasma etching, ozone, vacuum UV, or microwave[53] treatment. A tailor-made surfactant may also be removed by the selective chemical transformation, such as hydrolysis, of an incorporated chemical group. A characteristic feature of a structurally well-defined template-free mesoporous material is an open network of monodispersed pores or channels with a surface area that can be more than 1000 m^2/g and pore volume larger than 1 cm^3/g. These are attractive attributes for numerous applications[54] like large molecule catalysis[55,56] and separations,[57] electrodes in photovoltaics,[58] batteries and fuel cells, low dielectric constant microelectronic packaging materials,[59] chemical sensing,[60] controlled release of chemicals, encapsulation of proteins, and templates for nanostructures.

The choice of structure-directing agent, solvent, co-solvent, temperature, pH, added salts, precursor, and swelling agents afford considerable control over the static pore size and pore spacing of these materials. However, it is also possible to achieve dynamic control over the pore size by applying a stimulus such as light to a pre-formed mesoporous material. This has been demonstrated by using a pre-formed mesoporous silica, and grafting to the outside of it some photoactive coumarin groups.[61] Irradiating with long-wavelength UV light caused the photodimerization of the coumarin groups and the closure of the external pores, while irradiation with short-wavelength UV light photocleaved these dimers. Using this system, biologically active molecules such as the steroid progesterone could be trapped inside the mesopores and subsequently released using short-wavelength UV.

8.10 Mesostructure and Dimensionality

The composition-temperature phase diagram of different kinds of surfactants in water as well as polar solvents indicates the existence of regions of mesophases with hexagonal, cubic, lamellar, or other structures.[62–65] More exotic solvents for surfactants, such as supercritical carbon dioxide are also found to display a variety of ordered mesophases.[66] This guides but does not guarantee the synthesis conditions that favor the formation of a particular mesostructure.[67,68] A more reliable approach necessitates low-angle X-ray diffraction, neutron diffraction, and NMR studies of the "inorganotropic mesophases" that form in an "Surfactant-Inorganic-Temperature (S-I-T)" phase diagram and monitoring its conversion to the mesostructured product. Careful diffraction and microscopy analyses are required to distinguish well-ordered mesostructures from less ordered ones like the wormhole form, different cubic mesostructures like bicontinuous, gyroid and double diamond forms, and lamellar mesostructures with planar or corrugated, non-porous or porous sheets.

8.11 Mesocomposition – Nature of Precursors

The accessible framework compositions of the mesostructure depends on finding inorganic precursors that co-assemble with micelle or liquid crystal templates and remain intact throughout the nucleation and growth process leading to the mesoporous product material. Synthetic pathways to mesostructures with diverse compositions have been described in the literature.[69,70] For example, metallic

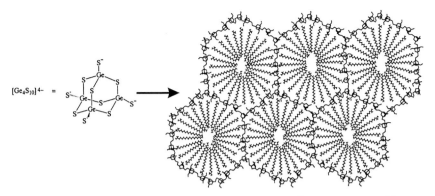

Figure 8.10 *Modular self-assembly of adamantanoid $M_4S_{10}^{4-}$ clusters with M^{2+} ions to form a mesostructured metal sulfide*
(Reproduced with permission from Ref. 79)

frameworks[71] have been made by chemical, electrochemical, sonochemical, and photochemical reduction,[72,73] semiconductors by salt precipitation[74] and semiconductor cluster assembly,[75] and insulators by sol–gel chemistry[76] using metal alkoxides[77,78] and metal glycolates. Figure 8.10 shows the structure of a semiconducting mesoporous framework assembled from germanium sulfide adamantane-like clusters.[79] This flexibility of framework composition enables the design of materials which combine the advantages of ordered porosity with a myriad of solid-state properties exemplified by electronic and ionic charge transport,[80] electroluminescence, electrochromism,[81] and catalysis.[82,83]

8.12 Mesotexture

The process of co-assembling a micelle or liquid crystal with an inorganic precursor is conducted under mild conditions to ensure the structural integrity of the "inorganotropic mesophase" remains intact. This is a necessary prerequisite for the formation of a well-defined mesostructure. Because of the mild synthesis (soft chemistry) conditions, mesostructures with glassy walls generally result due to insufficient thermal energy to reach thermodynamic equilibrium. A glassy texture may be desirable for certain purposes, however, in order to combine a particular solid-state property with mesoporosity, it might be important to obtain walls that are nanocrystalline or crystalline. In terms of Bloch–Wilson electronic bands and Schottky–Frenkel defects in crystalline solids, walls that have the right composition, structure, and long-range order may generate semiconductivity, photoconductivity, ionic conductivity, or an increase in thermal stability.[84] To this end, thermal post-treatment may facilitate conversion of an as-synthesized mesostructure with glassy channel walls to one composed of a contiguous arrangement of nanocrystals (Figure 8.11). Alternatively, it may be possible to synthesize a nanocrystalline mesostructure directly by co-assembling a dispersion of nanocrystals with a micelle or liquid crystal template. This has been achieved, for instance, in a periodic mesoporous nanocrystalline anatase material.[85,86]

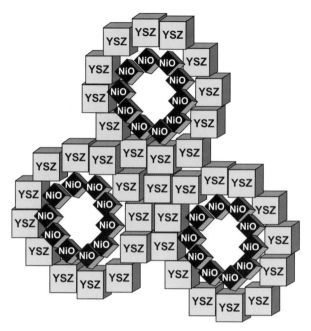

Figure 8.11 *Nanocrystalline mesoporous nickel oxide yttria-stabilized-zirconia, oxide ion conductor used as the anode in a solid oxide fuel cell*

8.13 Periodic Mesoporous Silica–Polymer Hybrids

Polymers are playing an increasingly important role both in the templated self-assembly of periodic mesoporous inorganic materials, as well as control of the structures and properties, function, and perceived utility of these materials. Four main scenarios have been explored that exploit this paradigm.

The first situation involves the tri- and di-block co-polymers based on ethylene and propylene oxide units that can form micelles and liquid crystals, as mentioned previously, and template the self-assembly with a range of inorganic precursors. In the case of silicates this results in silicas with crystalline mesoporosity, in which the polymeric template is trapped within the void spaces. One can imagine mesoporous silica–polymer hybrids of this type functioning as a permselective membrane or with imbibed lithium salts in the channels, being able to function as a solid electrolyte in a lithium solid-state battery.

The second situation takes the as-synthesized mesoporous silica–polymer hybrid, from which the polymeric template is removed by one of a collection of methods described earlier, and subsequently introduces other polymers into the void spaces. This can be done by either polymerizing a monomer in the space, or by incorporating the polymer directly into the space through a solution or melt technique. In this way, a myriad of mesoporous silica–polymer hybrids can be synthesized and their electrical or mechanical properties tailored through the choice of the imbibed polymer.[87,88]

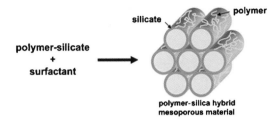

silicate
polymer

polymer-silicate
+
surfactant

polymer-silica hybrid
mesoporous material

Figure 8.12 *Polyelectrolytes can be trapped in the silicate walls of a PMS, conferring to it additional function*
(Reproduced with permission from Ref. 92)

The third situation makes use of the fact that the silica material mesoporous host can act as a mesomold to be dissolved in dilute hydrofluoric acid to release polymer nanowires for a range of uses.[89,90] Polymer nanofibers can also be formed when the mesoporous material is used as a nano-extrusion mold.[91]

The fourth situation uses a surfactant to template the self-assembly of a mixture of a silicate and a polyelectrolyte to create a periodic mesoporous silica–polyelectrolyte hybrid, in which the polyelectrolyte is incorporated in the silica walls of the pores.[92] This concept is illustrated in Figure 8.12. Both cationic polyacrylates and anionic polystyrenesulfonates have been found to work well in this situation, resulting in acidic sites dotting the pore walls. This simple fact provides opportunities for innovative organic catalytic transformations, new designs of proton conducting membranes for hydrogen–oxygen fuel cells, electrolytes for batteries and electrochromic devices to name but a few. One can also imagine performing layer-by-layer electrostatic self-assembly within these hybrids (see Chapter 3) to control the dimensions, shapes and surface properties of the mesopores and hence their electrical, optical, and mechanical properties.

8.14 Guests in Mesopores

Mesoporous hosts enable chemistry to be conducted in confined mesoscale void spaces.[93,94] They provide mesoscale reaction chambers in which to include a wide range of mesoscale guests.[95] Examples include carbon nanotubes,[96] conducting and non-conducting polymers,[97] metal porphyrins,[98] proteins,[99] organometallic, and coordination compounds,[100–103] semiconductor,[104] and metal clusters and wires[105,106] to name a few. It should be possible to encapsulate almost any type of species based on well-developed surface chemistry and geometrical considerations. The ordered nature of these porous materials makes them ideal for evaluating the properties of various supported chemical species such as catalysts.[107,108] Figure 8.13 shows a schematic of the growth of an organometallic polymer, polyferrocenylsilane, inside the channels of mesoporous silica.[109,110] The monodisperse channels or pores in mesoporous materials can serve for the generalized synthesis of various nanoparticles.[111] The surfactant can also be exchanged for other functional amphiphiles after material synthesis.[112] Mesoscale channels and

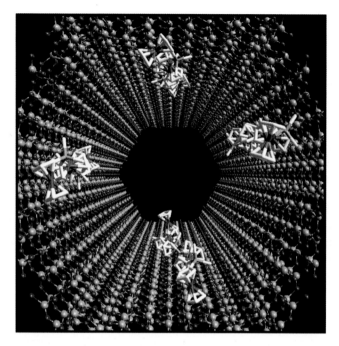

Figure 8.13 *Growth of polyferrocenylsilane within the channels of hexagonal mesoporous silica*

pores in the size range 2–35 nm provide a unique opportunity for studying confinement effects on the chemical and physical properties of diverse kinds of molecular, polymeric, cluster, and solid-state materials. In addition, mesoporous materials provide a readily recyclable scaffold on which to anchor catalysts or reagents for chemical transformations.[113]

8.15 Capped Nanocluster Meets Surfactant Mesophase

When a surfactant is dissolved in water, we know that hydrophobic molecules like cyclohexane can cause surfactant micelles to swell by essentially dissolving in the core of alkane chains. We have also seen that homogenous size nanoclusters can be synthesized bearing hydrophobic capping ligands (Chapter 6). Therefore, why not use capped nanoclusters as a nanoscale swelling agent for a surfactant micelle and surfactant mesophase?

This is a simple idea with a number of interesting consequences. First, a surfactant sheath coating the capped nanocluster would create a heterobilayer shell, making the hydrophobic nanocluster hydrophilic and hence water-soluble. As with other micelle systems, these nanocluster-loaded micelles could at the proper concentration self-assemble into an ordered mesophase. The resulting nanocluster-surfactant mesophase could then co-assemble with inorganic precursors, such as silica or titania, to create periodic composite materials comprised of a 3D array of capped nanoclusters in an inorganic matrix. A systematic investigation of such composites could afford control over their structure and lattice dimensions, and hence their electrical, optical, magnetic, and mechanical properties would be under chemical control. Fine-tuning of function would hinge on selecting the composition and diameter of the nanocluster core and the number of carbons in the alkane chains of the capping ligand and surfactant. Finally, the nanocomposites would be mechanically stable, enabling engineering routes to electronic, optical, and magnetic devices based on robust nanocluster composites.

A proof-of-concept example of the above idea is shown in Figure 8.14, which depicts the synthesis procedure as well as a TEM image of a {100} lattice plane of such a periodic, cubic gold nanocluster silica composite.[114] To amplify a little upon the synthesis details, alkanethiol capped nanoclusters are dissolved in hexane and stirred with surfactant in water to make an oil-in-water microemulsion. Transfer of the nanoclusters from the non-aqueous to the aqueous phase generates interdigitated bilayer surfactant–alkanethiol capped nanoclusters, which depending on concentration can exist as either micelles or a liquid crystal mesophase and in the presence of silicates co-assemble to make a gold nanocluster–silica composite in the form

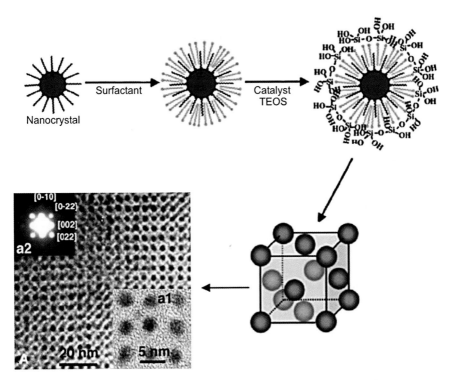

Figure 8.14 *Schematic showing the assembly of a periodic nanocrystal–silica composite. A TEM image is shown on the bottom left, highlighting the order of the material*
(Reproduced with permission from Ref. 114)

of a powder or film. Particularly noteworthy is the highly ordered structure of these nanocluster arrays, a feature that permitted evaluation of the current–voltage scaling law and temperature-dependent conductance for a 3D array of Coulomb islands, with nanometer precision command over their separation and surrounding matrix.

8.16 Marking Time in Mesostructured Silica – New Approach to Optical Data Storage

A number of the aforementioned attributes of mesostructured silica, namely the ability to form high optical quality films and synthesize in a straightforward fashion diverse kinds of host–guest nanocomposites, have been creatively amalgamated to generate a new kind of optical data storage material. This material offers the possibility to enhance the information storage capacity using a multibit-per-site digital format compared to competing materials.

The approach is based on a color change of a pH-sensitive dye, made light-sensitive *via* protons liberated from a photoacid generator (PAG), both of which are housed within the spatial confines of a tri-block co-polymer templated mesostructured silica.[115] The distinctiveness of this strategy compared to CDs or DVDs for light-based information storage is that the absorbance (transmittance or fluorescence) obtained for the protonated dye imbibed within the composite mesostructured silica film is a function of the time (or intensity) of the illumination. This special feature enables "write once read many times" multibit-per-site storage of optical information where the zeros and ones in digital code can be associated with the recording of the change in absorbance of the dye for a discrete illumination time or intensity.

A representative example showing the light-driven color change of the PAG-Dye couple embedded in a mesostructured film is shown in the top part of Figure 8.15, where can be seen two examples of digital images stored within the composite mesostructured silica film. An illustration of the means by which the time of illumination and associated color change of the photo-protonated dye can be utilized to store multiple bits of digital information, is shown in the bottom part of the figure. Here it can be observed that four distinct writing times give rise to four different absorbance values of the protonated dye that when read optically would correspond to a 00, 01, 10, 11 binary code.

The storage capacity of this mesostructured silica nanocomposite film can in principle be increased by decreasing the laser illumination spot size and exploiting the fact that the density of information scales as 2^n where n is the number of bits per site. This new class of optical data storage material can be made in a one-pot co-assembly synthesis involving all the aforementioned components, making the fabrication of films rather straightforward and cost effective. This is an important consideration if it is going to successfully compete with other enhanced optical data storage systems based on, for example, holographic or multi-layer 3D storage, which are difficult to implement mainly because the ideal material has yet to be discovered.

Optical Images

Spectra

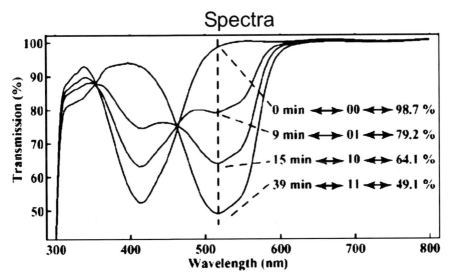

Figure 8.15 *Two digital pictures of patterns inscribed into a photoactive mesostructure at the centimeter and micron scale, top left and right, respectively. The bottom graph shows the measured absorbance change in the material after the denoted irradiation time*
(Reproduced with permission from Ref. 115)

8.17 Sidearm Mesofunctionalization

The majority of research that has been aimed at providing the channel walls of mesoporous materials with surface functionality has focused mainly on silica-based materials using $(EtO)_3SiR$ reagents.[116] These can either be co-assembled with $(EtO)_4Si$ and surfactants to directly form a mesoporous organosilica[117,118] or tethered to the channel surface of a pre-formed mesoporous silica *via* anchoring silanol groups[119,120] to indirectly from a mesoporous organosilica.[121] The former approach tends to give a more homogeneous distribution of terminally bound organic groups with a maximum loading of about 25%. In this way, it has been proven possible to functionalize the surface of mesoporous silica with, for example, alkyl amine, phosphine, thiol, vinyl,[122] sulfonic acid,[123] and aromatic terminal organic groups. A schematic depicting the functionalization of mesoporous silica

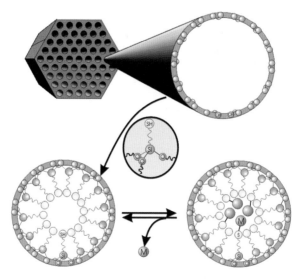

Figure 8.16 *Mesoporous silica with channel walls functionalized with terminally bound alkanethiol groups that chelate toxic heavy metals*

with terminal thiol groups, and its use in heavy metal remediation is shown in Figure 8.16. This endows the mesoporous silica materials with a range of properties that may be useful for base, acid, and metal catalysis,[124] and removal of heavy metal pollutants from wastewater streams.

8.18 Organics in the Backbone

To integrate organic functional groups into the channel wall of mesoporous silica a silsesquioxane precursor [(EtO)$_3$Si]$_n$R, containing two or more triethoxysilane groups bonded to a bridging organic group,[125] is co-assembled with a surfactant micelle or lyotropic liquid crystal. This provides a class of periodic mesoporous organosilicas, called PMOs,[126–128] in which bridging organic groups, like methylene, ethene, ethylene, 1,4-benzene, thiophene, methyne, and 1,3,5-benzene become an integral part of the channel wall. By judicious selection of the organic group and the synthesis conditions, it is possible to template PMOs in which the organic and silicate materials comprising the channel walls are either randomly oriented (glassy) or partially ordered where just the organic constituent is oriented but the silicate is disordered (crystalline-like). A case in point is periodic meso-porous benzenesilica,[129] where at the molecule scale incipient order is created through interaction between aromatic groups in the benzene silicate precursor, for example hydrogen-bonding between silanols and/or pi-bonding between aromatics. This ordering occurs simultaneously with ordering at the mesoscale induced by the co-assembly of pre-organized benzenesilicate construction units with micellar templates. After hydrolytic poly-condensation of the silicate moieties this results in hexagonal close-packed channels where bridge bonded benzene groups are locked

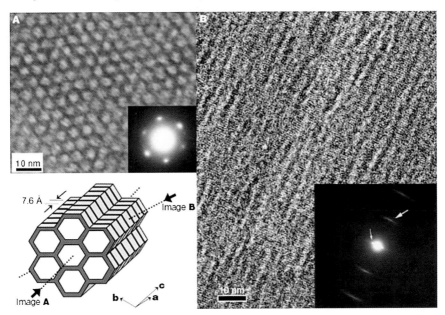

Figure 8.17 *Periodic mesoporous benzene silica in which the organization of the bridging benzene groups in the channel walls is ordered. TEM images perpendicular, left, and parallel, right, to the hexagonal channels are shown. In the latter image, an electron diffractogram shows weak peaks confirming an ordered arrangement of benzene rings*
(Reproduced with permission from Ref. 129)

in an ordered state within a glassy matrix of silica (Figure 8.17). There is great opportunity for this new class of mesoscale composite materials in a range of applications as diverse as chiral chromatography and microelectronics.

One application for PMO films that presents itself immediately is as low dielectric constant (k) materials for microelectronics. The increasing density of devices and interconnects demands insulating dielectric materials with k sufficiently low to prevent crosstalk and limit the problem of signal delays caused by intra- and interlayer capacitance. A new generation of low-k materials must be less than silica ($k \sim 3.8$), and ultimately, ultra-low dielectric constant layers with $k < 2.0$ are required very soon to keep pace with Moore's law. There are three important ways to achieve a low k value. The first is to incorporate a large porous volume fraction, since $k_{air} \sim 1.0$, though very porous or disordered networks like aerogels can become highly fragile. The second is to reduce the atomic weight and polarizability of the matrix network as k scales with electron density. For silica-based materials, there are many examples incorporating terminal organic groups in place of Si–O–Si bridges as organosilicas. However, many of these sacrifice mechanical and thermal stability as a result of disrupting the network connectivity with so-called dangling bonds. Finally, the material must effectively resist moisture adsorption, despite being highly porous. As a result, surface silanol groups are generally capped using hexamethyldisilazane (HMDS) or trimethylsilylchloride

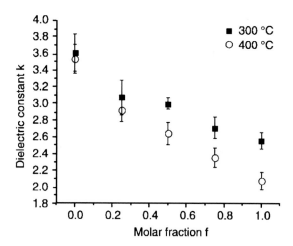

Figure 8.18   *Dielectric constant as a function of the loading of three-ring precursor, f, for*
materials treated at 300 and 400 °C
(Reproduced with permission from Ref. 59)

(TMSCl), for example. Considering these three approaches in combination, PMO
materials, as ordered, porous, bridged polysilsesquioxanes, are therefore well suited
in principle to achieve low k properties.

A recent example of a PMO material using a high organic content monomer
has shown how useful these materials may be as low k dielectrics in microelectronic packaging.[59] In this example, a mesoporous silica is assembled using a
cyclohexane analog with di(methoxy)silyl groups substituted for carbon at the
1,3,5 positions. Following the removal of the template, it was found that these
PMOs were exceptionally well ordered and demonstrated a dielectric constant
$k < 2$, which is low enough for next generation microelectronic packaging materials.
A plot of the dielectric constant of these materials as a function of the organic
molar fraction in the organosilica framework and post-treatment temperature is
shown in Figure 8.18, where pure silica has a dielectric constant of about 4.1.
A thermally driven, mesostructure maintaining, silanol consuming, bridge to
terminal organic group transformation in the channel walls, has been found to be
responsible for the humidity resistance of the PMO – self-hydrophobization
without the need for silanol group capping! Overall it seems that PMOs may have
an ideal combination of dielectric, mechanical, and humidity resistance properties
for next generation low-k microelectronic applications.

8.19 Mesomorphology – Films, Interfaces, Mesoepitaxy

Mesoporous silica spontaneously nucleates and grows at air–liquid,[130] liquid–
liquid,[131] and solid–liquid[132,133] interfaces in the form of oriented free standing
and supported films.[134–136] The most practical methods for making supported
mesoporous silica films involve dip coating[137,138] and spin coating[139,140] strategies.

The best-studied liquid phase in regards to film growth is water.[141,142] Surfactant micelles are well documented to assemble at air–water, oil–water, and solid–water interfaces. Co-assembled silicate micelles and lyotropic liquid crystals behave in a similar way, and hydrolytic poly-condensation results in the formation of hexagonal mesoporous silica film in which the channels are oriented parallel to the growth interface. The structure and hydrophobicity–hydrophilicity of the interface has a profound influence on the growth kinetics and channel architecture of the mesoporous silica film. Atomic force microscopy with gentle electrical double layer imaging with a surfactant-coated tip, has permitted *in situ* observation of the assembly of cetyltrimethylammonium surfactants at the interface between graphite–water and mica–water.[143,144] To expand, under dilute surfactant conditions parallel arrays of well-organized micelle hemi-cylinders and cylinders form monolayers on graphite and mica, respectively (Figure 8.19).[133] At higher concentrations of surfactant, AFM studies have shown the nucleation and growth

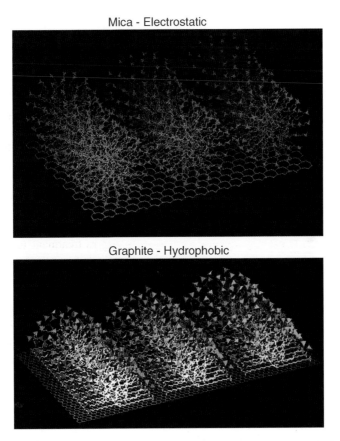

Figure 8.19 *Self-assembly of cationic surfactants into parallel arrays of cylinders and hemi-cylinders aligned with the hexagonal symmetry axes of the underlying mica and graphite substrates, respectively*
(Reproduced with permission from Ref. 133)

of well-ordered hexagonal liquid crystal films, comprised of a multi-layer stack of about 20 cylindrical micelles on hydrophobic graphite up to a thickness of around 1000 Å. Alkane chains of the surfactant are known to adsorb in a lying down position to the hydrophobic graphite surface with a head-to-head and tail-to-tail conformation. Methylene groups of the alkane chain provide an epitaxial match with the C–C bonds of the aromatic six-rings of the graphite. The alkane tail of the surfactant is hydrophobically bonded to the graphite. Together with ion-pair and image dipole interactions between the graphite, cationic surfactant, and counter anions, this induces ordering of a surfactant monolayer lying flat on the surface and aligned orthogonally to the hexagonal unit cell axes of graphite. Hydrophobic interactions between surfactants in solution and this organized monolayer drive the system to one of aligned hemi-micelle cylinders. In contrast, the formation of ordered cylindrical micelles in the case of hydrophilic mica is determined mainly by electrostatic interactions. These involve the cationic head group of a surfactant that has exchanged with charge-balancing potassium ions on the mica surface interacting with anionic six-ring sites of the hexagonal mica surface. Complementary interactions involving surface charge and geometry at the surfactant–mica and surfactant–graphite interfaces are considered responsible for "mesoepitaxy" of hexagonal mesoporous silica films with their channels registered with the hexagonal lattice of the underlying substrate.

In the case of mesoporous silica film growth at the air–water interface, neutron and X-ray reflection studies together with large-scale molecular dynamics calculations have shown that cetyltrimethyl-ammonium surfactants form a hemi-micellar overlayer at the interface between air and water (Figure 8.20). Under quiescent acidic aqueous conditions silicate micelles assemble at this interface and polymerize to an oriented free standing, mesoporous silica film in which the channels run strictly parallel to the air–water interface.[145] Further level of control can be imposed, for instance, by the application of a magnetic field, which causes a preferred axial alignment of silicate micelles.[146]

8.20 Stand Up and Be Counted

One of the greatest challenges in the field of periodic mesoporous materials since their invention around 15 years ago has been the synthesis of hexagonal symmetry films in which the orientation of the channels is orthogonal to the substrate surface. Experience has proven that the channels in these films prefer to lie parallel to the surface of the substrate on which they are grown. Templating cylindrical micelles (dilute synthesis conditions) or lyotropic liquid crystal hexagonal mesophase (concentrated synthesis conditions) are responsible for directing the co-assembly of inorganic precursors like silicates to form periodic mesoporous silica (PMS) films on substrates like mica, graphite, silicon, glass, and gold. It seems that these units prefer energetically to align in a planar rather than a homeotropic fashion on the substrate surface. Attempts to modify the alignment of the channels in a PMS film through the control of micelle or liquid crystal anchoring energies, using surface modification or electric fields, for example, have met with limited success.

Air

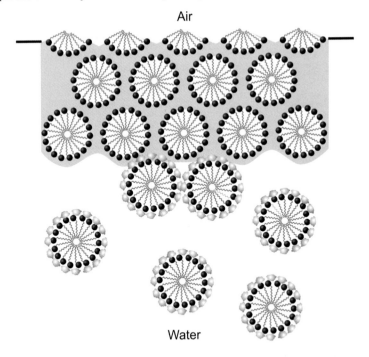

Water

Figure 8.20 *Silicate micelles self-assemble at the interface between air and water to form oriented free-standing, hexagonal mesoporous silica film*

A creative twist that enables vertical channel alignment curiously takes advantage of the preferred parallel channel alignment with respect to the substrate surface.[147,148] The approach makes use of the simple yet elegant idea of using large channels to organize small channels. In essence, the channels in an anodized aluminum oxide (AAO) membrane (see Chapter 5) are utilized as a template for the synthesis of PMS, and its channel surface directs the orientation of the resulting PMS channels. Because the cylindrical micelles or lyotropic liquid crystal hexagonal mesophase prefer planar alignment on the channel surfaces of the AAO, the outcome of this kind of "double-templating" procedure is a composite AAO–PMS membrane with a "channels within channels" architecture shown in Figure 8.21. In this particular structure, both large and small channels are orthogonally oriented with respect to the AAO membrane surface.

This is a natural construct for a practical size-selective molecule separation membrane because the AAO provides the mechanical stability for supporting the PMS, while the large channels of the AAO orient the small channels of the PMS perpendicular to the membrane surface. This configuration is ideally suited to enable molecular transport through the PMS channels since molecules can easily diffuse from one side of the membrane to the other while being in constant contact with a permselective mesophase. An attractive feature of this new kind of membrane, besides just the favorable orientation of the large and small channels, is that one has control over the composition of the periodic mesoporous material

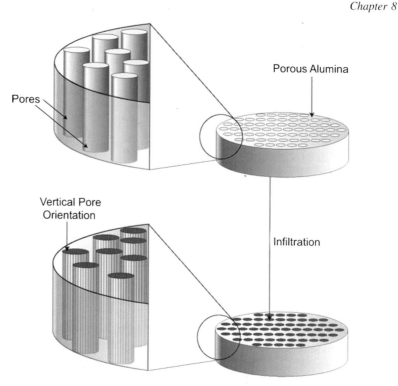

Figure 8.21 *Porous alumina templating method used to make mesoporous oxides with vertically aligned channels*
(Reproduced with permission from Ref. 147)

grown within the channels of the AAO as well as command over the relative diameters of large and small channels. Permselective composite AAO–PMS membranes of this type are anticipated to be useful for size-selective macromolecule and biomolecule separations and biosensing devices.

8.21 Mesomorphology – Spheres, Other Shapes

For mesoporous materials to find applications in a variety of fields, they must be synthesized with a correct form. A material with a potentially useful chemical property and meso-structure could be virtually useless if synthesized as a non-descript powder, which cannot be processed. A particularly attractive morphology for mesoporous materials is nanometer or micrometer-sized spheres, which would facilitate their application in chromatography columns,[149] or to colloidal photonic crystals with optical sensing capabilities if sufficiently monodisperse.

Several strategies have been developed to generate spheres of mesoporous oxides, some with surprisingly good control over ensemble monodispersity. A common way to make monodisperse spheres is to use the finely dispersed droplets in an emulsion as nano-reaction vessels[150] and nuclei for further growth. Correspondingly, it was found that a water-in-oil emulsion could template the formation of relatively monodisperse microspheres.[151,152] Other interesting templates for mesoporous spheres are liquid droplets in an aerosol.[153] Aerosols with good monodispersity can be produced by forcing a liquid through a small nozzle. The liquid jet ejected at high speed experiences Rayleigh instability and breaks into spherical drops. Particle size can be controlled by changing the nozzle diameter, and a very respectable monodispersity can be achieved by using a vibrating orifice aerosol generator.[154] A great variety of spheres with different internal structures and morphologies are available by aerosol templating, with just a few of these shown in Figure 8.22. While these types of spheres have not been prepared sufficiently monodisperse to form colloidal photonic crystals, such a construct can be made by templating the growth of mesoporous silica inside a polymer inverse opal (see Chapter 7). Micromolding in inverse polymer opals, MIPO, allows one to form a perfect lattice of silica spheres and transfer this order to a mesoporous silica through a double-replication step.[155]

Spheres of mesoporous oxides have also been synthesized as hollow shells, which increases their surface area as well as reduces the mass of material per unit volume. This would be advantageous to reduce costs if mesoporous-based chromatography devices were produced on an industrial scale. Hollow mesoporous shells can be synthesized by templating the oil–water interface in an oil–water

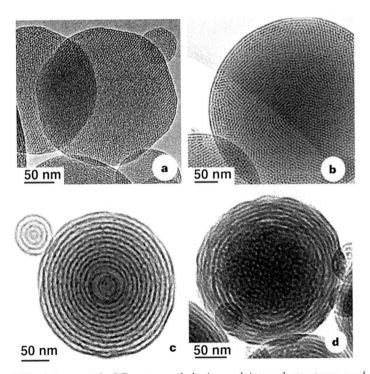

Figure 8.22 *Spheres with different morphologies and internal structures produced by aerosol-templating of mesoporous oxides*
(Reproduced with permission from Ref. 153)

emulsion,[156] and can also be produced by modifying the conditions in an aerosol particle synthesis.[157] Perhaps the most general way to form hollow mesoporous spheres is by templating against pre-formed spheres of another material. As mentioned in Chapter 7, a multitude of spheres are commercially available or available through published synthetic protocols. Such spheres can be incorporated into a mesosynthesis to either form core–shell composite particles,[158] or hollow shells if the templating sphere can be dissolved selectively from the porous oxide shell.[159]

8.22 Mesomorphology – Patterned Films, Soft Lithography, Micromolding

Patterns in mesoporous silica film have been intentionally designed using soft-lithographic techniques, such as microcontact printing with alkanethiolate monolayers (SAMs) on gold (Figure 8.23). A three-way templating mechanism has been proposed to explain the selective growth of mesoporous silica on the SAM defined regions of the substrate.[160] Langmuir adsorption of the cetyltrimethyl ammonium surfactant occurs preferentially on the SAM to form a heterobilayer.[161] Then negatively charged silicate micelles accrete from solution mainly to the positively charged headgroup region of adsorbed surfactants. This favors oriented growth of the mesoporous silica only on the SAM delineated areas of the gold, allowing the intentional design of micron scale patterned films of mesoporous silica.

Micromolding in capillaries (MIMIC) also provides an effective way for patterning mesoporous silica.[162,163] Capillary tension provides the driving force for the flow of the silicate liquid crystal precursor into the channels of a PDMS

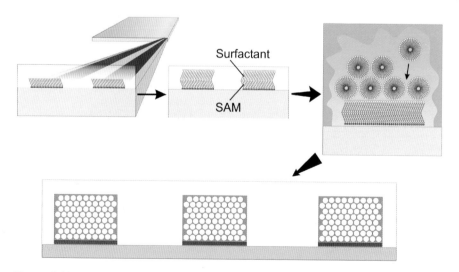

Figure 8.23   *Three-way templating of patterned mesoporous silica film*
(Reproduced with permission from Ref. 160)

elastomeric stamp that is pressed against a solid substrate. Simultaneous imposition of an electric field along the capillaries causes rapid infiltration by electro-osmotic flow[164] and increases the rate of polymerization due to localized Joule heating. The preferred alignment of cylindrical silicate micelles parallel to the capillary axis together with the flow of liquid in the capillary aligns the director field of the imbibed silicate liquid crystal along the capillary to minimize the shear energy on the silicate micelles. Polymerization of the silicate and removal of the elastomeric mold creates a replica pattern out of mesoporous silica in which the channel axes are organized along the length of the capillaries.

Micromolding in capillaries has enabled rhodamine 6G doped tri-block copolymer templated, photoluminescent mesoporous silica waveguide arrays to be patterned on low refractive index (1.15) mesoporous silica thin film cladding, as seen in the left panel of Figure 8.24.[165] These integrated optical mesoporous silica components have facilitated the practical realization of amplified spontaneous emission at a relatively low pump laser power threshold, depicted in the right panels of the figure. This property has been attributed to the ability of mesoporous silica to suppress dye molecule agglomeration even at relatively high dye loading levels inside the mesochannel waveguides. Mesoporous silica cladded mesoporous silica waveguides may lead to the development of integrated optical circuits.[166]

A silicatropic mesophase containing an imbibed PAG when exposed to UV light undergoes acid catalyzed condensation-polymerization of silicate oligomers to silica. This concept of creating the silicate polymerization catalyst photolytically has been utilized to pattern a thin film of the silicatropic mesophase by a creative combination of photolithography and selective etching.[167] The methodology used to achieve this goal is expressed in Figure 8.25. In essence photolithographic patterning allows the optically unexposed film of a silicatropic mesophase to be selectively removed from the mesoporous silica-exposed regions by an aqueous NaOH etchant. In addition, thermal post-treatment of the optically exposed silicatropic mesophase enables refractive index differentiation of the two optically exposed regions of the mesoporous silica originating from either (i) hexagonal mesoporous silica patterns with different channel diameters (lower surfactant

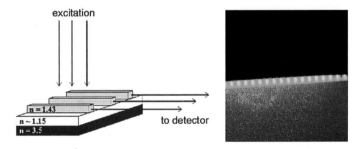

Figure 8.24 *Mirrorless lasing from mesoporous silica waveguides patterned by micromolding in capillaries. Above the lasing threshold the fluorescence narrows significantly, and yellow laser light can be seen by optical microscopy* (Reproduced with permission from Ref. 165)

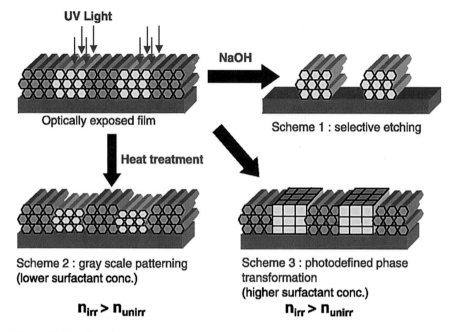

Figure 8.25 *Photolithographic patterning of a silicatropic mesophase film containing an imbibed photoacid generator*
(Reproduced with permission from Ref. 167)

concentration), left lower panel (Figure 8.25) or (ii) mesoporous silica patterns with different structures, cubic, and hexagonal (higher surfactant concentrations), right lower panel (Figure 8.25).

Mesoporous silica has also been patterned using ink jet printing.[168] The method is based upon evaporation induced self-assembly (EISA) of a dilute aqueous acidic ethanolic solution of tetraethoxyorthosilicate within microdroplets or microlines written on a substrate *via* an ink jet chemical delivery system as depicted in the top part of Figure 8.26. The synthesis ink emerges from the pen orifice and deposits the desired mesophase micropattern on the substrate, like the dots shown by optical microscopy on the bottom part of the figure. Silicate micelles co-assemble and polymerize at the air–liquid and liquid–solid interfaces (section 8.19) of printed microdroplets and microlines, to form mesoporous silica microreplicas with droplet and line shapes.

8.23 Mesomorphology – Morphosynthesis of Curved Form

In the absence of interfaces and under quiescent aqueous conditions mesoporous silica grows with uniquely curved shapes.[169–172] Depending mainly on the choice of pH, mesoporous silica can form sphere, gyroid, discoid, fiber, rod, hollow cylinder, and hollow helicoids shapes. These shapes often display periodic surface patterns.

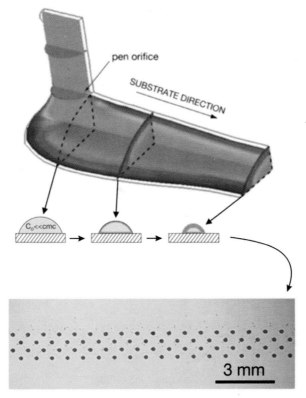

Figure 8.26 *Evaporation induced self-assembly (EISA) within the spatial confines of ink jet printed patterns of silicate-micelle microdrops or microlines can result in arbitrary patterns of mesoporous silica, such as the dots shown in the optical microscope image below*
(Reproduced with permission from Ref. 168)

To amplify, curved forms of mesoporous silica can be synthesized by a rational control of the interplay of defect-design and complex soft matter order.[173] Nanoscopic nuclei of mesoporous silica can be captured by quenching the early stages of a surfactant templated synthesis. They can also be trapped by evaporation-induced self-assembly of surfactants and silicates within the spatial confines of aerosol nanodroplets.[174] The growth of siliceous mesostructures begins with the formation of a seed, a co-assembly of surfactants and silicates. What is special about this situation is the nature and temporal evolution of the "viable nucleus" that is initially created from the synthesis nutrients. In the "incubation" of the seed, "soft silicatropic matter" is generated whose order and shape is not dictated by crystallographic physics, space groups and the underlying symmetry of a rigid unit cell that creates Platonic form. Instead, its shape is determined by a nano-scale deformable mesophase that assumes a form predicated by the theory of elasticity.[175]

The critical size of such an elastically deformable ordered nucleus, which is able to sustain growth rather than simply re-dissolve, is governed by the equality of surface and bulk free energy. This soft seed appears to be around 50 nm in size, which interestingly can be an association of about 10 surfactant micellar aggregates in each direction of order. Of special interest is the structure of this "nano–meso" unit as well as the evolution of this kind of complex soft matter nucleus in time, because it is expected to involve a rather intricate interplay of subtle factors each of which can have a profound effect on the final form.

Not a great deal is currently understood about this important aspect of morphogenesis with its biosilicification connotations. From the little knowledge that does exist, it is clear that growth is intimately connected to colloidal interactions[176] (electrical double layer repulsion and van der Waals attraction forces) operating between the evolving seed and accreting surfactant–silicate micelles for the particular choice of synthesis conditions (ionic strength, dielectric constant, temperature). It is also apparent that topological defects of particular strength assemble within the emerging nanoscale mesophase, driving it to assume a specific internal structure and shape that depends on the synthesis conditions as well as surface and bulk elastic-free energy. It is this intricate interplay of colloidal and elastic forces that lead to the mesostructure organization and impressive "biomimetic" silicified curved shapes (rods, gyroids, toroids, spirals, spheres) observed for mesoporous silica and organosilica materials templated by surfactant micelles (Figure 8.27).

Another unusual aspect of this growth process is the gradual rigidification, induced by hydrolytic poly-condensation of a silicatropic mesophases. This too is a complex topic and much remains to be learned about the details. What is presently known is that polymerization of silicate species in an emerging silicatropic nanomesophase leads to contraction-induced stresses in the channels of the mesostructure, which results in fascinating buckling patterns observed on mesoporous silica curved forms[177] (Figure 8.28). The design of these buckling patterns is related to the organization of the mesostructure and the anisotropy of contraction-induced stresses within the curved form. Curved patterned shapes are definitely a distinctive feature of mesoporous silica.

8.24 Chiral Surfactant Micelles – Chiral Mesoporous Silica

Section 8.23 hinted at a strategy for synthesizing hexagonal mesoporous silica hollow helicoids with chiral form, in which the channel direction appears to track the helical axis of the helicoids. The creation of the hollow helicoids was thought to originate from free-standing achiral hexagonal mesoporous silica film formed at the air–water interface. Silicate polymerization induced differential contraction stresses in the channels along longitudinal and transverse channel directions forces the film to curl and twist into the observed chiral form.

A more direct approach for the preparation of chiral hexagonal mesoporous silica involves the use of a specially designed surfactant with a chiral amino acid head

Figure 8.27 *Morphosynthesis of mesoporous silica curved shapes*
(Reproduced with permission from Ref. 169)

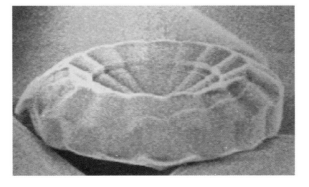

Figure 8.28 *Buckling patterns in a mesoporous silica discoid shape*
(Reproduced with permission from Ref. 177)

group that is able to self-assemble into chiral micelles and chiral lyotropic liquid crystals. These chiral assemblies can then direct the self-assembly of silicate building blocks into a chiral material in which the silica channels so formed replicates the structure of the chiral template.[178]

The surfactant template (chiral sodium *N*-acyl-*L*-alanate) and silica precursors (equimolar mixture of tetraethoxyorthosilicate and aminopropyltriethoxysilane or trimethylammonium–propyltriethoxysilane) that were employed under aqueous acid catalyzed synthesis conditions are depicted in Figure 8.29. The chiral surfactant micelle-silicate co-assembly believed to be responsible for the formation of chiral hexagonal mesoporous silica involves electrostatic interactions between negatively charged carboxyls of the chiral amino acid and positively charged amines on the silicate.

The material formed in this way displays fascinating well-defined spiral hexagonal rod morphology with a helical pitch estimated to be about 1.5 μm as shown in the figure. An elegant computer simulation analysis of high-resolution TEM images of the spiral hexagonal rods provides strong evidence that chiral mesoscale channels track the helical axis and helical pitch of the rods. Especially interesting is the observed ratio of right to left handed rods determined from an analysis of 500 rods which turns out to be between 6.5:3.5 and 7.5:2.5 depending on the precursor used. Noteworthy, the racemic amino acid surfactant is found to template achiral hexagonal mesoporous silica, which implies that factors other than just geometric packing of chiral surfactants and silicate in the co-assembly depicted

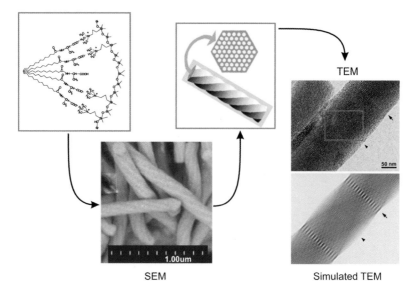

Figure 8.29 *Precursors used for the assembly of a chiral mesoporous silica shown in the top left. The chirality is readily apparent in SEM images, 3D models, and TEM images*
(Reproduced with permission from Ref. 178)

in the figure are responsible for the creation of the chiral hexagonal mesoporous silica.

A direct and intentional synthesis of hexagonal mesoporous silica with chiral channels and spiral rod shapes provides a number of exciting opportunities for templating chiral metal and semiconductor mesoscopic wires, for performing size and shape selective asymmetric separations and catalytic chemistry in a chiral compartment, and maybe even orchestrating the assembly of a spiral rod photonic crystal for replication in silicon, a photonic lattice that has been predicted to have a very large omnidirectional photonic bandgap immune to intrinsic defects (see Chapter 7).

8.25 Mesopore Replication

The synthesis of well-ordered porous inorganics can be adapted to a variety of materials, as we have seen. However, it may be quite difficult to find the ideal conditions to form a well-ordered mesophase from materials with a more exotic composition. In these cases, a double-templating method may be used. A structurally well-ordered mesoporous silicate is formed, and the pores in this material used to template the formation of another material. Removal of the silicate by an acidic (dilute HF) or basic (concentrated NaOH) etchant leaves behind the templated material having an inverted topology. This technique is conceptually similar to the formation of "inverted opals" from colloidal crystal synthetic opals (see Chapter 7).

Amongst the most interesting and promising materials made by double-templating are mesoporous carbons,[179–182] another addition to the carbon zoo recently re-stocked with Buckyball, Buckytubes, and Carbon Opals. A mesoporous silicate is infiltrated with a carbon source, an organic species that may be molecular or polymeric. It is advantageous to choose a precursor which retains the greatest possible mass (carbon yield) upon pyrolysis, as this will ensure an optimally filled mold. If the precursor used has a low carbon conversion, multiple infiltrations can be performed to fully fill the mesopores. Treatment at high temperature carbonizes the organic component, and etching using dilute HF in water affords a mesoporous carbon replica. This method could be adapted to a variety of materials, which are formed under prohibitive conditions for mesostructure assembly, such as those deposited by chemical vapor deposition (CVD). The silica mold is stable enough to withstand a variety of experimental deposition conditions, and in the case of carbon, conditions can be modified to obtain graphitic instead of amorphous walls.[183] Graphitic and non-graphitic carbons are widely used as conductors, fillers, anodes in lithium batteries, and supports for various heterogeneous catalysts. The formation of carbon into a structure with regular mesopores can greatly facilitate the understanding of carbon-supported catalysis by simplifying the structural models considered for porosity.

A great illustration of the potential utility of mesoporous carbons is in the synthesis of ordered carbon-supported Pt particles.[184] The mesoporous silica chosen for this study was so-called SBA15. This silicate is templated by non-ionic

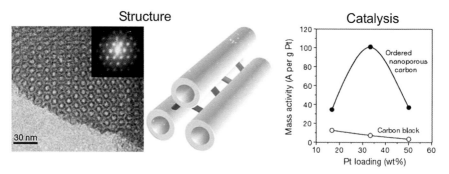

Figure 8.30 *TEM image of a mesoporous carbon, left, and a proposed structural model, center. Catalytic activity of Pt loaded ordered carbon compared to Pt loaded carbon black for different Pt loadings, right* (Reproduced with permission from Ref. 184)

PEO–PPO–PEO tri-block copolymers, and is permeated with 1D channels on a hexagonal lattice, interconnected by smaller through-pores.[185] In order to increase the thermal stability of this material, it is subjected to post-synthesis exchange, which replaces about 20% of the silicon atoms in the structure by aluminum atoms.[186] This modified template is then infiltrated with furfuryl alcohol, and treated at 80 °C to polymerize the alcohol as a coating on the aluminosilicate walls. Removal of excess furfuryl alcohol under vacuum followed by pyrolysis at 1100 °C gives the carbon–aluminosilicate composite, and etching of the template in aqueous HF results in the formation of mesoporous carbon. Such carbons are highly ordered, as shown in Figure 8.30. After its synthesis, the carbon support is loaded with platinum particles by simply exposing it to an acetone solution of hexachloroplatinate. The ordered channels and confined geometry of the mesoporous host results in nanoscale platinum particles with a narrow size distribution, which remain very well dispersed even at quite high Pt loading. The performance of these metal-loaded carbons was compared to commercially available materials such as platinum black (Pt supported on amorphous carbon). It was found that the catalytic activity of the materials for electro-reduction of oxygen far surpassed conventional materials, as shown in the right part of Figure 8.30. They were also found to be superior material for the storage of hydrogen, likely the fuel of the future.

The promise of mesoporous carbons in catalysis, electrocatalysis, and hydrogen storage is quite clear. Further research into these materials will definitely uncover more surprises.

8.26 Mesochemistry and Topological Defects

Micron scale defects in organic liquid crystals control their director field patterns, optical birefringence texture, electrical, and magnetic properties.[187] It should therefore come as no surprise that such defects exist in silicatropic liquid crystals and can exert profound effects on the growth and channel structure of mesoporous silica shapes and films.[188] As mentioned earlier, silicatropic liquid crystal droplets are believed to provide the seeds for viable growth of hexagonal mesoporous silica curved shapes. The stability of a particular director field pattern in the droplet is expected to be extremely sensitive to the solution environment. This is because the stability of a specific type of defect depends on a balance of bulk and surface elastic constants, where the surface-free energy or tension is expected to be most influenced by solution conditions. Once a director field pattern is established, the surface mesostructure and surface charge of the silicatropic liquid seed serves to direct the further accretion of silicate micelles, and can therefore be considered to template its own growth to a singular kind of mesoporous silica shape. Such a shape is shown in Figure 8.28. Similarly, the defects and director field patterns of a silicatropic liquid crystal seed or liquid crystal film located at the boundary between water and a solid, liquid, or gaseous surface will be mainly determined by interfacial interactions. If these forces are strong then the structure of the surface will be reflected in the channel architecture of the mesoporous silica film. Conversely, if the forces are weak then topological defects and director field patterns will be stabilized in the silicatropic liquid crystal film and replicated in the channel design of the resulting mesoporous silica film. Polarization optical microscopy, using a sample placed between crossed polarizer and analyzer in an optical microscope, is a good way to identify topological defects in uniaxial hexagonal mesoporous silica shapes and films.[189] In this way, diagnostic optical birefringence patterns that arise from constructive–destructive interference between ordinary and extraordinary rays traversing an optically anisotropic sample, can be identified in mesoporous silica shapes and films (Figure 8.31). These birefringence patterns or optical textures can provide detailed information concerning the strength and type of topological defects responsible for the nucleation, growth, form, and texture of mesoporous materials made under different synthesis conditions. Transmission electron microscopy images[190] of the architecture of the hexagonal channels within these mesoporous silica shapes and films reveal director field patterns and topological defects (Figure 8.32) that are consistent with observed optical birefringence patterns.

8.27 Mesochemistry – Synthesis in "Intermediate" Dimensions

The purpose of this chapter has been to document the key physicochemical design principles that underpin the template-directed synthesis of inorganic mesoporous materials with particular structure, composition and form, and to appreciate how the properties of materials fashioned at this length scale can be orchestrated to create a purposeful function that is directed to a specific field of use. It should be clear from the examples presented that surfactant and block copolymer templates enable *morphosynthesis,* or *shape synthesis,* of mesostructured materials with uniquely curved form and patterned surfaces. This stands in sharp distinction to Platonic form, which is more typical of materials structured at the microscale.

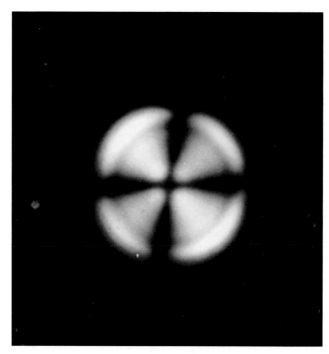

Figure 8.31 *Maltese cross optical birefringence pattern observed for a single hexagonal mesoporous silica discoid shape between a crossed polarizer and analyzer in an optical microscope*
(Reproduced with permission from Ref. 173)

The attribute of curved form is accomplished through judicious command of topological defects in "inorganotropic liquid crystal seeds" formed in the template-directed nucleation and growth process.

Synthesis conditions can be adjusted to control the texture of the material comprising the mesostructure to be either "glassy or crystalline." Chemical and physical properties of the mesostructure can be tailored through the inclusion of specific guests within channel spaces or the incorporation of organic functional groups either on the surface of channels or within channel walls. This capability enables chemistry *in* the channels, chemistry *of* the channels, and a chemical control over physical properties and function. Matching a characteristic dimension of the templating mesophase to the length scale at a growth interface, for example, air–liquid, liquid–liquid and liquid–solid, called *mesoepitaxy*, can be used as a synthesis tool to direct the orientation of and to pattern a mesostructured film. This can be done for example by soft lithography, micromolding, photolithography, and ink jet printing, for a range of perceived applications that include battery, fuel, and solar cell electrodes and electrolytes, sensing and chemical delivery, low k microelectronics packaging and acoustic insulation, and permselective membranes and catalyst supports.

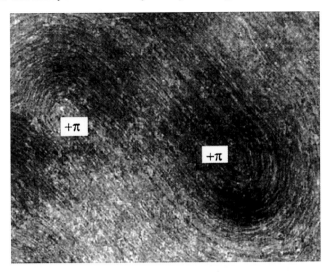

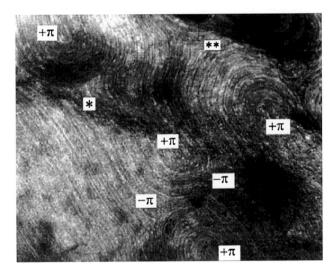

Figure 8.32 *TEM images of oriented free standing hexagonal mesoporous silica film that displays swirling and curling channel director field patterns and topological defects*
(Reproduced with permission from Ref. 173)

In the framework of materials chemistry, inorganic materials structured at the mesoscale represent an "intermediate form of matter," with unique properties. Mesomaterials of the genre described in this chapter appear well poised to spawn new mesoscience and mesotechnology. It will be interesting to watch the field unfold in the years ahead.

8.28 References

1. M.E. Davis, Ordered porous materials for emerging applications, *Nature*, 2002, **417**, 813.
2. C.J. Brinker, Porous inorganic materials, *Curr. Opin. Solid State Mater. Sci.*, 1996, **1**, 798.
3. A. Falciatore and C. Bowler, Revealing the molecular secrets of marine diatoms, *Annu. Rev. Plant Biol.*, 2002, **53**, 109.
4. T. Armbruster and M.E. Gunter, Crystal structures of natural zeolites, *Rev. Miner. Geochem.*, 2001, **45**, 1.
5. E. Passaglia and R.A. Sheppard, The crystal chemistry of zeolites, *Rev. Miner. Geochem.*, 2001, **45**, 69.
6. S. Polarz and M. Antonietti, Porous materials via nanocasting procedures: innovative materials and learning about soft-matter organization, *Chem. Commun.*, 2002, **22**, 2593.
7. K.J.C. van Bommel, A. Friggeri and S. Shinkai, Organic templates for the generation of inorganic materials, *Angew. Chem. Int. Ed.*, 2003, **42**, 980.
8. H.A. Lowenstam, Minerals formed by organisms, *Science*, 1981, **211**, 1126.
9. R.M. Barrer, Zeolite synthesis: an overview, *NATO ASI Ser., Ser. C: Math. Phys. Sci. (Surf. Organomet. Chem.: Mol. Approaches Surf. Catal.)*, 1988, **231**, 221.
10. S. Mann, *Biomimetic Materials Chemistry*, VCH, New York, NY, 1996.
11. D.W. Breck, *Zeolite Molecular Sieves: Structure, Chemistry, and Use*, Wiley-Interscience, New York, NY, 1974.
12. R.L. Bedard, L.D. Vail, S.T. Wilson and E.M. Flanigen, Crystalline microporous metal sulfide compositions, US Patent 4, 880, 761, 1989, **880**, 761.
13. H. Ahari, C.L. Bowes, T. Jiang, A. Lough, G.A. Ozin, R.L. Bedard, S. Petrov and D. Young, Nanoporous tin(IV) chalcogenides. Flexible open-framework nanomaterials for chemical sensing, *Adv. Mater.*, 1995, **7**, 375.
14. T. Jiang, A.J. Lough, G.A. Ozin, D. Young and R.L. Bedard, Synthesis and structure of the novel nanoporous tin(IV) sulfide material TPA-SNS-3, *Chem. Mater.*, 1995, **7**, 245.
15. C.W. Jones, K. Tsuji and M.E. Davis, Organic-functionalized molecular sieves as shape-selective catalysts, *Nature*, 1998, **393**, 52.
16. M. Eddaoudi, D.B. Moler, H. Li, B. Chen, T.M. Reineke, M. O'Keeffe and O.M. Yaghi, Modular chemistry: secondary building units as a basis for the design of highly porous and robust metal–organic carboxylate frameworks, *Acc. Chem. Res.*, 2001, **34**, 319.
17. O.M. Yaghi, M. O'Keeffe, N.W. Ockwig, H.K. Chae, M. Eddaoudi and J. Kim, Reticular synthesis and the design of new materials, *Nature*, 2003, **423**, 705.
18. C.L. Bowes and G.A. Ozin, Self-assembling frameworks: beyond microporous oxides, *Adv. Mater.*, 1996, **8**, 13.
19. M. Eddaoudi, D.B. Moler, H.L. Li, B.L. Chen, T.M. Reineke, M. O'Keeffe and O.M. Yaghi, Modular chemistry: secondary building units as a basis for the design of highly porous and robust metal–organic carboxylate frameworks, *Acc. Chem. Res.*, 2001, **34**, 319.

20. N.F. Zheng, X.H. Bu and P.Y. Feng, Synthetic design of crystalline inorganic chalcogenides exhibiting fast-ion conductivity, *Nature*, 2003, **426**, 428.

21. S.T. Batten and R. Robson, Interpenetrating nets: ordered, periodic entanglement, *Angew. Chem. Int. Ed.*, 1998, **37**, 1460.

22. D.T. Tran, P.Y. Zavalij and S.R.J. Oliver, $Pb_3F_5NO_3$, a cationic layered material for anion-exchange, *J. Am. Chem. Soc.*, 2002, **124**, 3966.

23. J.S. Seo, D. Whang, H. Lee, S.I. Jun, J. Oh, Y.J. Jeon and K. Kim, A homochiral metal–organic porous material for enantioselective separation and catalysis, *Nature*, 2000, **404**, 982.

24. O.M. Yaghi, G. Li and H. Li, Selective binding and removal of guests in a microporous metal–organic framework, *Nature*, 1995, **378**, 703.

25. N.L. Rosi, J. Eckert, M. Eddaoudi, D.T. Vodak, J. Kim, M. O'Keeffe and O.M. Yaghi, Hydrogen storage in microporous metal–organic frameworks, *Science*, 2003, **300**, 1127.

26. J.L.C. Rowsell, A.R. Millward, K.S. Park and O.M. Yaghi, Hydrogen sorption in functionalized metal–organic frameworks, *J. Am. Chem. Soc.*, 2004, **126**, 5666.

27. Y.P. Zhou, K. Feng, Y. Sun and L. Zhou, A brief review on the study of hydrogen storage in terms of carbon nanotubes, *Prog. Chem.*, 2003, **15**, 345.

28. M. Eddaoudi, J. Kim, N. Rosi, D. Vodak, J. Wachter, M. O'Keefe and O.M. Yaghi, Systematic design of pore size and functionality in isoreticular MOFs and their application in methane storage, *Science*, 2002, **295**, 469.

29. A. Corma, State of the art and future challenges of zeolites as catalysts, *J. Catal.*, 2003, **216**, 298.

30. S. Bhatia, *Zeolite Catalysis: Principles and Applications*, CRC Press, Boca Raton, FL, 1990.

31. T.E. Gier, X.H. Bu, P.Y. Feng and G.D. Stucky, Synthesis and organization of zeolite-like materials with three-dimensional helical pores, *Nature*, 1998, **395**, 154.

32. M.O. Adebajo, R.L. Frost, J.T. Kloprogge, O. Carmody and S. Kokot, Porous materials for oil spill cleanup: a review of synthesis and absorbing properties, *J. Porous Mater.*, 2003, **10**, 159.

33. D. Kallo, Applications of natural zeolites in water and wastewater treatment, *Rev. Miner. Geochem.*, 2001, **45**, 519.

34. C.T. Kresge, M.E. Leonowicz, W.J. Roth, J.C. Vartuli and J.S. Beck, Ordered mesoporous molecular sieves synthesized by a liquid-crystal template mechanism, *Nature*, 1992, **359**, 710.

35. J.S. Beck, J.C. Vartuli, W.J. Roth, M.E. Leonowicz, C.T. Kresge, K.D. Schmitt, C.T.W. Chu, D.H. Olson, E.W. Sheppard, S.B. McCullen, J.B. Higgins and J.L. Schlenker, A new family of mesoporous molecular sieves prepared with liquid crystal templates, *J. Am. Chem. Soc.*, 1992, **114**, 10834.

36. Y.H. Sakamoto, M. Kaneda, O. Terasaki, D.Y. Zhao, J.M. Kim, G. Stucky, H.J. Shim and R. Ryoo, Direct imaging of the pores and cages of three-dimensional mesoporous materials, *Nature*, 2000, **408**, 449.

37. S. Hafner and D.E. Demco, Solid-state NMR spectroscopy under periodic modulation by fast magic-angle sample spinning and pulses: a review, *Solid State NMR*, 2002, **22**, 247.

38. A. Firouzi, D. Kumar, L.M. Bull, T. Besier, P. Sieger, Q. Huo, S.A. Walker, J.A. Zasadzinski, C. Glinka, J. Nicol, D. Margolese, G.D. Stucky and B.F. Chmelka, Cooperative organization of inorganic-surfactant and bio-mimetic assemblies, *Science*, 1995, **267**, 1138.

39. Q.S. Huo, D.I. Margolese, U. Ciesla, P.Y. Feng, T.E. Gier, P. Sieger, R. Leon, P.M. Petroff, F. Schuth and G.D. Stucky, Generalized synthesis of periodic surfactant inorganic composite-materials, *Nature*, 1994, **368**, 317.

40. Y. Bouligand, Mesoporous solids of biological interest, *Actualite Chim.*, 2003, **7**, 4.

41. N. Coombs, D. Khushalani, S. Oliver, G.A. Ozin, G.C. Shen, I. Sokolov and H. Yang, Blueprints for inorganic materials with natural form: inorganic liquid crystals and a language of inorganic shape, *J. Chem. Soc. Dalton Trans.*, 1997, **21**, 3941.

42. G.S. Attard, J.C. Glyde and C.G. Göltner, Liquid-crystalline phases as templates for the synthesis of mesoporous silica, *Nature*, 1995, **378**, 366.

43. B.P. Binks, Particles as surfactants – similarities and differences, *Curr. Opin. Colloid Interface Sci.*, 2002, **7**, 21.

44. D.Y. Zhao, P.D. Yang, Q.S. Huo, B.F. Chmelka and G.D. Stucky, Topological construction of mesoporous materials, *Curr. Opin. Solid State Mater. Sci.*, 1998, **3**, 111.

45. P.T. Tanev and T.J. Pinnavaia, A neutral templating route to mesoporous molecular-sieves, *Science*, 1995, **267**, 865.

46. E. Leontidis, Hofmeister anion effects on surfactant self-assembly and the formation of mesoporous solids, *Curr. Opin. Colloid Interface Sci.*, 2002, **7**, 81.

47. M. Seul and D. Andelman, Domain shapes and patterns – the phenomenology of modulated phases, *Science*, 1995, **267**, 476.

48. D.Y. Zhao, J.L. Feng, Q.S. Huo, N. Melosh, G.H. Fredrickson, B.F. Chmelka, G.D. Stucky, Triblock copolymer syntheses of mesoporous silica with periodic 50 to 300 Angstrom pores, *Science*, 1998, **279**, 548.

49. G.J.D.A. Soler-Illia, E.L. Crepaldi, D. Grosso and C. Sanchez, Block copolymer-templated mesoporous oxides, *Curr. Opin. Colloid Interface Sci.*, 2003, **8**, 109.

50. Y.C. Liu and S.H. Chen, Analysis of the structure, interaction, and viscosity of pluronic micelles in aqueous solutions by combined neutron and light scatterings, *Scattering Polym.; ACS Symp. Ser.*, 2000, **739**, 270.

51. T.P. Lodge, Block copolymers: past successes and future challenges, *Macromol. Chem. Phys.*, 2003, **204**, 265.

52. S. Forster, Amphiphilic block copolymers for templating applications, *Top. Curr. Chem.*, 2003, **226**, 1.

53. B.Z. Tian, X.Y. Liu, C.Z. Yu, F. Gao, Q. Luo, S.H. Xie, B. Tu and D.Y. Zhao, Microwave assisted template removal of siliceous porous materials, *Chem. Commun.*, 2002, **11**, 1186.

54. C.J. Brinker, Porous inorganic materials, *Curr. Opin. Colloid Interface Sci.*, 1996, **1**, 798.
55. D.T. On, D. Desplantier-Giscard, C. Danumah and S. Kaliaguine, Perspectives in catalytic applications of mesostructured materials, *Appl. Catal. A- Gen.*, 2001, **222**, 299.
56. A.P. Wight and M.E. Davis, Design and preparation of organic–inorganic hybrid catalysts, *Chem. Rev.*, 2002, **102**, 3589.
57. S. Rubio and D. Perez-Bendito, Supra-molecular assemblies for extracting organic compounds, *Trends Anal. Chem.*, 2003, **22**, 470.
58. A. Hagfeldt and M. Gratzel, Molecular photovoltaics, *Acc. Chem. Res.*, 2000, **33**, 269.
59. K. Landskron, B.D. Hatton, D.D. Perovic and G.A. Ozin, Periodic mesoporous organosilicas containing interconnected $[Si(CH_2)]_3$ rings, *Science*, 2003, **302**, 266.
60. K. Domansky, J. Liu, L.Q. Wang, M.H. Engelhard and S. Baskaran, Chemical sensors based on dielectric response of functionalized mesoporous silica films, *J. Mater. Res.*, 2001, **16**, 2810.
61. N.K. Mal, M. Fujiwara and Y. Tanaka, Photocontrolled reversible release of guest molecules from coumarin-modified mesoporous silica, *Nature*, 2003, **421**, 350.
62. D.Y. Myers, *Surfactant Science and Technology*, Wiley, New York, 1988.
63. S.T. Hyde and G.E. Schroder, Novel surfactant mesostructural topologies: between lamellae and columnar (hexagonal) forms, *Curr. Opin. Colloid Interface Sci.*, 2003, **8**, 5.
64. G.G. Chernik, Phase studies of surfactant-water systems, *Curr. Opin. Colloid Interface Sci.*, 1999, **4**, 381.
65. A. Khan, Phase science of surfactants, *Curr. Opin. Colloid Interface Sci.*, 1996, **1**, 614.
66. R.A. Pai, R. Humayun, M.T. Schulberg, A. Sengupta, J.N. Sun and J.J. Watkins, Mesoporous silicates prepared using preorganized templates in supercritical fluids, *Science*, 2004, **303**, 507.
67. A.E.C. Palmqvist, Synthesis of ordered mesoporous materials using surfactant liquid crystals or micellar solutions, *Curr. Opin. Colloid Interface Sci.*, 2003, **8**, 145.
68. S.J. Candau and F. Lequeux, Self-assembling surfactant systems, *Curr. Opin. Colloid Interface Sci.*, 1997, **2**, 420.
69. P.D. Yang, D.Y. Zhao, D.I. Margolese, B.F. Chmelka and G.D. Stucky, Generalized syntheses of large-pore mesoporous metal oxides with semi-crystalline frameworks, *Nature*, 1998, **396**, 152.
70. T. Sun and J.Y. Ying, Synthesis of microporous transition-metal-oxide molecular sieves by a supramolecular templating mechanism, *Nature*, 1997, **389**, 704.
71. G.S. Attard, C.G. Goltner, J.M. Corker, S. Henke and R.H. Templer, Liquid-crystal templates for nanostructured metals, *Angew. Chem. Int. Ed.*, 1997, **36**, 1315.

72. G.S. Attard, P.N. Bartlett, N.R.B. Coleman, J.M. Elliott, J.R. Owen and J.H. Wang, Mesoporous platinum films from lyotropic liquid crystalline phases, *Science*, 1997, **278**, 838.

73. A.H. Whitehead, J.M. Elliott, J.R. Owen and G.S. Attard, Electrodeposition of mesoporous thin films, *Chem. Commun.*, 1999, **4**, 331.

74. A. Stein, M. Fendorf, T.P. Jarvie, K.T. Mueller, A.J. Benesi and T.E. Mallouk, Salt gel synthesis of porous transition-metal oxides, *Chem. Mater.*, 1995, **7**, 304.

75. B.T. Holland, P.K. Isbester, C.F. Blanford, E.J. Munson and A. Stein, Synthesis of ordered aluminophosphate and galloaluminophosphate meso-porous materials with anion-exchange properties utilizing polyoxometalate cluster/surfactant salts as precursors, *J. Am. Chem. Soc.*, 1997, **119**, 6796.

76. C.J. Brinker, Sol–gel processing of silica, *Colloid Chem. Silica; Adv. Chem. Ser.*, 1994, **234**, 361.

77. S.A. Bagshaw, E. Prouzet and T.J. Pinnavaia, Templating of mesoporous molecular-sieves by nonionic polyethylene oxide surfactants, *Science*, 1995, **269**, 1242.

78. T. Sun and J.Y. Ying, Synthesis of microporous transition metal oxide molecular sieves with bifunctional templating molecules, *Angew. Chem. Int. Ed.*, 1998, **37**, 664.

79. M.J. MacLachlan, N. Coombs and G.A. Ozin, Non-aqueous supramolecular assembly of mesostructured metal germanium sulphides from $(Ge_4S_{10})^{4-}$ clusters, *Nature*, 1999, **397**, 681.

80. M. Mamak, N. Coombs and G. Ozin, Self-assembling solid oxide fuel cell materials: mesoporous yttria–zirconia and metal–yttria–zirconia solid solutions, *J. Am. Chem. Soc.*, 2000, **122**, 8932.

81. C. Sanchez, B. Lebeau, F. Chaput and J.P. Boilot, Optical properties of functional hybrid organic–inorganic nanocomposites, *Adv. Mater.*, 2003, **15**, 1969.

82. P.T. Tanev, M. Chibwe and T.J. Pinnavaia, Titanium-containing mesoporous molecular-sieves for catalytic-oxidation of aromatic-compounds, *Nature*, 1994, **368**, 321.

83. M. Valden, X. Lai and D.W. Goodman, Onset of catalytic activity of gold clusters on titania with the appearance of nonmetallic properties, *Science*, 1998, **281**, 1647.

84. A. Davidson, Modifying the walls of mesoporous silicas prepared by supramolecular-templating, *Curr. Opin. Colloid Interface Sci.*, 2002, **7**, 92.

85. S.Y. Choi, M. Mamak, N. Coombs, N. Chopra and G.A. Ozin, Thermally stable two-dimensional hexagonal mesoporous nanocrystalline anatase, meso-nc-TiO_2: bulk and crack-free thin film morphologies, *Adv. Funct. Mater.*, 2004, **14**, 335.

86. T.A. Ostomel and G.D. Stucky, Free-standing mesoporous titania films with anatase nanocrystallites synthesized at 80 degrees C, *Chem. Commun.*, 2004, **8**, 1016.

87. K.M. Coakley, Y.X. Liu, M.D. McGehee, K.L. Frindell and G.D. Stucky, Infiltrating semiconducting polymers into self-assembled mesoporous titania films for photovoltaic applications, *Adv. Funct. Mater.*, 2003, **13**, 301.

88. Y.F. Lu, Y. Yang, A. Sellinger, M.C. Lu, J.M. Huang, H.Y. Fan, R. Haddad, G. Lopez, A.R. Burns, D.Y. Sasaki, J. Shelnutt and C.J. Brinker, Self-assembly of mesoscopically ordered chromatic polydiacetylene/silica nanocomposites, *Nature*, 2001, **410**, 913.

89. G.T. Li, S. Bhosale, T.Y. Wang, Y. Zhang, H.S. Zhu and K.H. Fuhrhop, Gram-scale synthesis of submicrometer-long polythiophene wires in mesoporous silica matrices, *Angew. Chem. Int. Ed.*, 2003, **42**, 3818.

90. Z.S. Zhang, S.Y. Zhang and W.R. Li, Templates and their applications in synthesis of nanomaterials, *Prog. Chem.*, 2004, **16**, 26.

91. K. Kageyama, J. Tamazawa and T. Aida, Extrusion polymerization: catalyzed synthesis of crystalline linear polyethylene nanofibers within a mesoporous silica, *Science*, 1999, **285**, 2113.

92. Y.S. Kang, H.I. Lee, Y. Zhang, Y.J. Han, J.E. Yie, G.D. Stucky and J.M. Kim, Direct synthesis of ordered mesoporous materials constructed with polymer–silica hybrid frameworks, *Chem. Commun.*, 2004, 1524.

93. G.A. Ozin, E. Chomski, D. Khushalani and M.J. Maclachlan, Mesochemistry, *Curr. Opin. Colloid Interface Sci.*, 1998, **3**, 181.

94. K. Moller and T. Bein, Inclusion chemistry in periodic mesoporous hosts, *Chem. Mater.*, 1998, **10**, 2950.

95. A. Stein, B.J. Melde and R.C. Schroden, Hybrid inorganic–organic mesoporous silicates – Nanoscopic reactors coming of age, *Adv. Mater.*, 2000, **12**, 1403.

96. B.Z. Tian, S.N. Che, Z. Liu, X.Y. Liu, W.B. Fan, T. Tatsumi, O. Terasaki and D.Y. Zhao, Novel approaches to synthesize self-supported ultrathin carbon nanowire arrays templated by MCM-41, *Chem. Commun.*, 2003, **21**, 2726.

97. P.L. Llewellyn, U. Ciesla, H. Decher, R. Stadler, F. Schueth and K.K. Unger, MCM-41 and related materials as media for controlled polymerization processes, *Stud. Surf. Sci. Catal.*, 1994, **84**, 2013.

98. C.J. Liu, S.G. Li, W.Q. Pang and C.M. Che, Ruthenium porphyrin encapsulated in modified mesoporous molecular sieve MCM-41 for alkene oxidation, *Chem. Commun.*, 1997, **1**, 65.

99. J. Deere, E. Magner, J.G. Wall and B.K. Hodnett, Mechanistic and structural features of protein adsorption onto mesoporous silicates, *J. Phys. Chem. B*, 2002, **106**, 7340.

100. T. Maschmeyer, F. Rey, G. Sankar and J.M. Thomas, Heterogeneous catalysts obtained by grafting metallocene complexes onto mesoporous silica, *Nature*, 1995, **378**, 159.

101. W. Zhou, J.M. Thomas, D.S. Shephard, B.F.G. Johnson, D. Ozkaya, T. Maschmeyer, R.G. Bell and Q. Ge, Ordering of ruthenium cluster carbonyls in mesoporous silica, *Science*, 1998, **280**, 705.

102. R. Anwander, SOMC@PMS: surface organometallic chemistry at periodic mesoporous silica, *Chem. Mater.*, 2001, **13**, 4419.

103. D. Brunel, N. Bellocq, P. Sutra, A. Cauvel, M. Lasperas, P. Moreau, F. Di Renzo, A. Galarneau and F. Fajula, Transition-metal ligands bound onto the micelle-templated silica surface, *Coord. Chem. Rev.*, 1998, **178**, 1085.

104. T. Hirai, H. Okubo and I. Komasawa, Size-selective incorporation of CdS nanoparticles into mesoporous silica, *J. Phys. Chem. B*, 1999, **103**, 4228.

105. L. Bronstein, E. Kramer, B. Berton, C. Burger, S. Forster and M. Antonietti, Successive use of amphiphilic block copolymers as nanoreactors and templates: preparation of porous silica with metal nanoparticles, *Chem. Mater.*, 1999, **11**, 1402.

106. M.H. Huang, A. Choudrey and P.D. Yang, Ag nanowire formation within mesoporous silica, *Chem. Commun.*, 2000, **12**, 1063.

107. J. Panpranot, J.G. Goodwin and A. Sayari, Synthesis and characteristics of MCM-41 supported CoRu catalysts, *Catal. Today*, 2002, **77**, 269.

108. M. Vettraino, M. Trudeau, A.Y.H. Lo, R.W. Schurko and D. Antonelli, Room-temperature ammonia formation from dinitrogen on a reduced mesoporous titanium oxide surface with metallic properties, *J. Am. Chem. Soc.*, 2002, **124**, 9567.

109. M.J. MacLachlan, P. Aroca, N. Coombs, I. Manners and G.A. Ozin, Ring-opening polymerization of a [1]silaferrocenophane within the channels of mesoporous silica: poly(ferrocenylsilane)-MCM-41 precursors to magnetic iron nanostructures, *Adv. Mater.*, 1998, **10**, 144.

110. M.J. MacLachlan, M. Ginzburg, N. Coombs, N.P. Raju, J.E. Greedan, G.A. Ozin and I. Manners, Superparamagnetic ceramic nanocomposites: synthesis and pyrolysis of ring-opened poly(ferrocenylsilanes) inside periodic mesoporous silica, *J. Am. Chem. Soc.*, 2000, **122**, 3878.

111. L.M. Bronstein, Nanoparticles made in mesoporous solids, *Colloid Chem. I; Top. Curr. Chem.*, 2003, **226**, 55.

112. A.B. Bourlinos, T. Karakostas and D. Petridis, "Side chain" modification of MCM-41 silica through the exchange of the surfactant template with charged functionalized organosiloxanes: an efficient route to valuable reconstructed MCM-41 derivatives, *J. Phys. Chem. B*, 2003, **107**, 920.

113. M.S. Morey, J.D. Bryan, S. Schwarz and G.D. Stucky, Pore surface functionalization of MCM-48 mesoporous silica with tungsten and molybdenum metal centers: perspectives on catalytic peroxide activation, *Chem. Mater.*, 2000, **12**, 3435.

114. H. Fan, K. Yang, D.M. Boye, T. Sigmon, K.J. Malloy, H. Xu, G.P. Lopez and C.J. Brinker, Self-assembly of ordered, robust, three-dimensional gold nanocrystal/silica arrays, *Science*, 2004, **304**, 567.

115. J. Wang and G.D. Stucky, Mesostructured composite materials for multibit-per-site optical data storage, *Adv. Funct. Mater.*, 2004, **14**, 409.

116. R.J.P. Corriu and D. Leclercq, Recent developments of molecular chemistry for sol–gel processes, *Angew. Chem. Int. Ed.*, 1996, **35**, 1420.

117. S.L. Burkett, S.D. Sims and S. Mann, Synthesis of hybrid inorganic–organic mesoporous silica by co-condensation of siloxane and organosiloxane precursors, *Chem. Commun.*, 1996, **11**, 1367.

118. U. Schubert, N. Husing and A. Lorenz, Hybrid inorganic–organic materials by sol–gel processing of organofunctional metal alkoxides, *Chem. Mater.*, 1995, **7**, 2010.

119. X. Feng, G.E. Fryxell, L.-Q. Wang, A.Y. Kim, J. Liu and K.M. Kemner, Functionalized monolayers on ordered mesoporous supports, *Science*, 1997, **276**, 923.

120. D.S. Shephard, W. Zhou, T. Maschmeyer, J.M. Matters, C.L. Roper, S. Parsons, B.F.G. Johnson and M.J. Duer, Site-directed surface derivatization of MCM-41: use of high-resolution transmission electron microscopy and molecular recognition for determining the position of functionality within mesoporous materials, *Angew. Chem. Int. Ed.*, 1998, **37**, 2719.

121. M.H. Lim and A. Stein, Comparative studies of grafting and direct syntheses of inorganic–organic hybrid mesoporous materials, *Chem. Mater.*, 1999, **11**, 3285.

122. M.H. Lim, C.F. Blanford and A. Stein, Synthesis and characterization of a reactive vinyl-functionalized MCM-41: probing the internal pore structure by a bromination reaction, *J. Am. Chem. Soc.*, 1997, **119**, 4090.

123. M.H. Lim, C.F. Blanford and A. Stein, Synthesis of ordered microporous silicates with organosulfur surface groups and their applications as solid acid catalysts, *Chem. Mater.*, 1998, **10**, 467.

124. D.E. De Vos, M. Dams, B.F. Sels and P.A. Jacobs, Ordered mesoporous and microporous molecular sieves functionalized with transition metal complexes as catalysts for selective organic transformations, *Chem. Rev.*, 2002, **102**, 3615.

125. D.A. Loy and K.J. Shea, Bridged polysilsesquioxanes. Highly porous hybrid organic–inorganic materials, *Chem. Rev.*, 1995, **95**, 1431.

126. T. Asefa, M.J. MacLachlan, N. Coombs and G.A. Ozin, Periodic mesoporous organosilicas with organic groups inside the channel walls, *Nature*, 1999, **402**, 867; C. Yoshina-Ishii, T. Asefa, N. Coombs, M.J. MacLachlan and G.A. Ozin, Periodic mesoporous organosilicas, PMOs: fusion of organic and inorganic chemistry 'inside' the channel walls of hexagonal mesoporous silica, *Chem. Commun.*, 1999, **24**, 2539.

127. S. Inagaki, S. Guan, Y. Fukushima, T. Ohsuna and O. Terasaki, Novel mesoporous materials with a uniform distribution of organic groups and inorganic oxide in their frameworks, *J. Am. Chem. Soc.*, 1999, **121**, 9611.

128. B.J. Melde, B.T. Holland, C.F. Blanford and A. Stein, Mesoporous sieves with unified hybrid inorganic/organic frameworks, *Chem. Mater.*, 1999, **11**, 3302.

129. S. Inagaki, S. Guan, T. Ohsuna and O. Terasaki, An ordered mesoporous organosilica hybrid material with a crystal-like wall structure, *Nature*, 2002, **416**, 304.

130. H. Yang, N. Coombs, I. Sokolov and G.A. Ozin, Free-standing and oriented mesoporous silica films grown at the air–water interface, *Nature*, 1996, **381**, 589.

131. S. Schacht, Q. Huo, I.G. Voigt-Martin, G.D. Stucky and F. Schuth, Oil–water interface templating of mesoporous macroscale structures, *Science*, 1996, **273**, 768.

132. H. Yang, A. Kuperman, N. Coombs, S. MamicheAfara and G.A. Ozin, Synthesis of oriented films of mesoporous silica on mica, *Nature*, 1996, **379**, 703.

133. I.A. Aksay, M. Trau, S. Manne, I. Honma, N. Yao, L. Zhou, P. Fenter, P.M. Eisenberger and S.M. Gruner, Biomimetic pathways for assembling inorganic thin films, *Science*, 1996, **273**, 892.

134. S. Pevzner, O. Regev and R. Yerushalmi-Rozen, Thin films of mesoporous silica: preparation and characterization, *Curr. Opin. Colloid Interface Sci.*, 1999, **4**, 420.

135. C.J. Brinker, Oriented inorganic films, *Curr. Curr. Opin. Colloid Interface Sci.*, 1998, **3**, 166.

136. J.E. Martin, M.T. Anderson, J. Odinek and P. Newcomer, Synthesis of periodic mesoporous silica thin films, *Langmuir*, 1997, **13**, 4133.

137. Y. Lu, R. Ganguli, C.A. Drewien, M.T. Anderson, C.J. Brinker, W. Gong, Y. Guo, H. Soyez, B. Dunn, M.H. Huang and J.I. Zink, Continuous formation of supported cubic and hexagonal mesoporous films by sol–gel dip-coating, *Nature*, 1997, **389**, 364.

138. D. Zhao, P. Yang, N. Melosh, J. Feng, B.F. Chmelka and G.D. Stucky, Continuous mesoporous silica films with highly ordered large pore structures, *Adv. Mater.*, 1998, **10**, 1380.

139. I. Honma, H.S. Zhou, D. Kundu and A. Endo, Structural control of surfactant-templated hexagonal, cubic, and lamellar mesoporous silicate thin films prepared by spin-casting, *Adv. Mater.*, 2000, **12**, 1529.

140. C.-M. Yang, A.-T. Cho, F.-M. Pan, T.-G. Tsai and K.-J. Chao, Spin-on mesoporous silica films with ultralow dielectric constants, ordered pore structures, and hydrophobic surfaces, *Adv. Mater.*, 2001, **13**, 1099.

141. D. Blaudez, T. Buffeteau, B. Desbat and J.M. Turlet, Infrared and Raman spectroscopies of monolayers at the air–water interface, *Curr. Opin. Colloid Interface Sci.*, 1999, **4**, 265.

142. J.A. Zasadzinski, R. Viswanathan, L. Madsen, J. Garnaes and D.K. Schwartz, Langmuir–Blodgett-films, *Science*, 1994, **263**, 1726.

143. S. Manne, J.P. Cleveland, H.E. Gaub, G.D. Stucky and P.K. Hansma, Direct visualization of surfactant hemimicelles by force microscopy of the electrical double layer, *Langmuir*, 1994, **10**, 4409.

144. S. Manne and H.E. Gaub, Molecular organization of surfactants at solid–liquid interfaces, *Science*, 1995, **270**, 1480.

145. H. Yang, N. Coombs, I. Sokolov and G.A. Ozin, Free-standing and oriented mesoporous silica films grown at the air–water interface, *Nature*, 1996, **381**, 589.

146. S.H. Tolbert, A. Firouzi, G.D. Stucky and B.F. Chmelka, Magnetic field alignment of ordered silicate–surfactant composites and mesoporous silica, *Science*, 1997, **278**, 264.

147. A. Yamaguchi, F. Uejo, T. Yoda, T. Uchida, Y. Tanamura, T. Yamashita and N. Teramae, Self-assembly of a silica–surfactant nanocomposite in a porous alumina membrane, *Nat. Mater.*, 2004, **3**, 337.

148. C.R. Martin and Z. Siwy, Pores within pores, *Nat. Mater.*, 2004, **3**, 284.

149. K.W. Gallis, J.T. Araujo, K.J. Duff, J.G. Moore and C.C. Landry, The use of mesoporous silica in liquid chromatography, *Adv. Mater.*, 1999, **11**, 1452.

150. M. Hager, F. Currie and K. Holmberg, Organic reactions in microemulsions, *Chem. Colloid II; Top. Curr. Chem.*, 2003, **227**, 53.

151. M. Grun, I. Lauer and K.K. Unger, The synthesis of micrometer- and submicrometer-size spheres of ordered mesoporous oxide MCM-41, *Adv. Mater.*, 1997, **9**, 254.

152. Q.S. Huo, J.L. Feng, F. Schuth and G.D. Stucky, Preparation of hard mesoporous silica spheres, *Chem. Mater.*, 1997, **9**, 14.

153. Y.F. Lu, H.Y. Fan, A. Stump, T.L. Ward, T. Rieker and C.J. Brinker, Aerosol-assisted self-assembly of mesostructured spherical nanoparticles, *Nature*, 1999, **398**, 223.

154. G.V.R. Rao, G.P. Lopez, J. Bravo, H. Pham, A.K. Datye, H.F. Xu and T.L. Ward, Monodisperse mesoporous silica microspheres formed by evaporation-induced self assembly of surfactant templates in aerosols, *Adv. Mater.*, 2002, **14**, 1301.

155. S.M. Yang, N. Coombs and G.A. Ozin, Micromolding in inverted polymer opals (MIPO): synthesis of hexagonal mesoporous silica opals, *Adv. Mater.*, 2000, **12**, 1940.

156. S. Schacht, Q. Huo, I.G. Voigt-Martin, G.D. Stucky and F. Schuth, Oil–water interface templating of mesoporous macroscale structures, *Science*, 1996, **273**, 768.

157. P.J. Bruinsma, A.Y. Kim, J. Liu and S. Baskaran, Mesoporous silica synthesized by solvent evaporation: spun fibers and spray-dried hollow spheres, *Chem. Mater.*, 1997, **9**, 2507.

158. G. Buchel, K.K. Unger, A. Matsumoto and K. Tsutsumi, A novel pathway for synthesis of submicrometer-size solid core/mesoporous shell silica spheres, *Adv. Mater.*, 1998, **10**, 1036.

159. G.S. Zhu, S.L. Qiu, O. Terasaki and Y. Wei, Polystyrene bead-assisted self-assembly of microstructured silica hollow spheres in highly alkaline media, *J. Am. Chem. Soc.*, 2001, **123**, 7723.

160. H. Yang, N. Coombs and G.A. Ozin, Mesoporous silica with micrometer-scale designs, *Adv. Mater.*, 1997, **9**, 811.

161. G.B. Sigal, M. Mrksich and G.M. Whitesides, Using surface plasmon resonance spectroscopy to measure the association of detergents with self-assembled monolayers of hexadecanethiolate on gold, *Langmuir*, 1997, **13**, 2749.

162. M. Trau, N. Yao, E. Kim, Y. Xia, G.M. Whitesides and I.A. Aksay, Microscopic patterning of oriented mesoscopic silica through guided growth, *Nature*, 1997, **390**, 674.

163. P.D. Yang, A.H. Rizvi, B. Messer, B.F. Chmelka, G.M. Whitesides and G.D. Stucky, Patterning porous oxides within microchannel networks, *Adv. Mater.*, 2001, **13**, 427.

164. A.D. Stroock, M. Weck, D.T. Chiu, W.T.S. Huck, P.J.A. Kenis, R.F. Ismagilov and G.M. Whitesides, Patterning electro-osmotic flow with patterned surface charge, *Phys. Rev. Lett.*, 2000, **84**, 3314.

165. P.D. Yang, G. Wirnsberger, H.C. Huang, S.R. Cordero, M.D. McGehee, B. Scott, T. Deng, G.M. Whitesides, B.F. Chmelka, S.K. Buratto and

G.D. Stucky, Mirrorless lasing from mesostructured waveguides patterned by soft lithography, *Science*, 2000, **287**, 465.

166. B.J. Scott, G. Wirnsberger and G.D. Stucky, Mesoporous and mesostructured materials for optical applications, *Chem. Mater.*, 2001, **13**, 3140.

167. D.A. Doshi, N.K. Huesing, M.C. Lu, H.Y. Fan, Y.F. Lu, K. Simmons-Potter, B.G. Potter, A.J. Hurd and C.J. Brinker, Optically defined multifunctional patterning of photosensitive thin-film silica mesophases, *Science*, 2000, **290**, 107.

168. H.Y. Fan, Y.F. Lu, A. Stump, S.T. Reed, T. Baer, R. Schunk, V. Perez-Luna, G.P. Lopez and C.J. Brinker, Rapid prototyping of patterned functional nanostructures, *Nature*, 2000, **405**, 56.

169. H. Yang, N. Coombs and G.A. Ozin, Morphogenesis of shapes and surface patterns in mesoporous silica, *Nature*, 1997, **386**, 692.

170. H. Yang, N. Coombs, O. Dag, I. Sokolov and G.A. Ozin, Free-standing mesoporous silica films; morphogenesis of channel and surface patterns, *J. Mater. Chem.*, 1997, **7**, 1755.

171. G.A. Ozin, C.T. Kresge and H. Yang, Nucleation, growth and form of mesoporous silica: role of defects and a language of shape, *Stud. Surf. Sci. Catal.*, 1998, **117**, 119.

172. S.M. Yang, I. Sokolov, N. Coombs, C.T. Kresge and G.A. Ozin, Formation of hollow helicoids in mesoporous silica: supramolecular origami, *Adv. Mater.*, 1999, **11**, 1427.

173. H. Yang, G.A. Ozin and C.T. Kresge, The role of defects in the formation of mesoporous silica fibers, films, and curved shapes, *Adv. Mater.*, 1998, **10**, 883.

174. Y.F. Lu, H.Y. Fan, A. Stump, T.L. Ward, T. Rieker and C.J. Brinker, Aerosol-assisted self-assembly of mesostructured spherical nanoparticles, *Nature*, 1999, **398**, 223.

175. S. Singh, Curvature elasticity in liquid crystals, *Phys. Rep. – Rev. Sec. Phys. Lett.*, 1996, **277**, 284.

176. J. Israelachvili, *Intermolecular and Surface Forces: with Applications to Colloidal and Biological Systems*, Academic Press, New York, 1992.

177. I. Sokolov, H. Yang, G.A. Ozin and C.T. Kresge, Radial patterns in mesoporous silica, *Adv. Mater.*, 1999, **11**, 636.

178. S. Che, Z. Liu, T. Ohsuna, K. Sakamoto, O. Terasaki and T. Tatsumi, Synthesis and characterization of chiral mesoporous silica, *Nature*, 2004, **429**, 281.

179. S. Jun, S.H. Joo, R. Ryoo, M. Kruk, M. Jaroniec, Z. Liu, T. Ohsuna and O. Terasaki, Synthesis of new, nanoporous carbon with hexagonally ordered mesostructure, *J. Am. Chem. Soc.*, 2000, **122**, 10712.

180. J. Lee, S. Yoon, S.M. Oh, C.H. Shin and T. Hyeon, Development of a new mesoporous carbon using an HMS aluminosilicate template, *Adv. Mater.*, 2000, **12**, 359.

181. J.S. Lee, S.H. Joo and R. Ryoo, Synthesis of mesoporous silicas of controlled pore wall thickness and their replication to ordered nanoporous carbons with various pore diameters, *J. Am. Chem. Soc.*, 2002, **124**, 1156.

182. R. Ryoo, S.H. Joo, M. Kruk and M. Jaroniec, Ordered mesoporous carbons, *Adv. Mater.*, 2001, **13**, 677.

183. T.W. Kim, I.S. Park and R. Ryoo, A synthetic route to ordered mesoporous carbon materials with graphitic pore walls, *Angew. Chem. Int. Ed.*, 2003, **42**, 4375.

184. S.H. Joo, S.J. Choi, I. Oh, J. Kwak, Z. Liu, O. Terasaki and R. Ryoo, Ordered nanoporous arrays of carbon supporting high dispersions of platinum nanoparticles, *Nature*, 2001, **412**, 169.

185. M. Kruk, M. Jaroniec, C.H. Ko and R. Ryoo, Characterization of the porous structure of SBA-15, *Chem. Mater.*, 2000, **12**, 1961.

186. R. Ryoo, S. Jun, J.M. Kim and M.J. Kim, Generalized route to the preparation of mesoporous metallosilicates *via* post-synthetic metal implantation, *Chem. Commun.*, 1997, **22**, 2225.

187. P. Poulin, Novel phases and colloidal assemblies in liquid crystals, *Curr. Opin. Colloid Interface Sci.*, 1999, **4**, 66.

188. H. Yang, N. Coombs and G.A. Ozin, Thickness control and defects in oriented mesoporous silica films, *J. Mater. Chem.*, 1998, **8**, 1205.

189. C. Noel, Defects and textures in nematic MCLCPs, *Acta Polym.*, 1997, **48**, 335.

190. S.D. Hudson, High resolution transmission electron microscopy of liquid crystalline polymers, *Curr. Opin. Colloid Interface Sci.*, 1998, **3**, 125.

Nanofood for Thought – Soft Blocks Template Hard Precursors, Holey Materials

1. The surfactant templated synthesis of periodic mesoporous silica (PMS) in 1992 created a sea of change in the way that solid-state materials are synthesized and since that time a periodic table of such mesostructures have been reported, yet no real world applications has yet emerged from more than a decade of research in universities and industry on this novel class of materials. Why do you think this is so?

2. Devise a method for synthesizing a mesoporous silica film with an architecture based upon (a) a superlattice of large and small mesopores and (b) a gradient of large to small mesopores. What applications can you imagine for such novel mesoporous silica film architectures?

3. What approach would you adopt to synthesize (a) hexagonal symmetry mesoporous gold; (b) mesolamellar tungsten bronze; (c) cubic bicontinuous Keggin ion mesostructure; (d) hexagonal mesostructure of alkanethiolate capped gold clusters; and (e) hexagonal symmetry mesoporous carbon. Imagine what utility could be devised for these new materials.

4. Invent a purely synthetic approach for the assembly of an electrochromic mesostructure starting from $Na_4W_{10}O_{32}$ clusters.

5. Describe the origin and type of defects that are expected to occur in a lyotropic hexagonal silicate liquid crystal and the effect that they have on the growth and form of mesoporous silica.

6. What methods are available for synthesizing periodic and aperiodic mesoporous inorganic materials?

7. Give a synthetic method for patterning an oriented mesoporous silica film.

8. Describe the synthesis, characterization and possible utility of templated mesostructured metal germanium sulfides $(CTA)_2MGe_4S_{10}$ incorporating the $[Ge_4S_{10}]^{4-}$ cluster, where CTA^+ = Cetyltrimethyl ammonium cation.

9. Explain how a microphase separated block copolymer could be used to synthesize a poly(methylmethacrylate) membrane with well-ordered mesopores.

10. How would you synthesize Angstrom tunable diameter poly(phenyl-formaldehyde) mesoscale fibers? Where might such fibers prove to be useful?

11. What approach would you use to make a hexagonal mesoporous silica film with the channels oriented either (a) parallel or (b) perpendicular to the surface of the film?

12. How many ways can you devise, chemical or physical means, for removing the organic supramolecular template from as-synthesized PMS?

13. Devise a purely synthetic method for creating a periodic array of luminescent silicon nanoclusters embedded in a silica film on a silicon wafer. What uses could you imagine for such a material?

14. How and why would you try to make (a) a lithium solid-state battery, (b) a solar cell, and (c) an electrochromic device from hexagonal mesoporous nanocrystalline titania?

15. How would you expect a V-shaped microchannel to influence the structural order of a mesophase such as that found in a microphase separated, block copolymer? Why would an extension to an inorganic mesophase be interesting and potentially useful?

16. It is well known that as electronic circuits shrink in size towards 100 nm and less, capacitive and resistive effects cause the RC time constant of the circuit to increase. One solution to this problem is to invent a material that can replace silica currently used as the insulator between electrical interconnects and components in the circuit. Such a material must have a dielectric constant less than 2. How could you employ mesoporous materials to solve this materials problem?

17. How many ways can you think of making an achiral PMS chiral?

18. What methods could you use to increase or decrease the pore diameter of an as-synthesized PMS?

19. How could you synthesize a square shaped nanoring made of PMS?

20. Devise a synthesis of a Bragg mirror, a 1D photonic crystal built of alternating layers with equal thickness but different refractive index, that is comprised of PMS and periodic mesoporous titania layers and explain how you could tune the frequency of reflected or transmitted visible light with such a construct. Imagine an application where this property might prove to be useful.

21. Explain why hexagonal symmetry (PMS) synthesized inside the macroscale channels of an AAO membrane would form with its mesoscale channels aligned parallel to the macroscale channel direction – think of creative ways of using such a PMS–AAO composite membrane to achieve some useful functions. How would you expect the ratio of AAO to PMS channel diameters to affect the assembly of PMS?

22. What bottom–up strategy would you employ to make a photonic crystal heterostructure on a substrate that is comprised of a thin film of periodic mesoporous nanocrystalline anatase sandwiched between two thick films of silica colloidal crystal – what would be the most special properties of this kind of photonic crystal architecture and how could you exploit these characteristics to create a functional device?

23. It has recently been demonstrated (*Chem. Commun.*, 2004, **13**, 1460) that template-free, thin films of hexagonal symmetry mesoporous nanocrystalline anatase grown on plastic substrates can be bent without signs of cracking from the macro- to the mesoscale. Provide possible explanations for the origin of this kind of mechanical behavior of a purely inorganic material that is normally associated with polymer and polymer–inorganic elastomeric composite materials?

24. How could you use soft lithography to make a micron scale pattern of hexagonal mesoporous silica in which the channels are aligned orthogonally to the substrate?

25. Up until the time this book was completed there was no such thing as single crystal PMS films. Now they exist: what was the trick and why is this a significant development for this class of materials?

Block
Copolymers

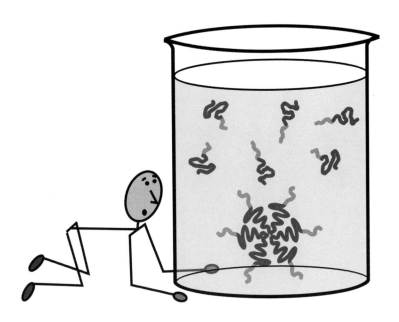

Self-Assembling Block Copolymers

*"Order is not pressure which is imposed on society from without, but an
equilibrium which is set up from within."*
Jose Ortega y Gasset (1883–1955)

9.1 Polymers, Polymers Everywhere in Nanochemistry

Throughout this textbook on nanochemistry, polymers have appeared in various
guises. These have been mainly homopolymers, polymer chains built of a single kind
of monomer unit. They range from elastomers to enable microcontact printing,
micromolding, and microfluidics, photoresists to enable 2D and 3D photolithography, cationic and anionic polyelectrolytes to facilitate layer-by-layer electrostatic
self-assembly, protective coatings to prevent protein adsorption, and conducting
polymers to drive the plastic electronics and optics revolution. However, homopolymers are only one facet of the polymer chemistry toolbox. Synthetic methods
now exist which can produce polymer chains with a custom-designed architecture
made out of two or more types of monomers, with unprecedented control over chain
length distribution. In this chapter, we will be taking a glance at the burgeoning area
of block copolymers, and how these complex macromolecules can undergo self-
assembly and lead to truly remarkable nanomaterials.[1]

9.2 Block Copolymer Self-Assembly – Chip Off the Old Block

What is a block copolymer and why does it self-assemble? Copolymers result from
the chemical integration of two or more monomer units into random, alternating, or
graft architecture chains. A block copolymer is comprised of alternating segments of
polymer chains synthesized using living polymerization,[2,3] illustrated in Figure 9.1.
In a living polymerization, a polymer chain is first grown while maintaining an active

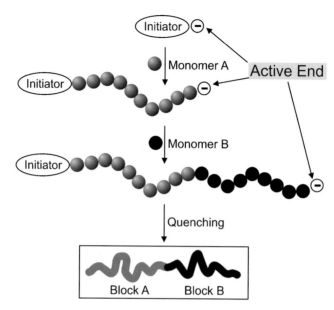

Figure 9.1 *A schematic illustration of anionic living polymerization, one of several living polymerization methods*

end, after which a second monomer is added to grow a second block onto the pre-existing chain. If care is taken during this synthesis, the process can be carried out several times to generate highly monodisperse, complex polymer architectures.

If two polymers are blended, they will almost inevitably undergo phase separation on the macroscale due to the unfavorable entropy of mixing two long-chain macromolecules. When a polymer chain is comprised of chemically connected, mutually immiscible or incompatible segments or blocks, the polymer can undergo what is termed microphase separation.[4-8] Due to the physical connectivity of the two blocks, they cannot form large phase-separated domains, so instead the individual segments self-organize into domains with nanoscale dimensions. At certain chain lengths and volume fractions, the self-organized structures can have well-defined 2D or 3D periodicity consisting, for example, of ordered spheres, cylinders, lamellae, and bicontinuous structures, Figure 9.2, resulting from the minimization of surface free energies.[9] Periodic structures are by no means limited to linear diblock copolymers. Many other morphologies have been observed in linear tri block or multi block copolymers[10] as well as in polymers having various branches, sidechains, or supramolecular complexation.[11]

The length scale of the separated domains and the architecture they adopt depends on the molecular weight, composition, interactions, and architecture of the segments and on the nature of any co-assembled additives, which may swell or cross-link the system. While there are many uses for block copolymers as liquid crystals in digital displays, micelle carriers for drug delivery, and thermoplastics for shoe soles, car tires, and adhesives, their exploitation in nanoscience is relatively new.

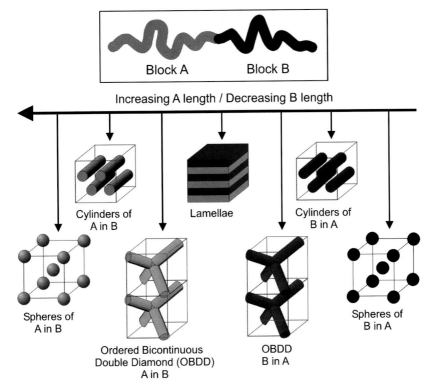

Block A   Block B

Increasing A length / Decreasing B length

Cylinders of
A in B

Lamellae

Cylinders of
B in A

Spheres of
A in B

Ordered Bicontinuous
Double Diamond (OBDD)
A in B

OBDD
B in A

Spheres of
B in A

Figure 9.2 *Typical self-assembly behavior of linear block copolymers*

In this context, the self-assembly and structural organization of a microphase separated block copolymer can also be directed by electric fields, shearing forces, surface control of wettability, chemical attachment to surfaces, epitaxial self-assembly on nanopatterned substrates, spatial confinement by surface relief patterns in substrates and molds, and voids in a range of porous hosts.[12] The ability of block copolymers to self-assemble into a myriad of nanoscale architectures has provided a fertile playground for innovative nanoscience and opportunities abound for making lithographic masks, porous membranes, scaffolds for biomaterials and photonic crystals, growing and organizing nanoclusters, nanowires, and nanotubes, to name but a few.[13,14]

9.3 Nanostructured Ceramics

Block copolymers can form templates whose length scale goes well beyond the 10 nm limit of most surfactant-based micellar and liquid crystal assemblies described in Chapter 8. An example of this block copolymer templating approach makes use of poly(isoprene)-b-poly(ethyleneoxide) (PI-PEO) for synthesizing different aluminosilicate mesostructures.[15] The synthesis begins with a PI-PEO block copolymer synthesized by living anionic polymerization, this polymer displaying the usual morphologies illustrated in Figure 9.2.[16] To this polymer is added a functionalized silane, (3-glycidyloxypropyl)-trimethoxysilane (GLYMO), which preferentially swells the PEO hydrophilic block. Various amounts of aluminum sec-butoxide can be added, which also dissolves in the PEO block aided by the GLYMO compatibilizer. The low $T_g \sim 213$ K of the PI block gives high RT mobility to the copolymer facilitating rapid formation of the aluminosilicate mesostructure with long-range order. This synthetic strategy is elaborated in Figure 9.3, which also shows hexagonal and lamellar phases viewed by TEM. Because the chain length, composition, and architecture of the block copolymer can be readily varied, it is easy to fine-tune the structure and properties of the inorganic–polymer composite. In addition, selective swelling of one of the blocks effectively raises the volume fraction of this block, and has the same effect on morphology as increasing its relative length, providing another synthetic handle on morphology. This methodology allows solvent casting of the block copolymer template and access to oriented mesostructured silica films. Cubic, hexagonal, and lamellar silica mesostructures, as well as other more exotic morphologies such as the cubic "plumber's nightmare",[17] have been synthesized in this way over a widely tunable length scale of 5–100 nm.[18]

The above method shows how the chemical affinities of a polymer segment can be tuned, allowing it to selectively dissolve a precursor for a desired material such as an aluminosilicate. But what if we want to obtain a nanostructured ceramic from a block copolymer without adding additional precursor? This seems a bit far-fetched, but has been demonstrated beautifully using a silicon-containing polymer as a ceramic precursor.[19] The procedure begins with a poly(isoprene)-b-poly(penta-methyl-disilylstyrene) (PI-PPDSS) polymer synthesized by anionic polymerization.

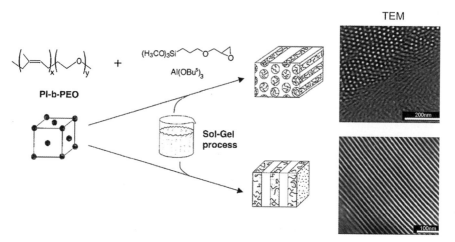

Figure 9.3 *The PEO phase of a PI-PEO block copolymer can be selectively swollen with aluminosilicate precursors, resulting in organic–inorganic composites after condensation*
(Reproduced with permission from Ref. 15)

Following self-assembly of this block copolymer into a desired morphology, such as an interpenetrating double gyroid (this is an exotic morphology only found in a small section of a diblock copolymer phase diagram), the material is treated with ozone in the presence of UV light. Such a treatment is very effective at breaking down organic compounds, and consequently degrades the PI block. However, instead of degrading the PPDSS block the ozone-UV treatment converts it into a silicon oxycarbide ceramic. The result is a nanoporous ceramic having the morphology of the PPDSS portion of the phase-separated block copolymer.

9.4 Nano-objects

In the previous example, we saw how a block copolymer can be modified in order to synthesize a nanoporous aluminosilicate. This same strategy can be employed to instead form isolated aluminosilicate nanoshapes (Figure 9.4).[20] Once again a PI-PEO block copolymer is used, and assembled in the presence of the alumino-silicate precursors GLYMO and $Al(O^sBu)_3$, which permeate the PEO segments. The difference in this case is that the length of the PEO segment is arranged to be quite a bit shorter than the PI segment. After condensation, the aluminosilicate consequently assumes three different morphologies: spheres, cylinders, and lamellae in a PI matrix. Because the ceramic phase is not interconnected, the PI can be dissolved with an organic solvent thereby releasing discrete alumino-silicate objects. In Figure 9.4 we can see isolated spheres, cylinders, and lamellae imaged by TEM. Not only are these nano-objects well defined and mono-dispersed, they are also surrounded by a sheath of PI polymer (chemically connected to the mineralized PEO segments), which confers to them excellent solubility in organic solvents, reminiscent of soluble capped nanoclusters seen in Chapter 6. In the case of cylinders the TEM clearly shows electron density from both the aluminosilicate (top portion) as well as from the polymer sheath (bottom portion), confirming the presence of the stabilizing polymer corona. Such aluminosilicate objects could be used in a variety of applications requiring spheres, wires, or sheets as well as self-assembled into functional architectures, plus the synthetic method should allow a diversity of compositions simply by changing the precursor absorbed into the PEO segments or the chemical nature of the block copolymer.

9.5 Block Copolymer Thin Films

As with most materials currently used in nanoscience, the study of block copolymer self-assembly started in the bulk phase. But again, as with a lot of other materials the bulk phase can be quite difficult to integrate into devices. Further processing is usually required to form the correct shape and architecture, and many samples must afterwards be discarded due to polycrystalline domain orientation. Many hurdles in the technological implementation of these materials can be circumvented by assembling them as thin films.[21] A thin film is subject to strong boundary conditions from the substrate surface and the air–polymer interface, and

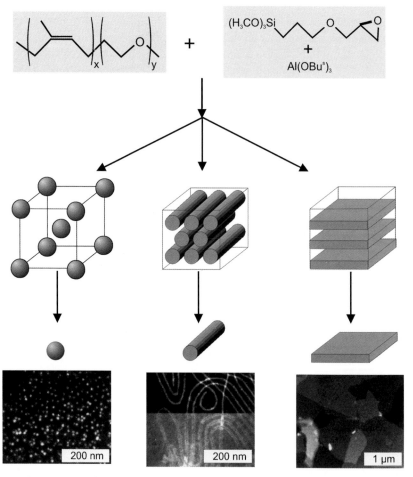

Figure 9.4 *Domains of PEO in self-assembled block copolymers can be swollen and polymerized with aluminosilicate precursors. These nanocomposites can then be dissolved to generate isolated nano-objects such as spheres, cylinders, and lamellae*
(Reproduced with permission from Ref. 20)

since block copolymer assembly is highly dependent on surface energetics these boundaries can effectively direct the orientation of nanoscopic polymer domains,[22,23] as illustrated in Figure 9.5. For instance let us say an A–B block copolymer, synthesized to assemble as cylinders of A in a B matrix, is cast as a thin film on a substrate. If the substrate wets B preferentially, the cylinders of A will be oriented parallel to the substrate such that a thin layer of B separates them from the substrate. If conversely the A block is wetting, the cylinders will orient vertically with the A domains in contact with the substrate. The wetting characteristics of a substrate can easily be controlled by its composition, hydrophobicity or hydrophilicity,

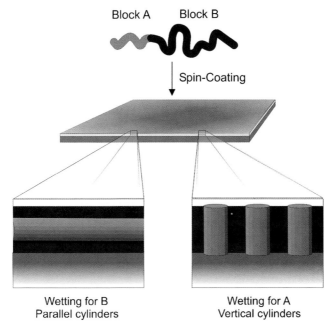

Figure 9.5 *Thin films of block copolymers can give different domain orientations depending on the wetting characteristics of the substrate*

charge, or by a surface treatment such as modification with a surface-active polymer.[24,25]

The usual methods for thin film fabrication are drop casting, dip coating, and spin coating, the latter usually giving the most homogenous films. Following deposition, the film is usually post-treated to increase the degree of ordering of the periodic mesostructure. This can be done for instance by thermally annealing for several days above the order–disorder transition (the temperature above which the copolymer is in a disordered isotropic state), which gives mobility to the polymer segments and allows a more extensive phase separation. Annealing in an atmosphere of solvent vapor, such as toluene, makes the polymer more mobile through a slight solvent swelling, and has a similar effect as thermal annealing. Although these techniques can create very good order, usually the periodicity is not perfect due to a random nucleation of ordered regions generating multiple domains on a substrate with defects at domain boundaries. Probably the easiest way to improve this situation is through a directional solidification strategy, such as using a temperature gradient, a type of zone refining.[26] If a film is brought to a temperature above its order–disorder transition, then cooled in the presence of such a gradient, the ordered phase nucleates at the cool end then gradually self-assembles as the rest of the film cools. As solidification proceeds the self-assembled region serves as a template and orients the rest of the film. Annealing a film in the presence of a gradient in solvent vapor can have a similar effect,[27] with a highly ordered region shown in Figure 9.6.

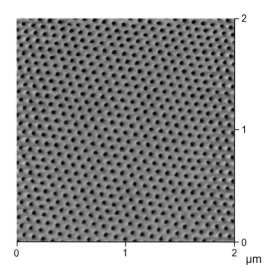

Figure 9.6   *A microphase-separated block copolymer film ordered by controlled solvent evaporation*
(Reproduced with permission from Ref. 27)

9.6 Electrical Ordering

One of the most conveniently tuned stimuli are electric fields. Electrodes can easily be patterned by various ways, which includes soft lithography (Chapter 2), microtransfer printing (Chapter 3), optical lithography, and electron lithography to name a few. There are also quite a few companies which sell substrate pre-patterned with electrodes, ready for use.[28,29] It is thus quite significant that the ordering of block copolymers can be directed by electric fields.[30] Let us consider an A–B block copolymer, which assembles as cylinders of A in a matrix of B, assembled on a substrate where B wets preferentially. As described earlier, the cylinders of A will, therefore, lay parallel to the substrate separated by a layer of B polymer. But how do we control their direction? The A and B segments of the polymer, being chemically different, will necessarily have different dielectric constants and polarizabilities. In the presence of an electric field, the lowest configurational energy of a diblock copolymer is that having the polymer interfaces parallel to the field. This effect, which can be calculated from electrostatic theory, causes all the cylinders to lie parallel and stretching between two counter-electrodes in the electric field direction, providing a kind of polymer map of electric flux lines. The difference between films annealed with and without an electric field present is dramatic, and shown in TEM images in Figure 9.7.

9.7 Spatial Confinement of Block Copolymers

From the discussion, thus far, it should be apparent that making homogenous large area block copolymer patterns could be quite tricky. This has an analogy in

Annealed with
Electric Field

Annealed
without Field

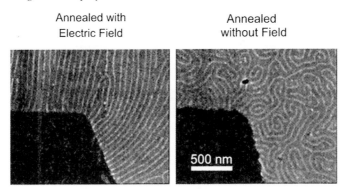

Figure 9.7 *TEM micrographs of block copolymer films annealed in the presence of an electric field and without a field, highlighting the electric field alignment near the electrode seen in the bottom left parts of the images*
(Reproduced with permission from Ref. 30)

the formation of self-assembled colloidal crystals (Chapter 7), where it is also difficult to make defect-free single crystals. One of the strategies employed to direct and perfect colloidal crystallization was confinement in topographic surface relief patterns, where the boundary conditions imposed by rigid geometrically well-defined surface features can nucleate crystalline regions with perfect alignment and orientation. Essentially the same strategy can be used to direct the assembly of block copolymers: local control over the positional order of microphase separation can be achieved by spatial confinement of the block copolymer in geometrically well-defined surface relief patterns that have been lithographically sculpted in a substrate.[31,32] The geometrical confinement can also be provided by a patterned stamp, brought into contact with a substrate creating a micromold (see Chapter 2). An impressive example involves poly(styrene)-b-poly(methylmethacrylate) in which the topography of a patterned micromold forces vertical alignment of hexagonally close-packed PMMA cylinders in the PS matrix.[33] The strategy for ordering the block copolymer assembly using nanoimprint lithography is illustrated in Figure 9.8.

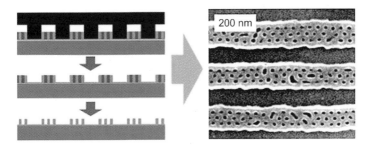

Figure 9.8 *Procedure used to control the orientation of block copolymer domains using nanoimprint lithography. The physical constraint of the PDMS stamp results in ordered structures as shown in the SEM image on the right*
(Reproduced with permission from Ref. 33)

A microchannel mold was employed to control the microphase separation and the outcome can be appreciated in the SEM image, which shows the selective removal of vertically aligned PMMA cylinders by oxygen plasma etching to create a hexagonal close-packed array of nanoscale channels. The resolution of this technique is not quite high enough to be on the length scale of the block copolymer domains. As such, the walls act as boundaries, which direct the orientation of the domains, but there are still quite a few defects due to roughness and non-commensurality (*i.e.*, the width of the channel and the block copolymer period do not match).

9.8 Nanoepitaxy

While directed self-assembly of block copolymer microphase separation through selective dewettability on chemically patterned surfaces can be utilized to order and position cylindrical and lamella domains, there still exist defects of various kinds, which preclude perfect domain ordering and thereby limit the use of block copolymers for advanced lithography applications. A way to surmount this difficulty is to integrate high-resolution lithography and self-assembled monolayers with block copolymer self-assembly, such that the length scale of patterns of chemically induced dewettability is commensurate with the dimensions of microphase separated domains.[34,35] The essence of the idea is to induce epitaxial self-assembly of the domains of a phase separated block copolymer on a patterned substrate to minimize formation of defects. Block copolymer epitaxy of this type has the ability to overcome the intrinsic defect limitations of the self-assembly process. Only when commensurability between the period of the block copolymer and the surface pattern was within a few percent was the resulting domain pattern found to be free of defects, oriented and registered with the underlying substrate over large areas, as displayed in Figure 9.9.

9.9 Block Copolymer Lithography

Optical lithography has come a long way, and next generation UV-lithography systems used by microchip makers such as Intel can now reach the order of 65 nm feature size. Electron beam lithography has also made great progress, with the capability of drawing feature sizes below 10 nm, albeit in a serial fashion. On the other hand, block copolymers microphase separate with highly controlled and tunable characteristic length scales on the order of 5–50 nm, an accuracy much greater than any available lithography technique. The question is, can we use the massively parallel ordering of block copolymers to lithographically pattern a substrate with the precision, perfection, and reproducibility that will be demanded for nanomanufacturing?

This strategy has successfully been accomplished, using an ordered block copolymer film to inscribe an ordered pattern into silicon nitride[36] (Figure 9.10). The procedure starts with a thin film of polystyrene–b–polybutadiene (PS–PB) block copolymer, which organizes on a substrate as a monolayer of PB spheres in a PS matrix. The next step is to subject such a self-organized film to reactive ion

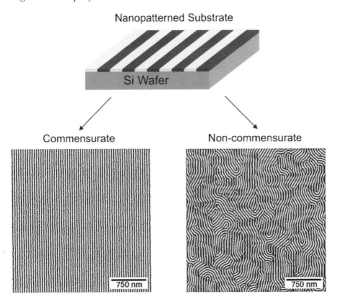

Figure 9.9 *A silicon wafer can be inscribed with a chemical line pattern by high-resolution electron lithography, above. If the line pattern has the same spacing as the block copolymer (commensurate) the order is perfect, but if the spacing is non-commensurate the block copolymer domains appear random* (Reproduced with permission from Ref. 34)

etching, a directional etching technique. However, before this is done the block copolymer film must be modified in some way, since the etching rates of the PS and PB blocks are too similar to pattern the underlying substrate. The first strategy is to react the film with ozone, which breaks down the PB blocks by cleaving the carbon–carbon double bonds, making the now-degraded PB domains much more rapidly etched than the PS domains. The second strategy is to treat the film with OsO_4 vapor, which reacts with the PB double bonds and makes these domains much more resistant to etching. After selective reactive ion etching of one of the polymer segments, and further reactive ion etching of the silicon nitride substrate, we can achieve the inscription of both holes and dots, respectively, using these two strategies. Furthermore, once the ordered block copolymer masks have been generated, it is also possible to transfer the periodicity using techniques such as metal deposition, allowing us to create either metal films with periodic nanoholes or a substrate decorated with metal nanoposts.[37] Such a strategy can also be carried out with other morphologies, resulting in a variety of inscribed patterns.

Notably, it is possible to carry out block copolymer lithography without either degrading or staining one of the blocks prior to etching. For instance, an iron-containing polymer segment such as polyferrocenylsilane (PFS) has an intrinsic resistance to reactive ion etching. Thus, if a block copolymer containing a PFS block is self-assembled on a substrate, reactive ion etching can directly transcribe the ordered morphology onto an underlying substrate.[38] For example,

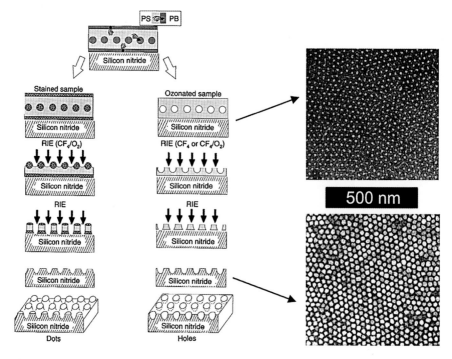

Figure 9.10 *Strategies for performing block copolymer lithography with a polystyrene–*
polybutadiene block copolymer. Staining or degradation of the PB block can
enable selective etching of polymer domains and underlying substrate
(Reproduced with permission from Ref. 36)

if the substrate consists of a thin film of cobalt, the etching process can ultimately give rise to an ordered array of cobalt dots (Figure 9.11). Such small magnetic nanostructures, produced in a massively parallel fashion, may well portend a next-generation technology for fabricating high-density magnetic data storage media.

9.10 Decorating Block Copolymers

A creative strategy has been employed to organize nanoparticles into block copolymer derived nanoscale pits using capillary forces.[39] A PS-b-PMMA polymer with 30% PMMA can be assembled as a thin film with PMMA cylinders oriented orthogonal to the substrate. Treatment with UV-light cross-links the PS domains, stabilizing them towards organic solvents, and at the same time degrades the PMMA domains, leaving nanoscale holes from 10–50 nm in diameter after washing with acetic acid. Such a nanoporous substrate can then be dipped into a toluene solution of capped CdSe nanocrystals and withdrawn at a controlled rate. During the withdrawal, solvent evaporation at the meniscus results in strong capillary forces, which act to drive the nanocrystals into the self-assembled surface

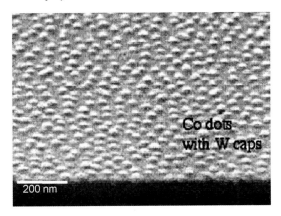

Figure 9.11 *Ordered phase-separated spherical domains of a polyferrocenylsilane block copolymer film can be used to lithographically pattern an underlying thin film of cobalt*
(Reproduced with permission from Ref. 38)

patterns. This strategy is not unlike the use of electron beam defined nanoscale surface relief patterns to achieve nanocrystal positional ordering described in Chapter 6. This evaporation-induced self-assembly procedure and an image showing the fidelity of the process are shown in Figure 9.12. Diverse polymer-nanocluster architectures have been synthesized using microphase separated block copolymers as templates for the macroscale patterning of metal and semiconductor clusters.[40–42]

9.11 A Case of Wettability

The self-assembly of block copolymer films displays a marked sensitivity on the surface properties of a substrate. This occurs because the two polymer blocks have different wetting characteristics, and it follows from this fact that the patterns generated by this route will have a periodic modulation in their wetting properties. For instance, if cylinders are assembled parallel to a substrate, this sample will display a pattern of parallel lines with distinct wettabilities. This contrast in surface properties can be used quite impressively to selectively decorate block copolymer domains with various types of vapor deposited metals.[43] This procedure begins with a PS-b-PMMA block copolymer with block ratios such that it self-assembles as cylinders of PMMA in a PS matrix. It is cast as a very thin film on a Si_3N_4 substrate, such that the block architecture consists of a very thin PMMA layer wetting the substrate, a layer of PS matrix, and a monolayer of PMMA cylinders laying parallel to the substrate and exposed to the air interface. Therefore, the surface of the film displays alternating stripes of PMMA (cylinders) and PS (matrix).

If a small amount of metal is thermally evaporated onto this self-assembled pattern, the different wetting characteristics of the two polymer blocks will cause

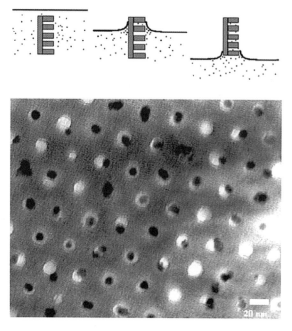

Figure 9.12 *Directed, evaporation-induced self-assembly of capped CdSe nanocrystals within nanoscale surface relief patterns created in a phase-separated PS-b-PMMA block copolymer by UV induced cross-linking of the PS domains and removal of the PMMA domains*
(Reproduced with permission from Ref. 39)

the metal atoms to preferentially accrete in one of the domains. By then annealing the sample above the glass transition of the polymer, the deposited metal can be driven exclusively to one block. The result of such an experiment in TEM images is shown in Figure 9.13, highlighting the effects of both gold and silver deposition. In the case of gold we see that even without annealing, the metal atoms have a distinct affinity for the PS phase. After annealing, it is evident that the gold has migrated exclusively to the PS domains, visualized as dark stripes in the SEM image. Further deposition and annealing increases the gold loading on the PS stripes, eventually forming nanowires composed of discrete nanoparticles. In the case of silver the metal also prefers the PS domains, but upon repeated deposition and annealing we see the formation of continuous silver nanowires, which have been found to display metallic conductivity. Such "wettability masks" can be applied to several different metals: Au and Ag as we have seen migrate to the PS domains, while In, Pb, and Sn instead decorate the PMMA phase. This strategy provides an excellent route to a variety of metal nanowires and nanocluster chains. If deposition of these two classes of metals were performed sequentially or maybe even simultaneously onto the polystyrene–b–polymethylmethacrylate (PS–b–PMMA) film it would be interesting to discover if one could synthesize, for example, coaxial Au/PS nanowires in a matrix of Sn/PMMA or other interesting combinations. If reduced to

Gold Evaporation

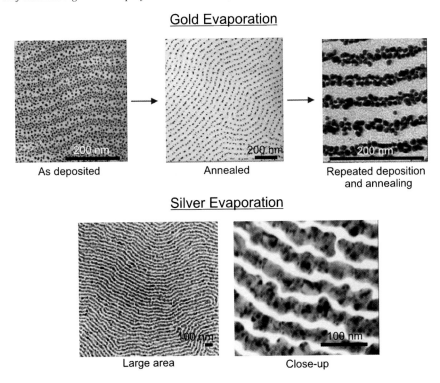

As deposited   Annealed   Repeated deposition and annealing

Silver Evaporation

Large area   Close-up

Figure 9.13 *Nanoscale decoration of phase-separated PS-b-PMMA block copolymer film on a silicon nitride substrate by selective wetting and accretion of a vapor deposited metals like gold and silver selectively in the PS phase* (Reproduced with permission from Ref. 43)

practice this would represent a novel phase-separated binary metal-polymer composite most likely with rather unusual properties.

9.12 Nanowires from Block Copolymers

In Chapter 5, we saw quite a few examples of templating methods for making nanowires. One of the most general ones involves the electrodeposition of a metal, semiconductor, or conducting polymer into the parallel pores of track-etched polymer or anodized alumina or silicon membranes. We have seen in previous examples that block copolymers can also self-assemble into a morphology of polymer A cylinders in a polymer B matrix. The difference here is the relative ease in which these phases can be formed, and the high degree of tunability of the cylindrical domains by molecular weight or swelling agent control. If the cylindrical domains in a self-assembled block copolymer can somehow be aligned in a parallel fashion and selectively etched, we have a highly tunable porous template in which to electrodeposit a high density of ultra-small parallel nanowires[44,45] (Figure 9.14).

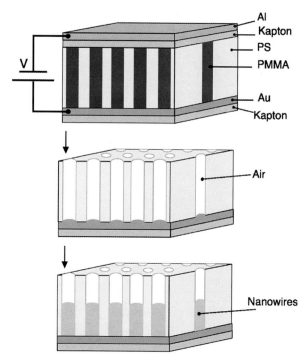

Figure 9.14 *An electric field aligns ordered cylindrical domains in a PS-b-PMMA block copolymer film, encased between two parallel-aligned electrodes, in a direction orthogonal to the electrode surfaces. Subsequent cross-linking of the PS and removal of the PMMA creates an ordered nanochannel membrane that can direct the electrodeposition of a metal in the pores to generate a high-density periodic nanowire array*
(Reproduced with permission from Ref. 44)

The first step to accomplish this goal is making a PS–PMMA block copolymer with a 71% fraction of styrene so that it self-assembles as 14 nm cylinders of PMMA in a PS matrix. A thin film of this block copolymer is spin cast on a conducting substrate, a flexible counter-electrode is laminated onto the top of the film, and a thermal annealing is carried out in the presence of an electric field. As we have seen previously, cylinders in a block copolymer are aligned parallel to an electric field making the PMMA cylinders in this example stand perpendicular to the substrate. After removing the top electrode, the film is treated with a deep UV irradiation, which simultaneously degrades the PMMA and cross-links the PS. An acetic acid wash removes the broken down PMMA blocks, leaving a regular array of cylindrical pores. Electrodeposition can now be performed to fill the pores with a metal, using an aqueous plating solution to which methanol has been added to better wet the hydrophobic membrane. When Co is plated inside the pores, what is obtained is a high-density array of magnetic nanowires. These wires have been found to display a marked magnetic anisotropy, with a magnetic coercivity parallel to the nanowires considerably higher than perpendicular to them. This is significant because a magnetic nanowire with a sufficiently high aspect ratio will be ferromagnetic, with a higher resistance to thermal demagnetization than a spherical nanoparticle with a comparable diameter, below the magnetization threshold (Weiss domain), which would be superparamagnetic. A nanowire with a diameter below the Weiss domain can, therefore, still be ferromagnetic: however, the wires can only be permanently magnetized in a direction parallel to their axes, the so-called easy axis of magnetization.

Such densely packed magnetic bits may well find uses in magnetic recording applications. Thus far these types of electrodeposited nanowires have not been isolated as individual units, though it should be feasible to do so. It is likely that block copolymer templates could be used to grow ultra-fine barcoded nanowires, as seen in Chapter 5, which could enable a whole new class of programmable self-assembly.

9.13 Making Micelles

Throughout the previous examples, we have seen how block copolymers can assemble in the bulk state and under boundary conditions. In all these examples, the polymer is in a concentrated state leading to periodic morphologies as predicted by theory. However, they can also display fascinating behavior in solution, akin to simple amphiphiles and surfactants in an aqueous medium.[46] A greasy surface can usually be readily cleaned by a dilute water solution of surfactant (soap). The surfactant, bearing a hydrophobic tail and a hydrophilic head, forms spherical assemblies called micelles with a solubilizing hydrophilic exterior and a hydrophobic interior, which can trap oils and dissolve them away. Similarly, block copolymers can also undergo this type of phase-separation behavior in solution.[47]

Let us consider a series of polystyrene-b-polyacrylic acid (PS-PAA) block copolymers with a very long PS block and short PAA block, dissolved in water. The PS block is obviously hydrophobic and does not dissolve in water, while the PAA block is quite soluble because of its polyanionic character. In this case, the PS segments will aggregate together through hydrophobic forces and these aggregates, will be terminated by the short soluble PAA segments, earning them the name "crew-cut" micelles. The assembly of such polymers is highly dependent on the relative molecular weights of the two blocks, as illustrated in Figure 9.15. At a 200:21 block ratio of PS to PAA the swollen PAA corona imposes a high curvature onto the self-assembled aggregates, causing them to assemble into PAA-terminated solid PS spheres. When the PAA segment is reduced to 15, the decrease in curvature results in the self-assembly of worm-like micelles. Further decreasing the PAA length to 8 the system assembles into hollow spheres or vesicles, having a PAA corona on the outside and inside of the capsule. Decreasing the PAA content even further to 4, simply causes the PS to macro-phase separate into large spherical clumps. The behavior of crew-cut aggregates is not only dependent on the block copolymer used, but is also sensitive to ionic additives, which can direct the assembly of interesting morphologies such as cell-mimics and cross-linked "pearl necklace" networks.[48]

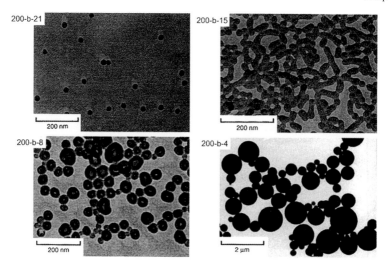

Figure 9.15 *Morphology control of PS-b-PAA micellar assemblies in water obtained by varying the relative molecular weights of the two blocks*
(Reproduced with permission from Ref. 47)

Block copolymer micelles can be made from a variety of different polymer compositions and solvents. The simple requirement is that one of the blocks is soluble in a given solvent, while the other is poorly soluble. The nature of the morphology so-formed is mainly dependent on the relative block lengths and volumes of the two polymers, but can also be tuned with a host of different additives. The resulting self-assembled micelles are quite tough, for instance block copolymer vesicles have been found to have an elastic modulus orders of magnitude higher than vesicles assembled from molecular surfactants.[49–51] A further enhancement in mechanical properties can be achieved by cross-linking either the core or the shell of a self-assembled micelle,[52] for instance by polymerizing the double-bonds in polybutadiene, as exemplified in the synthesis of rubber-like worm micelles.[53] The self-assembly of tri-block copolymers can result in even more complex architectures, such as worm-like core-shell-corona micelles, which can be converted into hollow tubes by post-synthetically cross-linking the shell block and then degrading the core block.[54]

One of the most promising areas of application for block copolymer micelles is in tailored drug delivery. Synthesis conditions can easily be manipulated to introduce biodegradable materials, such as polylactic acid, into the polymer blocks. The micelles can be injected into the body with a drug trapped inside their hydrophobic cores, allowing the drug release rate to be finely controlled without toxicity.[55] Active substances could also be trapped in the large void spaces of biodegradable polymer vesicles, possibly allowing a much larger loading. Micelles have also been observed to segregate within particular organelles in a living cell, which could potentially lead to ultra-specific drug targeting without any ill effects in non-affected areas.[56] It is evident that block copolymer micelles have a clear-cut role in future nanotechnologies.

9.14 Assembling Inorganic Polymers

Since its inception, living polymerization has gradually been expanded to include a great deal of polymerizable monomers. In particular, anionic polymerization has proven a viable strategy to synthesize block copolymers containing poly-ferrocenylsilane segments. This polymer, mentioned in previous chapters, consists of ferrocene units connected by substituted silicon atoms.[57,58] Its corresponding cyclic monomer, with the silicon atom bridging the two cyclopentadienyl ligands in ferrocene, can be ring-opened by anions, heat, or transition metal catalysts driven by the release of ring strain. Anionic polymerization can therefore afford narrowly dispersed PFSs, as well as block copolymers with other anionically polymerizable blocks, such as poly(dimethylsiloxane) (PDMS), poly(butadiene), or poly(isoprene).

The anionic ring-opening-polymerization synthesis of block copolymers that have an architecture based upon an inorganic PFS segment attached to a PDMS portion has provided a route to some interesting nanostructures. For example, in hexane hybrid organometallic–organic polymers can self-assemble into rod or wire-shaped cylindrical micelles.[59–61] The mesostructure and morphology of these assemblies are especially interesting as they have an iron-bearing polymer core and an outer sheath of PDMS. Controlled redox chemistry can provide an oxidized PFS core, known to be semiconducting, which could be interesting as active elements or interconnects in nanosize electronic devices. Post-treatment has also been used to transform the rod micelles into nanoscale magnetic wires with tailorable magnetic properties and high aspect ratios.[62] In Figure 9.16 is shown the chemical structure of a PDMS–PFS block copolymer used to make cylindrical micelles, as well as an AFM image after micelle alignment in lithographically defined trenches and conversion to nanoscopic ceramic lines using a hydrogen plasma. In the case of polyferrocenylsilane–polyisoprene micelles, shell cross-linking of the polyisoprene sheath provides enough stabilization to the micelles so that they can be converted to ceramic replicas through thermal pyrolysis.[63] Such magnetic nanowires may find many uses in future nanoscale magnetic data storage applications, and magnetically programmable self-assembly.

9.15 Harnessing Rigid Rods

Amongst the myriad of structural variants on block copolymers, an interesting one is the so-called rod–coil polymer.[64] True to their name, these polymers consist of a rigid segment or rod with low mobility, connected to a flexible segment which assumes a random coil conformation. Such specially designed diblock copolymers have been found to self-assemble in a solvent that has an affinity for the coil component to form micron diameter hollow spherical micellar-type constructs.[65] These giant assembles can further organize from solution to form a hexagonally close-packed macroporous film (Figure 9.17).[66] These films display iridescence characteristic of optical Bragg diffraction from a periodic dielectric photonic crystal lattice. The particular rod–coil diblock copolymer used in this study was poly(phenylquinoline)-b-polystyrene abbreviated PPQ_mPS_n, where m and n denote the number of repeat units of the respective blocks. Carbon disulfide was used as

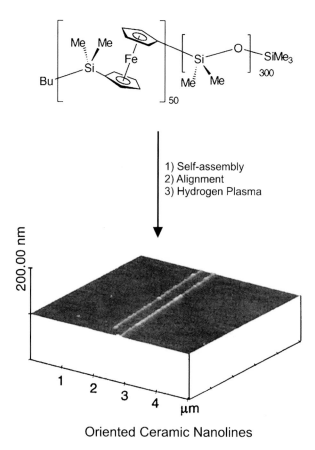

Oriented Ceramic Nanolines

Figure 9.16 *Evaporation induced self-assembly and hydrogen plasma treatment of PDMS-PFS block copolymer micelle cylinders in lithographically defined nanoscale trenches to form oriented ceramic nanolines*
(Reproduced with permission from Ref. 62)

the selective solvent for the PS block and casting of the films. For the representative case of $PPQ_{50}PS_{2000}$ the air holes were found to have a diameter of 3.4 ± 0.2 μm with a center-to-center hole-periodicity of 4.4 ± 0.2 μm and wall thickness about 1.0 μm. It was noted that the diameter of the holes, spacing between holes, and the wall thickness scaled linearly with the length of the PS block, reaching 2.6 ± 0.2, 2.8 ± 0.2, and 0.2 μm, respectively, on passing from $PPQ_{50}PS_{2000}$ to $PPQ_{10}PS_{300}$. The PPQ segment of the block copolymer has intrinsic fluorescence and non-linear optical behavior.[67] Additional tailoring of the optical properties of the micelles and films could be achieved by inclusion of C_{60} and C_{70} fullerene guests in the PPQ region of the walls.[68] Because of the diverse variations of composition and architecture that are feasible for this rod–coil block copolymer system, opportunities abound for development of novel photonic band gap materials, tissue engineering scaffolds and optically responsive coatings, to name but a few.

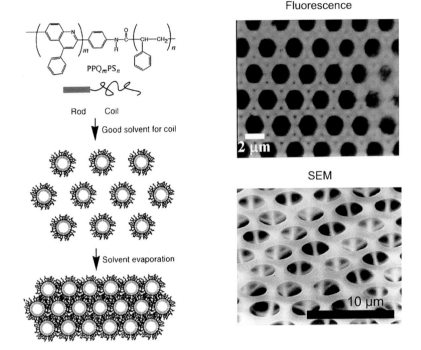

Fluorescence

2 μm

SEM

10 μm

Figure 9.17 *Casting and self-assembly of a rod–coil PPQ–PS block copolymer from carbon disulfide gives a macroporous polymer film periodic at the light scale, enabling it to function as a photonic crystal that Bragg diffracts light* (Reproduced with permission from Ref. 66)

9.16 Supramolecular Assemblies

Most polymers we mention in this text are linear, containing no branches, but branched polymers can also be quite interesting. If we take branching to the extreme, we get dendrimers: these are polymers which start from a central molecule with a certain number of branches.[69] The growth of a "generation" is then performed, which installs two or more branches on each of the initial branches. Further generations can be grown, each time adding a controlled number of new branch tips. If dendrimers are grown to a sufficiently high generation number, they usually adopt a spherical conformation because of the high steric repulsion at their periphery. A great deal of work has been expended in the study of dendrimers, with applications ranging from nanoscale reaction vessels for nanocluster synthesis to light-harvesting applications to drug delivery.

A single branch of a dendrimer is called a dendron,[70,71] and they can adopt interesting shapes such as flat tapers or cones.[72] When placed in a solvent, which interacts more favorably with the periphery than the interior of the dendron, these units can aggregate together into characteristic supramolecular architectures.[73] Flat tapers assemble into disks, which can then stack into columns, which in turn may organize into a hexagonally close-packed assembly. Cones on the other hand can

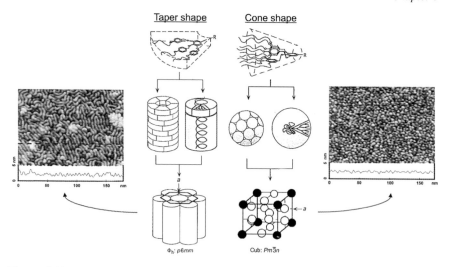

Figure 9.18 *Hierarchical assembly involving polymer attached taper and cone-shaped dendrons that self-assemble into supramolecular architectures like spheres and cylinders, which further assemble into periodic suprastructures* (Reproduced with permission from Ref. 75)

assemble into spheres, which can then organize into a bcc array for instance. In order to stabilize the thus-formed supramolecular architectures, dendrons can be anchored onto a linear polymer backbone with the self-assembly following the same building rules.[73–75] In Figure 9.18 is illustrated the structure of both taper and cone shaped dendrons substituted onto a polymer backbone. The configuration of both the dendron units and the backbone polymer chains in the self-assembled spheres and cylinders is also shown, as well as the organization of these suprastructures onto a periodic lattice. AFM images clearly confirm the presence of spheres and cylinders following self-assembly. Such supramolecular assemblies have given rise to very interesting materials, for instance incorporating electron donors and acceptors into self-assembled cylindrical stacks provided a route to high-electron mobility electronic materials.[76] With the ability to tailor the tethering backbone polymer as well as the chemical nature of each generation of the appended dendrons, there are indeed countless opportunities for new materials based on this type of hierarchical self-assembly.

9.17 Supramolecular Mushrooms

We have seen so far that block copolymers can organize into bulk periodic structures with characteristic morphologies. We have also seen that in a selective solvent they can form discrete micelles such as spheres, cylinders, vesicles, or microcapsules. Finally, we have seen that substituents, such as dendrons on a polymer backbone, can drive the assembly of a supramolecular mesophase. Considering this, there is another very interesting type of self-assembly which can occur with block copolymers: relatively ˙short block copolymers could self-assemble into discrete objects containing a finite amount of chains, and these could then assemble into a periodic assembly depending on the size and geometry of the intermediate object.[77] An example of such a situation is illustrated in Figure 9.19. The chemical units used in this study consist of quite short molecules, which can be described as miniaturized block copolymers. First, a short coil-like segment is prepared by anionic polymerization of a 9-unit polystyrene block then a 9-unit polyisoprene block. Because the chain is so short the molecular weight distribution is relatively broad, which coupled with the structural variability in the PI segment makes this section quite amorphous in the solid state. To this poly(styrene)-b-poly(isoprene) (PS-PI) polymer is chemically affixed a rod-like end consisting of a tri-biphenyl ester (TBE) segment, terminated with a polar hydroxy group, synthesized by well-defined organic reactions. The rod-like segments of this polymer, being highly rigid and aromatic, can pi-stack with each other to form a parallel assembly, much like molecular-sized nanowires packing parallel to each other. The coil-end, however, is much more loosely packed and takes up much more room. The result is a mushroom-shaped aggregate composed of about 100 molecules, where the base of the mushroom consists of the tightly packed aromatic segments, while the top of the mushroom is made up of the disordered coil-like segments. These supramolecular mushrooms stop growing at a particular size, because the top of the mushroom becomes too sterically bulky for more chains to

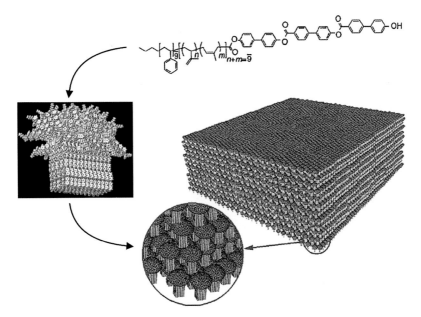

Figure 9.19 *Hierarchical self-assembly of a low molecular weight PS-PI-TBE tri-block copolymer into nanomushrooms that further assemble into sheets composed of parallel-aligned nanomushrooms, which themselves assemble into a layered supramolecular nanoarchitecture*
(Reproduced with permission from Ref. 78)

efficiently pack in the aggregate. Such a molecular arrangement is confirmed by experimental probes such as TEM and electron diffraction, and corroborated by theoretical calculations.

When a solution of this block copolymer is cast as a film about 1 μm in thickness, the supramolecular mushrooms in turn assemble into an ordered layered phase as shown in Figure 9.19. Each layer consists of a monolayer of mushroom shapes all standing up straight. The result is a material which is inherently polar: the rod segments are terminated with polar groups while the coil segments are terminated in non-polar groups. Thus one side of this film is hydrophilic, is readily wet by water, is tacky, and binds tenaciously to a glass slide. The other side is hydrophobic, not wet by water, and non-adhesive. What is effectively formed is a nanoscale, self-assembled strip of sticky tape! Because of the polar nature of the materials, they also display interesting non-linear optical behavior. This example is only one exciting prospect highlighting the multifunctionality and possible utility of self-assembling supramolecular block copolymers.

9.18 Structural Color from Lightscale Block Copolymers

With the ability of block copolymers to self-assemble into periodic nanostructures with tailored sizes and compositions of domains, why not use them as a universal platform for photonic crystal research?[79–81] To achieve this goal it is, however, mandatory to design and synthesize block copolymers that have microphase separated domains with good long-range order, length scales of the order of microns, suitable morphology, and high enough dielectric contrast to enable control over the flow of light at desirable wavelengths in the visible and near infrared spectral regions. In this context PS-PI block copolymer with 300,000–450,000 molecular weight and about 38% volume fraction of styrene has been cast into a film from a 10 wt% toluene solution over a period of two weeks, and found to display blue–violet iridescence.[82] An epoxy embedded, osmium tetroxide stained (addition of OsO_4 to double bonds increases the electron contrast of those regions) and cryo-microtomed thin section, when examined by transmission electron microscopy, defined the PS-PI film as a 3D photonic crystal having a cubic bicontinuous double gyroid morphology with a length scale of phase separated domains appropriate for the interaction of visible light with the microstructure. Sacrificial photo-oxidation of the PI blocks from the PS-PI double gyroid morphology by exposing the sample to UV light in an air atmosphere leaves behind a PS-air cubic single gyroid morphology in which the structure of the cubic interpenetrating network is maintained intact (Figure 9.20). The measured optical reflectivity spectra for the PI-PS double gyroid and PS-air single gyroid are in close accord with those expected from photonic crystal band diagram calculations, confirming that these microstructures are behaving as 3D photonic crystals at visible wavelengths. The 3D PS-air single gyroid can now function as a template or micromold for creating replicas with sufficient refractive index contrast to display a full photonic band gap in their photon density of states. In this context, band

Figure 9.20 *Cubic single gyroid photonic crystal that operates at visible wavelengths made by selective removal of the PI segments from a high molecular weight PS-PI block copolymer phase-separated cubic double gyroid morphology (Reproduced with permission from Ref. 82)*

diagram calculations show that if this PS-air microstructure can be copied in silicon that has a refractive index around 3.5, then the ratio of the gap to mid-gap frequency, the photonic band gap, for the silicon single gyroid replicate can be 25%. This is big enough to make it rather immune to intrinsic defects and, therefore, an appealing platform for photonic band gap materials research. Other ways of boosting refractive index contrast in block copolymer photonic crystals is to incorporate high electron density, large polarizability species into the phase separated domains like semiconductor and metal nanoclusters.

9.19 Block Copolypeptides

Proteins, polymerized amino acids also called polypeptides, are perhaps the most abundant polymers in the world. Every living organism synthesizes them based on a DNA or RNA coding sequence, and uses them for structural, catalytic, transport, and various other functions. We have also seen in chapters of this text that they are increasingly being used for various kinds of building block self-assembly. Given that they are so ubiquitous, it would be quite advantageous to develop a purely synthetic strategy to make polypeptides with controlled structure. In fact, this has been achieved using a metal catalyzed route permitting the synthesis of peptide block copolymers *in vitro* with a narrowly distributed molecular weight distribution.[83,84] The monomers used in this synthesis are not amino acids themselves but rather N-carboxyanhydrides (NCAs), which can be prepared in one step from commercially available amino acids. By using a zero-valent nickel catalyst, NCAs can be polymerized in a living fashion into well-defined homopolymers or block copolymers. Because natural amino acids are chiral the resulting polymer is also chiral, permitting the assembly of naturally occurring motifs, whereby the polymer backbone assumes a tight right-handed screw with the amino acid residues sticking outwards.

This controlled synthesis of polypeptides has resulted in pH-responsive block copolymer microcapsules.[85] A block copolypeptide was synthesized, and is illustrated in Figure 9.21, bearing a water soluble block with poly(ethylene oxide) side chains and a hydrophobic block with leucine side chains (iso-butyl). Both segments coil tightly into α-helices, and as a result the polymer is quite rigid. When dissolved in water, this polymer self-assembles into a low-curvature spherical microcapsule, with the membrane having a hydrophilic–hydrophobic–hydrophilic bilayer structure. If some of the isobutyl groups are replaced by aminobutyl groups, at pH 10.5 the block copolymer self-assembles into much the same morphology since the amine groups are deprotonated and hydrophobic. However, a large change occurs when the block copolymer vesicles are exposed to a lower pH:

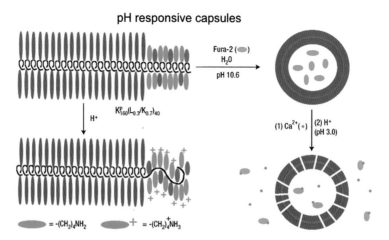

Figure 9.21 *pH controlled self-assembly and disassembly of spherical microcapsules comprised of block co-polypeptides*
(Reproduced with permission from Ref. 85)

once the medium is acidic, the amino groups become protonated, causing electrostatic repulsions and disruption in the hydrophobic domains. This causes the entire microcapsule to fall apart and release any molecules trapped within. This is shown schematically in Figure 9.21, and has been proven by a dye labeling experiment. The microcapsules can be assembled in a solution of Fura-2 dye, transferred to a dye-free solution, and then the pH can be lowered and the dye quenching by Ca^{2+} ions monitored by optical absorption spectroscopy. Such pH responsive microcapsules are completely biocompatible, and could be loaded with a drug, swallowed, and only released in the acidic environment of the stomach. Chemical modification of the block co-polypeptide sequences will likely open many novel avenues for targeted drug delivery.

9.20 Block Copolymer Biofactories

Nature is replete with examples of materials at a level of perfection we have not yet been able to approach. The same goes for block copolymers: natural block copolymers, such as collagen,[86] are under strict genetic control and their lengths are absolutely monodisperse, with every single biosynthesized macromolecule identical to one another.[87] Even in the best of living polymerizations, this is not the case – there is always a certain spread in the molecular weight distribution. If this is the case, can we somehow coerce an organism to produce our monodisperse polymers?

While reminiscent of science fiction, the use of bacteria or other cells as polymer biofactories is quickly gaining ground in the scientific community.[88] Genetic engineering has progressed immensely in the past decades, and it is now quite routine to insert arbitrary DNA sequences into, say, bacterial host. The DNA code we insert can code for a given polypeptide sequence, with each base-pair sequence coding for one of the 20 natural amino acids, and flanked by sequences which will activate biosynthesis as well as tell the organism to expel the polymer so we can collect it.

The problem with polypeptide biosynthesis is that substrates are typically limited to the naturally available amino acids, since organisms cannot process non-natural amino acids. However, in certain circumstances we can coerce an organism to make polypeptides out of amino acids that do not exist in nature.[89,90] It has been known for quite some time that derivatives sufficiently similar in chemical properties can be assimilated by organisms, for instance selenium can be substituted for sulfur in methionine (an amino acid with a mercaptoether residue) without ill effects. Such heavy atom substitutions have greatly facilitated the determination of protein crystal structures by X-ray diffraction, by increasing the X-ray scattering strength and the accuracy of the diffraction patterns.[91] The incorporation of such chemically similar analogues is facilitated by starving the organism of the natural amino acid, forcing it to use the surrogate. A more drastic method for synthesizing custom polypeptide involves modifying an organism's genetic decoding apparatus.[92,93] To make a protein, a DNA code is transcribed into RNA then three-by-three the RNA is decoded by transfer RNA (tRNA) molecules, bearing conjugated amino acids. These loaded tRNA molecules dock into the protein synthesizer, the ribosome, and add their amino acid cargo one at a time to the growing polypeptide chain. Therefore, by either introducing ^{t}RNAs with different cargo, or modifying the ^{t}RNA synthetases, it has proven possible to incorporate a diverse set of chemical functionalities into biosynthesized amino acid polymers. Unsaturated alkene groups, azides, aryl bromides, fluorinated groups, and precursors to conducting polymers have all been successfully integrated into biosynthesized polymers. While there are still some problems with this approach, such as scaling-up and improving the amino acid replacement selectivity, the biosynthetic approach will likely be a big part of future polymer synthesis and nanotechnology.[94]

9.21 References

1. M. Lazzari and M. Arturo López-Quintela, Block copolymers as a tool for nanomaterial fabrication, *Adv. Mater.*, 2003, **15**, 1583.
2. O.W. Webster, Living polymerization methods, *Science*, 1991, **251**, 887.
3. N. Hadjichristidis, M. Pitsikalis, S. Pispas and H. Iatrou, Polymers with complex architecture by living anionic polymerization, *Chem. Rev.*, 2001, **101**, 3747.
4. F.S. Bates and G.H. Fedrickson, Block copolymer thermodynamics: theory and experiment, *Annu. Rev. Phys. Chem.*, 1990, **41**, 525.
5. M.W. Matsen and F.S. Bates, Origins of complex self-assembly in block copolymers, *Macromolecules*, 1996, **29**, 7641.
6. F.S. Bates, Polymer–polymer phase behavior, *Science*, 1991, **251**, 898.
7. M.W. Matsen and F.S. Bates, Unifying weak- and strong-segregation block copolymer theories, *Macromolecules*, 1996, **29**, 1091.
8. I.W. Hamley, *The Physics of Block Copolymers*, Oxford University Press, Oxford, 1998.
9. E.L. Thomas, D.M. Anderson, C.S. Henkee and D. Hoffman, Periodic area-minimizing surfaces in block copolymers, *Nature*, 1988, **334**, 598.
10. P. Maniadia, R.B. Thompson, K.O. Rasmussen and T. Lookman, Ordering mechanisms in triblock copolymers, *Phys. Rev. E*, 2004, **69**, 031801.
11. J. Ruokolainen, R. Mäkinen, M. Torkkeli, T. Mäkela, R. Serimaa, G. Ten Brinke and O. Ikkala, Switching supramolecular polymeric materials with multiple length scales, *Science*, 1998, **280**, 557.
12. Z.-R. Chen, J.A. Kornfield, S.D. Smith, J.T. Grothaus and M.M. Satkowski, Pathways to macroscale order in nanostructured block copolymers, *Science*, 1997, **277**, 1248.
13. M. Muthukumar, C.K. Ober and E.L. Thomas, Competing interactions and levels of ordering in self-organizing polymeric materials, *Science*, 1997, **277**, 1225.
14. I.W. Hamley, Nanostructure fabrication using block copolymers, *Nanotechnology*, 2003, **14**, R39.
15. M. Templin, A. Franck, A. DuChesne, H. Leist, Y.M. Zhang, R. Ulrich, V. Schadler and U. Wiesner, Organically modified aluminosilicate meso-structures from block copolymer phases, *Science*, 1997, **278**, 1795.

16. G. Floudas, B. Vazaiou, F. Schipper, R. Ulrich, U. Wiesner, H. Iatrou and N. Hadjichristidis, Poly(ethylene oxide-b-isoprene) diblock copolymer phase diagram, *Macromolecules*, 2003, **34**, 2947.

17. A.C. Finnefrock, R. Ulrich, G.E.S. Toombes, S.M. Gruner and U. Wiesner, The plumber's nightmare: a new morphology in block copolymer–ceramic nanocomposites and mesoporous aluminosilicates, *J. Am. Chem. Soc.*, 2003, **125**, 13084.

18. S.M. De Paul, J.W. Zwanziger, R. Ulrich, U. Wiesner and H.W. Spiess, Structure, mobility, and interface characterization of self-organized organic–inorganic hybrid materials by solid-state NMR, *J. Am. Chem. Soc.*, 1999, **121**, 5727.

19. V.Z.-H. Chan, J. Hoffman, V.Y. Lee, H. Iatrou, A. Avgeropoulos, N. Hadjichristidis, R.D. Miller and E.L. Thomas, Ordered bicontinuous nanoporous and nanorelief ceramic films from self assembling polymer precursors, *Science*, 1999, **286**, 1716.

20. R. Ulrich, A. Du Chesne, M. Templin and U. Wiesner, Nano-objects with controlled shape, size, and composition from block copolymer mesophases, *Adv. Mater.*, 1999, **11**, 141.

21. M.J. Fasolka and A.M. Mayes, Block copolymer thin films: physics and applications, *Annu. Rev. Mater. Res.*, 2001, **31**, 323.

22. E. Huang, L. Rockford, T.P. Russell and C.J. Hawker, Nanodomain control in copolymer thin films, *Nature*, 1998, **395**, 757.

23. C. Harrison, D.H. Adamson, Z. Cheng, J.M. Sebastian, S. Sethuraman, D.A. Huse, R.A. Register and P.M. Chaikin, Mechanisms of ordering in striped patterns, *Science*, 2000, **290**, 1558.

24. E. Huang, T.P. Russell, C. Harrison, P.M. Chaikin, R.A. Register, C.J. Hawker and J. Mays, Using surface active random copolymers to control the domain orientation in diblock copolymer thin films, *Macromolecules*, 1998, **31**, 7641.

25. P. Mansky, Y. Liu, E. Huang, T.P. Russell and C. Hawker, Controlling polymer-surface interactions with random copolymer brushes, *Science*, 1997, **275**, 1458.

26. C.D. Rosa, C. Park, E.L. Thomas and B. Lotz, Microdomain patterns from directional eutectic solidification and epitaxy, *Nature*, 2000, **405**, 433.

27. S.H. Kim, M.J. Misner, T. Xu, M. Kimura and T.P. Russell, Highly oriented and ordered arrays from block copolymers via solvent evaporation, *Adv. Mater.*, 2004, **16**, 226.

28. Windsor Scientific: www.windsor-ltd.co.uk

29. Synkera Technologies: www.synkera.com

30. T.L. Morkved, M. Lu, A.M. Urbas, E.E. Ehrichs, H.M. Jaeger, P. Mansky and T.P. Russell, Local control of microdomain orientation in diblock copolymer thin films with electric fields, *Science*, 1996, **273**, 931.

31. R.A. Segalman, H. Yokoyama and E.J. Kramer, Graphoepitaxy of spherical domain block copolymer films, *Adv. Mater.*, 2001, **13**, 1152.

32. J.Y. Cheng, C.A. Ross, E.L. Thomas, H.I. Smith and G.J. Vancso, Templated self-assembly of block copolymers: effect of substrate topography, *Adv. Mater.*, 2003, **15**, 1599.

33. H.-W. Li and W.T.S. Huck, Ordered block-copolymer assembly using nanoimprint lithography, *Nano Lett.*, 2004, **4**, 1633.

34. S.O. Kim, H.H. Solak, M.P. Stoykovich, N.J. Ferrier, J.J. de Pablo and P.F. Nealey, Epitaxial self-assembly of block copolymers on lithographically defined nanopatterned substrates, *Nature*, 2003, **424**, 411.

35. E.W. Edwards, M.F. Montague, H.H. Solak, C.J. Hawker and P.F. Nealey, Precise control over molecular dimensions of block-copolymer domains using the interfacial energy of chemically nanopatterned substrates, *Adv. Mater.*, 2004, **16**, 1315.

36. M. Park, C. Harrison, P.M. Chaikin, R.A. Register and D.H. Adamson, Block copolymer lithography: periodic arrays of $\sim 10^{11}$ holes in 1 square centimeter, *Science*, 1997, **276**, 1401.

37. K. Shin, K.A. Leach, J.T. Goldbach, D.H. Kim, J.Y. Jho, M. Tuominen, C.J. Hawker and T.P. Russell, A simple route to metal nanodots and nanoporous metal films, *Nano Lett.*, 2002, **2**, 933.

38. J.Y. Cheng, C.A. Ross, V.Z.-H. Chan, E.L. Thomas, R.G.H. Lammertink and G.J. Vancso, Formation of a cobalt magnetic dot array via block copolymer lithography, *Adv. Mater.*, 2001, **13**, 1174.

39. M.J. Misner, H. Skaff, T. Emrick and T.P. Russell, Directed deposition of nanoparticles using diblock copolymer templates, *Adv. Mater.*, 2003, **15**, 221.

40. J.P. Spatz, S. Moessmer, C. Hartmann, M. Moeller, T. Herzog, M. Krieger, H.G. Boyen and B. Kabius, Ordered deposition of inorganic clusters from micellar block copolymer films, *Langmuir*, 2000, **16**, 407.

41. D.E. Fogg, L.H. Radzilowski, R. Blanski, R.R. Schrock and E.L. Thomas, Fabrication of quantum dot/polymer composites: phosphine-functionalized block copolymers as passivating hosts for cadmium selenide nanoclusters, *Macromolecules*, 1997, **30**(3), 417–426

42. K. Tsutsumi, Y. Funaki, Y. Hirokawa and T. Hashimoto, Selective incorporation of palladium nanoparticles into microphase-separated domains of poly(2-vinylpyridine)-block-polyisoprene, *Langmuir*, 1999, **15**, 5200.

43. W.A. Lopes and H.M. Jaeger, Hierarchical self-assembly of metal nano-structures on diblock copolymer scaffolds, *Nature*, 2001, **414**, 735.

44. T. Thurn-Albrecht, J. Schotter, G.A. Kästle, N. Emley, T. Shibauchi, L. Krusin-Elbaum, K. Guarini, C.T. Black, M.T. Tuominen and T.P. Russell, Ultrahigh-density nanowire arrays grown in self-assembled diblock copolymer templates, *Science*, 2000, **290**, 2126.

45. T. Thurn-Albrecht, R. Steiner, J. DeRouchey, C.M. Stafford, E. Huang, M. Bal, M. Tuominen and T.P. Russell, Nanoscopic templates from oriented block copolymer films, *Adv. Mater.*, 2000, **12**, 787.

46. S. Jain and F.S. Bates, On the origins of morphological complexity in block copolymer surfactants, *Science*, 2003, **300**, 460.

47. L.F. Zhang and A Eisenberg, Multiple morphologies of crew-cut aggregates of polystyrene-b-poly(acrylic acid) block-copolymers, *Science*, 1995, **268**, 1728.

48. L. Zhang, K. Yu and A. Eisenberg, Ion-induced morphological changes in "crew-cut" aggregates of amphiphilic block copolymers, *Science*, 1996, **272**, 1777.

49. B.M. Discher, Y.Y. Won, D.S. Ege, J.C.M. Lee, F.S. Bates, D.E. Discher and D.A. Hammer, Polymersomes: tough vesicles made from diblock copolymers, *Science*, 1999, **284**, 1143.

50. D.E. Discher and A. Eisenberg, Polymer vesicles, *Science*, 2002, **297**, 967.

51. H.W. Shen and A. Eisenberg, Control of architecture in block-copolymer vesicles, *Angew. Chem. Int. Ed.*, 2000, **39**, 3310.

52. H.P. Hentze and E.W. Kaler, Polymerization of and within self-organized media, *Curr. Opin. Colloid Interface. Sci.*, 2003, **8**, 164.

53. Y.-Y. Won, H.T. Davis and F.S. Bates, Giant wormlike rubber micelles, *Science*, 1999, **283**, 960.

54. S. Stewart and G. Liu, Block copolymer nanotubes, *Angew. Chem. Int. Ed.*, 2000, **39**, 340.

55. B. Jeong, Y.H. Bae, D.S. Lee and S.W. Kim, Biodegradable block copolymers as injectable drug-delivery systems, *Nature*, 1997, **388**, 860.

56. R. Savic, L.B. Luo, A. Eisenberg and D. Maysinger, Micellar nanocontainers distribute to defined cytoplasmic organelles, *Science*, 2003, **300**, 615.

57. K. Kulbaba and I. Manners, Polyferrocenylsilanes: metal-containing polymers for materials science, self-assembly and nanostructure applications, *Macromol. Rapid Commun.*, 2001, **22**, 711.

58. I. Manners, Materials science – putting metals into polymers, *Science*, 2001, **294**, 1664.

59. J. Massey, K.N. Power, I. Manners and M.A. Winnik, Self-assembly of a novel organometallic–inorganic block copolymer in solution and the solid state: nonintrusive observation of novel wormlike poly(ferrocenyldimethylsilane)-b-poly (dimethylsiloxane) micelles, *J. Am. Chem. Soc.*, 1998, **120**, 9533.

60. J.A. Massey, K.N. Power, M.A. Winnik and I. Manners, Organometallic nano-structures: self-assembly of poly(ferrocene) block copolymers, *Adv. Mater.*, 1998, **10**, 1559.

61. J.A. Massey, K. Temple, L. Cao, Y. Rharbi, J. Raez, M.A. Winnik and I. Manners, Self-assembly of organometallic block copolymers: the role of crystallinity of the core-forming polyferrocene block in the micellar morpho-logies formed by poly(ferrocenylsilane-b-dimethylsiloxane) in *n*-alkane solvents, *J. Am. Chem. Soc.*, 2000, **122**, 11577.

62. J.A. Massey, M.A. Winnik, I. Manners, V.Z.-H. Chan, J.M. Ostermann, R. Enchelmaier, J.P. Spatz and M. Möller, Fabrication of oriented nanoscopic ceramic lines from cylindrical micelles of an organometallic polyferrocene block copolymer, *J. Am. Chem. Soc.*, 2001, **123**, 3147.

63. X.-S. Wang, A. Arsenault, G.A. Ozin, M.A. Winnik and I. Manners, Shell cross-linked cylinders of polyisoprene-b-ferrocenyldimethylsilane: formation of magnetic ceramic replicas and microfluidic channel alignment and patterning, *J. Am. Chem. Soc.*, 2003, **125**, 12686.

64. S.I. Stupp, Self-assembly of rodcoil molecules, *Curr. Opin. Colloid Interface. Sci.*, 1998, **3**, 20.

65. X.L. Chen and S.A. Jenekhe, Supramolecular self-assembly of three-dimen-sional nanostructures and microstructures: microcapsules from electroactive and photoactive rod–coil–rod triblock copolymers, *Macromolecules*, 2000, **33**, 4610.

66. X.L. Chen and S.A. Jenekhe, Self-assembly of ordered microporous materials from rod–coil block copolymers, *Science*, 1999, **283**, 372.

67. J.A. Osaheni and A.A. Jenekhe, Electroactive and photoactive rod–coil copolymers – design, synthesis, and supramolecular regulation of photophysical properties, *J. Am. Chem. Soc.*, 1995, **117**, 7389.

68. S.A. Jenekhe and X.L. Chen, Self-assembled aggregates of rod–coil block copolymers and their solubilization and encapsulation of fullerenes, *Science*, 1998, **279**, 1903.

69. A.W. Bosman, H.M. Janssen and E.W. Meijer, About dendrimers: structure, physical properties, and applications, *Chem. Rev.*, 1999, **99**, 1665.

70. S.K. Grayson and J.M.J. Frechet, Convergent dendrons and dendrimers: from synthesis to applications, *Chem. Rev.*, 2001, **101**, 3819.

71. V. Percec, B. Barboiu, C. Grigoras and T.K. Bera, Universal iterative strategy for the divergent synthesis of dendritic macromolecules from conventional monomers by a combination of living radical polymerization and irreversible TERminator Multifunctional INItiator (TERMINI), *J. Am. Chem. Soc.*, 2003, **125**, 6503.

72. V. Percec, W.-D. Cho, G. Ungar and D.J.P. Yeardley, From molecular flat tapers, discs, and cones to supramolecular cylinders and spheres using Frechet-type monodendrons modified on their periphery, *Angew. Chem. Int. Ed.*, 2000, **39**, 1598.

73. S.D. Hudson, H.T. Jung, V. Percec, W.D. Cho, G. Johansson, G. Ungar and V.S.K. Balagurusamy, Direct visualization of individual cylindrical and spherical supramolecular dendrimers, *Science*, 1997, **278**, 449.

74. J.C.M. Vanhest, D.A.P. Delnoye, M.W.P.L. Baars, M.H.P. Vangenderen and E.W. Meijer, Polystyrene-dendrimer amphiphilic block-copolymers with a generation-dependent aggregation, *Science*, 1995, **268**, 1592.

75. V. Percec, C.-H. Ahn, G. Ungar, D.J.P. Yeardley, M. Moller and S.S. Sheiko, Controlling polymer shape through the self-assembly of dendritic side-groups, *Nature*, 1998, **391**, 161.

76. V. Percec, C.-H. Ahn and B. Barboiu, Self-encapsulation, acceleration and control in the radical polymerization of monodendritic monomers via self-assembly, *J. Am. Chem. Soc.*, 1997, 2978–2979.

77. V. Percec, M. Glodde, T.K. Bera, Y. Miura, I. Shiyanovskaya, K.D. Singer, V.S.K. Balagurusamy, P.A. Heiney, I. Schnell, A. Rapp, H.W. Spiess, S.D. Hudson and H. Duan, Self-organization of supramolecular helical dendrimers into complex electronic materials, *Nature*, 2002, **419**, 384.

78. S.I. Stupp, V. LeBonheur, K. Walker, L.S. Li, K.E. Huggins, M. Keser and A. Amstutz, Supramolecular materials: self-organized nanostructures, *Science*, 1997, **276**, 384.

79. M. Bockstaller, R. Kolb and E.L. Thomas, Metallodielectric photonic crystals based on diblock copolymers, *Adv. Mater.*, 2001, **13**, 1783.

80. A. Urbas, R. Sharp, Y. Fink, E.L. Thomas, M. Xenidou and L.J. Fetters, Tunable block copolymer/homopolymer photonic crystals, *Adv. Mater.*, 2000, **12**, 812.

81. A.C. Edrington, A.M. Urbas, P. DeRege, C.X. Chen, T.M. Swager, N. Hadjichristidis, M. Xenidou and E.L. Thomas, Polymer-based photonic crystals, *Adv. Mater.*, 2001, **13**, 421.

82. A.M. Urbas, M. Maldovan, P. DeRege and E.L. Thomas, Bicontinuous cubic block copolymer photonic crystals, *Adv. Mater.*, 2002, **14**, 1850.

83. T.J. Deming, Facile synthesis of block copolypeptides of defined architecture, *Nature*, 1997, **390**, 386.

84. T.J. Deming, Polypeptide materials: new synthetic methods and applications, *Adv. Mater.*, 1997, **9**, 299.

85. E.G. Bellomo, M.D. Wyrsta, L. Pakstis, D.J. Pochan and T.J. Deming, Stimuli-responsive polypeptide vesicles by conformation-specific assembly, *Nat. Mater.*, 2004, **3**, 244.

86. K.J. Coyne, X.X. Qin and J.H. Waite, Extensible collagen in mussel byssus: a natural block copolymer, *Science*, 1997, **277**, 1830.

87. J.C.M. van Hest and D.A. Tirrell, Protein-based materials, toward a new level of structural control, *Chem. Commun.*, 2001, **19**, 1897.

88. J.G. Tirrell and D.A. Tirrell, Synthesis of biopolymers: proteins, polyesters, polysaccharides and polynucleotides, *Curr. Opin. Solid State Mater. Sci.*, 1996, **1**, 407.

89. A.J. Link, M.L. Mock and D.A. Tirrell, Non-canonical amino acids in protein engineering, *Curr. Opin. Biotechnol.*, 2003, **14**, 603.

90. I. Kwon, K. Kirshenbaum and D.A. Tirrell, Breaking the degeneracy of the genetic code, *J. Am. Chem. Soc.*, 2003, **125**, 7512.

91. W.A. Hendrickson, J.R. Horton and D.M. LeMaster, Selenomethionyl proteins produced for analysis by multiwavelength anomalous diffraction (MAD): a vehicle for direct determination of three dimensional structure, *EMBO J.*, 1990, **9**, 1665.

92. K.L. Kiick, J.C.M. van Hest and D.A. Tirrell, Expanding the scope of protein biosynthesis by altering the methionyl-tRNA synthetase activity of a bacterial expression host, *Angew. Chem. Int. Ed.*, 2000, **39**, 2148.

93. D. Datta, P. Wang, I.S. Carrico, S.L. Mayo and D.A. Tirrell, A designed phenylalanyl-tRNA synthetase variant allows efficient in vivo incorporation of aryl ketone functionality into proteins, *J. Am. Chem. Soc.*, 2002, **124**, 5652.

94. R. Langer and D.A. Tirrell, Designing materials for biology and medicine, *Nature*, 2004, **428**, 487.

Nanofood for Thought – Block Copolymer Self-Assembling Nanostructures

1. Why might you wish to imbibe a block copolymer inside a capillary column? What would be the effect of changing the diameter and shape of the capillary as well as making the walls hydrophobic or hydrophilic?

2. Can you imagine what might happen if you coated the outside of a glass fiber with a continuous uniform film of a block copolymer with different thickness? Could there be something useful that could emerge from such adventurous research?

3. What would you expect to happen to different kinds of block copolymers when confined in the void spaces of an inverted silica opal? Could this knowledge be useful for making an optical device?

4. Can you think of a way of making a touch sensitive digital display from a block copolymer?

5. How would you expect different kinds of block copolymers to phase separate in the nanochannels of an anodized aluminum oxide membrane? Devise a means to separate the membrane from the polymer and think of some property–function relations that would enable utility of block copolymer nanorods or even barcode versions.

6. Devise a means of making a lithium ion conducting block copolymer and explain why such a nanocomposite would be useful.

7. A big question in block copolymer chemistry is whether it is possible to make, for example, a cubic bicontinuous morphology in which the non-intersecting portions of the network are electronically and ionically conducting? Can you dream of a way to rise to this challenge and find some interesting uses for this novel polymer nanocomposite.

8. What would be some of the benefits of making a phase-separated block copolymer morphology out of a heavily metalized polyferrocenylsilane?

9. Can you think why it is that many soft biomaterials and hard biominerals found in the natural world have microstructures that closely resemble morphologies of synthetic phase-separated block copolymers?

10. How and why would you want to integrate a liquid crystal forming building block into for example a columnar morphology block copolymer?

11. Imagine a method you could use to make a freestanding nanoporous membrane from a block copolymer. Could you think of ways to make a sensor from this membrane? Imagine various ways to modify this membrane to make a more selective sensor.

12. Devise and justify funding for the idea of DPN anionic living polymerization of block copolymers. Would you expect any surprises if you could

pattern block copolymers in this way at length scales smaller than the size of a phase-separated domain?

13. Devise several methods to make tough nano-objects with various morphologies starting from monodisperse block copolymers. Could any of these objects do nanoscale jobs that analogous shapes play in everyday macroscopic devices and machines?

14. What do you think would happen if a block copolymer melt was extruded from a narrow aperture or drawn on a spindle as a thin thread? What kind of morphologies and domain orientations would you expect at different block volume fractions?

15. Think about a way that you could make two monodispersed linear polymers such that they could reversibly link together to make a block copolymer.

16. Can you think of a way and a reason why you might want to synthesize CdS nanoclusters in a block copolymer?

17. How could you make a PI-b-PEO block copolymer in which nanocrystalline titania is exclusively located in the PEO phase-separated domains and why would this material be interesting?

18. Devise a way of converting the styrene groups in PS-b-PEO to bis-styrene chromium and think whether a phase-separated $(\eta^6\text{-PS})_2\text{Cr-b-PEO}$ bis-styrene chromium functionalized block copolymer would offer any opportunities for adventurous research relative to bis-arene chromium itself.

19. How and why would one increase the physical size of the phase-separated domains of a block copolymer?

20. Design a block copolymer in which you could synthesize titania nano-spheres, nanorods, and nanosheets and think of a function and use for these nano-objects.

21. Take a film of PS–PAA, ion exchange protons with Zn^{2+} to make PS-PAA/Zn^{2+}, react it with $(\text{Me}_3\text{Si})_2\text{Se}$ or H_2Se. What would one expect to form for various block copolymer morphologies?

22. Let us say you synthesized a block copolymer made of polystyrene–polyacrylamide (PS–PA), where the PA block can be cross-linked into a solvent swellable segment after self-assembly. What would be the effect of different solvents on the block copolymer morphology?

Bio Inspiration

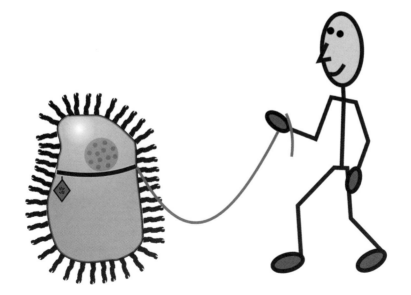

CHAPTER 10

Biomaterials and Bioinspiration

"There is nothing new under the sun, but there are lots of old things we don't know."
Ambrose Bierce (1842–1914)

10.1 Nature did it First

Nature has solved the problem of designing and synthesizing materials with structures that have been optimized to make them hard, tough and strong enough to house, protect and brace living organisms.[1] Optimization of these properties, refined by slow evolutionary engineering, serves to enhance the chances of survival. As well as having exceptional mechanical properties, biominerals usually end up being the least expensive in terms of raw materials, within biological limits, since the more energy the organism has to devote to material synthesis, the less energy it has to perform other important biological functions. Since organisms have spent millions of years optimizing structural materials for performance, durability and appearance, it is reasonable that materials chemists interested in designing functional materials should be curious about how nature has solved problems often encountered in science and technology.[2,3]

The challenge is one of technology transfer – the identification of questions in materials chemistry and connecting them with answers from nature.[4–6] This synergism between nature and materials chemistry should have considerable practical value, and this idea has had a long history. For instance, the Chinese around 3000 years ago looked for an artificial silk,[7] seeking to emulate silk's remarkable mechanical properties.[8] In the middle of the 20th century, the science of systems whose function is based upon or resembles living systems, termed bionics, emerged and inspired a surge of bioengineering research in systems diverse as animal sonar,[9] gravity devices, abalone shell[10] and insect cuticle.

Biominerals exemplified by microskeletons, biomagnets, teeth, shells and bones exhibit miraculous shapes, hierarchical structure and functional specificity – their form controls their function. Their formation and unique properties have inspired chemists to take a biomimetics approach to materials. The multilevel composite architecture of Haversian bone, where concentric layers of mineral surround a central blood channel, is an excellent example of biological hierarchy. It vividly

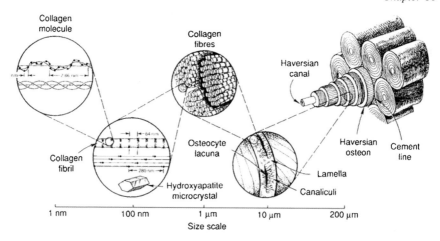

Figure 10.1   *Hierarchical structure of Haversian bone*
(Reproduced with permission from Ref. 12)

portrays the ability of a living organism to co-assemble the tropocollagen triple helix and hydroxyapatite crystal building blocks into progressively larger and complex microfibrils and osteons (Figure 10.1).[11,12] These are the construction units required for the final shaping of bone. Haversian bone is an excellent example of nature working with hierarchical assembly to create a finely tuned functional nanocomposite. It is an organic–inorganic hybrid material in which the bio-building blocks assemble from the lowest to the high level of construction in the hierarchy to create a material whose properties are greater than the sum of the parts. Each step in the process from site-selective nucleation and growth of oriented hydroxyapatite crystals on specific points of the α-triple helix to the ultimate placement of osteons in bone is genetically encoded[13] and the co-assembly strategy carries valuable information for a chemist attempting to synthesize a bone implant material.

10.2 To Mimic or to Use?

There are two ways we can think of bioinspired materials: to mimic those features found in Nature, or to directly use biological constructions in synthetic systems.

When we mimic Nature, we are attempting to emulate or duplicate features of natural systems using synthetic components and traditional fabrication techniques. One of the most impressive examples of mimicking is the Eiffel tower in Paris. It's conception and curved design was inspired by the microscopic diatom skeletons discovered by Haëckel. This grand emulation can be reduced in size several orders of magnitude right down to the molecular level.[14–17] A very important feature of biomaterials is their flawless integration of organic and inorganic components. For instance, proteins can be pre-organized as filaments, these filaments coated with nanocrystals produced in another part of the cell, and further crystal growth directed by the protein template. The result is a hierarchical construct with robust interfaces, curved and ordered shapes, and an optimized function using the least starting materials. In these types of examples, Nature teaches us that the interface between organics and inorganics is of the utmost importance.[18–20] It is important for making strong interfaces, for directing minerals into unusual morphologies, and the achievement of outstanding mechanical properties through the combination of both hard and soft elements. As we gain an increasing understanding of the molecular basis behind biomaterials, we become more adept at imitating both the macroscopic features of biological materials as well as the molecular basis of these features.

The second bioinspired strategy is to take Nature's self-assembled creations as they are and use their functionality in synthetic systems, perhaps in ways they were not intended to be used.[21] A prime example of this is the use of DNA in materials science and chemistry. Before the discovery of its structure and role in heredity, DNA was thought of as exotic but not very useful. Once this was disproved, research in the chemistry of DNA and its role in biological processes increased exponentially. Increasingly, the study of DNA has permeated from biochemistry into all areas of materials research and is now center stage in a lot of bionano-technology. The strong and selective bonds between complementary DNA sequences permit one to tag a material with a unique code, or a variety of materials with a variety of unique codes. These selective attractions can be used to link nanoparticles, assemble particles on surfaces and more. Preliminary experiments have appeared in the literature, but using the true power of DNA encoding has yet to materialize. Other biological components finding their way into materials are proteins, ubiquitous in biominerals where they direct synthesis, assembly, stabilize structures, and so forth. They are central in biological catalysis (enzymes), transport (motor proteins, transmembrane pores) and structural functions (actin filaments). In short, they are an indispensable part of any biological materials application. It thus follows that the most successful avenues of biomimetic research are likely to include the use of proteins in synthesis, assembly, and function of synthetic systems. Several common proteins are routinely used in materials applications, such as the use of the strong biotin–streptavidin interaction for the immobilization of a desired tag onto the surface of a material. Proteins can have an almost limitless variation of their sequence, which imparts them with a particular three-dimensional shape once folded. In materials, especially nanomaterials, shape is everything. It follows logically that if shape is everything, and proteins can adopt any shape, proteins are everything! Once again, the true power of proteins in materials synthesis lies somewhere in the near-future.

10.3 Faux Fossils

If you looked through a microscope at the complex patterned siliceous micron scale construct shown in Figure 10.2, it would be hard to tell if it was the fossilized remains of bacteria-like bodies formed during Earth's infancy, or rather a shape generated by a purely chemical process involving the hydrolytic polymerization of soluble silicates. Paleontologists looking for signs of early life on Earth have long been aware of this problem and confounded by a class of objects called "dubiofossils": these dubious microconstructions are formed by purely geological processes, and are difficult to distinguish from genuine microfossils.

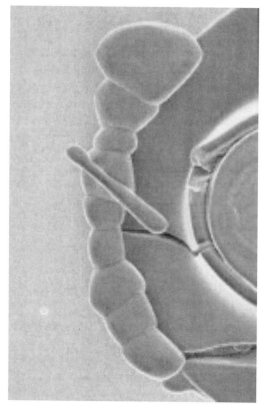

Figure 10.2 *Synthetic siliceous fossil. Are you fooled?*

The controversy over the Martian microfossils highlights the dilemma of describing fossil formation either to living or non-living causes.[22]

10.4 Nature's Siliceous Sculptures

The key to sculpting siliceous artificial fossils reminiscent in both size and shape to the fossilized remains of bacteria is the participation of organics in the processes responsible for their growth and form. What is remarkable about synthetic morphologies with complex and natural form of the type shown is that a living organism was not involved in their creation. As D'Arcy Thompson predicted,[23] chemical and physical forces alone can lead to siliceous constructions like Ernst Haeckel's radiolarians (Figure 10.3).[24] This occurs simply by depositing silica from solution into the Plateau border regions of a protoplasmic froth of vesicle-like bubbles called vacuoles. Almost a century later, synthetic morphologies resembling radiolarian and coccolith microskeletons are being synthesized from imprints of vesicles, microemulsions and latex spheres. The organics serve as templates for patterning deposited mineral in much the same way as vesicles do for the marine organisms.[25,26]

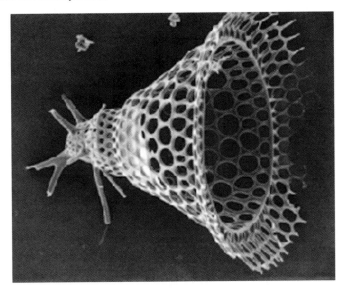

Figure 10.3 *Haekel's siliceous radiolarian microskeleton*
(Photo courtesy of the Museum of Science, Boston)

10.5 Ancient to Modern Synthetic Morphology

More than 50 years before the publication of D'Arcy Thompson's book "On Growth and Form", the Dutch zoologist and microscopist Pieter Harting succeeded in making facsimiles of some of nature's calcareous forms by reacting calcium and carbonate containing salts in biological media like albumen and gelatin (Figure 10.4). He discovered that his calcareous formations were not simply inorganic. They also contained organics as found in the shells of living organisms. It was not clear from Harting's experiments whether the reactions that were responsible for the growth of his synthetic shapes resembled in any way those which sculpted biological minerals. The answer to this question emerged almost a century after Harting's work when Heinz Lowenstam demonstrated that organics did participate in the growth of magnetite radula in the rasping tongue of the sea urchin.[27]

Harting's work can be considered to represent the beginning of biomimetic inorganic materials chemistry. At the time, Harting believed that his synthetic morphologies would one day be as important as Wöhler's synthetic organic chemistry. He wrote in French at the end of his seminal 1872 paper on synthetic morphology, roughly translated to English: *this is a big field, there is much to do, and we have taken the first step.* However, the importance of this work was only realized a century or so later upon the emergence of a unified theory on the formation of biominerals. Biomimetic inorganic materials chemistry in its modern guise essentially follows Harting's strategy of coercing organics and inorganics to co-assemble into materials and composites that have their structure, hierarchy and morphology designed to achieve a specific set of properties and functions.[28–30]

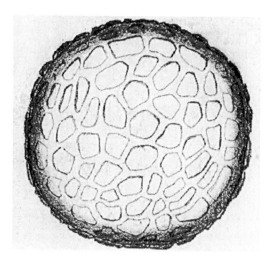

Figure 10.4 *One of Harting's synthetic calcareous concretions*

10.6 Biomimicry

Composite materials combine the properties of two or more components that are judiciously selected to maximize aspects of mechanical performance. Materials science evaluates different compositions and architectures to optimize properties like hardness, toughness, stiffness, strength and damage tolerance. Materials classes used in composites include metals, ceramics, glasses, elastomers and polymers. Fiber and laminate composite constructions are most common and designs are often based on the architecture, function and performance of composites found abundantly in animals and plants. In materials science, a composite is usually fabricated from preformed components in a process that organizes them in a matrix and with a particular arrangement. Since the integration of constituents is often a top-down physical procedure, the structure and composition of interfaces, which can have a profound effect on the properties of the composite, are not under control at the molecular scale.

A solution to this problem is through biomimetic materials chemistry, imitating Nature, which by contrast operates from the bottom-up. The idea is to co-assemble organic and inorganic precursors into composite materials with molecular level command over interfaces and structure at different length scales.[31] Organics can be molecules, assemblies of molecules, macromolecules or colloids, and they direct or template inorganics into a specific structure, pattern or shape. A goal of biomimetic materials chemistry is to utilize a synthetic approach to place organic and inorganic components in specific arrangements to create a composite with superior performance than those accessible through materials engineering methods. Consider for instance how cement, a brittle calcium aluminosilicate-based refractory material, can be made elastic and form a spring (Figure 10.5).[32,33] Although counterintuitive, a solution to this problem is to make some of the hard inorganic

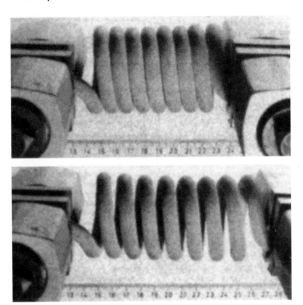

Figure 10.5 *A cement spring made by replacing "hard" inorganic cations in regular cement by "soft" organic cations or integrating polymers* (Reproduced with permission from Ref. 33)

cations in cement soft, by exchanging them for organics like tetrabutylammonium or integrating polymers into the synthesis of the cement. This is a case of intentional design of an organic–inorganic composite material for tailoring mechanical properties, a natural solution to a materials problem through a biomimetic way of thinking.

The benefits to be derived from molecular level control of interfaces between inorganics and organics are becoming apparent from the outpouring of research with a biomimetic chemistry theme. The remainder of this discussion will focus attention on how an appreciation of the synergy between biology and chemistry is leading to materials with remarkable structures and properties, expected by many to power the bio-nanotechnology industrial revolution.

10.7 Biomineralization and Biomimicry Analogies

A common thread that links biomineralization and biomimetic inorganic materials chemistry is an organic molecule, macromolecule, supramolecular assembly or matrix, which directs inorganics to a particular structure, pattern or shape. However, materials chemistry aimed at mimicking biomineralization operates without genetic regulation and control, and the results can deliver only limited mechanistic information about mineral formation in living organisms. On the other hand, molecular biology views biomineralization from a different perspective to materials chemistry. For example, to begin to understand the formation of the nacre outer mantle of molluscan shell, the spicule of sponge, the radula of chiton, the test of sea urchin or the biomagnetic compass of bacteria, one must identify which organics are actually involved in mineral formation. Consider the procedure for identifying proteins that initiate mineralization, direct the growth and sculpt aragonite into plates that comprise the "bricks-and-mortar" construction of molluscan shell[34] (Figure 10.6). This first involves the extraction of mineralizing proteins from the shell, a step which can cause significant loss of structure or activity, called denaturation. Then they must be purified, crystallized and sequenced to provide information about which amino acid residues participate in the process of engineering the nucleation, growth, size and orientation of calcium carbonate crystals and their arrangement into an overall shape.[35] Molluscan calcite structured in this way has a 1000× higher fracture toughness than regular calcite crystals.[36,37] Modification of specific regions on the protein by site selected mutagenesis helps to pinpoint amino acids that are thought to be involved in biomineralization. Nacre proteins can then be used for *in vitro* mineralization of calcium carbonate to discover whether they exert any kind of control over the growth of particular polymorphs (*i.e.*, calcite, aragonite, vaterite), habits and state of aggregation.[38–40] Elucidated protein structures can also be used to design purely synthetic polymers that can effect similar mineralization functions.[41]

Molecular biologists utilize biological templates such as phospholipids, polysaccharides and proteins that have been obtained from mineralizing organisms to discover the role they play in biomineralization. Understanding structure and dynamics at the interface between biological templates and minerals is one of the most challenging problems in the field of molecular biology. An example of a

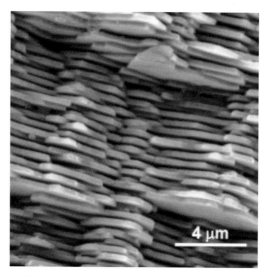

Figure 10.6   *Hierarchical laminated architecture of the nacre outer mantle of molluscan shell*
(Reproduced with permission from Ref. 34)

Figure 10.7 *Nacre mimic made of silicon carbide and graphite substituting for calcite crystals and protein. When subjected to a bending test after making an initial notch the stress is dissipated laterally by delamination, making the composite much less brittle than the bulk material*
(Reproduced with permission from Ref. 42)

biomimetic approach to materials with the fracture toughness of nacre would be to devise synthetic strategies to a highly ordered laminate microstructure. As nacre is built of hard thick plates of aragonite cemented together by thin and soft stress dissipating protein layers, an appealing materials solution to the problem is to co-assemble thick slabs of strong silicon carbide cemented together by thin soft layers of graphite (Figure 10.7).[42]

10.8 Learning from Nature

One can learn a lot about organic matrix mediated synthesis of inorganic materials and organic–inorganic composites by searching for analogies with biological design principles used in nature to process minerals. In this regard, molecular biology studies of biomineralizing systems are extremely valuable. Chemistry can utilize biological templates for synthesizing, patterning and shaping materials and composites. Numerous examples can be referenced to illustrate this approach. Ferritin is a large spherical cage self-assembled by the cell from many identical protein units, which the organism uses as an iron storage container by mineralizing iron oxide inside the central cavity.[43,44] Magnetic bacteria are able to synthesize aligned ferromagnetic nanoparticles inside vesicles, and line them up to form a bar magnet with a strong magnetic field.[45,46] Similarly, bacteria are able to survive toxic cadmium concentrations by sequestering the element as CdS inside vesicles, resulting in nanoparticle formation.[47] An analogy to the mineralization of ferritin, bacterial magnetite and cadmium sulfide bionanocrystallites has proven a powerful and general way to synthesize nanoclusters. These include metal oxide and sulfide nanoclusters within an empty ferritin protein cage (apoferritin), a viral coat devoid of internal RNA, and a phospholipid vesicle. Examples include oxides of silver, manganese, iron, uranium and tungsten, and sulfides of iron and cadmium. The spatial constraint of the internal void of the spheroidal organic template is found to impose a high degree of monodispersity on the nanoclusters formed within.

Monodispersed nanoclusters are fundamentally interesting because of the new science and technology that they promise (see Chapter 6).

10.9 Viral Cage Directed Synthesis of Nanoclusters

Consider a host–guest inclusion chemistry approach for growing tungstate nanoclusters inside viral cages.[48] The top of Figure 10.8 shows a cryo-TEM image reconstruction of cowpea chlorotic mottle virus (CCMV) in an unswollen (pH < 6.5) and swollen (pH > 6.5) condition viewed along the pseudo-threefold axis. Viral RNA is removed and ultra-centrifugation is used to purify the virus shell. This is followed by selective encapsulation of WO_4^{2-} inside the cationic virus cage at pH > 6.5. On reducing the pH < 6.5, the pores of the virus capsid close and the confined tungstate anions undergo oligomerization and mineralization to form $H_2W_{12}O_{42}^{10-}$ within the viral shell. The composite is subsequently purified by centrifugation. TEM images and measurements of the lattice spacing of unstained and negatively stained samples show that the mineral core is surrounded by the protein cage and is commensurate in size and shape with the 15 nm internal

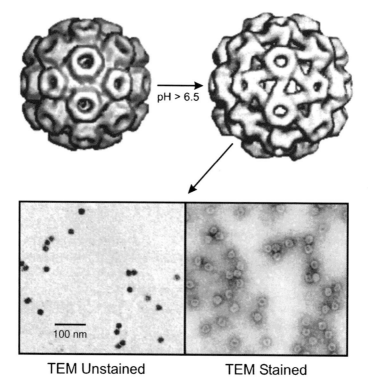

pH > 6.5

TEM Unstained   TEM Stained

Figure 10.8 *Minerizable virus capsids. The virus capsid swells at pH above 6.5, allowing it to be loaded by tungstate ions. The resulting virus-encapsulated particles are shown on the bottom in regular and negatively stained TEM images (Reproduced with permission from Ref. 48)*

diameter of the virus particle, bottom of Figure 10.8. With such a diversity of viruses in the natural world this idea could eventually lead to custom-made nanoparticles with any given size and shape, such as elongated nanocrystals grown in the cylindrical tobacco mosaic virus.[49] Virus capsids can be readily engineered to express particular surface or structural functionalities, giving us extra synthetic handles to further customize templated nanoparticles.[50,51]

An extension of the biological idea of restricting the size of nanoclusters by confining the growth process can be seen in numerous examples of materials chemistry attempts to mimic biology. The use of surfactant inverted micelles and alkanethiols to nucleate, arrest the growth and stabilize particular size clusters has been used successfully to prepare narrow size distribution capped magnetite, cadmium selenide and gold nanoclusters. Control over dispersity can be as fine as a single lattice plane, analogous to the remarkable monodispersity of nanoclusters grown in virus templates. For more details, see the examples outlined in Chapter 6.

10.10 Viruses that Glitter

The inside of a virus capsid provides an almost perfect chemical reactor for the synthesis of nanoscale particles, but one must remember that its exterior is by no means less perfect. Both the protein residues on the interior and exterior of a self-assembled virus shell are highly organized, and amenable to precise and pre-determined modifications.[52] By introducing cysteine amino-acid residues into spatially defined regions of the surface of cowpea mosaic virus (CPMV), it has proven possible to selectively anchor gold nanoclusters through Au–S bond formation on the exterior of the mutant capsid with pre-determined distances and patterns.[53] This is a nice example of how the chemical and structural specificity of a biological template can be used to organize inorganic building blocks into nanostructures where the positioning of the components is under strict geometrical and spatial control. With the help of molecular models and crystallographic data, together with site selective mutagenesis of CPMV, three different arrangements of cysteine units were explored for the anchoring of small 2 nm, and large 5 nm diameter gold nanoclusters. Light scattering is an effective tool for probing the anchoring dynamics of the nanoclusters to the virus as the scattering intensity increases with particle size. Further, the intrinsically high electron-contrast provided by the gold nanoclusters attached to the virus surface allows direct visualization of their spatial distribution by TEM for each cysteine pattern, where three examples are depicted in Figure 10.9. There appears to be an excellent correlation between the numbers and patterns of gold nanoclusters on the different mutant CPMV and those anticipated from the models. The fivefold arrangement of nanoclusters associated with the icosahedral fivefold symmetry sites can be seen in some of these images of gold decorated CPMV. It is found that the number of tethered gold nanoclusters to the capsid can be pre-arranged by working with mutant CPMV having larger numbers of cysteine units and smaller nanoclusters. Viral organization of gold nanoclusters in particular geometries and distances apart

TEM Images   Structural Models

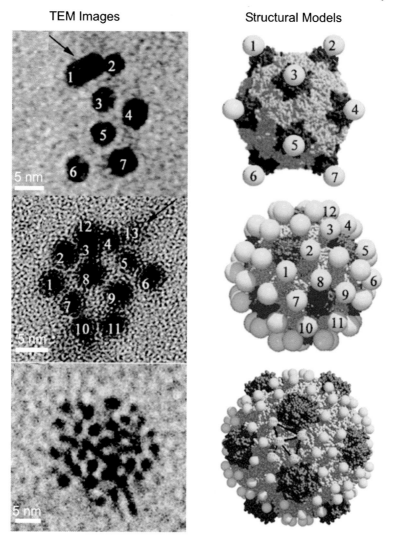

Figure 10.9 *TEM images of genetically engineered virus capsids decorated by gold nanoparticles. Beside each TEM image is a model of the decorated virus and the expected positions of the gold clusters based on the position of the inserted cysteine residues*
(Reproduced with permission from Ref. 53)

is a powerful means of controlling their optical and electrical properties potentially of value for biosensing.

10.11 Polynucleotide Directed Nanocluster Assembly

The properties of nanoclusters are inherently tied to their capping and stabilizing groups, which can impart solubility, processibility and functionality. A creative

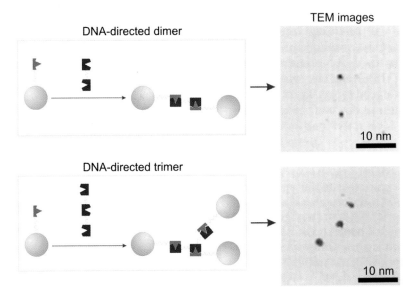

Figure 10.10 *Bioinspired strategy for interconnecting capped gold nanoclusters into assemblies with pre-determined size and shape using complementary DNA sequences*
(Reproduced with permission from Ref. 57)

extension of the capping concept utilizes polynucleic acid-substituted (DNA or RNA) alkanethiolates with complementary base pair sequences to synthesize capped gold nanocluster assemblies with pre-determined geometries and nuclearities.[54-56] The example shown in Figure 10.10 depicts the linking together of two and three capped nanoclusters at fixed distances apart using the base-pairing methodology.[57] These DNA connected gold nanocluster assemblies can be melted apart by heating, and linked again by cooling,[58] with profound and readily measurable differences in optical absorption spectra and electrical properties.[59] In addition, making use of the coded nature of DNA can result in the assembly of several types of particles into a single aggregate.[60]

The information content of DNA can be staggering. In the content of a single cell is contained all the information necessary to build and operate complex organisms, such as those writing this text. The engineering of nanocluster connectivity, charge transport and optical properties with DNA is pointing to a new generation of self-assembling integrated chemical systems, such as DNA chips that literally build themselves from components, or highly specific and multiplexed sensors.[61-64]

10.12 DNA Coded Nanocluster Chains

The information content and small size of DNA may make it the ultimate assembly module for self-assembling integrated systems. However, DNA by itself without any co-assembled functional units is of limited use. Several studies have shown that single molecules of DNA effectively function as insulators, while bundles and networks can have a finite resistance and behave as poor charge carriers.[65,66] The ultimate in the miniaturization of a variety of devices would hinge on the use of single DNA strands, and while DNA can be readily assembled onto surfaces and in solution it is clear that further steps are needed to make these assemblies functional.

One could imagine the first step in this direction being the conversion of the DNA strand into a metal. This does not happen through alchemy, but the strand is rather selectively coated with a metal sheath. The process used to achieve this goal is shown in Figure 10.11.[67] First, one gold electrode is coated with a SAM of 12 base-pair

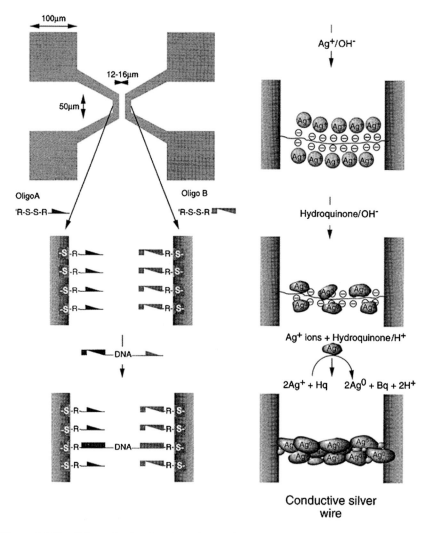

Figure 10.11 *Schematic showing the self-assembly of a silver wire, templated by a strand of DNA attached to opposite electrodes through selective recognition sequences*
(Reproduced with permission from Ref. 67)

single-stranded DNA, or sequence A. Then, a counter-electrode is coated with a different DNA sequence we call B. Another key component is a long piece of DNA having one end, A′, complementary to A and the other end, B′, complementary to B. Flowing a solution of this coded DNA strand over the electrodes results in the bridging by a single double-stranded helix, confirmed simultaneously by fluorescence and conductivity measurements. The bare strand is insulating, exhibiting hysteresis in the current–voltage plot. The DNA is then exposed to silver ions, which electrostatically bind to the backbone phosphate groups, and these nucleation sites are first reduced then grown into a continuous wire by electroless silver deposition. The linear I–V curve confirms the Ohmic conduction and metallic behavior. The thickness of the plated wire in this case is below 50 nm, and consists of connected silver grains. Silver clusters,[68] palladium[69] and platinum[70] wires have also been constructed from DNA for instance, as have conducting polymer lines,[71] and likely many more fascinating examples in the future.[72]

While metallization of DNA is a big step towards making tiny metal interconnects for future nanoelectronics, it is still a rather indiscriminate way of using DNA's exquisitely precise coding sequence. In order to make a diversity of architectures and compositions needed for integrated circuits, the sequence-specific modification of DNA becomes necessary.[73] One way to achieve this is by using the RecA protein, which biologically performs homologous recombination consisting of exchanging some genetic information between two strands of double-stranded DNA and increases an organism's genetic diversity. If RecA monomers are reacted with a 2000 base-pair single-stranded probe DNA, they will bind and polymerize along this strand. This probe can then be hybridized in a sequence-specific manner to the middle of a longer 70,000 base-pair double-stranded DNA. The resulting strand can then be stretched on a substrate, or bound to two electrodes as per the previous example. The big difference is that now there is a small section of the DNA bridge that is coated with RecA, which behaves as a chemical resist. For example, the DNA can be coated with silver clusters and these clusters connected by electroless gold deposition, but the RecA segment remains unmodified. A sequence of AFM and SEM images of the steps in the process is shown in Figure 10.12, demonstrating the highly specific and selective DNA modification. Such sequence-specific localization bodes well for future DNA-based devices, such as DNA templated carbon nanotube transistors[74] for instance, and further steps in this direction will likely result in even more impressive constructions.[75,76]

10.13 Building with DNA

It should be clear by now that DNA has tremendous potential as a template for the construction of functional architectures at length scales inaccessible to photolithography. It was mentioned previously that for most applications, the template DNA must be post-modified to endow it with a particular function. This is not always the case, and DNA in its own right can display very interesting behavior.[77–79] For instance, this biomolecule has been coaxed to assemble into cubes[80] and

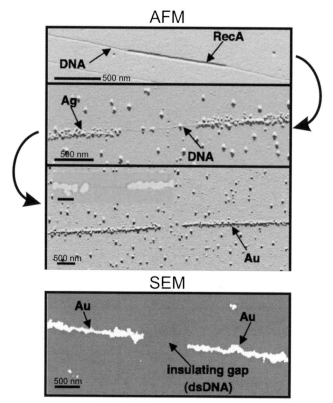

Figure 10.12 *AFM images depicting the coating of a sequence of DNA with RecA protein which protects it from chemical transformation. Silver clusters are coated with gold through electroless deposition to make a continuous wire with a narrow gap. An SEM image is shown on the bottom clearly delineating the engineered gap in the resulting wire*
(Reproduced with permission from Ref. 73)

other shapes with varied topology.[81] The assembly of DNA is quite flexible, since one simply has to define "sticky ends" on the tips of given strands to make them pop into a pre-determined position. However, DNA can be quite floppy, and this can lead to assemblies with ill-defined shapes or periodicities. To improve the stiffness of these constructs we can make use of a so-called double-crossover (DX) motif, which consists of a junction point between four single-stranded DNA.[82] Using the DX motif allows one to make shapes and lattices which follow much more closely a predicted assembly pathway. This strategy has been most successful in the construction of two-dimensional lattices and patterns.[83,84] Schematic steps to form such an ordered DNA array are shown in Figure 10.13, along with AFM images confirming the results of the assembly. The top-left image depicts the cross-shaped construction unit, which can then be linked together in various ways: on the left is shown self-assembled nanoribbons consisting of a square lattice of DNA crosses. Right, the positions of the sticky ends have been changed, resulting in a

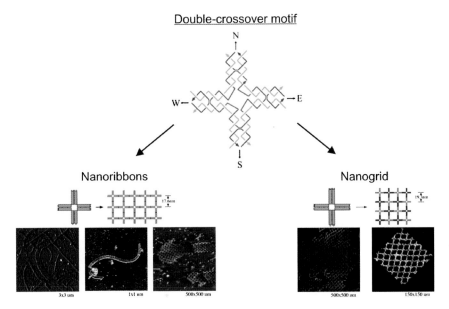

Figure 10.13 *Well-defined nanoribbons and nanogrids can be synthesized from stiff DNA segments called double-crossover motifs*
(Reproduced with permission from Ref. 84)

corrugated DNA mesh. These DNA ribbons and meshes can be straightforwardly modified with proteins, for instance with streptavidin at the mesh cross-points, or converted to metallic grids with high conductivity.

Another exciting prospect for DNA is in nanomechanical devices.[85] DNA is very stable when kept apart from digestion enzymes, can adopt a three-dimensional shape and can be prompted to change this shape, giving us the basis of durable DNA nanomechanics. Once DNA is assembled into a structure it can be actuated in various ways: supercoiling or phase-change induced by ions or intercalation agents, hybridization or dehybridization events, selective denaturation of A–T *vs.* G–C segments using temperature, and more.[86–88] A prime example of such a nano-mechanical system is a recently reported DNA switch,[89] made up of loops of double-stranded DNA, shown schematically in Figure 10.14. The addition of a target DNA strand causes a rotational motion in the supra-strand, which has been monitored by AFM measurements. This type of rotational actuator is much smaller than anything we can microfabricate, and has the recognition capabilities which could even-tually lead to highly multiplexed and addressable arrays of nanomechanical units.

10.14 Bacteria Directed Materials Self-Assembly

Organized arrays of the filamentous superstructure of *B. subtilis* have been utilized as proteinaceous templates for directing the infiltration of preformed inorganic

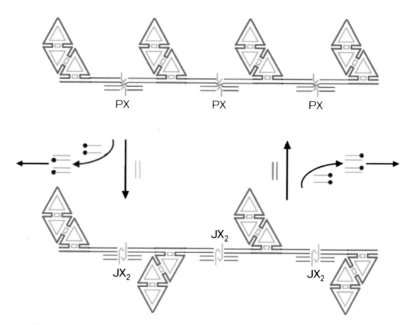

Figure 10.14 *Schematic depicting a DNA-based nanomechanical device*
(Reproduced with permission from Ref. 89)

colloids to form macroscopically ordered semiconductor and magnetic fiber composites, appropriately coined "brittle bacteria".[90–92] To amplify, different levels of cellular organization in *B. subtilis* bacterial thread can be obtained. They range from individual rod-shaped cells, constant diameter multi-cellular filaments, intertwined filaments, web structures and multi-cellular filaments. By drawing bacterial cultures at the air–water interface it is possible to produce macroscopic bacterial thread with organized internal superstructure[93] that can be reversibly swollen in aqueous inorganic colloidal suspensions. This allows for the infiltration of particles to form inorganic–organic fibrous composite materials. Preformed magnetic Fe_3O_4[94] and semiconducting CdS nanoparticles have been incorporated into *B. subtilis* organized bacterial superstructures (Figure 10.15). The air-dried materials consist of a close packed array of multi-cellular bacterial filaments, each 0.5 μm in diameter and coated with a 30–70 nm thick layer of aggregated colloidal particles. The surface charge on the bacterial filaments is negative (acidic proteins) allowing negatively charged colloidal nanoparticles to infiltrate into swollen inter-filament spaces. Similarly, silicate micelles have been imbibed into the voids between close packed bacterial filaments and polymerized to form siliceous fiber composites with hierarchical construction.[95] The macropores are packed with bacterial filament and the silica channel walls are permeated with ordered mesopores. The long-term goal is to integrate biotechnology and materials chemistry for fabrication of hybrid materials with patterned microstructures and morphological control.

Low
Magnification

High
Magnification

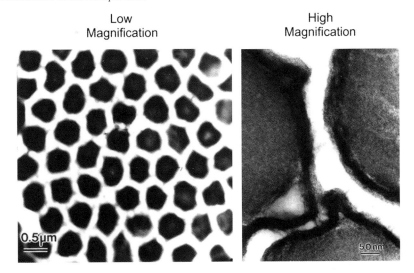

0.5μm

50 nm

Figure 10.15 *Close packed bacterial threads direct the synthesis of macroporous CdS* (Reproduced with permission from Ref. 91)

10.15 Using a Virus that is Benign, to Align

We saw in Chapter 8 that ordered phases can be formed from the close packing of cylindrical silicate micelles. In fact, it seems that any type of rod with a large aspect ratio is predisposed to the formation of liquid crystalline phases.[96] The simplest of these is a nematic phase, where the collection of rods is oriented with a small angle to a common axis. Apart from well-known molecular liquid crystals, other systems displaying liquid crystallinity are becoming increasingly prevalent. Widely different units such as metallic nanowires[97] and anisotropically shaped viruses[98,99] can form materials with axially oriented order.[100] The liquid crystalline property of virus capsids can be used to form hierarchical materials, co-assembled with nanoparticles which would otherwise be unable to assemble in such a fashion.[101–103]

The binding properties of surface proteins[104] on phage viruses (bacterial virus) can be modified in pre-determined ways by a so-called phage-display technique. For instance, if we required a virus that selectively bound streptavidin, we would first expose a collection of streptavidin-coated beads (commercially available from distributors such as Polysciences Inc.) with a library of phages. Virus capsids are self-assembled from protein units, as seen in a previous example. If the amino acids exposed to the liquid medium in these proteins are randomly mutated by DNA modification, a large library is obtained with each individual displaying slightly different surface proteins. A library can also be assembled from a population of viruses with natural variations in amino acid sequences. When such a collection is exposed to streptavidin-coated beads, there will be some individuals whose particular surface protein happens to bind streptavidin. The binding viruses are isolated by washing off the non-binding ones, and finally released from the substrate by changing the ionic strength. The selected phages are allowed to multiply, and

Streptavidin-Coated Viruses

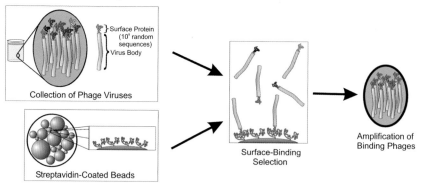

Virus Directed Self-Assembly

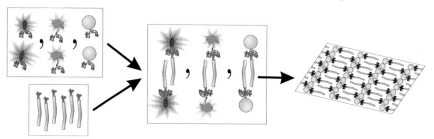

Figure 10.16 *Procedure used to make inorganic-binding viruses and using these to create virus–nanoparticle composites. On the top is shown the phage selection technique, while the bottom shows the co-assembly of the engineered phages with tagged nanoparticles*

the streptavidin binding step is repeated. Iterative cycles result in the selection of high affinity streptavidin-binding viruses. A scheme illustrating the procedure is shown in Figure 10.16.

On the other hand, it is relatively easy to coat a variety of materials with streptavidin. Commercially available streptavidin derivatives can conjugate to various groups such as amines, hydroxyl groups and carboxylic acids. Many streptavidin conjugates are commercially available, such as modified gold nanoparticles and fluorescent tags. When isolated phages were exposed to streptavidin-coated units such as gold nanoparticles, the nanoparticles were found to bind to exposed proteins on the virus tips. By concentrating the system through centrifugation, a solution displaying liquid crystalline behavior was obtained. This liquid crystalline phase could be dried to obtain free-standing, hierarchically ordered films of viruses with either metal particles, fluorescent molecules, or large fluorescent proteins.

10.16 Magnetic Spider Silk

Dragline spider silk has been mineralized with a preformed magnetite colloid to create a magnetic web[105] (Figure 10.17). Spider dragline silk is a semi-crystalline

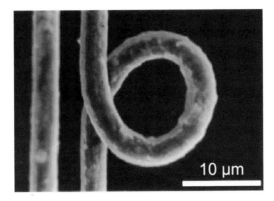

Figure 10.17   *Silk threads direct the synthesis of magnetic fibers, which retain many of their physical properties such as flexibility*
(Reproduced with permission from Ref. 105)

biopolymer with a unique combination of high tensile strength, high elasticity and high modulus.[106] Fibers that are 0.2–10 μm in diameter have a higher breaking energy than other natural or synthetic fibers. It far exceeds high tensile strength steel and Kevlar (a partially oriented crystalline polymer used for bullet-proof vests) on a weight-for-weight basis. The binding of 10–20 nm superparamagnetic magnetite particles to the surface of silk fibers using aqueous colloidal sols creates magnetic silk, which may be integrated into devices concerned with audio reproduction where strong fabrics that respond to magnetic fields are required. The methodology has also been applied to CdS, Au_n and polypyrrole (Fe^{3+} oxidative polymerization of pyrrole in the presence of silk fiber). Silk can also be used as a template to form hollow fibers of mesoporous oxides (see Chapter 8). Silk fiber that combines the natural strength and elasticity of the biopolymer with physical properties such as magnetism, electrical conductivity and semiconductivity could find interesting uses in the fabrication of a range of smart fabrics, for magnetic shielding amongst other things. This creative combination of biotechnology and materials chemistry could lead to a new class of fibrous composites for diverse applications.

10.17 Protein S-layer Masks

Bacterial protein S-layers are being used as templates for organizing semiconductor nanoclusters[107] and as lithographic masks for patterning silicon.[108] The outer membrane of certain classes of bacteria is composed of ordered assemblies of globular proteins that have impressive thermal and acid stability. The separation and spaces between proteins are in the mesoscopic length scale making them attractive for experiments in biomimetic materials chemistry.[109,110]

   Representative freeze fracture TEM images together with their image reconstruction can fully characterize S-layers that envelope archaea- and eubacterial cells. Schematic illustrations of different symmetry S-layers are shown in Figure 10.18. The coats show highly ordered arrays of proteins with tetragonal, hexagonal and

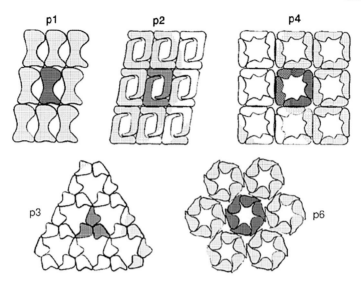

Figure 10.18   *Different symmetry bacterial S-layers*
(Reproduced with permission from Ref. 110)

trigonal mesh symmetries. The biological function of the S-layer is for molecular filtration, cell–cell recognition, immune response, and cell adhesion. In the context of biomimetics, the S-layer represents an interesting template for nanoscale patterning of metals and semiconductors. S-layers can be isolated from bacterial coats and deposited on a range of substrates. Just like crystals and liquid crystals, protein S-layers also display point and line defects, the effects of which have to be recognized in any materials application.

Two approaches to S-layer patterning of CdS nanoclusters have been explored. Re-crystallization of the S-layer on a TEM grid produces a mesoporous monolayer with a corrugated negatively charged inner face exposed to the external medium. The corresponding orientation of the S-layer formed over a LB monolayer of dipalmitoyl phophatidylethanol (DPPE) and transferred by horizontal dipping to a TEM grid, instead exposes a charge-neutral S-layer outer surface to the medium.

An anionic S-layer is anticipated to allow ion exchange of Cd^{2+} with specific surface sites whereas the neutral S-layer is expected to be less selective. Subsequent exposure of the S-layer-Cd^{2+} composite template to gaseous H_2S leads to the nucleation and growth of CdS nanoclusters, as illustrated in Figure 10.19. TEM images show the outcome of this experiment (Figure 10.20). The order present on the anionic surface of the bacterial S-layers can be clearly seen in images viewed after negative staining of the organic protein components, and this order is maintained after mineralization with CdS. The left panel in the figure shows the result of an experiment where CdS was generated on the anionic inner surface of a *B. Sphaericus* S-layer. Here the square symmetry inherited from the biological template is readily apparent in the inorganic replica. The right panel of the figure shows the mineralization onto the neutral outer surface of a *B. Stearothermophilus*

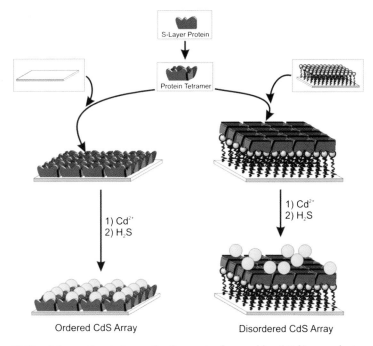

Figure 10.19 *S-layer directed growth of organized assembly of CdS nanoclusters*

S-layer, displaying weak periodicity. The effect of selective *vs.* non-selective ion exchange of the Cd^{2+} on the anionic compared to the neutral S-layer is a higher degree of order of the CdS nanocluster array in the former compared to the latter.

Parallel molecular nanolithography has also been achieved using S-layers as masks.[111] A sample procedure involves deposition of the protein S-layer onto the

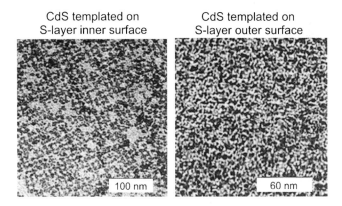

Figure 10.20 *TEM micrographs showing the S-layer directed growth of CdS nanoclusters which can either be organized or disorganized depending whether they are templated on the inner or outer surface, respectively, of the S-layer (Reproduced with permission from Ref. 107)*

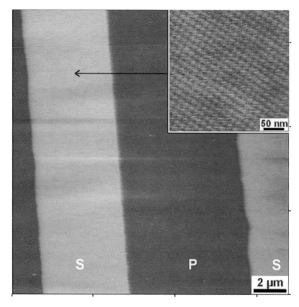

Figure 10.21 *Micron-scale patterns of nanopatterned S-layers made by micromolding in capillaries. Regions denoted "S" were exposed to the S-layer protein solution, while regions denoted "P" were covered with a PDMS stamp* (Reproduced with permission from Ref. 108)

substrate, followed by shadow metallization using e-beam vaporization of 12 Å of Ti, which becomes 35 Å of TiO_2 when exposed to air. Normal incidence Ar ion milling preferentially removes deposited material from the substrate to create a Ti or TiO_2 coated S-layer. The masking properties of S-layer templates can also be combined with soft lithographic techniques such as micromolding in capillaries (see Chapter 2). For instance, Figure 10.21 shows lines of S-layer monolayers patterned in this way, with the dark areas corresponding to the position of the PDMS stamp applied to the surface during S-layer crystallization. S-layers are clearly templates to be reckoned with.[112,113] Their self-assembly can routinely produce structures much smaller than the smallest achievable by conventional photolithography. New types of S-layers are being discovered constantly, and totally synthetic protein arrays are now appearing in the literature,[114] giving protein lithography a lively and bright future.

10.18 Morphosynthesis – Inorganic Materials with Complex Form

The hierarchical architecture and exquisite morphologies sculpted out of calcite in coccolithophore microskeletons and apatite in macroporous bone has inspired experiments in biomimicry. Consider the individual scales of the marine algae *Emiliania huxleyi* (Figure 10.22), which comprises an oval arrangement of

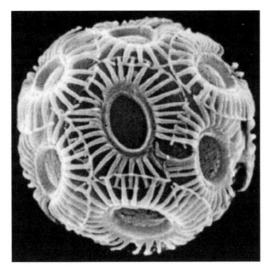

Figure 10.22 *Calcite coccolithophore*
(Reproduced with permission from Ref. 115)

30–40 oriented crystals of calcite with hammerhead morphologies. Calcite crystals are organized in radial arrangement of vesicles. Individual scales are exocytosed from the cell and assembled into the coccosphere, which surrounds the cell wall of the algae.[115] This biomineralization process represents a masterpiece of crystal tectonics involving supramolecular pre-organization, interfacial molecular recognition, vectorial regulation and cellular processing.

Bicontinuous microemulsions composed of oil, water and surfactant have been used to synthesize some interesting inorganic morphologies.[116–118] For instance, the interpenetrating non-intersecting network of mesoscale water conduits that exist in an oil-in-water microemulsion serve as a template for shaping calcitic and apatitic mesh structures (Figure 10.23). To amplify, surfactants can stabilize oil droplets in water or the reverse of water droplets in oil. At intermediate concentrations bicontinuous emulsions can form in which labyrinths of water interpenetrate an oil matrix, which is stabilized by interfacial surfactant. Although the structure is a dynamic one with water channels rapidly fluctuating, disconnecting and reforming, the time scales are short compared with measured bulk properties and can be used as a supramolecular template to shape inorganic mesh structures as well as other shapes. In this way, a microskeletal calcium phosphate has been formed in a bicontinuous microemulsion as well as a thin aragonite cellular film on a stainless steel substrate with an average pore size of 225 nm and roughly 115 cells μm^{-2} of surface. Reticulated mineralized open frameworks can have pore sizes that are tunable from the meso- to the macroscale.

An extension of the bicontinuous microemulsion templating method has led to biomimetic coccoliths,[119] micron size hollow shells of porous aragonite (Figure 10.24). The approach involves layering a $Ca(HCO_3)_2/Mg^{2+}$-tetradecane-didodecyldimethylammonium bromide bicontinuous microemulsion over micron

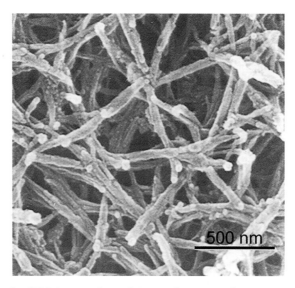

Figure 10.23 *An SEM image of a calcium carbonate mesh structure templated by a bicontinuous microemulsion*
(Reproduced with permission from Ref. 117)

dimension latex spheres. The spheres are washed with hot hexane to facilitate mineralization of the aragonite shell. The polymer sphere is removed from the mineral-coated beads with acetone resulting in porous hollow aragonite spheres with a form that superficially resembles the coccosphere of the marine algae. The morphology of these mesh structures and biomimetic coccoliths suggests possible uses as lightweight ceramics, magnetic microspheres, catalyst supports, membranes for high temperature separations, coatings for biological implants, separation of viruses and cells, encapsulation of DNA and targeted drug delivery.

By comparison, the architecture of the biomineralized calcite coccolith skeleton is far more intricate and the construction process responsible for its formation is much more complicated.[120,121] The steps involve intracellular nucleation, growth and assembly of differently shaped parts of the calcite scales. Component manufacture is followed by exocytosis and positioning of the scales, facilitated by microtubule scaffolds, into the recognizable shape of the coccolith microskeleton. Interestingly, the calcite scales are believed to function as a bio-optical device, namely a solar optical concentrator for coccolith photosynthesis in a marine environment.

10.19 Echinoderm vs. Block Copolymer

In the context of templates for calcite mineralization, the delicately curved surfaces and sculpted pores of calcite tests of the echinoderm (sea urchin) bear an appealing spatial relation to the architecture of microphase separated cubic bicontinuous block-copolymers in materials chemistry (see Chapter 9) and cubic bicontinuous membranes found in plant biology. Specifically, the domains of microphase separated

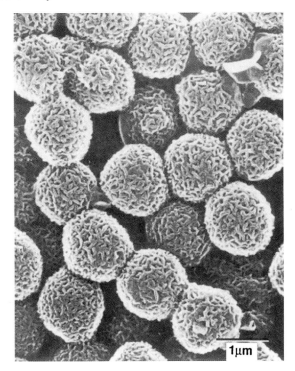

Figure 10.24 *Biomimetic coccoliths templated by a bicontinuous microemulsion coated latex microspheres*
(Reproduced with permission from Ref. 119)

linear diblock co-polymer poly(styrene)-*b*-poly(isoprene) have a topology that follows the Schwartz D and gyroid surfaces. They are strikingly similar to the ultra-structure of the calcium carbonate skeletal networks of the echinoderm, Figure 10.25, suggesting that their mode of formation could originate by the same mechanism. By comparison, cubic mesoporous silica films and crystals with cubo-octahedral morphologies having channel dimensions adjustable from 2 to 10 nm emerge from a surfactant-templating synthesis based upon silicification of a cubic bicontinuous lyotropic silicate mesophase (see Chapter 8). Although the templating length scales are different, a cubic bicontinuous lyotropic block copolymer has also been shown to form cubic mesoporous silica but with channel sizes that are tunable from 10 to 35 nm.

10.20 Fishy Top-Down Photonic Crystals

As we have seen in Chapter 7, nature has found ways to perfect structural color biologically and geologically through the creation of photonic crystal micro-structures in the butterfly wing, peacock feather, sea mouse spine, abalone shell, and gemstone opal. In principle, these natural photonic crystals can all be used as templates to create inverse versions that can display structural color contingent

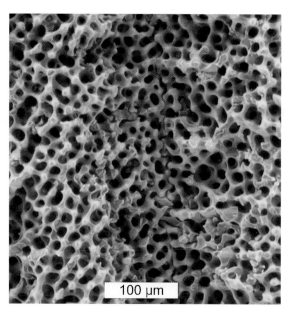

Figure 10.25 *Echinoderm calcite skeletal networks resemble the architecture of microphase separated cubic bicontinuous block copolymers in chemistry and biology*
(Reproduced with permission from http://www.glencoe.com/sec/science/cgi-bin/splitwindow.cgi?top = http://www.glencoe.com/sec/science/top2.html&link = http://www.ucmp.berkeley.edu/echinodermata/echinomm.html)

upon the lattice dimensions and refractive index contrast created in the replica. It turns out that there exist a number of other naturally occurring periodic structures that could also function as photonic crystals, like the sea urchin exoskeleton, but their lattice dimensions are far too large for them to operate in the visible or near infrared regions of the electromagnetic spectrum, of interest in optical telecommunication and other applications discussed in Chapter 7. A creative means of circumventing this problem is to find a means of linearly shrinking the lattice dimensions in natural photonic crystals, too big to be useful in these spectral regions, down to the right size.[122]

The size reduction procedure, applied to the periodic bicontinuous sea urchin calcite exoskeleton, begins by infilling the microstructure with the elastomer polydimethylsiloxane (PDMS). This step is followed by removing the biomineral calcite by aqueous HCl and then pyrolyzing the PDMS replica to create a 50% shrunken silicon oxycarbide (SiOC) inverse of the original exoskeleton. The shrinking is due to the loss of mass upon pyrolysis, and Figure 10.26 shows the original calcitic framework (left) as well as the SiOC replica (right). By infiltrating this shrunken SiOC microstructure with polystyrene (PS) and subsequently etching away the SiOC in dilute HF one obtains a reduced size PS facsimile of the original exoskeleton.

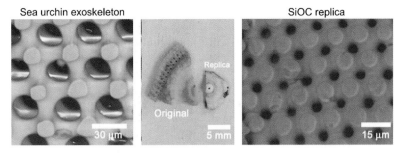

Sea urchin exoskeleton   SiOC replica

Figure 10.26 *The microstructure of the sea urchin exoskeleton can be linearly shrunk by replicating as a SiOC ceramic. The middle section shows optical images before and after the replication process*
(Reproduced with permission from Ref. 122)

This size reduction process can continue in a cyclic fashion by next infilling the PS with PDMS, removing the PS in nitrobenzene and pyrolyzing the so-obtained PDMS replica to SiOC. At this stage one obtains yet a second 50% linear size reduction in the dimensions of the original exoskeleton. If the SiOC microstructure obtained at any stage of this cyclic process is then replicated in a high refractive index material like tellurium, using the method of vacuum-assisted infiltration followed by HF etching of the SiOC template, then a photonic crystal with a desired lattice dimensions and refractive index contrast can be obtained. This is a kind of "natural" top-down way of transforming periodic structures in nature that are too large to smaller copies with lattice dimensions and refractive index contrast to show optical photonic crystal properties with stopbands that occur in a particular wavelength regime of the visible or infrared spectrum. Amusingly, the size reduction method could perhaps be applied to a hand made diamondoid lattice of glass marbles to make an inverse diamondoid silicon photonic crystal with an omnidirectional photonic band gap at 1.5 μm (see Chapter 7).

10.21 Aluminophosphates Shape Up

Intricate biomimetic inorganic forms have been synthesized from surfactant-templated aluminophosphate materials.[123–128] The synthesis comprises quite a concoction, namely alumina, phosphoric acid and decylamine in tetraethyleneglycol (TEG)–water co-solvent. Millimeter sized inorganic spheroids and hollow shells bearing micron dimension surface patterns (*e.g.*, pores, bowls, star dodecahedra) result from this biomimetic synthesis. The resemblance of the morphologies to radiolaria and diatom microskeletons is impressive (Figure 10.27). The morphologies are comprised of aluminophosphate lamellae with mesoscale but not microscale order. Two key aspects of the synthesis of these morphologies are noteworthy. One is the templating behavior of a decylammonium-dihydrogenphosphate (DDP) liquid crystal phase. A single crystal X-ray diffraction structure of DDP reveals a network of interdigitated decylammonium cations arranged orthogonally and intricately hydrogen bonded to a layer of dihydrogenphosphate

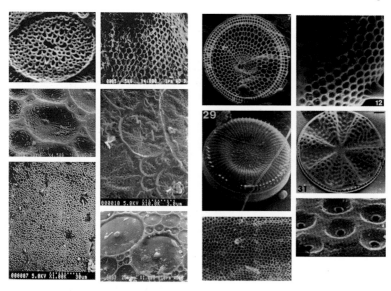

Figure 10.27   *Real and synthetic diatoms and radiolaria – can you tell the difference?*
(Reproduced with permission from Ref. 127)

counteranions. The second is the behavior of the solvent. TEG served a multifunctional role: as a co-solvent with water, a polydentate ligand for Al(III) and a co-surfactant for inducing bilayer curvature. It was established that DDP and DDP/TEG displayed smectic liquid crystal properties at the reaction temperatures required to template the aluminophosphate morphologies. Optical microscopy provided direct *in situ* images of the species responsible for the growth of the surface patterned morphologies. In essence, cellular-shaped objects with a smectic liquid crystal behavior spontaneously appear from an initially homogeneous solution under the conditions that yield the aluminophosphate morphologies. It is inferred that these objects are phase separated water–DDP–TEG and air–DDP–TEG liquid crystal microemulsions. These foam like structures behave as the precursors most likely responsible for the formation of solid and hollow multilayer aluminophosphate spheroidal shapes and surface patterns.

10.22 Better Bones Through Chemistry

Synthetic analogues of bone are being actively pursued as materials for biomedical applications in the field of bone replacement, augmentation and repair.[129,130] Numerous stringent criteria have to be met for a biomaterial to be considered as an acceptable bone implant, including the ability to integrate into bone and not cause any deleterious side effects. Bone itself consists of roughly 70% hydroxyapatite mineral embedded in an organic matrix consisting largely of collagen. Needle-like or plate-like morphologies of hydroxyapatite crystallize at regular intervals along the collagen fibers with their long *c*-axes oriented parallel to the fibrils. Moreover, hydroxyapatite crystals are found around the fibrils, randomly seeded in the collagen matrix and not intimately interacting with the collagen. The presence of HPO_4^{2-} groups in bone apatites has suggested that octacalcium phosphate (OCP) may be a precursor to hydroxyapatite (OHAp) formation in bone. OCP serves to establish the final morphology, composition, solubility, and interfacial energy of apatitic materials, as well as controlling the nucleation and growth of OHAp. They result from the fact that OCP and OHAp have similar crystal structures and are able to epitaxially grow together. As a consequence of this relationship, a prime consideration in the preparation and performance of a bone implant material is the selection of a surface that is able to induce nucleation of OCP. Osteogenic cells control the remodeling process. New bone is laid down on the resorbed surface of old bone, creating an interface with itself on a continual basis. One of the models of bone formation proposes that the sequence begins with the secretion of a collagen-free organic matrix, which serves as a source of the binding sites for nucleation of calcium phosphate. Uncalcified collagen fibers secreted by the osteogenic cells assemble on the mineral and then calcify by fiber mineralization and extracollagenous matrix mineralization by random seeding of crystallites in the non-collagenous proteins that are continually secreted by osteogenic cells.

In certain studies, metals or metal alloys are used directly as implants or these are coated with bioactive materials, such as hydroxyapatite,[131] to aid in adhesion to bone. Although metals or metal alloys meet biomechanical requirements of implants and some meet biocompatibility requirements, they exhibit poor interfacial bonding between the metallic surface and the surrounding bone.[132] Good biocompatibility of certain metals, for example titanium, is thought to occur as a result of spontaneous formation of a thin titanium oxide layer on the surface. In addition, titanium and titanium-based alloys have been shown to be bioactive because calcium phosphate readily nucleates and forms a film on these substrates. One of the most promising biomimetic approaches to bone implant materials is the use of composite organic–inorganic materials since they can be designed to be compositional, structural and functional analogues of bone. Different combinations of materials have been used with one of the most well-known examples being polymer–hydroxyapatite composites.

Another approach involves the spontaneous growth, under aqueous physiological pH conditions, of an oriented hydroxyapatite film with micron dimension porosity, on the surface of a layer of TiO_2 that has been sputter deposited on Ti metal.[133] This procedure creates desirable co-crystallized phases of hydroxyapatite (OHAp) and OCP with preferred orientation, respectively, along $\langle 001 \rangle$ and $\langle 101 \rangle$ directions. To mimic the hierarchical organic–inorganic composite architecture of bone, a calcium dodecylphosphate (CaDDP) ester mesolamellar phase has been incorporated synthetically into these oriented porous films to create a multilayer CaDDP–OHAp–OCP–TiO_2–Ti integrated chemical system. The process facilitates formation of a macroporous organic–inorganic composite film in which there is evidence of epitaxial growth of bilayers of the lamellar calcium dodecylphosphate ester on the apatitic phase. Complementarity of charge and geometry involving calcium ions and protonated phosphate located at the interface between ester and mineral is likely responsible for the observed intimate co-alignment of the organic and inorganic components of the composite film. The hydrophobic bilayer of the calcium dodecylphosphate ester component can be used to store and release bioeffector and drug molecules. These bioactive agents can be targeted directly to a site of bone surgery to promote osteoblast resorption and bone regeneration, and combat bone disease.

10.23 Mineralizing Nanofibers

In natural bones, collagen molecules serve as nucleation sites for oriented crystals of hydroxyapatite (HA). While many materials can be made which nucleate this mineral, it is difficult to match the precision with which collagen accomplishes it. The precise crystalline orientation of the HA seeds plays a key role in the mechanical properties of bone, since fusion of two crystallographically aligned crystals will result in less detrimental grain boundaries and defects.

The mineralizing properties of collagen have been reproduced using nano-fibers self-assembled from carefully constructed peptide amphiphiles, shown in Figure 10.28.[134] The large amphiphilic molecule has a conical shape, assembles as cylindrical micelles in aqueous solution (see Chapter 9), and is built of several sections (**1–5** in the figure) each designed for a different purpose. Section **1** is a long hydrophobic alkyl chain, gives the conical molecule a narrow tip, and provides the hydrophobicity necessary for efficient phase separation and self-assembly. In order to stabilize the self-assembled nanofibers, a section of cysteine residues

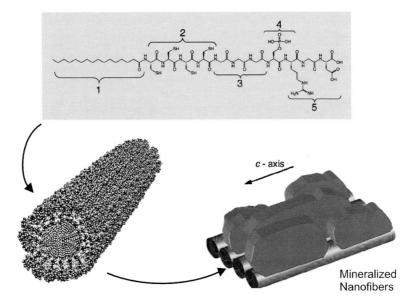

Figure 10.28 *Chemical structure of a peptide-based amphiphile which self-assembles into cylindrical micelles. These micelles can be mechanically stabilized, then used to nucleate oriented crystals of hydroxyapatite*
(Reproduced with permission from Ref. 134)

was added (**2**), the thiol groups of which can be reversibly crosslinked as disulfide bonds. Section **3** is a stretch of glycine amino acids, inserted as an inert buffer. It is known from many studies of biological materials that calcium phosphate minerals are often associated with phosphorylated proteins, therefore a phosphoserine group was inserted (**4**) in order for the nanofibers to display surface phosphate groups. If these synthetic collagen mimics were to be used *in vivo*, it would be beneficial that they can be able to promote the adhesion of bone-forming cells. The final section **5** therefore contains an arginine-glycine-aspartate (Arg-Gly-Asp) sequence, known to be key in many cell-adhesion promoting proteins. At the same time, the terminal anionic groups can attract calcium ions to the nanofiber periphery, and create a local supersaturation facilitating crystal nucleation. All these factors together result in a robust, biodegradable self-assembled system capable of precisely nucleating hydroxyapatite in exactly the same way as collagen, Figure 10.28, providing a powerful and tunable way to imitate a natural process.

10.24 Biological Lessons in Material Design

Composite fiber and laminate morphologies abound in the biomineral world. Protein β-sheets serve to facilitate formation and organization of calcite and aragonite crystal tablets into brick and columnar constructions that constitute the fabric of the hard and tough abalone shell. Protein α-helices and silica deposition vesicles direct the formation and shaping of silica spicules, called spiculogenesis, that provide support and protection to the soft body of the marine sponge. The stinging nettle uses vesicle compartments to mineralize silica spines that protect the plant from damage by intruders. The loricae silica basket of the unicellular protozoan *Stephanoeca diplocostata* provides the choanoflagellate with cell protection. Intracellular silica deposition vesicles are organized by protein microtubule scaffolds in an extracellular construction process, that leads to the open-ended, intricately woven loriate silica framed basket. The loricae assembly is considered as a masterpiece of microscale structural engineering that involves depositing, transporting and gluing together up to 300 costae silica rods in a highly specific organic templated sculpting process. There are countless more examples of well-characterized biominerals, and many more left to be discovered. Nature's biominerals can teach us many lessons in the design of new materials.[135,136]

10.25 Surface Binding Through Directed Evolution

Nanotechnology is as much about interfaces as materials, maybe more.[137] In just about any system, interfaces between materials have the greatest free energy, and are most prone to failure, be it mechanical, electrical or chemical. The surface area of a material divided into roughly spherical particles scales with the inverse square of the particle diameter, giving very small particles a huge high-energy surface. Therefore, when working at the nanoscale it is imperative to have a very good understanding and control of surface properties. For instance, one can make nanometer dimension metal clusters, but these are only stable if protected with a monolayer whose one end adheres to the metal and whose other end is miscible in the surrounding medium (see Chapter 6). However, it is not always so easy to find the perfect agent for surface modification. Often these molecules have poor selectivity (modify more than the intended surface) and poor stability. This makes them adequate for some applications, but increasingly problematic as the complexity of nanosystems increases. Theoretical methods do exist to simulate binding sequences for surfaces, however, these require unduly computing time and cost.[138]

A fascinating and potentially universal method to surface modification uses biology to do the binding for us.[139] When a virus enters our body, we may initially get sick. However, after a little while our body starts to make antibodies against the virus. These antibodies are those which can very efficiently bind the virus, tagging them for disposal by our white blood cells. Even though there were no such antibodies against the virus before infection, the body generates millions or billions of diverse antibodies and amplifies those that happen to be virus binding.

This selection–amplification strategy can work *in vitro* by using a phage-display technique,[140] or cell-surface display technique.[141] In a previous example we showed how this technique, can result in virus capsids that bind streptavidin, a well-studied biological molecule. It is not a far stretch to imagine that a virus can adapt to binding a biomolecule, but it is quite surprising to see that it can also bind non-biological metals[142] and semiconductors.[143]

The first step of the process again starts with a large population of viruses, each displaying a distinct surface protein. These are then allowed to bind to a single-crystal substrate of a given material in order to isolate those individuals with binding properties. The non-binding viruses are washed off, and the binding ones amplified and carried through a few more rounds of selection–amplification. In the end are obtained a collection of virus particles that can bind strongly and selectively to the given material, and are even selective enough to bind only one family of crystal planes. In Figure 10.29 is shown one of the results of the binding experiment. The top panel shows a substrate with exposed GaAs lines, which have been treated with a rhodamine–streptavidin derivative that binds to the engineered virus capsids. The image on the right shows strong and selective fluorescence, and the left image shows weak background fluorescence when the virus is not added. The middle left panel shows an SEM of bound virus particles decorated with gold nanoparticles, and the right panel shows a schematic of viral binding. The bottom two panels show the general procedure for conjugating chemical species to the virus-decorated surface.

A very interesting use for these inorganic binding protein sequences displayed on virus surfaces is related to inorganic nanocluster growth. In Chapter 6 we saw that the growth of nanoparticles relies on having a ligand which binds strongly to the growing nanoparticle, stabilizing the particles in a solvent and preventing agglomeration. While it is easy to find a proper ligand for the growth of gold particles, such as thiols, a ligand for GaAs surfaces is much harder to find. A solution is to select a protein sequence, identified through the phage-display technique, which binds strongly to a material, and use this sequence as a capping ligand in nanocluster growth. Since these ligands are material as well as crystal face selective, the resulting nanoparticles can have a variety of shapes due to preferential growth on

Fluorescence images

Background fluorescence   **Virus added**

Au decorated viruses   **Viruses standing up**

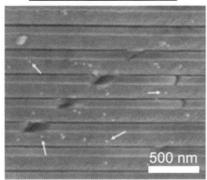

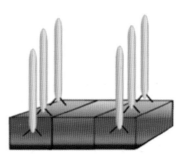

500 nm

Conjugating chemical species

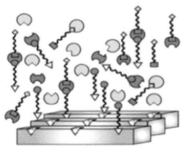

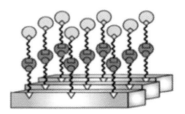

Figure 10.29 *Viruses can be engineered to bind inorganics such as metals and semiconductors by a process of directed evolution*
(Reproduced with permission from Ref. 143)

the uncapped crystal planes.[144] This technique can also be elaborated with RNA segments instead of proteins, with comparable selectivities for crystal growth.[145]

10.26 Nanowire Evolution

Another fascinating prospect for peptide sequences identified through evolutionary methods is the biologically mediated growth of nanowires.[146] To achieve this goal one must first identify a virus or cell with a wire-like profile, such as the filamentous M13 bacteriophage. This virus, like most viruses, consists of a highly ordered array of protein subunits, which self-assemble to give the final construct. These subunits are readily amenable to genetic manipulation, allowing one to self-assemble a mutant virus bearing inorganic-recognition peptide sequences on its outside coat. For instance, peptide sequences can be identified, which bind to the ZnS {100} crystal planes, and these expressed on the outside of the M13 bacteriophage. When this phage is placed in a solution of ZnS precursors, normally too dilute to nucleate particles, the peptides act as nucleation sites resulting in the growth of a homogenous coat of ZnS along the virus. This same process can be carried out for other semiconductors such as CdS, and magnetic materials like CoPt and FePt. The as-synthesized nanowires are polycrystalline, made up of interconnected grains with a rough texture, but having preferred crystalline orientations. Annealing at temperatures between 400 and 500 °C removes the organic constituents of the virus, and consolidates the grains into a single-crystal nanowire. Figure 10.30 shows TEM images, at different magnifications, of semiconductor nanowires assembled using this technique: panels A to D are ZnS wires, while E and F are CdS. The first ZnS image shows the wire before annealing, highlighting the grainy texture, while all the other images are single-crystals obtained after thermal annealing.

As we have seen in these examples, it is possible to use the diversity in biology to help solve challenging materials problems. With so many biological "starting materials", it is possible to screen a multitude of candidates for a particular physical property even if we do not know exactly how this property arises. Once this is achieved, the structural identification of the unit of interest can help us design more selective biological assays, as well as apply the biological design principles in purely synthetic systems.

10.27 Biomolecular Motors – Nanomachines Everywhere

Nanotechnology has many goals: smaller circuits, more efficient drug delivery, efficiency leaps in manufacturing. In fact, if nanotechnology attains but a fraction of its promise we are set to behold a revolution in almost all aspects of science and the world around us. Of all these possible applications the one that has most grasped the attention (and aversion) of the general public is the concept of a nanomachine. It is the dream of taking all the mechanical machines we use for building, fixing, moving, and shrinking all their parts until they are smaller than a

ZnS nanowires - before annealing

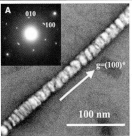

ZnS nanowires - after annealing

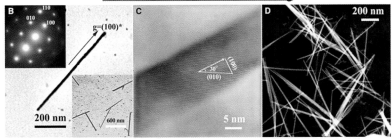

CdS nanowires - after annealing

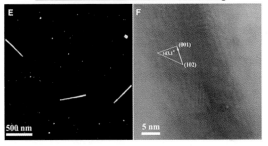

Figure 10.30 *Semiconductor nanowires formed by growth onto viruses coated with peptide recognition sequences identified by directed evolution*
(Reproduced with permission from Ref. 146)

cell, smaller even than the individual parts of the cell. The advantages of such machines and the number of them we could fit in a small space are obvious, but the thought of having impossible-to-see machines running amok around us and inside our bodies is one that makes many view nanotechnology with skepticism and worry.[147,148] Much talk has focused on nanotechnology doomsday scenarios: an army of nanorobots gradually disassembling the whole world and reassembling it into copies of themselves, the so-called "grey goo" scenario. While this may be a good topic for science fiction writers, such as Michael Crichton's book *Prey*, it is quite far from the truth. We are nowhere near making a nanomachine that can assemble arbitrary solids atom by atom, let alone pick up a single atom. In addition, an atom-by-atom assembly scheme would be hopelessly inefficient: we would likely benefit much more from a chemical self-assembly approach where a strict control of synthetic parameters can build up nanostructures in a highly parallel fashion.

Now that we have heard the truth, the real truth is that there are in fact nanomachines all around us. They are in the plants we grow, in our pets, and yes, they can be found in almost every cell in our bodies! We are not speaking of a secret government conspiracy to control our minds, but creations of the first and last nanotechnologist, Nature. Over millions of years of evolutionary refinement, Nature has crafted nanomachines capable of much more than our synthetic achievements, even viewed through the lens of our wildest dreams. We are discovering greater numbers of these biomechanical machines everyday, learning about their purpose and how they work. These biomolecular motors not only give us clues on how to design synthetic systems, but they can be directly used with or without modification in our own original constructions. In the following sections, we will attempt a glancing overview of different types of biological nanomotors and nanomachines, their structure–function relationships, and examples of how they have been creatively integrated into materials systems.

Biomolecular motors and machines are mechanical molecules, and truly are nanoscale versions of machines we know and use everyday.[149–153] They are primarily composed of proteins, but also nucleic acids and other organic molecules, and are quite possibly the prototypical example of programmed self-assembly. DNA contains the raw code, which is transcribed as a protein chain by the ribosome. The produced proteins then spontaneously fold into well-defined three-dimensional structures, and many of these folded structure associate with each other to form a functional multi-component system. These machines do not use electrical energy, but rather convert chemical energy stored in chemical bonds or gradients across membranes into mechanical energy. They are involved in a myriad of essential biological processes, and without them life as we know it would cease to exist. Transport of molecules, intracellular and cellular cargo, decontamination, sensing, actuation, synthesis, pumping, organism motility, viral infection, are but a few of the essential task these mechanical molecules accomplish.

10.28 How Biomotors Work

One of the main classes of biomolecular motors is linear motors whose function is summarized in Figure 10.31. These are molecules or molecular assemblies which move in a linear fashion along a track of some kind. Nature has devised a plethora of elegantly self-assembled systems, and there exist many types of "tracks" made of organic molecules as well as many different types of motors that move along them. The first type of linear motor is a processive motor, that is a motor which is constantly in contact with the track it moves along. Processive motors are exemplified by the kinesin protein super-family, which move along microtubules.[154–158] RNA polymerase, which synthesizes new RNA from a single strand RNA template,[159] and DNA helicase, which translates along and unwinds DNA in preparation for new DNA synthesis,[160] are also linear processive motors.

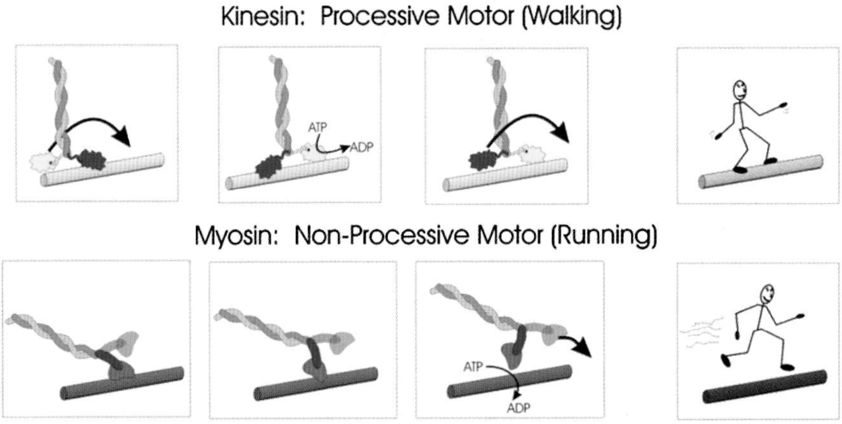

Kinesin: Processive Motor (Walking)

Myosin: Non-Processive Motor (Running)

Figure 10.31 *Linear biomolecular motors can move in either a processive or non-processive fashion*

In addition to processive motors, there are also non-processive motors, which detach from the track and subsequently re-attach, and therefore can be seen as hopping along the track instead of walking. Non-processive linear motors include myosin,[161–163] which binds to actin filaments and generates the contractile force in muscle tissue,[164] as well as the dynein protein family, which transports cargo along microtubules in the opposite direction to kinesin.[165]

One of the most common types of motors we encounter in everyday life is the rotary motor. A jet engine on aircraft is a marvel of modern engineering, which employs a rotary motor. However advanced, these behemoths pale in comparison to motors such as the bacterial flagella,[166] shown in Figure 10.32, where the rotation of a rigid, helical propeller is driven by a transmembrane chemical gradient. Flagella can spin at an incredible 100,000 rotations per second, and efficiently propel bacteria by a still poorly understood mechanism. Another well-studied bio-rotary motor is ATP synthase, also called ATPase,[167] which is an assembly of proteins anchored in the cell lipid bilayer. ATPase is responsible for the synthesis of ATP, the biological energy currency, and is powered by a proton gradient across the membrane created by nearby proton pumps. Every time an ATP molecule is synthesized the head of ATPase rotates by 120°, making this protein a very small and very efficient nanopropeller.

In addition to linear and rotary motors, Nature has devised many other types of mechanical molecules. Some of these perform tasks analogous to machines we know, some are based on principles we do not understand, and an unknown number of them are still lurking in the biological unknown waiting to be discovered.

Springs: One of the more elegant biological versions of the spring is the spasmoneme supra-molecular spring.[168] Upon binding and removal of calcium these filaments cause reversible contraction and extension, which the organism uses to protect itself, using the contractile avoidance reaction first described by Leeuwenhoek in 1676.

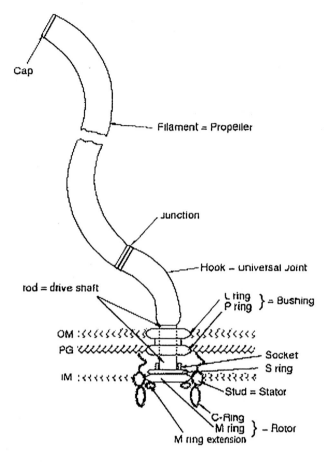

Figure 10.32 *Propeller-like structure of a bacterial flagellum*
(Reproduced with permission from Ref. 166)

Hinges: Some proteins, such as the maltose-binding protein, undergo hinge-like conformational changes upon binding a ligand. This principle has been used *in vitro* to construct a very sensitive maltose sensor.[169]

Spindles: Some viruses have devised ways of packaging their DNA into the viral capsid using a mechanism which is analogous to a spindle used for spinning yarn.

Electrostrictive materials: Prestin is a motor protein, which resides in the inner ear, whose shape responds to changes in electrical potential across membranes.[170,171] Prestin's electrostrictive mechanism is responsible for sound amplification and results in a 1000-fold enhancement of sound detection.

10.29 Kinesin – Walk Along

Nature has ever been and will continue to be a source of inspiration for materials chemistry. Many materials that are now becoming familiar to us are made by methods which seek to imitate biology. This is a strategy of "imitate the process and make it better". When we look at biomolecular motors, however, it is easy to get discouraged. These machines are so small and so complex that they are well beyond what is currently achievable by synthetic means. Now comes the other strategy: "If you can't beat 'em, join 'em!" That is, enough is known about biochemistry and protein modification that we can actually use Nature's self-assembled marvels directly in our own synthetic constructions. In this way, we can make biomolecular motors perform their natural tasks out of their natural environment, and we can even make them perform functions which they were not intended to perform.[172]

Conventional kinesin is a 400 kDa (in other words having a molecular mass of 400,000 g mol^{-1}) protein assembly whose total size is approximately 80 nm. It is composed of two larger (heavy) protein chains, which are involved in microtubule binding, mobility, ATP hydrolysis and protein dimerization, as well as two smaller (light) protein chains, which regulate heavy chain activity and binding to cargo (it is therefore a hetero-tetramer). Kinesin transports cargo along microtubules, self-assembled from monomeric proteins, using a "walking" motion with 8 nm steps.[173] Each of these steps is coupled to the hydrolysis of an ATP molecule, which provides the chemical energy for motion.[174] It moves with a high speed of about 1.8 μm s^{-1}, and can move against loads of 6 pN giving it a greater horsepower per unit weight than a jet engine. Kinesin is one of the most widely studied motor proteins, and can be easily modified by genetic engineering. Its rate of movement, binding site, binding affinity, total size, and so forth, can all be predictably modified resulting in finely tailored mechanical motion. Because it is so well studied, kinesin has been successfully incorporated into a variety of synthetic systems. This 80 nm large molecular motor is smaller and more efficient than any motor we can make, and therefore the use of this motor to move around artificial things could be the beginning of a new paradigm in bionanotechnology.

As mentioned, it has become fairly straightforward to chemically modify kinesin in essentially any way we would like. X-ray crystallography as well as other probes have been used on this motor at various stages in its movement,[175–179] allowing a detailed understanding of where chemical bonds should be made to attach kinesin to different surfaces. In this way, kinesin can be tethered to various substrates using well-known surface chemistry. In many cases, motors are found to retain biological activity after the surface-binding step allowing them to bind and move microtubules. However, while kinesin usually moves along fixed microtubules, bound motors function in the opposite way to glide free microtubules across a surface. Simply providing the system with ATP causes the motors to "turn on" and begin their walking motion. Such microtubule motion can even be guided along microlithographically defined tracks.[180,181]

In a similar fashion, microtubules may be bound to a substrate while retaining their structure and function.[182] Molecules or objects functionalized with kinesin can then be shuttled along these tracks by the addition of ATP.[183] It is possible to align microtubules, for instance by fluid flow, allowing the kinesin-powered cargo to move in a directional fashion. Other known linear motors can also be used to the same effect, such as the myosin–actin system,[184] which has similarly been used to perform biologically unconventional tasks *in vitro*.[185] Myosin is larger, and actin filaments are more flexible than microtubules, allowing freedom in the design of synthetic systems.

Many self-assembled systems with biochemical propulsion can be imagined using such biochemically powered trains. For illustration purposes, we will examine one such example that uses the motility generated by kinesin to stretch individual DNA molecules.[186] In this study the authors are using a strand of DNA from a phage virus, which is short and has a well defined fully stretched length of 16.5 μm. However, the force of one kinesin unit, 6 pN, is not enough on its own to compensate for the loss in entropy of a fully stretched DNA molecule. Therefore, a collection of kinesin molecules (1000 motors per square micrometer), were tethered to a substrate in order to get several of these units working in conjunction. Onto this surface were deposited microtubules, labeled with streptavidin, which bind to the exposed kinesin. These were then treated with DNA whose end had been modified with biotin. Due to the strong streptavidin–biotin complementary interactions, the microtubules became decorated with coiled DNA. When ATP was added to the system, the microtubules started gliding along the surface bearing DNA cargo. These were visualized by fluorescence microscopy using red and green fluorescent labeling of the microtubules and DNA, respectively. If the pH was lowered to 6, DNA tips were found to adhere to the substrate. With one end stuck to the substrate, and the other to a microtubule, DNA molecules gradually got stretched to their full length as the microtubules glided along the surface. In Figure 10.33 we can see the result of such an experiment where a DNA in a coil state (Green dot) gets stretched into a linear chain.

The production of aligned and stretched DNA is significant, since such templates can be used to generate nanoparticle chains,[187] conductive nanowires,[67] or even to spatially localize deposition onto the DNA using its nucleotide base code.[73]

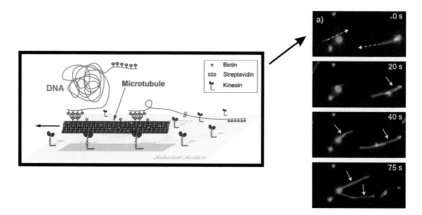

Figure 10.33   *Stretching of coiled DNA segments by microtubules gliding across a kinesin-decorated substrate*
(Reproduced with permission from Ref. 186)

10.30 ATPase – Biomotor Nanopropellers

As mentioned previously, ATPase belongs in the class of rotary biomolecular motors. This motor usually rotates in conjunction with the synthesis of ATP molecules for use in cell metabolism, with one of these molecules produced for every 120° right-handed turn. It is notable, however, that the motor can be operated in reverse by providing it with ATP. Instead of forming ATP, the enzyme hydrolyzes it to ADP with a concomitant 120° left-handed turn. ATPase is a difficult motor to study, since it is hard to separate it from a lipid bilayer without significant denaturation and structural collapse. Enough is known about it, however, that we can conduct well-defined coupling chemistry to different protein chains on this enzyme. In this way, *Escherichia coli* ATPase has been used to construct an inorganic–biological hybrid device bearing an ATP powered nickel propeller.[188]

The first part of this hybrid system is an engineered substrate designed to hold ATPase motors at precisely defined locations. To achieve this, a silicon substrate was coated with photoresist and 50–120 nm wide holes were made by electron beam lithography. Evaporation of nickel followed by photoresist dissolution leads to an array of nickel posts with a width of the original holes and a height of about 200 nm.

The next step was the production of engineered propellers. These were made similarly to the nickel posts, by first drawing the propeller shape into photoresist with an electron beam, developing, evaporating nickel, and finally the release of these features into solution. The resulting nickel propellers had a diameter of about 150 nm and a length of about 1000 nm, and were subsequently coated with streptavidin.

The final step was the production of mutant ATPase, made by modifying *E. coli* DNA so that the produced protein had histidine (an amino acid with a cyclic diamine residue) tags (10 consecutive histidines) on the bottom part of the molecule.[189–191] Histidine units are well-known metal ligands having a particular affinity for nickel, causing these ATPase units to bind strongly to the engineered Ni posts. The ATPase binds such that the rotating head is facing upwards, and this subunit can subsequently be tagged with biotin. When the biotin-coated ATPases are exposed to the streptavidin-coated propellers, the two parts stick strongly together. Feeding ATP to this system activates the ATPases whose γ–subunits begin to rotate, pulling its nickel cargo with it in concerted rotation. A convincing series of optical micrographs, Figure 10.34, clearly shows the nickel propeller rotating in a counterclockwise manner when ATP is provided to the system. This example demonstrates in a striking fashion how biological constructions can be used far out of biological context, and how they can allow us to reach goals never before possible.

Biomolecular motors are of great interest to nanoscientists, either as models for synthetic systems or sources of mobility with unmatched power density. As we learn more and more about these fascinating bioconstructions, and increasingly apply them to nanotechnological tasks,[192–197] we will see whether or not they are capable of truly revolutionary nano-advances. In Appendix G some very recent developments on chemically powered nanomotors are presented.

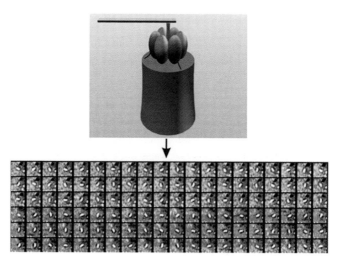

Figure 10.34 *A schematic of a hybrid organic–inorganic propeller powered by the ATPase enzyme, and consecutive frames of a video image showing the synthetic propeller fed ATP as an energy source*
(Reproduced with permission from Ref. 188)

10.31 (Bio)Inspiration

Are we now (bio)inspired by what we have read in this chapter? What did we learn from the above examples of materials synthesis with an underlying biomimetic theme? Clearly organics did organize inorganics into impressive nanocomposites with amazing structures, patterns and shapes, and over multiple length scales but would this knowledge have helped us to reach a synthetic target more efficiently than we would have without that knowledge? How much inspiration is really under the cover of this bio-umbrella? Is it science fiction or reality? So these are some of the tough questions being asked these days about a biomimetic approach to materials. It is certainly interesting to observe analogies between nanocomposite materials that emerge from biomineralization and biomimicry. The (bio)inspiration comes from recognizing the connections between the way biology assembles nanocomposite materials with a particular function and end use, and the way materials chemistry can do similar things to create nanocomposite materials with structure, properties and functions designed for the needs of a materials society. Often times scientists will claim that they have a biomimetic structure when what comes out of their reaction flasks bears a superficial resemblance to biological structures. The truth is that Nature does not intend for its composites to look a particular way: these structures are a means to an end and a result of the organism's necessity to survive. Nature worked over evolutionary time scales to create our soft and hard biomaterials, and like Pieter Harting, we have only taken the first step and there is still much to do!

10.32 References

1. B.L. Smith, T.E. Schaffer, M. Viani, J.B. Thompson, N.A. Frederick, J. Kindt, A. Belcher, G.D. Stucky, D.E. Morse and P.K. Hansma, Molecular mechanistic origin of the toughness of natural adhesives, fibres and composites, *Nature*, 1999, **399**, 761.
2. L. Addadi and S. Weiner, Control and design principles in biological mineralization, *Angew. Chem. Int. Ed.*, 1992, **31**, 153.
3. S.I. Stupp and P.V. Braun, Molecular manipulation of microstructures: biomaterials, ceramics, and semiconductors, *Science*, 1997, **277**, 1242.
4. P. Ball, *The Self-Made Tapestry: Pattern Formation in Nature*, Oxford University Press, Oxford, 2001.
5. S.A. Wainwright, *Mechanical Design in Organisms*, Princeton University Press, Princeton, 1982.
6. S. Vogel, *Cats' Paws And Catapults*, W.W. Norton & Company, New York, 1998.
7. D. Kaplan, W.W. Adams, B. Farmer and C. Viney, Silk – biology, structure, properties, and genetics, *Silk Polym.; ACS Symp. Ser.*, 1994, **544**, 2.
8. J.M. Gosline, P.A. Guerette, C.S. Ortlepp and K.N. Savage, The mechanical design of spider silks: from fibroin sequence to mechanical function, *J. Exp. Biol.*, 1999, **202**, 3295.
9. J.D. Pye, Animal sonar in air, *Ultrasonics*, 1968, **6**, 32.
10. G. Mayer and M. Sarikaya, Rigid biological composite materials: structural examples for biomimetic design, *Exp. Mech.*, 2002, **42**, 395.
11. National Materials Advisory Board, *Hierarchical Structures in Biology as a Guide to New Materials Technology*, National Academy Press, Washington, 1994.
12. R. Lakes, Materials with structural hierarchy, *Nature*, 1993, **361**, 511.
13. D.D.Y. Ryu and D.H. Nam, Recent progress in biomolecular engineering, *Biotechnol. Prog.*, 2000, **16**, 2.
14. S. Mann and P. Calvert, Synthesis and biological composites formed by *in-situ* precipitation, *J. Mater. Sci.*, 1988, **23**, 3801.
15. M. Sarikaya and I.A. Aksay, *Biomimetics: Design and Processing of Materials*, American Institute of Physics, College Park, MD, 1993.
16. S. Mann, *Biomimetic Materials Chemistry*, VCH, New York, 1996.
17. P. Ball, Life's lessons in design, *Nature*, 2001, **409**, 413.
18. M. Sarikaya, Biomimetics: materials fabrication through biology, *Proc. Natl Acad. Sci. USA*, 1999, **96**, 14183.
19. N.C. Seeman and A.M. Belcher. Emulating biology: building nanostructures from the bottom up, *Proc. Natl Acad. Sci. USA*, 2002, **99**, 6452.
20. A.H. Heuer, D.J. Fink, V.J. Laraia, J.L. Arias, P.D. Calvert, K. Kendall, G.L. Messing, J. Blackwell, P.C. Rieke, D.H. Thompson, A.P. Wheeler, A. Veis and A.I. Caplan, Innovative materials processing strategies – a biomimetic approach, *Science*, 1992, **255**, 1098.
21. I. Willner, Biomaterials for sensors, fuel cells, and circuitry, *Science*, 2002, **298**, 2407.

22. S.L. Cady, J.D. Farmer, J.P. Grotzinger, J.W. Schopf and A. Steele, Morphological biosignatures and the search for life on Mars, *Astrobiology*, 2003, **3**, 351.

23. D.W. Thompson, *On Growth and Form*, Cambridge Press, Cambridge, 1942.

24. P. Ball, Spheres of influence, *New Scientist*, 1995, **148**, 42.

25. N. Kroger, R. Deutzman and M. Sumper, Polycationic peptides from diatom biosilica that direct silica nanosphere formation, *Science*, 1999, **286**, 1129.

26. M. Sumper, A phase separation model for the nanopatterning of diatom biosilica, *Science*, 2002, **295**, 2430.

27. H.A. Lowenstam, Lepidocrocite an apatite mineral and magnetite in teeth of chitons (polyplacophora), *Science*, 1967, **156**, 1373.

28. S. Oliver, G.A. Ozin and L.A. Ozin, Skeletons in the cupboard – rediscovery in science, *Adv. Mater.*, 1995, **7**, 948.

29. G.A. Ozin and S. Oliver, Skeletons in the beaker – synthetic hierarchical inorganic materials, *Adv. Mater.*, 1995, **7**, 943.

30. G.A. Ozin, Bones about skeletons, *Adv. Mater.*, 1996, **8**, 184.

31. S. Mann, D.D. Archibald, J.M. Didymus, T. Douglas, B.R. Heywood, F.C. Meldrum and N.J. Reeves, Crystallization at inorganic–organic interfaces – biominerals and biomimetic synthesis, *Science*, 1993, **261**, 1286.

32. J.D. Birchall, A.J. Howard and K. Kendall, A cement spring, *J. Mater. Sci. Lett.*, 1982, **1**, 125.

33. J.D. Birchall, A.J. Howard and K. Kendall, New cements – inorganic plastics of the future, *Chem. Brit.*, 1982, **18**, 860.

34. X.D. Li, W.C. Chang, Y.J. Chao, R.Z. Wang and M. Chang, Nanoscale structural and mechanical characterization of a natural nanocomposite material: the shell of red abalone, *Nano Lett.*, 2004, **4**, 613.

35. X.Y. Shen, A.M. Belcher, P.K. Hansma, G.D. Stucky and D.E. Morse, Molecular cloning and characterization of lustrin A, a matrix protein from shell and pearl nacre of *Haliotis rufescens*, *J. Biol. Chem.*, 1997, **272**, 32472.

36. J.D. Currey, Mechanical properties of mother of pearl in tension, *Proc. R. Soc. London Ser. B – Biol. Sci.* 1977, **196**, 443.

37. A.P. Jackson, J.F.V. Vincent and R.M. Turner, The mechanical design of nacre, *Proc. R. Soc. London Ser. B – Biol. Sci.* 1988, **234**, 415.

38. G. Falini, S. Albeck, S. Weiner and L. Addadi, Control of aragonite or calcite polymorphism by mollusk shell macromolecules, *Science*, 1996, **271**, 67.

39. A.M. Belcher, X.H. Wu, R.J. Christensen, P.K. Hansma, G.D. Stucky and D.E. Morse, Control of crystal phase switching and orientation by soluble mollusc-shell proteins, *Nature*, 1996, **381**, 56.

40. M. Fritz, A.M. Belcher, M. Radmacher, D.A. Walters, P.K. Hansma, G.D. Stucky, D.E. Morse and S. Mann, Flat pearls from biofabrication of organized composites on inorganic substrates, *Nature*, 1994, **371**, 49.

41. J.N. Cha, G.D. Stucky, D.E. Morse and T.J. Deming, Biomimetic synthesis of ordered silica structures mediated by block copolypeptides, *Nature*, 2000, **403**, 289.

42. W.J. Clegg, K. Kendall, N.M. Alford, T.W. Button and J.D. Birchall, A simple way to make tough ceramics, *Nature*, 1990, **347**, 455.

43. N.D. Chasteen, Ferritin. Uptake, storage, and release of iron, *Met. Ions Biol. Syst.* 1998, **35**, 479.

44. A.K. Powell. Ferritin. Its mineralization, *Met. Ions Biol. Syst.* 1998, **35**, 515.

45. T. Matsunaga and T. Sakaguchi, Molecular mechanism of magnet formation in bacteria, *J. Biosci. Bioeng.*, 2000, **90**, 1.

46. D. Schuler and R.B. Frankel, Bacterial magnetosomes: microbiology, biomineralization and biotechnological applications, *Appl. Microbiol. Biotechnol.*, 1999, **52**, 464.

47. C.T. Dameron, R.N. Reese, R.K. Mehra, A.R. Kortan, P.J. Carroll, M.L. Steigerwald, L.E. Brus and D.R. Winge, Biosynthesis of cadmium sulfide quantum semiconductor crystallites, *Nature*, 1989, **338**, 596.

48. T. Douglas and M. Young, Host–guest encapsulation of materials by assembled virus protein cages, *Nature*, 1998, **393**, 152.

49. W. Shenton, T. Douglas, M. Young, G. Stubbs and S. Mann, Inorganic–organic nanotube composites from template mineralization of tobacco mosaic virus, *Adv. Mater.*, 1999, **11**, 253.

50. T. Douglas and M. Young, Virus particles as templates for materials synthesis, *Adv. Mater.*, 1999, **11**, 679.

51. T. Douglas, E. Strable, D. Willits, A. Aitouchen, M. Libera and M. Young, Protein engineering of a viral cage for constrained nanomaterials synthesis, *Adv. Mater.*, 2002, **14**, 415.

52. P.S. Arora and K. Kirshenbaum, Nano-tailoring: stitching alterations on viral coats, *Chem. Biol.*, 2004, **11**, 418.

53. A.S. Blum, C.M. Soto, C.D. Wilson, J.D. Cole, M. Kim, B. Gnade, A. Chatterji, W.F. Ochoa, T. Lin, J.E. Johnson and B.R. Ratna, Cowpea mosaic virus as a scaffold for 3-D patterning of gold nanoparticles, *Nano Lett.*, 2004, **4**, 867.

54. C.A. Mirkin, R.L. Letsinger, R.C. Mucic and J.J. Storhoff, A DNA-based method for rationally assembling nanoparticles into macroscopic materials, *Nature*, 1996, **382**, 607.

55. J.J. Storhoff, R. Elghanian, R.C. Mucic, C.A. Mirkin and R.L. Letsinger, One-pot colorimetric differentiation of polynucleotides with single base imperfections using gold nanoparticle probes, *J. Am. Chem. Soc.*, 1998, **120**, 1959.

56. J.J. Storhoff, A.A. Lazarides, R.C. Mucic, C.A. Mirkin, R.L. Letsinger and G.C. Schatz, What controls the optical properties of DNA-linked gold nanoparticle assemblies? *J. Am. Chem. Soc.*, 2000, **122**, 4640.

57. A.P. Alivisatos, K.P. Johnsson, X.G. Peng, T.E. Wilson, C.J. Loweth, M.P. Bruchez and P.G. Schultz, Organization of 'nanocrystal molecules' using DNA, *Nature*, 1996, **382**, 609.

58. R.C. Jin, G.S. Wu, Z. Li, C.A. Mirkin and G.C. Schatz, What controls the melting properties of DNA-linked gold nanoparticle assemblies? *J. Am. Chem. Soc.*, 2003, **125**, 1643.

59. S.J. Park, A.A. Lazarides, C.A. Mirkin, P.W. Brazis, C.R. Kannewurf and R.L. Letsinger, The electrical properties of gold nanoparticle assemblies linked by DNA, *Angew. Chem. Int. Ed.*, 2000, **39**, 3845.

60. C.J. Loweth, W.B. Caldwell, X.G. Peng, A.P. Alivisatos and P.G. Schultz. DNA-based assembly of gold nanocrystals, *Angew. Chem. Int. Ed.*, 1999, **38**, 1808.

61. Y.W.C. Cao, R.C. Jin and C.A. Mirkin, Nanoparticles with Raman spectroscopic fingerprints for DNA and RNA detection, *Science*, 2002, **297**, 1536.

62. S.J. Park, T.A. Taton and C.A. Mirkin, Array-based electrical detection of DNA with nanoparticle probes, *Science*, 2002, **295**, 1503.

63. R. Elghanian, J.J. Storhoff, R.C. Mucic, R.L. Letsinger and C.A. Mirkin, Selective colorimetric detection of polynucleotides based on the distance-dependent optical properties of gold nanoparticles, *Science*, 1997, **277**, 1078.

64. G.P. Mitchell, C.A. Mirkin and R.L. Letsinger, Programmed assembly of DNA functionalized quantum dots, *J. Am. Chem. Soc.*, 1999, **121**, 8122.

65. S.O. Kelley and J.K. Barton, Electron transfer between bases in double helical DNA, *Science*, 1999, **283**, 375.

66. D. Porath, G. Cuniberti and R. Di Felice, Charge transport in DNA-based devices, in "Long-range Charge Transfer in DNA II", *Top. Curr. Chem.*, 2004, **237**, 183.

67. E. Braun, Y. Eichen, U. Sivan and G. Ben-Yoseph, DNA-templated assembly and electrode attachment of a conducting silver wire, *Nature*, 1998, **391**, 775.

68. J.T. Petty, J. Zheng, N.V. Hud and R.M. Dickson, DNA-templated Ag nanocluster formation, *J. Am. Chem. Soc.*, 2004, **126**, 5207.

69. J. Richter, R. Seidel, R. Kirsch, M. Mertig, W. Pompe, J. Plaschke and H.K. Schackert, Nanoscale palladium metallization of DNA, *Adv. Mater.*, 2000, **12**, 507.

70. W.E. Ford, O. Harnack, A. Yasuda and J.M. Wessels, Platinated DNA as precursors to templated chains of metal nanoparticles, *Adv. Mater.*, 2001, **13**, 1793.

71. Y.F. Ma, J.M. Zhang, G.J. Zhang and H.X. He, Polyaniline nanowires on Si surfaces fabricated with DNA templates, *J. Am. Chem. Soc.*, 2004, **126**, 7097.

72. J. Richter, Metallization of DNA, *Phys. A – Low Dimens. Syst. Nanostruct.*, 2003, **16**, 157.

73. K. Keren, M. Krueger, R. Gilad, G. Ben-Yoseph, U. Sivan and E. Braun. Sequence-specific molecular lithography on single DNA molecules, *Science*, 2002, **297**, 72.

74. K. Keren, R.S. Berman, E. Buchstab, U. Sivan and E. Braun, DNA-templated carbon nanotube field-effect transistor, *Science*, 2003, **302**, 1380.

75. K. Keren and E. Braun, Sequence-specific molecular lithography towards DNA-templated electronics, *Chem. Eng. Technol.*, 2004, **27**, 447.

76. K. Keren, R.S. Berman and E. Braun, Patterned DNA metallization by sequence-specific localization of a reducing agent, *Nano Lett.*, 2004, **4**, 323.

77. X.D. Chen, S.G. Jiang and M.H. Liu, Progress in DNA-based molecular assemblies, *Prog. Chem.*, 2003, **15**, 367.

78. J.J. Storhoff and C.A. Mirkin, Programmed materials synthesis with DNA, *Chem. Rev.*, 1999, **99**, 1849.

79. N.C. Seeman, DNA in a material world, *Nature*, 2003, **421**, 427.

80. J. Chen and N.C. Seeman, The synthesis from DNA of a molecule with the connectivity of a cube, *Nature*, 1991, **350**, 631.

81. N.C. Seeman, Nucleic acid nanostructures and topology, *Angew. Chem. Int. Ed.*, 1998, **37**, 3220.
82. X. Li, X. Yang, J. Qi and N.C. Seeman, Antiparallel DNA double crossover molecules as components for nanoconstruction, *J. Am. Chem. Soc.*, 1996, **118**, 6131.
83. E. Winfree, F.R. Liu, L.A. Wenzler and N.C. Seeman, Design and self-assembly of two-dimensional DNA crystals, *Nature*, 1998, **394**, 539.
84. H. Yan, S.H. Park, G. Finkelstein, J.H. Reif and T.H. LaBean, DNA-templated self-assembly of protein arrays and highly conductive nanowires, *Science*, 2003, **301**, 1882.
85. C.M. Niemeyer and M. Adler, Nanomechanical devices based on DNA, *Angew. Chem. Int. Ed.*, 2002, **41**, 3779.
86. C. Mao, W. Sun, Z. Shen and N.C. Seeman, A DNA nanomechanical device based on the B–Z transition, *Nature*, 1999, **397**, 144.
87. B. Yurke, A.J. Turberfield, A.P. Mills, F.C. Simmel and J.L. Neumann. A DNA-fuelled molecular machine made of DNA, *Nature*, 2000, **406**, 605.
88. W.B. Sherman and N.C. Seeman, A precisely controlled DNA biped walking device, *Nano Lett.*, 2004, **4**, 1203.
89. H. Yan, X. Zhang, Z. Shen and N.C. Seeman, A robust DNA mechanical device controlled by hybridization topology, *Nature*, 2002, **415**, 62.
90. S.A. Davis, S.L. Burkett, N.H. Mendelson and S. Mann, Bacterial templating of ordered macrostructures in silica and silica-surfactant mesophases, *Nature*, 1997, **385**, 420.
91. S.A. Davis, H.M. Patel, E.L. Mayes, N.H. Mendelson, G. Franco and S. Mann, Brittle bacteria: a biomimetic approach to the formation of fibrous composite materials, *Chem. Mater.*, 1998, **10**, 2516.
92. B.J. Zhang, S.A. Davis, N.H. Mendelson and S. Mann, Bacterial templating of zeolite fibres with hierarchical structure, *Chem. Commun.*, 2000, **9**, 781.
93. J.J. Thwaites and N.H. Mendelson, Biomechanics of bacterial walls – studies of bacterial thread made from *Bacillus subtilis*, *Proc. Natl Acad. Sci. USA*, 1985, **82**, 2163.
94. R. Massart, Preparation of aqueous magnetic liquids in alkaline and acidic media, *IEEE Trans. Magn.*, 1981, **17**, 1247.
95. S.A. Davis, S.L. Burkett, N.H. Mendelson and S. Mann, Bacterial templating of ordered macrostructures in silica and silica-surfactant mesophases, *Nature*, 1997, **385**, 420.
96. S. Chandrasekhar, *Liquid Crystals*, Cambridge University Press, Cambridge, 1992.
97. L.S. Li and A.P. Alivisatos, Semiconductor nanorod liquid crystals and their assembly on a substrate, *Adv. Mater.*, 2003, **15**, 408.
98. Z. Dogic and S. Fraden, Smectic phase in a colloidal suspension of semiflexible virus particles, *Phys. Rev. Lett.*, 1997, **78**, 2417.
99. M. Adams, Z. Dogic, S.L. Keller and S. Fraden, Entropically driven microphase transitions in mixtures of colloidal rods and spheres, *Nature*, 1998, **393**, 349.
100. N.R. Jana, Shape effect in nanoparticle self-assembly, *Angew. Chem. Int. Ed.*, 2004, **43**, 1536.

101. S.-W. Lee, C. Mao, C.E. Flynn and A.M. Belcher, Ordering of quantum dots using genetically engineered viruses, *Science*, 2002, **296**, 892.

102. S.-W. Lee, S.K. Lee and A.M. Belcher, Virus-based alignment of inorganic, organic, and biological nanosized materials, *Adv. Mater.*, 2003, **15**, 689.

103. C. Mao, C.E. Flynn, A. Hayhurst, R. Sweeney, J. Qi, G. Georgiou, B. Iverson and A.M. Belcher, Viral assembly of oriented quantum dot nanowires, *Proc. Natl Acad. Sci. USA*, 2003, **100**, 6946.

104. L. Weissbuch, L. Addadi, M. Lahav and I. Leiserowitz, Molecular recognition at crystal interfaces, *Science*, 1991, **253**, 637.

105. E.L. Mayes, F. Vollrath and S. Mann, Fabrication of magnetic spider silk and other silk-fiber composites using inorganic nanoparticles, *Adv. Mater.*, 1998, **10**, 801.

106. A.H. Simmons, C.A. Michal and L.W. Jelinski, Molecular orientation and two-component nature of the crystalline fraction of spider dragline silk, *Science*, 1996, **271**, 84.

107. W. Shenton, D. Pum, U.B. Sleytr and S. Mann, Synthesis of cadmium sulphide superlattices using self-assembled bacterial S-layers, *Nature*, 1997, **389**, 585.

108. E.S. Gyorvary, A. O'Riordan, A.J. Quinn, G. Redmond, D. Pum and U.B. Sleytr, Biomimetic nanostructure fabrication: nonlithographic lateral patterning and self-assembly of functional bacterial S-layers at silicon supports, *Nano Lett.*, 2003, **3**, 315.

109. T.J. Beveridge, Bacterial S-layers, *Curr. Opin. Struct. Biol.*, 1994, **4**, 204.

110. U.B. Sleytr, P. Messner, D. Pum and M. Sara, Crystalline bacterial cell surface layers (S layers): from supramolecular cell structure to biomimetics and nanotechnology, *Angew. Chem. Int. Ed.*, 1999, **38**, 1035.

111. K. Douglas, G. Devaud and N.A. Clark, Transfer of biologically derived nanometer-scale patterns to smooth substrates, *Science*, 1992, **257**, 642.

112. D. Pum and U.B. Sleytr, The application of bacterial S-layers in molecular nanotechnology, *Trends Biotechnol.*, 1999, **17**, 8.

113. D. Pum, A. Neubauer, E. Gyorvary, M. Sara and U.B. Sleytr, S-layer proteins as basic building blocks in a biomolecular construction kit, *Nanotechnology*, 2000, **11**, 100.

114. P. Ringler and G.E. Schulz, Self-assembly of proteins into designed networks, *Science*, 2003, **302**, 106.

115. J.R. Young, S.A. Davis, P.R. Bown and S. Mann, Coccolith ultrastructure and biomineralisation, *J. Struct. Biol.*, 1999, **126**, 195.

116. D. Walsh, J.D. Hopwood and S. Mann, Crystal tectonics – construction of reticulated calcium phosphate frameworks in bicontinuous reverse microemulsions, *Science*, 1994, **264**, 1576.

117. D. Walsh and S. Mann, Chemical synthesis of microskeletal calcium phosphate in bicontinuous microemulsions, *Chem. Mater.*, 1996, **8**, 1944.

118. S.D. Sims, D. Walsh and S. Mann, Morphosynthesis of macroporous silica frameworks in bicontinuous microemulsions, *Adv. Mater.*, 1998, **10**, 151.

119. D. Walsh and S. Mann, Fabrication of hollow porous shells of calcium carbonate from self-organizing media, *Nature*, 1995, **377**, 320.

120. E. Paasche, Biology and physiology of coccolithophorids, *Annu. Rev. Microbiol.*, 1968, **22**, 71.

121. E. Paasche, A review of the coccolithophorid *Emiliania huxleyi* (Prymnesio-phyceae), with particular reference to growth, coccolith formation, and calcification–photosynthesis interactions, *Phycologia*, 2001, **40**, 503.

122. Y.-H. Ha, R.A. Vaia, W.F. Lynn, J.P. Costantino, J. Shin, A.B. Smith, P.T. Matsudaira and E.L. Thomas, Three-dimensional network photonic crystals via cyclic size reduction/infiltration of sea urchin exoskeleton, *Adv. Mater.*, 2004, **16**, 1091.

123. S. Oliver, A. Kuperman, N. Coombs, A. Lough and G.A. Ozin, Lamellar aluminophosphates with surface patterns that mimic diatom and radiolarian microskeletons, *Nature*, 1995, **378**, 47.

124. G.A. Ozin and S. Oliver, Skeletons in the beaker – synthetic hierarchical inorganic materials, *Adv. Mater.*, 1995, **7**, 943.

125. S. Oliver, N. Coombs and G.A. Ozin. Synthetic hollow aluminophosphate microspheres, *Adv. Mater.*, 1995, **7**, 931.

126. S. Mann and G.A. Ozin. Synthesis of inorganic materials with complex form, *Nature*, 1996, **382**, 313.

127. G.A. Ozin, Morphogenesis of biomineral and morphosynthesis of biomimetic forms, *Acc. Chem. Res.*, 1997, **30**, 17.

128. S.R.J. Oliver and G.A. Ozin, Phosphate liquid crystals: novel supramolecular template for the synthesis of lamellar aluminophosphates with natural form, *J. Mater. Chem.*, 1998, **8**, 1081.

129. M. Cehreli, S. Sahin and K. Akca, Role of mechanical environment and implant design on bone tissue differentiation: current knowledge and future contexts, *J. Dent.*, 2004, **32**, 123.

130. N. Al-Saffar and P.A. Revell, Pathology of the bone–implant interfaces, *J. Long-Term Eff. Med. Implants*, 1999, **9**, 319.

131. K.A. Thomas, Hydroxyapatite coatings, *Orthopedics*, 1994, **17**, 267.

132. Y.H. An and R.A. Draughn, *Mechanical Testing of Bone and the Bone–Implant Interface*, 1st edn, CRC Press, Boca Raton, FL, 1999.

133. I. Soten and G.A. Ozin, Porous hydroxyapatite-dodecylphosphate composite film on titania–titanium substrate, *J. Mater. Chem.*, 1999, **9**, 703.

134. J.D. Hartgerink, E. Beniash and S.I. Stupp, Self-assembly and mineralization of peptide-amphiphile nanofibers, *Science*, 2001, **294**, 1684.

135. P. Ball, Does nature know best? *Nat. Mater.*, 2003, **2**, 510.

136. P. Ball, Natural strategies for the molecular engineer, *Nanotechnology*, 2002, **13**, R15.

137. M. Sarikaya, C. Tamerler, A.K.-Y. Jen, K. Schulten and F. Baneyx, Molecular biomimetics: nanotechnology through biology, *Nat. Mater.*, 2003, **2**, 577–585.

138. G. Schneider and P. Wrede, Artificial neural networks for computer-based molecular design, *Biophys. Mol. Biol.*, 1998, **70**, 175.

139. L. Giver and F.H. Arnold, Combinatorial protein design by *in vitro* recombination, *Curr. Opin. Chem. Biol.*, 1998, **2**, 335.

140. G.P. Smith, Filamentous fusion phage: novel expression vectors that display cloned antigens on the virion surface, *Science*, 1985, **228**, 1315; R.H. Hoess, Protein design and phage display, *Chem. Rev.*, 2001, **101**, 3205.

141. K.D. Wittrup, Protein engineering by cell surface display, *Curr. Opin. Biotechnol.*, 2001, **12**, 395.

142. S. Brown, Metal recognition by repeating polypeptides, *Nat. Biotechnol.*, 1997, **15**, 269.

143. S.R. Whaley, D.S. English, E.L. Hu, P.F. Barbara and A.M. Belcher, Selection of peptides with semiconductor binding specificity for directed nanocrystal assembly, *Nature*, 2000, **405**, 665.

144. R.R. Naik, S.J. Stringer, G. Agarwal, S.E. Jones and M.O. Stone, Biomimetic synthesis and patterning of silver nanoparticles, *Nat. Mater.*, 2002, **1**, 169.

145. L.A. Gugliotti, D.L. Feldheim and B.E. Eaton, RNA-mediated metal–metal bond formation in the synthesis of hexagonal palladium nanoparticles, *Science*, 2004, **304**, 850.

146. C.B. Mao, D.J. Solis, B.D. Reiss, S.T. Kottmann, R.Y. Sweeney, A. Hayhurst, G. Georgiou, B. Iverson and A.M. Belcher, Virus-based toolkit for the directed synthesis of magnetic and semiconducting nanowires, *Science*, 2004, **303**, 213.

147. E. Dexler, Mightier machines from tiny atoms may someday grow – molecular engineering, *Smithsonian*, 1982, **13**, 145.

148. E. Regis, Interview: Eric Drexler, *Omni*, 1989, **11**, 66.

149. J. Howard, *Mechanics of Motor Proteins and the Cytoskeleton*, Sinauer Associates, Sunderland, MA, 2001.

150. L.S.B. Goldstein, Molecular motors: from one motor many tails to one motor many tales, *Trends Cell Biol.*, 2001, **11**, 477.

151. C.D. Montemagno, Nanomachines: a roadmap for realizing the vision, *J. Nanopart. Res.*, 2001, **3**, 1.

152. R.D. Vale and R.A. Milligan, The way things move: looking under the hood of molecular motor proteins, *Science*, 2000, **288**, 88.

153. J.J. Schmidt and C.D. Montemagno, Using machines in cells, *Drug Discov. Today*, 2002, **7**, 500.

154. S.M. Block, Kinesin: what gives? *Cell*, 1998, **93**, 5.

155. G. Woehlke, A look into kinesin's powerhouse, *FEBS Lett.*, 2001, **508**, 291.

156. N. Hirokawa, Kinesin and dynein superfamily proteins and the mechanism of organelle transport, *Science*, 1998, **279**, 519.

157. L.S.B. Goldstein, Kinesin molecular motors: transport pathways, receptors, and human disease, *Proc. Natl Acad. Sci.*, 2001, **98**, 6999.

158. M. Schliwa and G. Woehlke, Switching on kinesin, *Nature*, 2001, **411**, 424.

159. J. Gelles and R. Landick, RNA polymerase as a molecular motor, *Cell*, 1998, **93**, 13.

160. T.M. Lohman, K. Thorn and R.D. Vale, Staying on track: common features of DNA helicase and microtubule motors, *Cell*, 1998, **93**, 9.

161. K. Kitamura, M. Tokunaga, A.H. Iwane and T. Yanagida, A single myosin head moves along an actin filament with regular steps of 5.3 nanometres, *Nature*, 1999, **397**, 129–134.

162. A. Yildiz, J.N. Forkey, S.A. McKinney, T. Ha, Y.E. Goldman and P.R. Selvin, Myosin V walks hand-over-hand: single fluorophore imaging with 1.5-nm localization, *Science*, 2003, **300**, 2061.

163. Y.E. Goldman, Wag the tail: structural dynamics of actomyosin, *Cell*, 1998, **93**, 1.

164. M. Irving, V. Lombardi, G. Piazzesi and M.A. Ferenczi, Myosin head movements are synchronous with the elementary force-generating process in muscle, *Nature*, 1992, **357**, 156.

165. H.C. Taylor and E.J. Holwill, Axonemal dynein – a natural molecular motor, *Nanotechnology*, 1999, **10**, 237.

166. D.J. DeRosier, The turn of the screw: the bacterial flagellar motor, *Cell*, 1998, **93**, 17.

167. K. Kinosita, R. Yasuda, H. Noji, S. Ishiwata and M. Yoshida, F-1-ATPase: a rotary motor made of a single molecule, *Cell*, 1998, **93**, 21.

168. L. Mahadevan and P. Matsudaira, Motility powered by supramolecular springs and ratchets, *Science*, 2000, **288**, 95–99.

169. D.E. Benson, D.W. Conrad, R.M. de Lorimier, S.A. Trammell and H.W. Hellinga, Design of bioelectronic interfaces by exploiting hinge-bending motions in proteins, *Science*, 2001, **293**, 1641.

170. J. Zheng, W. Shen, D.Z.Z. He, K.B. Long, L.D. Madison and P. Dallos, Prestin is the motor protein of cochlear outer hair cells, *Nature*, 2000, **405**, 149.

171. M.C. Liberman, J. Gao, D.Z.Z. He, X. Wu, S. Jia and J. Zuo, Prestin is required for electromotility of the outer hair cell and for the cochlear amplifier, *Nature*, 2002, **419**, 300.

172. H. Hess, G.D. Bachand and V. Vogel, Powering nanodevices with biomolecular motors, *Chem. Eur. J.*, 2004, **10**, 2110.

173. K. Svoboda, C.F. Schmidt, B.J. Schnapp and S.M. Block, Direct observation of kinesin stepping by optical trapping interferometry, *Nature*, 1993, **365**, 721.

174. D.L. Coy, M. Wagenbach and J. Howard, Kinesin takes one 8 nm step for each ATP that it hydrolyzes, *J. Biol. Chem.*, 1999, **274**, 3667.

175. M. Kikkawa, E.P. Sablin, Y. Okada, H. Yajima, R.J. Fletterick and N. Hirokawa, Switch-based mechanism of kinesin motors, *Nature*, 2001, **411**, 439.

176. A. Marx, M. Thormahlen, J. Muller, S. Sack, E.M. Mandelkow and E. Mandelkow, Conformations of kinesin: solution vs. crystal structures and interactions with microtubules, *Eur. Biophys. J. Biophys. Lett.*, 1998, **27**, 455.

177. F.J. Kull, E.P. Sablin, R. Lau, R.J. Fletterick and R.D. Vale, Crystal structure of the kinesin motor domain reveals a structural similarity to myosin, *Nature*, 1996, **380**, 550.

178. E.P. Sablin, F.J. Kull, R. Cooke, R.D. Vale and R.J. Fletterick, Crystal structure of the motor domain of the kinesin-related motor ncd, *Nature*, 1996, **380**, 555.

179. R.D. Vale, T. Funatsu, D.W. Pierce, L. Romberg, Y. Harada and T. Yanagida, Direct observation of single kinesin molecules moving along microtubules, *Nature*, 1996, **380**, 451.

180. Y. Hiratsuka, T. Tada, K. Oiwa, T. Kanayama and T.Q.P. Uyeda, Controlling the direction of kinesin-driven microtubule movements along microlithographic tracks, *Biophys. J.*, 2001, **81**, 1555.

181. S.G. Moorjani, L. Jia, T.N. Jackson and W.O. Hancock, Lithographically patterned channels spatially segregate kinesin motor activity and effectively guide microtubule movements, *Nano Lett.*, 2003, **3**, 633.

182. D.C. Turner, C.Y. Chang, K. Fang, S.L. Brandow and D.B. Murphy, Selective adhesion of functional microtubules to patterned silane surfaces, *Biophys. J.*, 1995, **69**, 2782.

183. L. Limberis and R.J. Stewart, Toward kinesin-powered microdevices, *Nanotechnology*, 2000, **11**, 47.

184. Y. Harada, A. Noguchi, A. Kishino, T. Yanagida, Sliding movement of single actin-filaments on one-headed myosin-filaments, *Nature*, 1987, **326**, 805.

185. H. Suzuki, A. Yamada, K. Oiwa, H. Nakayama and S. Mashiko, Control of actin moving trajectory by patterned poly(methyl methacrylate) tracks, *Biophys. J.*, 1997, **72**, 1997.

186. S. Diez, C. Reuther, C. Dinu, R. Seidel, M. Mertig, W. Pompe and J. Howard, Stretching and transporting DNA molecules using motor proteins, *Nano Lett.*, 2003, **3**, 1251.

187. A.P. Alivisatos, K.P. Johnsson, X.G. Peng, T.E. Wilson, C.J. Loweth, M.P. Bruchez and P.G. Schultz, Organization of 'nanocrystal molecules' using DNA, *Nature*, 1996, **382**, 609.

188. R.K. Soong, G.D. Bachand, H.P. Neves, A.G. Olkhovets, H.G. Craighead and C.D. Montemagno, Powering an inorganic nanodevice with a biomolecular motor, *Science*, 2000, **290**, 1555.

189. H. Noji, R. Yasuda, M. Yoshida and K. Kinosita, Direct observation of the rotation of F-1-ATPase, *Nature*, 1997, **386**, 299.

190. R. Yasuda, H. Noji, M. Yoshida, K. Kinosita and H. Itoh, Resolution of distinct rotational substeps by submillisecond kinetic analysis of F-1-ATPase, *Nature*, 2001, **410**, 898.

191. C. Montemagno and G. Bachand, Constructing nanomechanical devices powered by biomolecular motors, *Nanotechnology*, 1999, **10**, 225.

192. D.V. Nicolau, H. Suzuki, S. Mashiko, T. Taguchi and S. Yoshikawa, Actin motion on microlithographically functionalized myosin surfaces and tracks, *Biophys. J.*, 1999, **77**, 1126.

193. H. Hess, J. Clemmens, D. Qin, J. Howard and V. Vogel, Light-controlled molecular shuttles made from motor proteins carrying cargo on engineered surfaces, *Nano Lett.*, 2001, **1**, 235.

194. H. Hess, J. Clemmens, J. Howard and V. Vogel, Surface imaging by self-propelled nanoscale probes, *Nano Lett.*, 2002, **2**, 113.

195. H. Hess, J. Howard and V. Vogel, A piconewton forcemeter assembled from microtubules and kinesins, *Nano Lett.*, 2002, **2**, 1113.

196. K.J. Bohm, R. Stracke, P. Muhlig and E. Unger, Motor protein-driven unidirectional transport of micrometer-sized cargoes across isopolar micro-tubule arrays, *Nanotechnology*, 2001, **12**, 238.

197. C. Niemeyer and C.A. Mirkin, *Nanobiotechnology: Concepts, Applications and Perspectives*, Wiley-VCH, New York, 2003.

Nanofood for Thought – Organic Matrix, Biomineralization, Biomimetics, Bioinspiration

1. Describe how technology transfer from biomineralizing systems in Nature might enable biomimetic synthesis of new materials that could function as an improved performance (a) optical fiber, (b) solar concentrator, (c) data storage medium, (d) military armor, (e) waste water heavy metal sponge, (f) abrasion resistant film, (g) bone implant, (h) bacterial or viral filter and (i) magnetic cloth.

2. Nature has learned over evolutionary time scales how and when to use color derived from structure rather than pigmentation. Expand upon this statement and where the knowledge could prove useful in today's materials world.

3. Nature is parsimonious in her choice of inorganic and organic building blocks for making incredible biomineral constructions like bone, teeth and shells. What developments are still needed for the chemist to be able to match Nature's skills in the materials laboratory?

4. What kinds of biological molecules can be used to form nanoparticles with unconventional shape, size or size distribution?

5. List different types of hierarchical biological systems that can be "turned into stone" by self-assembly.

6. How would you distinguish a synthetic fossil, or *faux fossil*, from the real thing?

7. Describe various ways of making tougher materials by self-assembly, using inspiration from mechanical structures in biology.

8. Describe possible ways in which biomolecular motors could be integrated into synthetic systems in order to power the motility of microfabricated gears.

9. Imagine how one could use a rotary biomolecular motor such as ATPase to generate a nanoscale fluorescent light source attached exclusively to the tip of a rotating propeller. Why would such a structure be useful?

10. The biological function of the regular nanoscale filigree patterns that decorate the silicified microskeletons of the single cell diatom, to this day remain an enigma for biologists and a delight for artists – in the context of nanoscience and nanotechnology can you think of ways in which these biomaterials could be orchestrated to provide a useful function?

11. How and why would you want to make a biophotonic crystal made of a protein or DNA?

12. Biosilicified spicules of the marine sponge support the soft tissue of the organism in ocean currents and act as a protective coat to keep predators at bay, however, recent research has suggested they also have an optical function as waveguides.

Look into this suggestion and see if there is anything unusual about optical fibers that have evolved biologically rather than abiologically.

13. Devise a chip-based method to assay for 10 different unique strands of DNA, with precise readouts as fluorescence in the visible light range.

14. Devise a biomimetic way to synthesize and pattern mother of pearl (nacre) and think of some property–function connections that suggest a particular use for what you have made.

15. Wood is a biomaterial with remarkable mechanical properties. Using synthetic approaches, how could you improve the mechanical properties of natural wood? Can you think of a way to make a totally synthetic wood analogue with similar or superior properties?

Large Building Blocks

CHAPTER 11

Self-Assembly of Large Building Blocks

"When you look at yourself from a universal standpoint, something inside always reminds or informs you that there are bigger and better things to worry about."
Albert Einstein (1879–1955)

11.1 Self-assembling Supra-micron Shapes

By viewing self-assembly as a map of forces acting over multiple length scales, it is possible to imagine a wide range of interactions that could auto-construct building blocks larger than a micron into functional architectures.[1,2] This way of looking at what forces drive objects of different size, shape and surface functionality to self-assemble into functional architectures was introduced in the opening chapter of this book and amplified upon in subsequent chapters in the context of objects with dimensions in the range of nanometers to microns.

In this chapter, we have larger thoughts in mind: self-assembling supra-micron-shaped objects. Instead of organizing the smallest possible atomic-scale building blocks into molecules, polymers and solids through covalent, ionic, hydrogen-bonding and metal–ligand chemical bonding interactions, here we will explore recent experiments, which seek to utilize predominantly physical forces operating between building blocks larger than a micron, to create new structures. These include electric, magnetic, colloidal and capillary forces, not usually considered particularly useful in the chemical approach to manipulating matter. However, because of their strength, selectivity and long-range interactions, they are proving to be a powerful means for organizing and assembling large building blocks in a fluid phase, at interfaces and in confined spaces.

The assembly of large building blocks is not as abstract as it may initially seem.[3] Countless common phenomena are based on self-assembly, such as the crystallization of molecular materials, growth of epitaxial layers in microchip manufacture and the folding of proteins after their manufacture in living cells. The use of large units, which mimic the behavior of atoms and molecules in these systems, can allow us to test various assembly models with easily detectable and rigorously defined responses.[4–8]

11.2 Synthesis Using the "Capillary Bond"

A series of studies have focused on capillary interactions between large building blocks, where the driving force for self-assembly is the minimization of interfacial free energy.[9–13] Capillary forces between surfaces can be tailored through chemical control of their hydrophobic and hydrophilic properties. It is possible to design the surfaces of large building blocks to cause them to either attract or repel when floating at a fluid–fluid or fluid–air interface, or suspended in an isodense fluid. These concepts are illustrated in Figure 11.1.

When objects float at an interface, denoted by the dotted lines in the figure, they will sink into this interface to a degree depending on their density. In doing so, they induce a surface deformation which increases the free energy of the system. If two of these shapes get close to one another, and both have a similar effect on the interface, they will attract one another thereby reducing the area of the interface. If they have dissimilar effects on the interface, the interaction will be repulsive. The magnitude of the attractive or repulsive force depends, in turn, on the degree of interfacial deformation.[14] When objects are suspended in an isodense liquid, particular faces of these objects can be wet by another immiscible liquid if the surface properties are properly chosen. Surfaces coated with this immiscible liquid

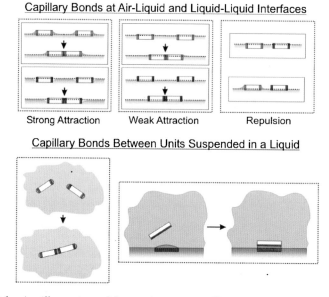

Figure 11.1 *An illustration of how micron- or millimeter-scale building blocks can assemble at air–liquid and liquid–liquid interfaces, as well as in solution, through capillary forces*

can stick together through capillary bonds, or can stick to surface patterns which have been similarly functionalized.

All these principles are based on the minimization of interfacial energy by reducing the area of a particular interface. Because these interactions usually do not result in irreversible sticking, the building blocks are free to explore the energy landscape and arrive to an assembly which is at thermodynamic equilibrium. Therefore, large building blocks can be coerced to self-assemble into predetermined geometries driven by capillary interactions by simply controlling their shape and the properties of specific surfaces.

Capillary interactions can also be tailored to operate between large building blocks and substrates, planar or curved, that have been patterned with hydrophobic and hydrophilic domains to create functional constructs. The use of patterns of low melting solder on specific surfaces of building blocks provides a large interfacial free energy to guide self-assembly and electrical interconnection to form electrically functional constructs.[15,16]

11.3 Crystallizing Large Polyhedral-Shaped Building Blocks

Capillary-driven self-assembly has provided a means of "crystallizing" millimeter-sized building blocks into regular, 3D arrays with open structures.[17–19] Polyhedral building blocks are made from molded polyurethane and particular faces are covered with copper tape wet by a film of low-melting solder. The polyhedra are shaken around in an isodense potassium bromide solution heated above the solder melting point to cause self-assembly into equilibrium structures. Capillary forces between molten solder surfaces on the polyhedral construction units cause them to self-assemble into a regular open structure, polyhedral crystals, the structure of which is controlled by the particular polyhedral shape and location of solder-coated surfaces (Figure 11.2).[19]

11.4 Self-Assembling 2D and 3D Electrical Circuits and Devices

Capillary-guided self-assembly of large building blocks has been expanded to include polyurethane polyhedra with surfaces patterned with solder dots, wires and light emitting diodes (LEDs) and self-assembled to electrically functional 3D networks.[15] The design of dots and wires determined the network structure of parallel and serial networks that formed. A similar approach has been used for the self-assembly of large building blocks on a patterned substrate.[20] The experiment involved capillary self-assembly of semiconductor cube-shaped building blocks onto a solder patterned substrate.

To make a functional device, a number of steps were needed to organize and interconnect the cubes and align a top electrode with the 2D array of cubes to make them individually electrically addressable. A dramatic illustration of the power of this approach for making functional electrical devices is shown in Figure 11.3,

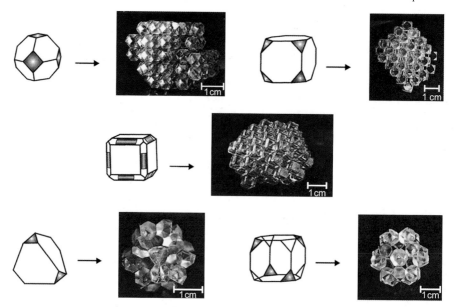

Figure 11.2 *Crystallization of patterned polyurethane polyhedra. The dark patches on the schematic diagrams represent the surfaces wet by solder and determine the connectivity of the self-assembled array*
(Reproduced with permission from Ref. 19)

where GaAlAs cube-shaped LEDs were organized into a prototype cylindrical display.

The auto-constructed display is comprised of an interleaved, fully addressable array of 113 LEDs constructed through hierarchical self-assembly steps.

11.5 Crystallizing Micron-Sized Planar Building Blocks

The examples highlighted above have mainly concerned the assembly of millimeter-scale units, problematic because their fabrication was serial and time-consuming. A similar approach has been used to self-assemble metal plates with hexagonal shapes made by photolithography into regular 3D arrays.[21] The procedure for making the shapes and assembling them is shown in Figure 11.4. The example concerns 10 μm Cr(OH)–Au(CH$_3$)–Cr(OH) plates where this notation describes the terminal functional groups on the bottom, side and top surfaces of the plates, respectively.

The process to make gold plates begins with electrodeposition of Au through a patterned photoresist on an electrode made of Si/SiO$_2$/Cr/Ag/Cr/Au. Next, evaporation of Cr and dissolution of the photoresist in acetone and of the Ag film with nitric acid served to release the plates as a sheet from the underlying Si substrate. Ultrasound treatment of the plates in ethanol and subsequent filtration removed residual thin films of Cr that held the plates together. The plates were then treated in an ethanolic solution of HS(CH$_2$)$_{15}$CH$_3$ to form a hydrophobic SAM on the Au leaving the Cr/Cr$_2$O$_4$ faces hydrophilic. The plates with their selectively

Figure 11.3 *A fully addressable cylindrical display made by capillary driven patterned assembly. The figures shows the activation of LEDs in various configurations on this display*
(Reproduced with permission from Ref. 20)

functionalized top, side and bottom surfaces are now primed for self-assembly. This involved dispersing approximately 10^5 plates in ethanol in a glass cuvette, plus 30 μL of either a photo- or thermally curable adhesive that served to lubricate the plates during self-assembly and join them together permanently afterwards. Next, water is slowly added to dilute the ethanol, which caused the adhesive to precipitate selectively onto the hydrophobic faces of the plates. The axial rotation of the cuvette provided the agitation that allowed self-assembly of the plates to

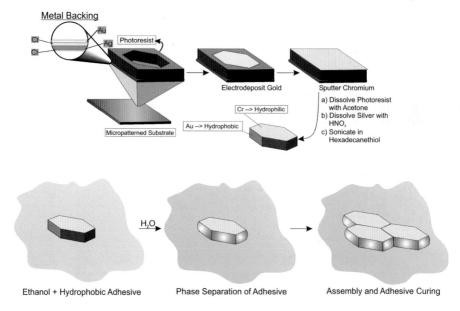

Figure 11.4   *Process for making and assembling micron-scale hexagonal-shaped plates into ordered 3D arrays*

proceed to the equilibrium structure. Finally, the adhesive is cured by ultraviolet (UV) or thermal means to create mechanically stable arrays. This type of shape-selective self-assembly driven by hydrophobic and hydrophilic forces between selectively functionalized surfaces of hexagonal-shaped $Cr(OH)$–$Au(CH_3)$–$Cr(OH)$ plates has been expanded to include hexagons with holes, hexagon-shaped stars, elongated hexagons and indented hexagons. Some examples of multilayered ordered arrays formed by these large building blocks are seen in Figure 11.5.

From this work, it was established that the shape of the resulting array is governed by the geometry and pattern of edge and face functionalization of the hexagonal plates. By manipulating the lateral area and thickness of the hexagonal plates as well as their functionality, with variations such as $Au(CH_3)$–$Au(CH_3)$–$Au(CH_3)$, $Cr(OH)$–$Au(PO_3H_2)$–$Cr(OH)$, $Pt(CH_3)$–$Au(CO_2H)$–$Pt(CH_3)$, it proved possible to manipulate the shape assembly process to favor different geometry arrays as seen in Figure 11.6. These hexagonal-shaped metal plates can be guided by capillary forces to self-assemble on the surface of a drop of perfluorodecalin in water to create a synthetic geodesic dome,[18] or can be assembled in shape-complementary templates.[22]

11.6 Polyhedra with Patterned Faces that Autoconstruct

It can be appreciated from the above examples that making a large population of desired polyhedral shapes with predetermined surface functionality and getting

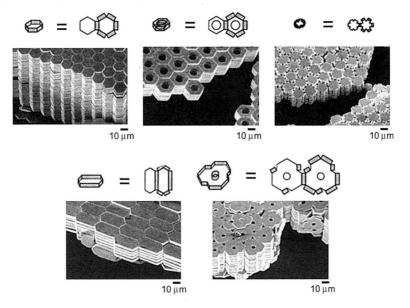

Figure 11.5 *Shape self-assembly of different geometries of hexagonal plates* (Reproduced with permission from Ref. 21)

them to spontaneously assemble into a particular geometry require a creative fusion of top-down microfabrication with bottom-up self-assembly methodologies. An interesting idea is to find a way of getting polyhedral shapes with selectively patterned faces to build themselves in the first place, making that step easier.[23] A clever way of doing this is illustrated in Figure 11.7.

The process begins with the deposition of a sacrificial silica layer on a silicon layer. The required pattern is defined by a combination of photolithography and an evaporated Cr/Au metallic layer. The first layer of photoresist is dissolved and next a seed layer of Cr–Au is evaporated to facilitate electrodeposition, and patterning of a second layer of photoresist is performed with the boundary of each face of the polyhedron in registry with the already present pattern. A Ni–Cu electrode is then deposited to build each face followed by solder deposition by dip coating. Dissolution of the photoresist, the metallization layers and the sacrificial layer releases the desired structure from the substrate. The final step is a capillary-driven self-construction of the released 2D shape into the desired 3D polyhedral shape with patterned faces. This is achieved by heating the 2D structure above the solder melting point, driving it to fold into a 3D polyhedron by minimizing the surface area of the molten solder.

Examples of this auto-construction strategy for making 200 μm dimension polyhedra are shown in Figure 11.8. This series of SEM images clearly show how the shape and geometry of the patterned faces can lead to a rational folding into a desired 3D shape. While this highlighted study is a proof-of-concept experiment, it could prove a great asset in next generation, auto-constructing, MEMS devices.[24]

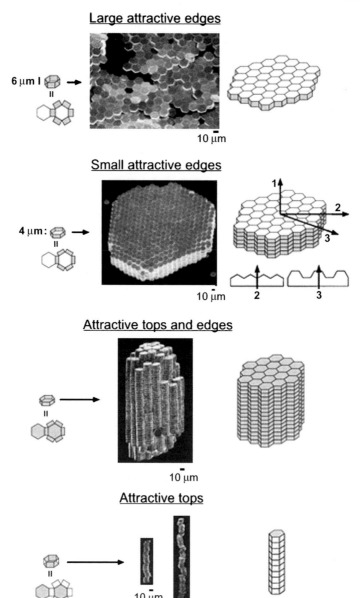

10 μm

Figure 11.6 *Chemical control of capillary-driven self-assembly of hexagonal-shaped plates through selective edge and face (top) functionalization to give arrays with different geometries*
(Reproduced with permission from Ref. 21)

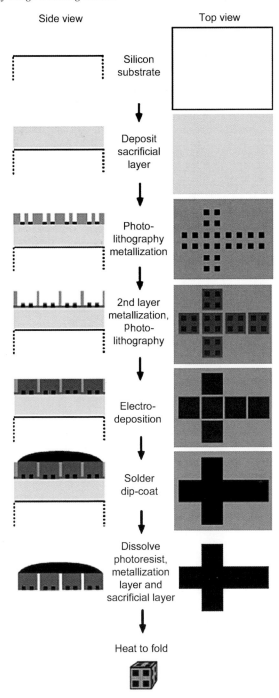

Figure 11.7 *Fabrication of polyhedra with patterned faces that build themselves through capillary-driven auto-folding and self-construction of a 2D structure* (Reproduced with permission from Ref. 23)

Metallic ➜ Solder covered ➜ Folded
2D Precursors   2D Precursors   Polyhedra

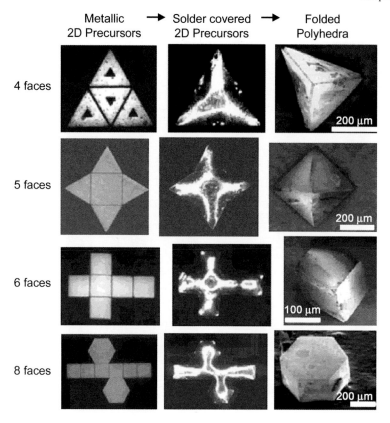

4 faces

5 faces

6 faces

8 faces

Figure 11.8   *Auto-folding of 2D patterned shapes into 3D patterned polyhedra*
(Reproduced with permission from Ref. 23)

11.7   Large Sphere Building Block Self-Assemble into 3D Crystals

A strategy has been formulated for generating regular 3D arrays of large millimeter dimension spherical building blocks by using confined crystallization in columnar wells.[25] The orientation and arrangement of the spheres in the columns can be predetermined, by tailoring the sphere diameter with respect to the column cross-sectional size and shape. The process is shown schematically in Figure 11.9. Columns are then filled with a polymer to form rod encapsulated sphere arrays, which upon release can then be self-assembled using capillary interactions between solder-covered sphere columns to form 3D sphere lattices. This methodology has provided a useful means for making high symmetry sphere lattices (cubic, hexagonal, tetragonal or orthorhombic) with lattice dimensions that could find utility as microwave photonic crystal devices. While it will be challenging to precisely adapt this method to micron diameter spherical building blocks for visible wavelength photonic crystal applications, it does operationally bear a resemblance

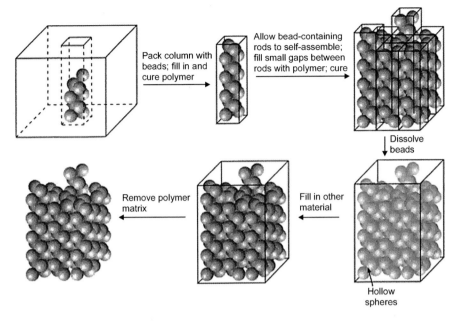

Figure 11.9 *Confined assembly of sphere lattices*
(Reproduced with permission from Ref. 25)

to the growth of light scale colloidal crystals by confined self-assembly of micron-scale microspheres within surface relief patterns on chips[26] as described in Chapter 7.

11.8 Synthetic MEMS?

The self-assembly of sub- and supra-micron-shaped objects like those described in this brief introduction to the subject and other places in this text inspires the notion of a synthetic approach to microelectromechanical systems, known as MEMS. While the fabrication and integration of 3D topologically complex objects into microsensor, microfluidic, microsynthesis and microdelivery systems is an emerging field of research, the new-found ability to synthesize and arrange similar size and shape objects into a synthetic MEMS has only just become a possibility. One can imagine a future in which synthetic micron-scale gears and screws auto-construct into tiny machines driven through on-board miniature battery, fuel or solar cell power sources – see Appendix G.

11.9 Magnetic Self-Assembly

If we can do electrostatic self-assembly can we do it magnetically? An elegant recent demonstration shows that indeed this is a feasible strategy for achieving programmable assembly of different kinds of colloidal dimension microspheres.[27] This begins with patterned cobalt micromagnets on a substrate made by standard lithography of an e-beam deposited, 100 nm thick cobalt film. Next, an array of square SU-8 microwells are positioned over the poles of the micromagnets with a well size and external field magnetization direction selected to trap, through designed attractive and repulsive magnetic forces, just a single kind of colloidal microsphere in predefined spatial locations on a substrate.

To amplify upon the magnetic self-assembly strategy, the direction and magnitude of the magnetization of the micromagnets were driven to saturation by a permanent magnet and checked by magnetic force microscopy (MFM). Then dispersions of super-paramagnetic fluorescent microspheres, red and green color coded with biotin–streptavidin Texas Red and fluorescein isocyanate conjugates, were exposed to the magnetized micromagnets with external magnetic field bias, respectively, up then down. The resulting magnetic assembly was effectively probed by fluorescence microscopy.

The first stage in magnetic self-assembly involves organizing one set of magnetic microspheres into rows, say the red ones, with field bias upwards. That this is an achievable objective is dramatically demonstrated by the images in Figure 11.10. Reversing the field bias and introducing the green microspheres cause them to selectively self-assemble in the adjacent microwells as seen in the figure. Especially noteworthy is the achievement of zero cross-contamination of red and green microspheres.

All of this unequivocally demonstrates the power of magnetic self-assembly for organizing magnetic building blocks into nanoscale patterns that can be targeted for particular applications, where electric or optical fields, high temperature or UV processing, harmful solvents and molecular and ionic solutes could have a deleterious effect on the outcome of, for example, patterning of delicate species such as proteins, oligonucleotides and cells. A particular asset of this system is that magnetic information can be stored in a power-free mode of operation.

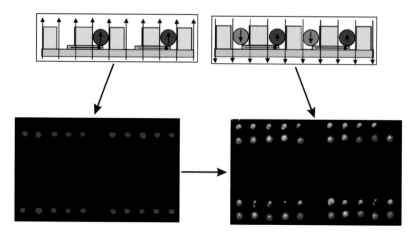

Figure 11.10 *Red and green magnetic beads can be selectively assembled into pits covering magnetic micropatterns*
(Reproduced with permission from Ref. 27)

11.10 Dynamic Self-Assembly

Most examples of self-assembly reported in the literature are static. Building blocks with predetermined "sticky" and "slippery" spots come together through a variety of metastable states to form the structure which resides at thermodynamic equilibrium. If the building blocks have an element of symmetry, most often the equilibrium structure is ordered with a symmetry reflecting that of the individual components. What may be more interesting than such static assembly is the so-called dynamic assembly. The difference here is that components can only achieve or maintain an ordered structure while dissipating energy. A prime example of this can be seen in living cells, containing a variety of self-assembled structures which only stay together as long as the cell is alive and consuming energy. If the intricate chain of chemical reactions driven by the release of energy stored in chemical bonds is broken, the cell dies and its self-assembled structures fall apart.

Biological systems are perhaps the most complicated example of dynamic self-assembly, of which most aspects are relatively unknown. Dynamic assembly in synthetic systems, on the other hand, is much more rudimentary but conversely much easier to understand. One of the most elegant examples of dynamic self-assembly to appear in the literature involves magnetic disks spinning at an air–water interface.[28] Millimeter-scale disks containing magnetic particles can easily be made of various polymers such as PDMS, and the density of these disks adjusted with fillers such that they float on water or other solvents. In the absence of a magnetic field, the disks freely float and form disordered arrays, since there is no driving force for self-assembly. If a collection of these floating disks is placed above a rotating magnetic field produced by a magnetic stir plate, the disks start to rotate with an angular frequency equal to the applied field.

On one hand, the disks are attracted to the axis of rotation of the magnet, but as they spin they also repel each other due to hydrodynamic shear between the fluid flow lines produced by neighboring disks. The balance of attractive and repulsive forces, which are necessary requirements for self-assembly, causes the formation of ordered structures. The fluid flow caused by disk spinning can be directly visualized by placing a drop of dye solution at the interface and watching how it distributes over time (Figure 11.11, top). Two examples of ordered aggregates formed at the interface of an ethylene glycol–water mixture using a rotation frequency of 1100 rpm are shown in the bottom part of the figure. Several studies have since been performed investigating this assembly process in detail.[29–31]

An even greater range of attractive and repulsive forces can be induced, by giving the floating magnetic disks an asymmetric shape.[32] To illustrate this, epoxy shapes doped with magnetite particles were fabricated, consisting of a central circular disk with an arm jutting tangentially from it. Such a shape can be made in two distinct forms, which are mirror images of each other. These two "isomers" are named R and S with regards to the spinning direction of the external magnetic field: the R isomer has an arm pointing in the rotational direction of the field, while the S isomer has opposite symmetry. The behavior of these plates is illustrated in Figure 11.12. At a low spin rate (low energy input) a mixture of R and S isomers forms disordered aggregates, highlighting the fact that such dynamic systems need a sufficient

Vortex visualization

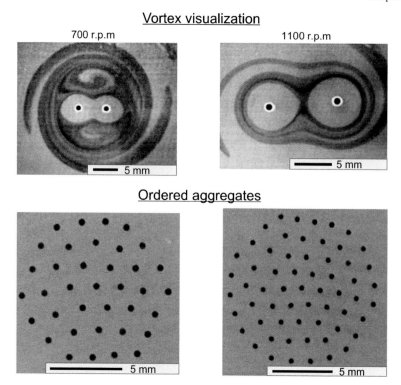

Figure 11.11 *Self-assembly of magnetic disks rotating at the air–water interface* (Reproduced with permission from Ref. 28)

constant energy influx to induce assembly. If the spin rate is increased above ~200 rpm, interesting behavior is observed. As the R isomers spin, their arms impede their rotation by causing considerable hydrodynamic drag. Thus, it is energetically favorable for two R isomers to dimerize through their arms and spin together, minimizing the drag. However, R isomers will not dimerize with S isomers since they are not complementary and such an aggregate does not minimize the free energy of the system. The arms of the S isomers, on the other hand, do not impede fluid flow and there is no advantage for these shapes to dimerize with others. Thus, when spinning, S isomers are repulsive to both R and S shapes. Finally, when the rotational frequency is increased to <450 rpm, the fluid flow in the vortexes around each shape is strong enough for all combinations to be repulsive. In this case, a mixture of R and S isomers forms ordered assemblies due to the balance of an attraction to the magnet rotation axis and mutual vortex repulsions.

11.11 Autonomous Self-Assembly

By and large, synthetic systems intended for self-assembly are built by rationally designing the components such that they form a particular intended structure. If we are seeking to emulate Nature in our constructions, such a rational approach may

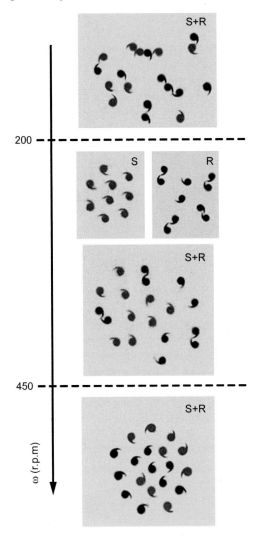

Figure 11.12 *Self-assembly of chiral magnetic disks rotating at the air–water interface. We can see the effect of chirality as well as the evolution of aggregation behavior as the magnet rotation rate is increased (top to bottom)* (Reproduced with permission from Ref. 32)

not be the most beneficial one. Nature, having no consciousness to speak of, does not plan for its constructions to achieve a certain goal. The interaction of a multitude of autonomously moving components ultimately results in systems much more complex, and much more ordered, than we could ever imagine. Consequently, the study of synthetic components having autonomous movement capabilities and the observation of their interaction may shed clues onto the emergence of complexity from simple systems,[33] a trait characteristic of living organisms.

One of the first examples directed towards this goal has been demonstrated using shapes floating on an air–water interface.[34] These shapes were made asymmetric, and bear hydrophilic and hydrophobic regions such that they can aggregate due to capillary interactions. In order to endow these shapes with movement, a small coating of platinum was applied to the end of these shapes. Such a shape is shown schematically in the top of Figure 11.13, while a diagram of the attractive capillary forces is shown on the bottom part of this figure. Platinum is a very good catalyst for the decomposition of hydrogen peroxide, which breaks down to liquid water and gaseous oxygen $(2H_2O_2 \rightarrow 2H_2O + O_2)$. When floated onto an aqueous solution of hydrogen peroxide, the evolution of oxygen from the platinum turns these metallic coatings into miniature motors. When allowed to motor themselves around this aqueous interface, the shapes move forward almost directionally until they bump into another shape with a complementary hydrophobic surface. The shapes then form dimers with two motors. If the two shapes are the same isomer, the dimer formed has motors at opposite ends and the aggregate starts making loops on the interface. If, however, the shapes are mirror images of each other, the two motors in the dimer are facing the same direction and it thus begins to move in a straight line. This behavior is illustrated with series of optical photographs shown in Figure 11.14, where A shows a heterodimer while B shows a homodimer.

Although simple, this example shows that unexpected behavior can occur when simple shapes are allowed to randomly explore space. Such a deceivingly simple system, where a homodimer "swims" in circles and the heterodimer "swims" away, has a tantalizing analogy in biological evolution. Were these organisms, the homodimer would stay in place and quickly deplete its food supply, giving the heterodimer an evolutionary advantage. Further studies in this area, and increasing the complexity of the components, may well uncover systems with emerging order and patterns at hierarchical length scales. While we may never achieve synthetic life, we may yet learn about how our own came to be.

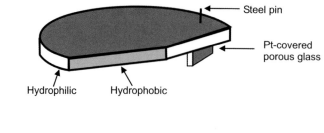

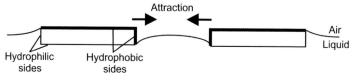

Figure 11.13 *The construction of a self-propelled chiral floater, with attractive and repulsive interactions*
(Reproduced with permission from Ref. 34)

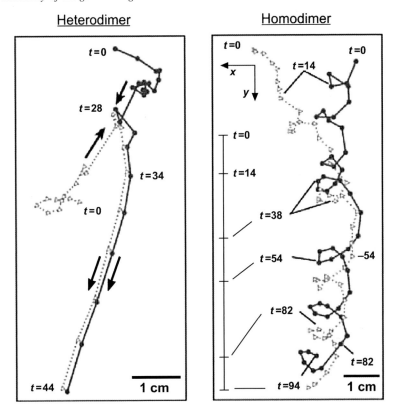

Figure 11.14 *Autonomous movement and self-assembly of self-propelled chiral floaters* (Reproduced with permission from Ref. 34)

11.12 Self-Assembly and Synthetic Life

Studies of both natural and synthetic systems are shedding clues to the rules of self-assembly where multiple forces and components are involved. While natural systems produce a multitude of self-assembled structures by poorly understood means, synthetic systems produce a few structures whose assembly can be rigorously described by physical theory. What may be most interesting of all is the combination of the two. Studies of the interaction of synthetic systems with dynamic living systems would increase both our understanding of the complexity of biological processes, and the diversity achievable with well-understood synthetic units. One of the ultimate goals of self-assembly is to create systems with "living" characteristics: those that acquire energy and nutrients, respond to the environment, transport and recycle materials, and are capable of self-replication. Rather than trying to create synthetic life, the adaptability of life could be combined with unnatural components to make a whole greater than the sum of its parts. It will be fascinating to watch this field unfold in the near future.

11.13 References

1. L. Isaacs, D.N. Chin, N. Bowden, Y. Xia and G.M. Whitesides, Self-assembling systems on scales from nanometers to millimeters: design and discovery, *Perspect. Supramol. Chem.*, 1999, **4** (Supramolecular Materials and Technologies), 1.

2. G.M. Whitesides and M. Boncheva, Beyond molecules: self-assembly of mesoscopic and macroscopic components, *Proc. Natl Acad. Sci. USA*, 2002, **99**, 4769.

3. M. Boncheva, D.A. Bruzewicz and G.M. Whitesides, Millimeter-scale self-assembly and its applications, *Pure Appl. Chem.*, 2003, **75**, 621.

4. I.S. Choi, M. Weck, N.L. Jeon and G.M. Whitesides, Mesoscale folding: a physical realization of an abstract, 2D lattice model for molecular folding, *J. Am. Chem. Soc.*, 2000, **122**, 11997.

5. N.B. Bowden, M. Weck, I.S. Choi and G.M. Whitesides, Molecule-mimetic chemistry and mesoscale self-assembly, *Acc. Chem. Res.*, 2001, **34**, 231.

6. V.R. Thalladi, A. Schwartz, J.N. Phend, J.W. Hutchinson and G.M. Whitesides, Simulation of indentation fracture in crystalline materials using mesoscale self-assembly, *J. Am. Chem. Soc.*, 2002, **124**, 9912.

7. M. Weck, I.S. Choi, N.L. Jeon and G.M. Whitesides, Assembly of mesoscopic analogues of nucleic acids, *J. Am. Chem. Soc.*, 2000, **122**, 3546.

8. T.D. Clark, M. Boncheva, J.M. German, M. Weck and G.M. Whitesides, Design of three-dimensional, millimeter-scale models for molecular folding, *J. Am. Chem. Soc.*, 2002, **124**, 18.

9. N. Bowden, A. Terfort, J. Carbeck and G.M. Whitesides, Self-assembly of mesoscale objects into ordered two-dimensional arrays, *Science*, 1997, **276**, 233.

10. N. Bowden, I.S. Choi, B.A. Grzybowski and G.M. Whitesides, Mesoscale self-assembly of hexagonal plates using lateral capillary forces: synthesis using the "capillary bond", *J. Am. Chem. Soc.*, 1999, **121**, 5373.

11. N. Bowden, S.R.J. Oliver and G.M. Whitesides, Mesoscale self-assembly: capillary bonds and negative menisci, *J. Phys. Chem. B*, 2000, **104**, 2714.

12. N. Bowden, J. Tien, W.T.S. Huck and G.M. Whitesides, Mesoscale self-assembly: the assembly of micron- and millimeter-sized objects using capillary forces, *Supramolecular Organization and Materials Design*, Cambridge University Press, Cambridge, 2002, p. 103.

13. N. Bowden, F. Arias, T. Deng and G.M. Whitesides, Self-assembly of microscale objects at a liquid/liquid interface through lateral capillary forces, *Langmuir*, 2001, **17**, 1757.

14. D.B. Wolfe, A. Snead, C. Mao, N.B. Bowden and G.M. Whitesides, Mesoscale self-assembly: capillary interactions when positive and negative menisci have similar amplitudes, *Langmuir*, 2003, **19**, 2206.

15. D.H. Gracias, J. Tien, T.L. Breen, C. Hsu and G.M. Whitesides, Forming electrical networks in three dimensions by self-assembly, *Science*, 2000, **289**, 1170.

16. M. Boncheva, D.H. Gracias, H.O. Jacobs and G.M. Whitesides, Biomimetic self-assembly of a functional asymmetrical electronic device, *Proc. Natl Acad. Sci. USA*, 2002, **99**, 4937.

17. A. Terfort, N. Bowden and G.M. Whitesides, Three-dimensional self-assembly of millimeter-scale components, *Nature*, 1997, **386**, 162.
18. W.T.S. Huck, J. Tien and G.M. Whitesides, Three-dimensional mesoscale self-assembly, *J. Am. Chem. Soc.*, 1998, **120**, 8267.
19. T.L. Breen, J. Tien, S.R.J. Oliver, T. Hadzic and G.M. Whitesides, Design and self-assembly of open, regular, 3D mesostructures, *Science*, 1999, **284**, 948.
20. H.O. Jacobs, A.R. Tao, A. Schwartz, D.H. Gracias and G.M. Whitesides, Fabrication of a cylindrical display by patterned assembly, *Science*, 2002, **296**, 323.
21. T.D. Clark, J. Tien, D.C. Duffy, K.E. Paul and G.M. Whitesides, Self-assembly of 10 μm-sized objects into ordered three-dimensional arrays, *J. Am. Chem. Soc.*, 2001, **123**, 7677.
22. T.D. Clark, R. Ferrigno, J. Tien, K.E. Paul and G.M. Whitesides, Template-directed self-assembly of 10 μm-sized hexagonal plates, *J. Am. Chem. Soc.*, 2002, **124**, 5419.
23. D.H. Gracias, V. Kavthekar, J.C. Love, K.E. Paul and G.M. Whitesides, Fabrication of micrometer-scale, patterned polyhedra by self-assembly, *Adv. Mater.*, 2002, **14**, 235.
24. R.R.A. Syms, E.M. Yeatman, V.M. Bright and G.M. Whitesides, Surface tension-powered self-assembly of microstructures – the state-of-the-art, *J. Microelectromech. Syst.*, 2003, **12**, 387.
25. H.K. Wu, V.R. Thalladi, S. Whitesides and G.M. Whitesides, Using hierarchical self-assembly to form three-dimensional lattices of spheres, *J. Am. Chem. Soc.*, 2002, **124**, 14495.
26. G.A. Ozin and S.M. Yang, The race for the photonic chip: colloidal crystal assembly in silicon wafers, *Adv. Funct. Mater.*, 2001, **11**, 95.
27. B.B. Yellen and G. Friedman, Programmable assembly of colloidal particles using magnetic microwell templates, *Langmuir*, 2004, **20**, 2553.
28. B.A. Grzybowski, H.A. Stone and G.M. Whitesides, Dynamic self-assembly of magnetized, millimeter-sized objects rotating at a liquid–air interface, *Nature*, 2000, **405**, 1033.
29. B.A. Grzybowski, X. Jiang, H.A. Stone and G.M. Whitesides, Dynamic, self-assembled aggregates of magnetized, millimeter-sized objects rotating at the liquid–air interface: macroscopic, two-dimensional classical artificial atoms and molecules, *Phys. Rev. E*, 2001, **64**, 011603.
30. B.A. Grzybowski and G.M. Whitesides, Three-dimensional dynamic self-assembly of spinning magnetic disks. Vortex crystals, *J. Phys. Chem. B*, 2002, **106**, 1188.
31. B.A. Grzybowski, H.A. Stone and G.M. Whitesides, Dynamics of self assembly of magnetized disks rotating at the liquid–air interface, *Proc. Natl Acad. Sci. USA*, 2002, **99**, 4147.
32. A.A. Grzybowski and G.M. Whitesides, Dynamic aggregation of chiral spinners, *Science*, 2002, **296**, 718.
33. R. Singh, V.M. Maru and P.S. Moharir, Complex chaotic systems and emergent phenomena, *J. Nonlinear Sci.*, 1998, **8**, 235.
34. R.F. Ismagilov, A. Schwartz, N. Bowden and G.M. Whitesides, Autonomous movement and self-assembly, *Angew. Chem. Int. Ed.*, 2002, **41**, 652.

Nanofood for Thought – Static and Dynamic, Capillary Bond, Shape Assembly

1. How would you synthesize a collection of nanorods with a nanopower source located at one end of each nanorod? Think of something creative you could do with this novel motorized building block.

2. Can you think of a way of making a collection of micron-scale cubes and tetrahedra with compositions based upon silver, silica, silicon and titania?

3. Provided with a beaker of monodispersed micron-scale cubes develop a chemical strategy that permits opposite or adjacent faces to be selectively modified (a) with the same and (b) with different functionalities and having achieved this goal think of something innovative that you could accomplish with these novel cubic micron-scale building blocks that is not feasible with the pristine cubes.

4. Try to come up with a way of organizing micron-scale rods and tetrahedra to form a diamondoid photonic lattice and explain why this would be a good thing to accomplish.

5. How would you make micron-scale single crystal silicon cubes with {100} faces and then convert them into cubo-octahedra and octahedra.

6. Is it possible to selectively modify the {111} and {100} faces of micron-scale single crystal silicon cubo-octahedra and self-assemble them into a zeolite-type photonic crystal with the structure of sodalite, zeolite A and zeolite Y. If possible, why would this be interesting?

7. Starting with millimeter-scale silica spheres how could you make a diamond lattice photonic crystal made of silicon that operates in the near infrared?

8. Draw building blocks with the correct shape and surface functionality in order to assemble a: (i) simple cubic array, (ii) body-centered cubic array, (iii) face-centered cubic array, (iv) diamond-topology array. Think about other types of arrays and the building block necessary to self-assemble them.

9. Can you think of a toy based on capillary assembly of centimeter size building blocks that would keep a child amused in the bath?

10. Large building blocks have been coerced through the capillary bond to self-assemble into a micro-geodesic dome, could you come up with a way to assemble a micro-Eiffel tower?

11. Let us say you fabricated magnetic PDMS shapes which can be floated at an air–water interface and rotated with a magnetic field. How would you expect a collection of shapes to behave if the shapes were fashioned in the form of a right hand? What would be the effect of making a hand with open fingers versus one with closed fingers? What would happen if you mixed right hand and left hand shapes?

Nano Future

CHAPTER 12

Nano and Beyond

"Any smoothly functioning technology will have the appearance of magic."
Arthur C. Clarke (1917–)

12.1 Assembling the Future

It seems that in just a fleeting moment of time, materials self-assembly has changed our entire way of thinking about making new matter. We have seen a shift in materials research from familiar chemistry approaches based on synthesizing molecules, polymers and solids, and engineering physics means of making planar and lateral structures, to a materials self-assembly paradigm, more akin to a Lego construction process, where pre-formed and pre-programmed building blocks, with respect to size, shape, and surface functionality, are self-guided or directed to automatically assemble into some sort of device or machine.

Materials self-assembly works with a few relatively simple building blocks based on wires, rods, tubes, layers, rings, solid and hollow spheres and hemispheres, spirals and polyhedrons, yet devises rational ways to put them together into a diversity of structures to achieve function. This can be likened to the way Nature uses a few materials to make a myriad of biological constructs with complex form designed to achieve a specific function. Lehn's paradigm of complexity in biology and diversity in chemistry seems to be converging in the new and exciting field of materials self-assembly.

After reading this introductory text on how to synthesize building blocks with a specific size, shape, bulk, and surface structure and composition and make materials through building block self-assembly, it will hopefully become clear that we have a new toolbox for designing and making materials with a purpose. This has happened because of the extraordinary advances in the synthesis of structurally well-defined building blocks over the last few years and the concomitant development of creative strategies for directing these building blocks into integrated systems.

The field is changing on a day-to-day basis and it has not been possible to include every exciting development in this brief look at the field of nanochemistry through the eye of materials self-assembly over "all" scales.

Figure 12.1 *Microfluidic large-scale integration*
(Reproduced with permission from Ref. 1)

However, in what follows it is worthwhile mentioning some thought provoking breakthroughs and try to imagine how the field will evolve over the next few years.

12.2 Microfluidic Computing

Elastomeric microfluidic chips made by soft-lithography can have micropumps, microvalves, and microchambers integrated on a large scale, making them ideal for miniaturized labs-on-chips. The available complexity can produce microfluidic "circuits" (Figure 12.1), analogous to electronic integrated circuits, enabling liquids on a chip to be individually compartmentalized, mixed pair-wise and selectively purged with solutions.[1] This could provide a means of using liquids for memory storage, displays and maybe a new type of computing based on one of the most plentiful substances on our planet, water!

12.3 Fuel Cells – Hold the Membrane

When fluid flows in a thin enough channel, it undergoes flow without turbulence, otherwise known as laminar flow.[2] Laminar flow in microfluidic channels has led to a miniature redox fuel cell which operates without a membrane.[3] This could provide local power supplies to drive electronic, optoelectronic, and photonic devices on chips, hand held devices and laptop computers.

12.4 Curved Prints

Nanoscale soft-lithography patterning using a thin PDMS elastomeric mask draped conformally over photoresist-coated sphere surfaces has facilitated the direct

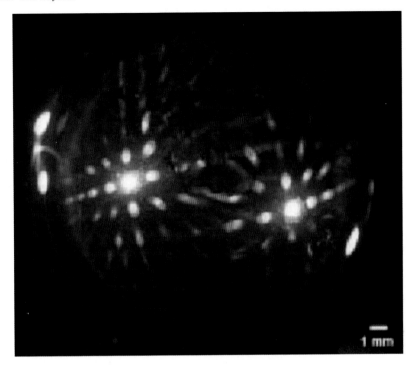

Figure 12.2 *Nanoscale soft-lithography on curved surfaces*
(Reproduced with permission from Ref. 4)

transfer of 2D patterns with features as small as 175 nm to the surface of a sphere.[4]
The diffraction from a patterned sphere is shown in Figure 12.2. This flexible
printing technique suggests new applications in wide-angle optics and sensors.

12.5 Beating the Ink Diffusion Dilemma

Dip-pen nanolithography (DPN, Chapter 4) is an ideal way to pattern substrates
with arbitrary patterns on the nanoscale. It is also possible to locally transform a
functional group on a self-assembled monolayer (SAM) to make micron scale
patterns using catalytic transformations mediated by elastomeric stamps or molds.
These two concepts can be combined into one: by supporting a catalyst on an AFM
tip, ultra-fine patterns can be drawn on a SAM by catalytic transformation. For
instance, a sulfonic acid modified tip can locally hydrolyze ether groups to create
25 nm features of alcohol groups.[5] Since no chemicals are being transported,
deleterious blurring of written features *via* diffusive transport and spread does not
occur. There are a myriad more opportunities for catalytic lithography, such as
reducing nitro groups to amines using anchored silver nanoparticles,[6] and
photocatalytic oxidation with a UV-exposed nanocrystal anatase tip.

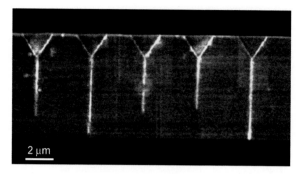

Figure 12.3 *30 nm diameter channels anodically etched in silicon, templated on an array of anisotropically etched pyramidal pits*
(Reproduced with permission from Ref. 7)

12.6 Tip of the Pyramid

It is quite straightforward to create well-defined square pyramidal-shaped pits in silicon using a base such as potassium hydroxide in isopropanol. The nanoscale apices of such pits can then be used to focus the electric field during anodic oxidation of silicon in an aqueous HF, in order to make nanochannels as small as 30 nm with an aspect ratio as high as 250[7] (Figure 12.3). These open channels benefit from the well-known surface chemistry of silicon: such ultra-fine nanochannels could thus be of great interest in bio-separations, encapsulation of biomolecules, and controlled drug release. The funnel-shaped channels could also be used to template tip-bonded nanowires or funnel-attached nanotubes for nanoimaging and nanocircuitry, nanosyringes, and nanojets.

12.7 Biosensing Membranes

Making ultra-small nanochannel membranes could provide the ultimate route to high accuracy biosensors. By using a 30 nm pore polycarbonate membrane, the channel diameter can be reduced to 12 nm by coating the walls of the channels by gold electroless deposition. These 12 nm channels can then be coated with a DNA recognition sequence, which only lets a particular DNA sequence through the channel. This has allowed the recognition of DNA strands accurate to a single coding base.[8] Such results are extremely promising for all types of biological and non-biological sensing applications.

12.8 Crossing Nanowires

VLS synthesis has enabled a rational way of making heterostructured nanowire superlattices with control over the length, composition, and surface functionalization. As a next step, functional molecules could be placed on specific segments of the nanowire and orthogonal assembly could provide a means to self-assemble them into a crosspoint array of memory or logic devices.[9] Nanochannel membranes have allowed larger diameter barcoded nanorods to be synthesized, and orthogonal assembly could provide a way of organizing them into nanorod electronic lattices or photonic crystals, with the barcodes auto-arranged to form point and line functional defects.

12.9 Complete Crystallographic Control

Crystal engineering methods have been used with great success to control the diameter and length of VLS grown single-crystal nanowires, as well as their spatial location in a nanowire array. However, command over the crystallographic growth direction of the individual nanowires has remained elusive. The key to achieve this goal, as for the successful growth of single-crystal films, has been epitaxy.[10] It was found that the {100} faces of a γ-LiAlO$_2$ single-crystal substrate favored the <110> growth direction of triangular cross-section single-crystal GaN nanowires using MOCVD. In contrast, on the {111} faces of a MgO single-crystal substrate, the selective growth of hexagonal cross-section GaN nanowires along the <001> growth direction was observed instead (Figure 12.4). Such crystallographic control can result in nanowires with finely tuned electronic, mechanical, and optical properties needed for the future development of nanowire devices and circuitry.

12.10 Down to the Wire

Is there more than one way of rising to the challenge of making flexible nanowire macroelectronics? We have seen how purely top-down fabrication and bottom-up self-assembly strategies can integrate the disparate worlds of small devices and large electronic circuitry but why not fuse the attributes of both strategies? This has

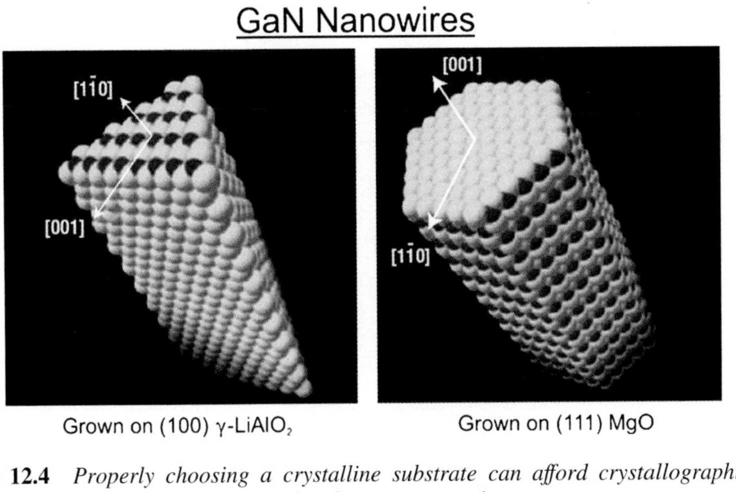

GaN Nanowires

Grown on (100) γ-LiAlO$_2$ Grown on (111) MgO

Figure 12.4 *Properly choosing a crystalline substrate can afford crystallographic and cross-section control of VLS-grown nanowires*
(Reproduced with permission from Ref. 10)

been realized through an ingenious amalgamation of traditional photolithography and anisotropic etching (top-down) with dry transfer printing (bottom-up) to create a parallel array of oriented single-crystal GaAs nanowires on a plastic substrate.[11] If silica lines are patterned on a GaAs substrate, this semiconductor can be anisotropically etched such that triangular GaAs nanowires remain affixed to the silica. These nanowires, with a width down to 50 nm and a width variation of less than 9%, can be subsequently transfer printed onto a piece of PDMS, and then embedded into a polyurethane (PU) film supported on a flexible polyethylene terepthalate (PET) sheet. A three layer GaAs nanowire array is shown in Figure 12.5, illustrating the accuracy of the process. This strategy gives embedded wires that have excellent electrical characteristics, can withstand repeated flexure, and which could be made of other materials such as InP with a similar etching strategy. Flexible nanoelectronics may be on the market sooner than we think!

12.11 Shielded Nanowires

Electromagnetic radiation (EMR) with a wide range of frequencies is ubiquitous in our environment, and more often than not wires in electronic devices must somehow be shielded from this undesirable EMR. The practical construct designed

Figure 12.5 *A three-layer GaAs nanowire array formed by lithography, anisotropic etching, and transfer printing*
(Reproduced with permission from Ref. 11)

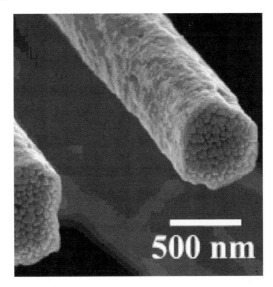

Figure 12.6 *A multi-coaxial bundle of crystalline SiC nanowires, produced by direct thermal synthesis*
(Reproduced with permission from Ref. 12)

to achieve this end is the well-known coaxial cable, but what if we need to safeguard nanowires in a nanoelectronic circuit? Believe it or not, this remarkable feat has been accomplished in a "one-pot" nanowire synthesis.[12] By passing a flow of silicon vapor and methane over a heated copper foil, what results is a multi-coaxial nanocable consisting of a parallel-aligned bundle of oriented single-crystal β-SiC nanowires, enclosed in a sheath consisting of copper oxide, graphite, and amorphous carbon (Figure 12.6). Having seen this, clearly the future of nanowire electronics is now in safe hands with this kind of defense from extraneous EMR!

12.12 Writing 3D Nanofluidic and Nanophotonic Networks

We have seen in Chapter 5 how nanoscale solid and hollow polymer fibers can be electrospun into mats and trapped between electrodes on substrates to give organized arrays but what if one needs them to be drawn and patterned in a serial or parallel fashion into 3D suspended nanobeams or nanocapillary networks? An ingenious solution to this intriguing problem, with technological implications in flexible nanofluidics and nanophotonics, is to employ the tip of a scanning probe microscope to draw and knit hollow individual polymer nanofibers, from and between polymer microdroplets in an array.[13] This feat can also be accomplished in parallel by dragging the edge of a polymer film across the surface of an array of fine tips to create a network of suspended polymer nanofibers.

12.13 Break-and-Glue Transistor Assembly

One of the challenges in the quest for molecular electronics is to find a way to practically and reproducibly make devices like a FET with molecular size in all spatial dimensions and ideal electrical contacts between metal electrodes and molecules. If a carbon nanotube is aligned across two electrodes, the application of a sufficiently high current can create a tiny break as small as ~2 nm wide. The broken nanotube now becomes two perfectly aligned counter-electrodes, which can then be glued together with a molecular semiconductor such as pentacene grown from the vapor phase, or poly(3-hexylthiophene) deposited from solution.[14] This break junction breakthrough is a promising development towards the practical realization of high-performance ballistic FETs. The dream of a molecular-FET with molecular width and channel length now seems feasible, for example, by exploiting the aforementioned break-and-glue assembly and figuring out how to coerce a semiconducting molecule to chemically bridge functional groups at the ends of the SWCN source and drain electrodes.

12.14 Turning Nanostructures Inside-out

The Kirkendall effect is a well-known phenomenon discovered in the 1930s. It occurs during the reaction of two solid-state materials and involves the diffusion of reactant species, like ions, across the product interface usually at different rates to make the product. In the special case when the movement of the fast-diffusing component cannot be balanced by the movement of the slow component, the net mass flow is accompanied by a net flow of atomic vacancies in the opposite direction. This effect leads to Kirkendall porosity, formed through the super-saturation of vacancies into hollow pores. However, when starting with perfect building blocks such as cobalt nanocrystals,[15] a reaction meeting the Kirkendall criteria can lead to supersaturation of vacancies exclusively in the center of the nanocrystal. This provides a general route to hollow nanocrystals out of almost any given material,[16] as shown in Figure 12.7 for the synthesis of a cobalt sulfide nanoshell starting from a cobalt nanocluster and sulfur in a high-boiling solvent. The nanoscale Kirkendall effect can be applied to the diversity of compositions and shapes of nano-building blocks described in this book, leading to chemically synthesized materials with mind-boggling complexity.

12.15 Confining Spheres

Confined assembly within surface relief patterns on planar chips is a general route to all kinds of nanocrystal and microsphere patterns with packing and geometries. In the case of microspheres, it enables the formation of planarized microphotonic crystals as shown in Figure 12.8. The control over the crystalline orientation, and thus optical properties, in this approach could pave the way to synthetic all-optical devices and circuits of light.[17]

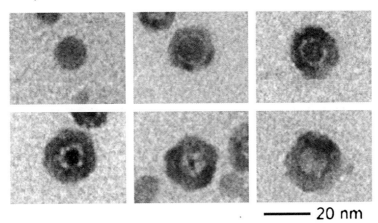

Figure 12.7 *Time evolution of a hollow cobalt sulfide nanocrystal grown from a cobalt nanocrystal via the nanoscale Kirkendall effect*
(Reproduced with permission from Ref. 15)

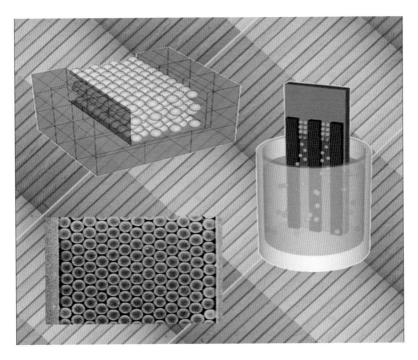

Figure 12.8 *Confined microsphere assembly*
(Reproduced with permission from Ref. 16)

12.16 Escape from the Silica and Polystyrene Prison

Colloidal photonic crystals (CPCs) with a photonic band gap are most often obtained by inverting the structure of a silica or latex colloidal crystal template with a high-refractive index material using a suitable precursor and infiltration method followed by sacrificial removal of the template as described in Chapter 7. But why not make the CPC directly by self-assembling monodispersed microspheres of high-refractive index semiconductors and metals? While the quest for single size semiconductor microspheres and their colloidal assembly into colloidal lattices has met with some success for the case of CdS, ZnS, TiO_2 and a-Se, monodisperse metal microspheres have until very recently remained elusive. This state-of-affairs changed suddenly with the recent discovery that low melting point metals like Bi, Pb, In, Cd and their alloys can be coerced into microspheres with dispersity better than 5% and diameters in the range 100–600 nm.[18] This accomplishment was achieved by either reducing a metal acetate precursor or breaking up the molten metal by rapid stirring in a high-boiling glycol-based solvent and in the presence of a polymeric stabilizer like poly(vinyl pyrrolidone). The monodispersed metal microsphere genie is out of the bottle!

12.17 Smart Dust

Modulated current anodic oxidation of lithographically patterned silicon wafers is a route to an array of porous silicon photonic crystals.[19] Here the diffraction in each microscale feature on the wafer is caused by the alternating regions of high and low porosity caused by a high and low anodization current, respectively. The ability to ultrasonically shake these microscale photonic crystals from the wafer and control their Bragg diffracted color and surface functionality through electrochemistry and chemistry, may provide a means of tracking molecules in microbiology due to the ease of modifying the silicon surface – very smart dust indeed!

12.18 Light Writing for Light Guiding

Holographic photolithography and two-photon polymerization patterning of polymer resists leads to 3D microstructures that replicate the interference patterns of intersecting and writing laser beams, respectively. This may provide a new generation of periodic macroporous materials, as shown in Figure 12.9, that can function as templates for making high-refractive index photonic crystals with an omnidirectional photonic band gap for applications in optical telecommunications.[20] Templates made in this way can be totally perfect, and can be arbitrarily dotted with designer defects to introduce a desired functionality. With respect to laser patterning methods, it should also be possible to use an inorganic photoresist to obtain a high-refractive index, full photonic band gap photonic crystal.[21] This idea has been reduced to practice using well-known inorganic phase change materials, whereby following 3D laser patterning of the inorganic photoresist the desired 3D PBG PC is obtained in a single chemical etching step!

Figure 12.9 *An SU-8 diamond-type woodpile lattice, with supporting frame, made by direct-write two-photon polymerization. The lattice spacing is designed for this crystal to have a photonic band gap in the optical telecommunications range*
(Reproduced with permission from Ref. 20)

12.19 Nanoring Around the Collar

If one can create nanoscale patterns by diffusing chemical and biological inks to a substrate *via* the tip of a scanning probe microscope, then why not flow them *via* a contacting microsphere to create nanorings? Such a strategy is called edge-spreading lithography (ESL), as illustrated in Figure 12.10.[22] First, a monolayer of silica spheres is deposited on a gold- or silver-coated substrate, then a flat PDMS stamp inked with an alkanethiol is pressed against this monolayer. The ink flows from the stamp *via* the microsphere to the substrate, to create a ring-shaped SAM at the contact point with the substrate. The longer the stamp is in contact, the thicker the ring grows, with a diffusion-controlled rate. Wet etching of the unprotected regions of the gold generates an array of gold nanorings where the dimensions are related to the microsphere diameter, type and concentration of alkanethiol, solvent carrier, and contact time. Although not demonstrated, lift-off from the substrate and subsequent alkanethiol functionalization of the nanorings provides a means for performing LbL electrostatic self-assembly of alternately charged nanorings, either in solution or on a chemically patterned substrate, to form novel architectures like a nanoring nanocylinder. This method could be diversified, by self-assembling nanorings of different metals and opposite surface charges to form a barcoded nanoring nanocylinder, for example $\{Au–Ag\}_n$ or $\{Pd–Ag\}_n$. They could then be thermally transformed into barcoded nanocylinders!

12.20 A Meso Rubbed Right

The synthesis of periodic mesoporous silica in the early 1990s literally broke the mold, and provided a means to move beyond single molecule templating of microporosity. However, several issues remained to be ironed out: although

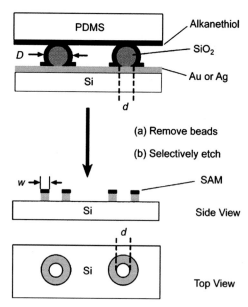

Figure 12.10 *Edge-spreading lithography: using diffusion along a monolayer of silica spheres to make metallic rings on a substrate*
(Reproduced with permission from Ref. 22)

the material was periodic at the mesoscale, at the microscale it was thoroughly disordered and confounding to analyze. Also, the liquid crystal templates used in the synthesis were pervaded by topological defects making it impossible to create a film in which the channels were ordered over macroscopic length scales. We saw in Chapter 8 that a periodicity at the microscale can be obtained by introducing pi-stacking aromatic groups into sol–gel precursors. So how can one make the channels crystalline-like over large areas? Rubbing the substrate does the job![23] It is a trick commonly used for aligning liquid crystals on polyimide surfaces for liquid crystal displays, and works beautifully for silicatropic liquid crystal precursors. Microgrooves in the rubbed polyimide substrate facilitate alignment of the silicatropic mesophase and long-range order of the channels in the so-formed periodic mesoporous silica film. Many technological objectives that were impeded by channel disorder in conventional periodic mesoporous silica now seem achievable in new generation "single-crystal" periodic mesoporous silica films.

12.21 Fungus with the Midas Touch

Gold nanoclusters functionalized with single stranded DNA have been assembled into ordered hierarchical architectures using not just biological templates, but living filamentous fungus strands (hypha)[24] (Figure 12.11). Although the precise details of the interaction between the hypha surface and the oligonucleotide-functionalized gold are not understood (*i.e.*, electrostatic, chemical, or hydrogen bonding), the fungi

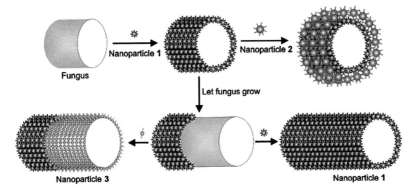

Figure 12.11 *Nanoparticles self-assembled on growing fungus*
(Reproduced with permission from Ref. 24)

hypha co-assembly enables additional gold nanoclusters to be coated on the first layer by exploiting complementary oligonucleotide interactions between the first and second layer of DNA-functionalized gold nanoclusters. The truly living nature of fungal hypha as templates is demonstrated by the continued growth of the hypha after co-assembly with the gold nanoclusters. This allows a higher level of hierarchical complexity to be achieved through the co-assembly of the newly grown bare hypha with a different kind of oligonucleotide-functionalized gold nanocluster, thereby providing command of the structure along the axial direction of the filament. There are currently unprecedented opportunities for studying the interaction of nanoparticles and nanosystems with organisms.

12.22 Self-assembled Electronics

3D assembly of polyhedral shapes having surfaces patterned with solder dots, wires, and light emitting diodes can self-assemble, on the basis of capillary interactions involving liquid solder, into electrically functional 3D networks with parallel and serial connections.[25] Patterned assembly of this type could provide a means of organizing micro-devices into electronic, optoelectronic, and photonic circuits. 3D electrical networks, in turn, can provide the means for a wholly different type of computation, a neural net, approaching the sort of connections that we have in our brains.

12.23 Gears Sink Their Teeth into the Interface

An interesting energy dissipating dynamic system involves an auto-constructing simple machine based upon magnetically driven, magnetically doped PDMS gears that self-assemble and self-level at the interface between air and liquid perfluorodecalin[26] (Figure 12.12). Torque between millimeter to centimeter scale gears is developed by mechanical, hydrodynamic, or capillary interactions. Capillary-driven gears for example, develop their torque from the interaction

Figure 12.12 *The evolution in time of magnetic self-assembling gears at an air–perfluorodecalin interface*
(Reproduced with permission from Ref. 26)

between scalloped menisci associated with hydrophobic and hydrophilic patches (teeth) on the rim of the gears. Simple and complex systems of self-assembling gears have been studied using this strategy. A curious example shown involves a diamagnetic gear with an asymmetric shape around which a smaller circular magnetic gear travels but in an opposite sense of rotation. One can imagine the paradigm being extended to include much more complex assemblies of gears in 3D and reduction of the dimension of the gears to the micron scale with potential utility in the field of microelectromechanical systems.

These are just a sampling of surprising breakthroughs that are taking our breath away on a regular basis in the scientific literature. They are profoundly novel concepts, which provoke creative ideas that would have seemed fanciful not so long ago. The possibilities are clearly enormous for making new materials in static and dynamic systems of self-assembling building blocks. Hopefully, we have convinced the reader that we can all look forward to a very exciting nanochemistry future.

12.24 Materials Retro-assembly

In the field of organic chemistry, one of the great breakthroughs was the elaboration of the technique of retro-synthesis.[27] This concept is extremely simple yet powerful, and consists of identifying a target molecule and elaborating all the ways the molecule could be disassembled into smaller and smaller components. This process leads to a tree diagram, where at the branch tips are located the simplest, commercially available building units. The development of retro-synthesis was truly a paradigm shift in organic synthesis, for which E.J. Corey was awarded the 1990 Nobel Prize in chemistry.

The field of self-assembling materials can be seen as being in its infancy, much as organic synthesis in the beginning of the 20th century. In other words, we have not yet approached the synthesis of these materials from a target-oriented point of view although this approach is becoming more prevalent. We are mostly putting building blocks together and seeing how they assemble, and concentrating on the assembly process rather than starting from the final structure and working backwards. Due to its inherent multidisciplinary nature, materials chemistry has a long history of borrowing ideas from other fields. Many concepts from organic synthesis have been borrowed, such as the chemical modification of microbead surfaces, and combinatorial chemistry, which allows us now to easily screen for new materials from a very large pool of materials with gradually changing compositions.[28,29] Retro-synthesis is applicable in all of organic synthesis, and we would like to present it as a concept which may be equally applicable to materials self-assembly.

To illustrate, suppose that we required a nanometer-sized train to haul small cargo from one side to the other of a silicon chip. This would consist of something like that shown in the top of Figure 12.13. We see a series of trains headed by a motor car. In theory, the first thing we can de-assemble in this structure are the bonds

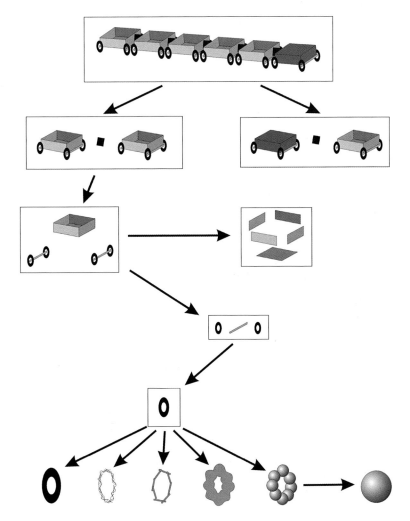

Figure 12.13 *Hypothetical retro-synthesis of a nanoscale train*

between cars, then within a single car we can separate the wheel-axel from the box, after which the wheel can then be separated from the axel. A wheel may consist of a nanoring made by soft-lithography, an annulus of entangled 1D strands, a ring of joined nanorods, or a ring of joined or fused nanospheres. If we then look at the isolated nanosphere, we have a wealth of compositions and surface properties available from simple chemistry. Although this example may not be possible as of yet, the more building blocks we uncover, the more branches we will be able to draw. Eventually, building such a nanoscale train may just be a matter of identifying the right blocks.

12.25 Matter that Matters – Materials of the "Next Kind"

Writing this textbook on nanochemistry has been a breathtaking experience. A good question is "what next"? When pondering this issue, it is really hard to believe that in just a fleeting moment in time, so much has changed in our ability to synthesize, fashion, manipulate, and visualize matter at length scales and into forms that matter. As a matter-of-fact, for more than a century of chemistry, "Molecule was king"! Diversity overpowered complexity and function was scarce. Biology, in its characteristically parsimonious manner, with only a few well-chosen pieces of matter, "was the pre-eminent builder of materials with diverse composition, multiple length scales, complex form, and myriad function." How could such frugality yield materials with so much sophistication?

Towards the end of the last century, a unification of chemical and biological construction principles blossomed forth as chemistry of the supra molecular kind – "beyond the molecule was its centerpiece." The new supra molecular chemistry inspired a movement particularly amongst chemists, to fuse concepts and methods in materials chemistry, materials science, engineering, physics, and biology, to take a ground-upward chemistry approach to make nanomaterials from which grew the field of nanochemistry with self-assembly as a central theme, the subject of this book.

Within this textbook an attempt has been made to organize more-or-less all the key building blocks in the new "materials tool kit" and the main construction principles in the "new solid-state instruction manual" under one all-encompassing umbrella, which we call nanochemistry. In this still emerging field, entrenched boundaries between large and small, soft and hard, surface and bulk, synthesis and micro-fabrication have become blurred and borders between traditional science disciplines have been bridged, irreversibly so it seems.

To the student and practitioner of nanomaterials research, anything now seems possible – how exciting it is to realize that "there is plenty of room at the top, bottom and middle"! Practitioners of materials research have been unleashed from the prison of serial or parallel synthesis of matter with a particular composition and structure. Proof-of-concept dynamic and static self-assembly of equilibrium and non-equilibrium forms have been demonstrated in a spectacular eruption of materials and biomaterials architecture and visual imagery that has taken an experienced materials community by surprise.

The future challenge for this emerging field is not just what incremental technological extensions are now possible, but also to make a concerted effort to unravel how to utilize the strength of a chemistry approach to nanomaterials and self-assembly to create "matter of a new kind" with properties that we have not yet anticipated. Here we are not thinking of a new composition or structure or modification thereof, but rather an integration of known matter in the form of building blocks in new and surprising ways to give new matter, new properties, new function, new uses.

When we talked about assembling the future, a potpourri of some very leading-edge developments in nanochemistry, which we believe will have lasting impact on the field, were highlighted. This was one way of conveying to the reader that exciting breakthroughs are occurring on a daily basis and it is really difficult to keep abreast of such a highly competitive bourgeoning field. On concluding this textbook, it is instructive and slightly adventurous to try to imagine some of the ways in which the field may develop from hereon in, as a type of blueprint for future research.

It seems likely that self-assembling building blocks of various shapes and sizes could become very "smart," beyond those frozen in the words and images of this text. Next generation building block assembly, besides being motion driven by some power source in a dynamic purposeful mode (see Appendix G), and designed to interact cooperatively either with each other or structure directing templates or patterned surfaces to emerge in some new, complex, and surprising form, could also be orchestrated through chemistry to undergo dynamical bending and twisting motions as well as changes in size and shape. To illustrate the point, clever rod, ring, sphere, ellipsoid, spiral and disc forms, could auto-assemble into a myriad of machines, small and not so small, with dynamic metamorphic parts, purposeful scale, and function. With internal or external energy sources, heat, light, chemical, or electrochemical, they might do the work of existing machines or might even be the basis of machines of a new kind. The unison of self-constructing abiological and biological machines could be one such adventure into the materials unknown and unimaginable. Assemblers, nanobots, gray goo will be left to the imagination of writers of science fiction!

Chapter 7 said much about CPCs, the new "light scale materials" based on spherical building blocks. At the micron length scale characteristic of this class of materials, non-spherical constructs could self-assemble into shape-changeable lattices not previously encountered or even imagined. They would be of interest in the drive to miniaturize optical components, develop new chemical and biological sensors and devise polychrome digital display and print media. A dielectric construct made of periodic arrays of shape-adjustable building blocks, besides being able to wavelength tune optical diffraction of light may provide new platforms for creatively integrating other materials functionality with tunable coloration.

Imagine batteries, solar and fuel cells that report their electrical states, output, or performance through electrodes and electrolytes that diffract light according to their refractive index contrast. Contemplate gas membranes, stationary phases, storage, and delivery materials whose permeability and loading determine the color of optically diffracted light. Think about sensory materials whose diffracted color can be used to detect and discriminate molecules within the confines of its lattice. Ponder a new type of magneto-optic material where it is the altered color of diffracted light rather than changes of polarization with magnetic field strength that is the basis of data storage. The theme of combining two or more properties of matter in a single material is nothing new in solid-state, the idea being to create a multi-functional material or device whose integrated properties are superior to the component individual properties. Self-assembled photonic crystals have a myriad of possibilities that exploit this basic concept – we have seen the light!

A nanoring made of a swellable–shrinkable polymer gel would change diameter in response to a change of solvent, temperature, pH, or charge. A collection of smaller spheres contained within the annulus of the nanoring could be moved between ordered and disordered states as the nanoring swells and shrinks in size. This could give rise to a detectable change in reflected or diffracted light incident on the system. Pixilated matrix addressable nanorings might form the basis of a new type of imaging or printing system. Similarly swellable–shrinkable nanorods, nanotubes, and nanowires could be made to function as nanoscale actuator and switches, fluidic, and optical circuits. An electrically powered swellable–shrinkable polymer gel coating within an inverse silicon opal, of the polyferrocenylsilane type described in Chapter 7 under the guise of photonic ink (P-Ink), could also serve to switch the refractive index contrast of the photonic lattice and hence its ability to reflect or transmit light – an optical transistor.

Alkanethiol SAMs, renown for their aurophilicity and argentophilicity, on simple planar and curved surfaces could yield unexpected multi-functionality when confined to the 3D topologically complex surface of a gold or silver photonic lattice. One idea is to exploit the well-known sensitivity of gold and silver plasmons to a SAM and what is adsorbed onto the SAM, with the twist that the SAM is now sculpted in the topology of a colloidal photonic lattice whose optical properties will be sensitive to adsorbate and adsorbents. Plasmonics can be imagined whereby biomolecular recognition events facilitate molecule detection and discrimination.

Why not use the multi-tip capability of DPN to write chemical patterns on the surfaces of nanoscale building blocks, such as spheres, rods, rings, and various polyhedra. Imagine a monolayer of silica microspheres with DPN written patches arranged at the vertices of a tetrahedron. These DPN patches can be surface anchored onto functional groups or deposited metals. The idea is to mimic the tetrahedrally hybridized sp^3 orbitals of carbon by tetrahedrally configured chemical patches of microspheres. With such spheres one could devise ways of coercing them to self-assemble into the Holy Grail of photonic crystals – a diamondoid periodic lattice. When made of silicon, this connectivity is predicted to have the largest omnidirectional photonic band gap, essentially immune to structural imperfection, what perfection that would be!

DPN chemical patterns written on rods could facilitate the self-assembly of side-to-side bonded nanorod dimers, trimers, tetramers, and so forth. This raises the possibility of simply making bipolar shapes for assembling novel constructs. For instance, a monolayer of silica microspheres, random or ordered on a substrate, with gold deposited on one side by DPN of gold nanoclusters or thermal/sputtering of gold, creates a situation where selective functionalization of the gold side with terminally functionalized alkanethiols, would enable self-assembly of novel microsphere aggregates. These novel construction units could be organized into devices with distinctive properties.

Consider mixing sphere (SAM1) with sphere (SAM2) to give sphere (SAM1)(SAM2) sphere dimers held together by say hydrophobic, hydrogen bonding, or acid–base interactions. Sphere dimers could for instance be assembled into novel sphere dimer lattices with distinctive optical properties. It is not difficult to imagine how this methodology can be engineered to yield sphere trimers, tetramers, chains and so forth – maybe even surprising microsphere lattice architectures could emerge.

Bipolar building blocks offer a myriad of new opportunities. For instance, a new kind of digital imaging or printing system could be devised based on the self-assembly of a close packed monolayer of cationic–anionic bipolar microspheres on a matrix addressable micropatterned electrode surface. The latter could be made by microsphere lithography. This architecture provides a pixilated microsphere surface that is electrostatically switchable between charge states thereby creating color contrast at the micron scale and possibly the basis of a new digital imaging or printing system. Bipolar microspheres with Cu, Ag, or Au half-caps might undergo rotational motion at different voltages to reflect different colors. Similar thinking but utilizing instead of spheres a parallel-aligned array of bipolar nanorods, could enable a rolling nanorod display, which makes one think about new nanorod machines like a nanorod conveyer belt, nanorod rotary motors, nanorod windmills, and even nanorod rollers smoothing surfaces and running up and down chemically graded slopes.

The field of nanochemistry can be seen in the way one might view a scientific breakthrough. First comes the "discovery pregnant with promise," and then, "the hard graft begins"! It is an exciting time for the field of nanochemistry, which is clearly an integral and inseparable component of nanoscience and nanotechnology. The possibilities for innovation and invention in this emerging field seem endless, only limited by imagination.

Nanochemistry is a big and exciting field in which there is much to discover and do. Hopefully this first look at the subject in its entirety and as it stands today will convince the reader that a first important step has been taken to the realization of "matter that matters – materials of the next kind."

12.26 References

1. T. Thorsen, S.J. Maerkl and S.R. Quake, Microfluidic large-scale integration, *Science*, 2002, **298**, 580.
2. P.J.A. Kenis, R.F. Ismagilov, S. Takayama, G.M. Whitesides, S.L. Li and H.S. White, Fabrication inside microchannels using fluid flow, *Acc. Chem. Res.*, 2000, **33**, 841.

3. R. Ferrigno, A.D. Stroock, T.D. Clark, M. Mayer and G.M. Whitesides, Membraneless vanadium redox fuel cell using laminar flow, *J. Am. Chem. Soc.*, 2002, **124**, 12930.

4. K.E. Paul, M. Prentiss and G.M. Whitesides, Patterning spherical surfaces at the two-hundred-nanometer scale using soft lithography, *Adv. Funct. Mater.*, 2003, **13**, 259.

5. M. Péter, X.-M. Li, J. Huskens and D.N. Reinhoudt, Catalytic probe lithography: catalyst-functionalized scanning probes as nanopens for nano-fabrication on self-assembled monolayers, *J. Am. Chem. Soc.*, 2004, **126**, 11684.

6. K. Kim and I. Lee, Chemical lithography by Ag-nanoparticle-mediated photo-reduction of aromatic nitro monolayers on Au, *Langmuir*, 2004, **20**, 7351.

7. S.E. Létant, T.W. van Buuren and L.J. Terminello, Nanochannel arrays on silicon platforms by electrochemistry, *Nano Lett.*, 2004, **4**, 1705.

8. P. Kohli, C.C. Harrell, Z. Cao, R. Gasparac, W. Tan and C.R. Martin, DNA-functionalized nanotube membranes with single-base mismatch selectivity, *Science*, 2004, **305**, 984.

9. N.I. Kovtyukhova and T.E. Mallouk, Nanowires as building blocks for self-assembling logic and memory circuits, *Chem. Eur. J.*, 2002, **8**, 4355.

10. T. Kuykendall, P.J. Pauzauskie, Y. Zhang, J. Goldberger, D. Sirbuly, J. Denlinger and P. Yang, Crystallographic alignment of high-density gallium nitride nanowire arrays, *Nature Mater.*, 2004, **3**, 524.

11. Y. Sun and J.A. Rogers, Fabricating semiconductor nano/microwires and transfer printing ordered arrays of them onto plastic substrates, *Nano Lett.*, 2004, **4**, 1953.

12. W.H. Ghim, A.S.W. Wong, A.T.S. Wee and M.E. Welland, Self-assembled growth of coaxial crystalline nanowires, *Nano Lett.*, 2004, **4**, 2023.

13. S.A. Harfenist, S.D. Cambron, E.W. Nelson, S.M. Berry, A.W. Isham, M.M. Crain, K.M. Walsh, R.S. Keynton and R.W. Cohn, Direct drawing of suspended filamentary micro- and nanostructures from liquid polymers, *Nano Lett.*, 2004, **4**, 1931.

14. K. Tsukagoshi, I. Yagi, Y. Aoyagi, Pentacene nanotransistor with carbon nanotube electrodes, *Appl. Phys. Lett.*, 2004, **85**, 1021.

15. V.F. Puntes, K.M. Krishnan and A.P. Alivisatos, Colloidal nanocrystal shape and size control: the case of cobalt, *Science*, 2001, **291**, 2115.

16. Y. Yin, R.M. Rioux, C.K. Erdonmez, S. Hughes, G.A. Somorjai and A.P. Alivisatos, Formation of hollow nanocrystals through the nanoscale Kirkendall effect, *Science*, 2004, **304**, 711.

17. S.M. Yang, H. Miguez and G.A. Ozin, Opal circuits of light – Planarized microphotonic crystal chips, *Adv. Funct. Mater.*, 2002, **12**, 425.

18. Y. Wang and Y. Xia, Bottom-up and top-down approaches to the synthesis of monodispersed spherical colloids of low melting-point metals, *Nano Lett.*, 2004, **4**, 2047

19. J.R. Link and M.J. Sailor. Smart dust: self-assembling, self-orienting photonic crystals of porous Si, *Proc. Natl Acad. Sci. USA*, 2003, **100**, 10607.

20. M. Deubel, G. von Freymann, M. Wegener, S. Pereira, K. Busch and C.M. Soukoulis, Direct laser writing of three-dimensional photonic-crystal templates for telecommunications, *Nature Mater.*, 2004, **3**, 444.
21. S.H. Wong, G. von Freymann, M. Dubel, M. Wegener and G.A. Ozin, Patent Filed, 2004.
22. J.M. McLellan, M. Geissler and Y. Xia, Edge spreading lithography and its application to the fabrication of mesoscopic gold and silver rings, *J. Am. Chem. Soc.*, 2004, **126**, 10830.
23. H. Miyata, T. Suzuki, A. Fukuoka, T. Sawada, M. Watanabe, T. Noma, K. Takada, T. Mukaide and K. Kuroda, Silica films with a single-crystalline mesoporous structure, *Nature Mater.*, 2004, **3**, 651.
24. Z. Li, S.W. Chung, J.M. Nam, D.S. Ginger and C.A. Mirkin. Living templates for the hierarchical assembly of gold nanoparticles, *Angew. Chem. Int. Ed.*, 2003, **42**, 2306.
25. D.H. Gracias, J. Tien, T.L. Breen, C. Hsu and G.M. Whitesides, Forming electrical networks in three dimensions by self-assembly, *Science*, 2000, **289**, 1170.
26. J.M.K. Ng, M.J. Fuerstman, B.A. Grzybowski, H.A. Stone and G.M. Whitesides, Self-assembly of gears at a fluid/air interface, *J. Am. Chem. Soc.*, 2003, **125**, 7948.
27. E.J. Corey. Retrosynthetic thinking-essentials and examples, *Chem. Soc. Rev.*, 1988, **17**, 111.
28. D.R. Liu and P.G. Schultz, Generating new molecular function: a lesson from nature, *Angew. Chem. Int. Ed.*, 1999, **38**, 36.
29. P.G. Schultz and X.D. Xiang, Combinatorial approaches to materials science, *Curr. Opin. Solid State Mater. Sci.*, 1998, **3**, 153.

Nanofood for Thought – Nano Potpourri

1. Explain the principles that underpin a materials self-assembly chemistry approach to a better (a) polymer light emitting diode; (b) biological fluorescent tag; (c) polymer electrolyte; (d) third harmonic generation thin film; (e) mask for nanolithography; and (f) microlens array for projection lithography.

2. Discuss structure-property design principles that underpin the synthesis of a self-assembled material that functions as a (a) 1D metal; (b) 2D semiconductor; (c) 2D metal; (d) 2D superconductor; (e) optically transparent thermal insulator; (f) low dielectric constant film; (g) anode in a solid oxide fuel cell; (h) cathode in a dye sensitized solar cell; (i) selective CO sensor; and (j) air-bridge between photonic crystal components on a photonic chip.

3. How might self-assembly techniques contribute to the manufacture of miniature components required for the construction of a micron size satellite, for stealth monitoring of the surface of the earth.?

4. Describe the different ways of engineering purely through materials chemistry and self-assembly (a) a lattice of periodic mesoporous silica microspheres; (b) a square array of nanorings made of aurothiol nanoclusters; (c) a photonic crystal made of nanorods; (d) a nanocylinder made of stacked polymer nanorings; and (e) an inverse opal made of a carbon nanotube-polyphenylenevinylene nanocomposite. Describe where these novel forms of matter might find utility.

5. Defects control the properties of materials and there is no such thing as a perfect crystal and if it did exist it would be useless. Defend this statement in the context of perfecting imperfection at the microscopic, mesoscopic, and macroscopic length scales.

6. Describe how materials self-assembly is contributing to improved performance LEDs and FETs made of polymers, inorganics, and hybrids of these materials.

7. Contrive of a chemical lithography process for making ohmic contact to (a) multilayer metal phosphonate LB film; (b) a colloidal photonic crystal; and (c) a SAM.

8. Why does the T_c of a BiSrCuO defect perovskite ceramic superconductor not change on intercalating a 5 nm thickness (cetylpyridinium)$_2$HgI$_4$ bilayer between the BiO layer-planes. How would you utilize this knowledge to make a synthetic Josephson junction?

9. How would you expect the properties of the polymer electrolyte PEO–LiCF$_3$SO$_3$ to compare with PEO–ncTiO$_2$–LiCF$_3$SO$_3$ and PEOSH–Au$_n$–LiCF$_3$SO$_3$.

10. What materials self-assembly strategy could you devise to thread a nanorod through a nanoring.

11. Imagine a nanoscale automobile built by self-assembly and directed assembly. Describe the possible building blocks to construct this car, and ways these can be linked together in a functional fashion. How could one then make a nanoscale track for this nanocar?

12. Explain how synthetic command over diffusion lengths in solid-state nanomaterials can be used to advantage to build a better lithium solid-state secondary battery cathode. Give a real example.

13. Can you devise a way of assembling a chain of nanorings?

14. Can you think of a means of making nanorings in the form of the logo for the Olympic games – what would you name such a nanoring construct?

15. Three laser-based techniques are currently vying for supremacy in 3D photo-lithography of polymer resists, describe the chemistry and physics that underpin these methods and think how you might apply them to an inorganic photoresist and explain why if reduced to practice this would represent a significant breakthrough?

16. How could you use carbon black and an organic polymer to make a molecule discriminating chemical sensor?

17. Through precision injection of fluorocarbon oil and water *via* separate inlets into a microfluidic channel of a PDMS chip, it is possible to create a stream of phase-separated single size microdroplets of either water in oil or oil in water, in which the diameter, separation and content of the droplets are under strict control. With real examples, how could you make use of this simple concept to conduct combinatorial chemistry that would be difficult to perform by other methods?

18. Can you think of a way of orchestrating the microfluidic phase separation microdroplet generator to make a photonic crystal?

19. It is said by those who live on the coast that sea air has the smell and taste of salt, how can nanochemistry account for this natural phenomenon and think how you could use this knowledge to develop an original program of research?

20. Concoct a means of synthesizing free standing, micron diameter, single-crystal silicon rings, and provide a reason for doing this in the first place?

21. Can you come up with a microfluidic chip design that allows you to play pinball with microspheres?

22. Could a microfluidic chip be used to optimize the monodispersity of silica microspheres made by Stöber chemistry?

23. Let us say you modified a gold surface with a 1:1 mixture of two distinct alkanethiols. Then, spin-coat a block copolymer film onto this substrate and thermally anneal so it self-assembles into an ordered mesophase. Would the self-assembly occur differently than on a non-modified surface? What do you think would happen to the underlying substrate?

Nanolabs

CHAPTER 13

Nanochemistry Nanolabs

"By far the best proof is experience"
Francis Bacon (1561–1626)

It occurred to us as the text of *Nanochemistry* began to self-assemble and take shape, so-to-speak, that a knowledge of how to make, touch, see, arrange, measure, find function and determine a use for nanomaterials would be rather sterile and incomplete without the opportunity for the student of nanoscience, chemists, physicists, materials scientists, engineers and biologists, to experience these things first hand. There is no better way to remedy this situation than to actually do it hands-on in a nanochemistry laboratory setting, and get a first hand experience of the methods of working with things very small. In what follows, a first step in this direction is presented – an outline of the undergraduate nanolab of the future!

This laboratory is designed to introduce the student to the emerging field of nanochemistry through some carefully chosen experiments that collectively illustrate the practice of synthesizing, organizing, visualizing, measuring and utilizing a range of nanomaterials, the underpinnings of which have been presented in the textbook *Nanochemistry – a chemical approach to nanomaterials*.

The nanochemistry experiments outlined in this laboratory manual have been devised to mirror the content of the nanochemistry textbook, the goal being to illustrate the bottom–up building block philosophy for making nanomaterials using the methods of chemical synthesis combined with self-assembly, sometimes directed by porous templates and soft lithographic patterned surfaces.

As the text of *Nanochemistry* describes, building blocks can be made out of most known organic, inorganic, polymeric and hybrid materials. A key objective of the nanochemistry laboratory is to train the student in (i) synthetic methods for making and organizing different kinds of building blocks, with a particular composition, size and shape and surface functionalization, (ii) imaging techniques for their visualization, (iii) diffraction, microscopy, spectroscopy, electrical, optical, electrochemical and photochemical strategies for determining the structure and properties and (iv) to creatively imagine and practically implement how this knowledge can be orchestrated to establish a function and utility for the building blocks and their self-assembled systems.

From these hands-on experiences in the nanochemistry laboratory the student will learn the challenges of synthesizing building blocks having a particular size, identical size and shape, and will realize that the consequences of polydispersity will be manifest in the achievable degree of structural perfection and nature of defects in a self-assembled system, and ultimately how this will affect its usefulness. Another learning experience for the student will be to discover that it is very demanding to make building blocks with a particular volume and surface structure and composition, surface charge, hydrophobicity or hydrophilicity, and functional groups. These properties will control interactions between building blocks as well as with solvents, interfaces and substrates, which ultimately determine the geometry and distances at which building blocks come to equilibrium in a static or dynamic self-assembled system.

Experiments performed by the student in the nanochemistry laboratory will show how building blocks can self-assemble amongst themselves or co-assemble with structure directing templates or chemically patterned substrates, over length scales from nanometers to centimeters, to make new kinds of electrical, optical, photonic and magnetic materials with hierarchical structures and complex form. This is the central theme running throughout the book *Nanochemistry*, and the goal of the experiments in the nanochemistry laboratory is to illustrate these guiding principles through a hands-on and state-of-the-art experience in making nanomaterials in a chemistry laboratory setting. Any student of nanoscience with a background in chemistry, physics, materials science, engineering or biology will greatly benefit from the nanochemistry laboratory experience.

In the following section we will outline a series of 20 experiments taken from the different topics covered in *Nanochemistry*. Although a set of detailed procedures will be published separately from this text in the future, these examples will undoubtedly guide students and lab instructors intent on exploring nanochemistry for themselves.

1-Luminescent Nanoring Array

The soft lithographic patterning of luminescent nanorings will be accomplished by microcontact printing wettability patterns with self-assembled monolayers. Nanorings will be made by sol–gel chemistry, which will provide rare-earth doped photoluminescent oxides, which can then be probed.

2-Ferromagnetic Nanocrystal Array

Soft lithography will be used to make hydrophilic squares on a hydrophobic substrate. Solution phase precursors will then be patterned by selective wetting, followed by solid-state reaction to form a magnetic ferrite nanocrystal array. The magnetic properties of this array will be directly visualized by the magnetic assembly of magnetic microspheres and nanoparticles.

3-Zeolite Membrane

Zeolite A cubic morphology crystals will be synthesized by hydrothermal synthesis. Cation and anion surface functionalization of zeolite crystals and glass slide will enable the layer-by-layer electrostatic self-assembly of designer zeolite crystal films.

4-Electrochromic Device

In this lab experiment, Prussian-blue nanocrystals will be synthesized, and will be used in the layer-by-layer electrostatic self-assembly of electrochromic films and patterns. Both the soft lithography patterning of surface charge and microtransfer printing of polyelectrolyte multilayers will be used for this purpose.

5-Size Reduction Soft-Lithography

In this lab experiment, various techniques for reducing the feature size of soft-lithographically printed patterns will be investigated. Amongst the techniques will be: (i) reactive spreading of an alkanethiol, (ii) printing with a high pressure applied to the stamp, (iii) multiple printing on a single substrate and (iv) electrochemical "whittling" of patterns.

6-Self-assembly of Barcoded Magnetic Nanorods

Nanowires will be made by electrochemical and electroless deposition of metals inside nanochannel templates. Modulating the metal composition will provide barcoded nanorods with ferromagnetic and diamagnetic segments, allowing magnetization of nanorods and magnetic self-assembly on their own or with magnetic nanoparticles.

7-Carbon Nanotube Field Emitting Display

Arrays of carbon nanotubes will be formed by chemical vapor deposition growth from a catalytic metal nanoluster array, formed by soft lithography. The field-emission properties of this array will be investigated.

8-Photoconducting Selenium Nanowires

Selenium nanowires will be synthesized in the solution phase by nucleation and anisotropic nanowire growth. Nanowires will be electrostatically self-assembled across electrical contacts, enabling photocurrent measurements.

9-Metal Colloids

The synthesis of gold nanoparticles will be performed using citric acid reduction in aqueous solution. Hydrophobic nanoparticles of gold and silver will be prepared using phase-transfer synthesis. Sizes of nanoparticles can be probed by plasmon absorbance and microscopy.

10-Metal–Nonmetal Transition

In this lab experiment, the plasmon-coupling of metal nanoparticles will be investigated. Coupling between nanoparticles can be probed by spectroscopy on self-organized nanoparticle superlattices. Monolayers of nanoparticles on an air–water interface will be compressed until the onset of the nonmetal–metal transition, identified as a metallic reflectivity.

11-Near Infrared Emitting Quantum Dots

Lead sulfide nanocrystals will be synthesized by nucleation and arrested growth in a high temperature solvent and capping ligand. Size-selected crystallization, imaging of quantum dots, and IR photoluminescence spectroscopy will be investigated.

12-Nanocrystals in Nanobeakers

Nanoscale chemical beakers will be synthesized by embossing indents into an aluminum surface using a self-assembled film of silica microspheres. Different types of nanocrystals will then be grown in these indents: Au nanocrystals by dipping the embossed aluminum in $AuCl_4^-$, then reducing with BH_4^-, CdS nanocrystals by dipping in $Cd(ac)_2$, then treating with H_2S. The optical properties of these nanocrystals, such as absorption and fluorescence, will be probed.

13-Colloidal Photonic Crystal Fingerprinting

Silica microspheres will be synthesized, and self-organized as single-crystal films. The structure will be infiltrated with a polymeric elastomer, followed by etching of the silica from the polymer–silica colloidal crystal composite to create an elastomeric inverse colloidal crystal. Pressure-sensitive optical Bragg diffraction measurements on films will be performed, including their response to fingerprint pressure.

14-Colloidal Crystal Capillary Column

Microspheres will be synthesized and packed in capillaries using pressure-assisted assembly. The optical Bragg diffraction of the capillary columns will be used to separate octane, nonane and decane by optical chromatography.

15-Low Dielectric Constant Film

Periodic mesoporous organosilica films made by evaporation-induced self-assembly will be made, using organically modified silsesquioxane precursors and tri-block co-polymer templates. Extraction of the template will result in nanoporous films, which can be post-modified by high-temperature treatment, and whose humidity dependent dielectric constant will be probed.

16-Block Copolymer Lithography

In this lab experiment, students will explore the patterning opportunities offered by polystyrene–polymethylmethacrylate (PS–PMMA) block copolymers. Films of this commercially available polymer will be spin-coated on silicon and glass substrates, forming different morphologies: PMMA cylinders or spheres in a PS matrix, PS cylinders or spheres in a PMMA matrix, or PS–PMMA lamellae. Treatment of the film with UV light and oxygen, followed by an acetic acid rinse, removes the PMMA and crosslinks the PS. Different patterning methods can then be performed and evaluated: etching the underlying substrate (alcoholic KOH for silicon, aqueous HF for glass), evaporating or sputter deposition of metals into the porous regions, or organizing nanoclusters (such as commercially available gold) into the nanopits by capillary forces.

17-Virus Mineralization

The tobacco mosaic virus (TMV) is a commercially available, self-assembled rod-shaped virus. In this lab experiment, the mineralization of the TMV will be investigated by sol–gel chemistry to form metal oxide and semiconductor nanowires.

18-Biological Structures and Templates

In this lab experiment, the structure of various biological materials will be probed by microscopy. The materials investigated may include evergreen wood, deciduous tree wood, bamboo, oyster shell, silk, bacterial filaments, sea urchin, amongst others. The potential of these materials as templates will be investigated by infiltrating them with nanoparticles and sol–gel precursors.

19-Mesoscopic Self-Assembly

Millimeter scale polydimethylsiloxane (PDMS) hexagonal shapes will be prepared, and specific faces of the hydrophobic hexagons will be functionalized to be hydrophilic. The capillary self-assembly of the patterned hexagonal shapes will be performed at liquid–liquid, and air–liquid interfaces.

20-Colloidal Crystal Shapes

Drops of a dispersion of silica microsphere from a syringe will be injected onto the surface of perfluorodecalin in a Petri dish held at a temperature that enables evaporation-induced self-assembly of the microspheres within the spatial confines of the droplets. Different colloidal crystal morphologies that form under a variety of synthesis conditions will be investigated by microscopy and optical spectroscopy.

Appendix A: Origin of the Term "Self-Assembly"

"Thereafter, I showed how the greatest part of the matter of this chaos must, in accordance with these laws, dispose and arrange itself in such a way as to present the appearance of heavens; how in the meantime some of its parts must compose an earth and some planets and comets, and others a sun and fixed stars"
– Rene Descartes, Discourse on Method (1637)

The meaning of the term *self-assembly* is not so easy to explain, as attested by reading the chapters of this book. *Self-assembly* as a fundamental principle, which creates structural organization from disordered components in a system, pervades many fields and suffuses the boundaries of many disciplines. These days it is virtually impossible to read a paper in the chemistry literature without being confronted with the term *self-assembly*. One gets the impression from the ubiquity of this expression that *matter of all kinds and over all length scales*, be they atoms or molecules, clusters or colloids, polymers or biopolymers, can undergo *self-assembly* to a higher level of structural organization and complexity. But where did the expression *self-assembly* actually originate?

It turns out that the derivation of the term *self-assembly* is not so easy to unearth in an unambiguous way! Nevertheless, it is a pertinent contemporary enquiry to define the origin of the term because *self-assembly* as a synthetic methodology is omnipresent in chemistry. Specifically in the context of the burgeoning field of Nanochemistry, synthetic *self-assembly* has emerged as a powerful means of making nanoscale materials and organizing them into functional constructs designed for a specific purpose.[1] So where did the term *self-assembly* have its roots?

The ancient Greek philosopher Democritus (Figure A.1), around 400 BC, expounded the idea that atoms and voids organized in different arrangements constitute all matter. His theory of how the universe began, from the minutest atomistic building blocks to stars and galaxies, can be considered to represent the oldest recorded vision of matter undergoing self-assembly over all scales, an underlying theme running throughout the book *Nanochemistry*. Although much of the writing of Democritus was destroyed in the fall of Rome, could it be that this Greek philosopher foresaw *self-assembly* but without explicitly using the words?

Figure A.1 *Democritus, forefather of self-assembly. The image shows Democritus Laughing*, a painting by Hendrick ter Brugghen (1628)
(Reproduced with permission from Wikipedia, The Free Encyclopedia, http://en.wikipedia.org)

More than two thousand years after Democritus, the French philosopher and mathematician René Descartes (Figure A.2) in his *Discourse on Method*, imagined the organization of the universe arising out of chaos, according to the laws of nature, from the coming together of the smallest objects to form larger and larger aggregations displaying order over different length scales. It would appear that René Descartes thinking about the natural laws, which govern the formation of the universe, set the stage for the theories underpinning self-organizing systems in physics (*complex systems*), biology (*growth and form*) and mathematics (*cellular*

Figure A.2 *René Descartes, oft called the father of modern philosophy and founder of modern mathematics, also made possible key developments in the understanding of self-assembly. Shown here is a portait painted by Frans Hals (ca. 1649)*
(Reproduced with permission from Wikipedia, The Free Encyclopedia, http://en.wikipedia.org)

automata). In all these fields, the language seems to be synonymous with *self-assembly* in chemistry by which the internal organization of a system spontaneously increases without being directed by an external influence. But nowhere in the writings of René Descartes can one find direct mention of *self-assembly*. So where exactly does this simple, yet inherently complex, hyphenated combination of the words *self* and *assembly* first appear in the context of a synthetic methodology in chemistry?

The 20th century biomathematician D'Arcy Wentworth Thompson in his 1912 book *On Growth and Form* used these ancient ideas, combined with 18th century theories of the naturalist movement, as a springboard to try to explain the biological form problem through the use of physical, mathematical and mechanical principles. From single cells to organic form and from simple inorganic chemicals to biomineral form, Thompson identified numerous correlations between phenomena and systems in the physical and biological world that supported his theory of morphogenesis (from the Greek *morphê,* shape and *genesis,* creation). A search of Thompson's classic book reveals no evidence that the term *self-assembly* is ever used as a descriptor for his theory, however it is implicit in his thesis that physical-mechanical forces drive all biological matter at all scales to an observed form.

Another clue to the origin of the term *self-assembly* emerges in 1935, when Blodgett working with Langmuir at General Electric Schenectady[2] recognized that collections of amphiphilic molecules, having hydrophilic head groups and hydrophobic tails, could be used to make molecular assemblies on solid surfaces. They achieved this by first forming a closely packed monolayer film of these molecules at the interface between air and water, accomplished by laterally compressing a random collection of the molecules localized at the air-water boundary. The system evolves from a chaotic to a well-organized state and the close-packed molecular assembly is subsequently transferred to a solid substrate by simply dipping and withdrawing the substrate into the water phase and through the interface. Repetitive transfer of these monolayers to the substrate enabled the creation of multilayer films. Although *self-assembly* is clearly operating in molecularly organized Langmuir-Blodgett monolayers formed at liquid and solid surfaces, one cannot find usage of the term *self-assembly* in any of this pioneering work.

In a similar vein, a much quoted paper published in 1946 entitled *Oleophobic Monolayers 1. Films Adsorbed from Solution in Non-Polar Liquids* by Zisman, Bigelow and Pickett[3] is often given the credit for the beginning of synthetic *self-assembly*. Curiously, nowhere in the paper is the terminology *self-assembly* to be found. There can be little doubt, however, in assuming the authors of this work recognized that a dilute solution of long chain alkylamines in a non-polar solvent were able to adsorb and self-organize onto the surface of platinum to form a densely packed monolayer. Clearly this assemblage of close-packed adsorbed molecules represents the dawn of a materials construct, analogous to the one discovered by Nuzzo and Allara some 37 years later. This latter, now affectionately called self-assembled monolayers (SAMs), are synthesized by chemisorbing alkanethiol molecules from a non-polar solvent onto the surface of gold.[4] This synthetic *self-assembly* strategy was first described in their 1983 paper entitled *Adsorption of Organic Disulfides on Gold Surfaces*.

Has the above information clarified where the term *self-assembly* originated? It seems the early notion that universal laws of nature coerce matter of all kinds and at all scales to organize into systems of increasing complexity provided the inspiration that set the scene for developing the language of *self-assembly*. In chemical *self-assembly* as we know today, it is second nature that ordinary chemical forces typified by ionic, covalent, hydrogen, non-covalent and metal-ligand coordinate bonding interactions drive molecules to self-assemble into higher order structures that can display properties distinct to the individual components. Lehn's pioneering supramolecular chemistry[5] emerged from application of this toolbox of molecule *self-assembly* forces.

The central tenet pervading the textbook *Nanochemistry* is that *materials self-assembly* can transcend Lehn's *molecular self-assembly*. The *Nanochemistry* thesis is that materials building blocks with designed sizes and shapes, bulk composition and surface functionalities, in contrast to the molecules and macromolecules of chemistry, can undergo *self-assembly* into higher order structures and at "all" scales[6,7] through the operation of forces beyond those of ordinary chemical bonds mentioned above. In this context, synthetic chemical *self-assembly* exploits capillary, colloidal, elastic, electric, magnetic, optical, and shear driving forces between building blocks, as well as traditional chemical interactions, to create new materials and functional architectures.

Arguing in this way, we have come full circle. It seems the ideas that underpin *self-assembly* are essentially encompassed in the thinking of Democritus and Descartes. While the hyphenated words *self* and *assembly* were not used explicitly by these early thinkers, *self-assembly* was clearly implicit in the thinking of their time.

References

1. G.A. Ozin, Nanochemistry: Synthesis in Diminishing Dimensions, *Adv.Mater.* 1992, **4**, 612.
2. A. Ulman, An Introduction to Ultrathin Organic Films from Langmuir-Blodgett to Self-Assemblies, Academic Press, San Diego, 1991.
3. W.C. Bigelow, D.L. Pickett, W.A. Zisman, Oleophobic Monolayers 1. Films Adsorbed from Solution in Non-Polar Liquids, *J. Colloid Sci.* 1946, **1**, 513.
4. R.G. Nuzzo, D.L. Allara, Adsorption of Bifunctional Organic Disulfides on Gold Surfaces, *J. Am. Chem. Soc.* 1983, **105**, 4481.
5. J.-M. Lehn, Supramolecular Chemistry, VCH, New York, 1995.
6. G.A. Ozin, Panoscopic Materials: Synthesis Over "All" Length Scales, *Chem. Commun.* 2000, 419.
7. G.M. Whitesides, B. Grzybowski, Self-Assembly at All Scales, *Science*, 2002, **295**, 2418.

Appendix B: Cytotoxicity Of Nanoparticles

"Out of this nettle, danger, we pluck this flower, safety."
– William Shakespeare (1564–1616)

In the introduction to *Nanochemistry*, a brief mention on the subject of health and safety of nanomaterials was presented. It was emphasized that no chemical is considered safe until proven otherwise, and it is therefore prudent to assume the same principle should be applied to nanochemicals.

Since this text was completed in the summer of 2004, a few papers have appeared dealing with the toxicity of nanomaterials towards cells (cytotoxicology). These begin to address the concern of potential hazards that could arise from the utilization of nanoparticles in future life science and biomedical applications. Such studies are timely considering some earlier trendsetting papers, demonstrating the value of fluorescent semiconductor nanoparticles in cell labeling, magnetic metal nanoparticles as contrast agents in magnetic resonance imaging, and high electron contrast gold nanoparticles for histoimmunological staining, to name just a few cases.

Key safety issues surrounding the use of nanopaticles in biomedical applications relate mainly to their biocompatibility with cells. While a material at the microscopic scale might be innocuous when introduced into an organism, it could become a health hazard on down sizing it to the nanoscale. This is because the minute size of nanoscale particles endows them not only with physicochemical properties that are distinct to the bulk form, arising from quantum size effects, but also greatly amplifies the concentration of reactive surface sites given the vastly increased surface-to-volume ratio of nanoparticles compared to the bulk form.

Current literature on this subject teaches the different ways in which nanoparticle-cell interactions may have a deleterious effect on cell function. The negative effects of nanoparticles on cells fall into three main categories: (i) nanoparticle corrosion within the cell membrane or in the proximity of cells could release toxic chemicals causing cell death, (ii) nanoparticle ingestion through cell membranes or adhesion to the cell surface may impair cell function, (iii) the particular shape of nanoparticles, like sphere versus fiber shapes, of the same composition may exhibit distinct toxic effects on cells.

In what follows some early cytotoxicology results will be briefly summarized in an attempt to throw some glimmer of light on the various ways ligand-capped CdSe and CdSe-ZnS semiconductor nanoparticles, composed of inherently toxic materials, can influence cells. Three studies have been reported on this topic using different cell types.

One of these studies[1] utilized primary hepatocytes (liver cells) as a liver model, and studied their cytocompatibility in the presence of CdSe nanoparticles. The results shows that CdSe nanoparticles in water release Cd^{2+}, the concentration of which shows a correlation with cytotoxic effects. These studies also revealed that air and UV exposure enhance nanoparticle deterioration and the extent of liberation of free Cd^{2+} ions into solution. This effect could be significantly impeded, but not eliminated, by either capping with an appropriate ligand like mecaptopropionic acid, wrapping with a biopolymer like bovine serum albumin or coating with a sheath of semiconductor like ZnS. With this knowledge, it proved possible to create relatively cytocompatible luminescent CdSe nanoparticles that were able to track the migration and reorganization of cells *in vitro* through their narrow band light emission These observations provided a means of establishing design criteria for minimizing toxic heavy metal release from nanoparticles, thereby paving the way to *in vivo* applications of nanoparticles in the life sciences and medicine.

A related study[2] employed luminescent CdSe-ZnS nanoparticles with thiolate capping ligands having terminal carboxylic acid, polyalcohol and amine groups as well as combinations thereof to probe nanoparticle cytotoxicity. This study arrived at similar conclusions, namely that the biological behavior of nanoparticles depend upon surface modification. The control of nanoparticles surfaces can enhance their stability, which is vital for any perceived future applications in biomedicine.

In another related study[3] nanoparticle coatings included mercaptopropionic acid, phosphonated-silica, polyethyleneglycol-silica, an amphiphilic polymer, as well as combinations thereof. Nanoparticle sizes were in the range 10–24 nm. In the case of the mecaptopropionic acid capping ligand, surface cadmium atom concentration limits were established for different CdSe and CdSe-ZnS nanoparticles at which fibroblast cells were poisoned from released Cd^{2+}. This occurred at critical surface concentrations ($[Cd]_{surface}$) of 0.65 ± 0.12 μM and 5.9 ± 1.3 μM for CdSe and CdSe-ZnS nanoparticles, respectively. This limiting value was increased by an order of magnitude on coating the nanoparticles with ZnS, a response that was attributed to diminished release of noxious Cd^{2+}. By applying a PEG-SiO$_2$ coating to CdSe and CdSe-ZnS nanoparticles, toxic effects were only observed when the $[Cd]_{surface}$ critical surface concentration reached $30\,\mu$M. This was ascribed to diminished nanoparticle uptake by the cell presumably due to the decreased cell adhesion effect of the PEG coat. Amphiphilic polymer coatings on CdSe nanoparticles acted as better Cd^{2+} diffusion barriers than mercaptopropionic acid capping ligands on the same nanoparticle, however this was not the case for CdSe-ZnS nanoparticles most likely because of their precipitation on the cell surface.

These initial studies of the cytotoxic behavior of nanoparticles on cells have shown that release of Cd^{2+} from the surface of the nanoparticles, as well as the precipitation of nanoparticles on the cell surface, can lead to cell death (apoptosis) and cause cell impairment. Such cytotoxic effects are highly dependent on the chemical composition of the nanoparticles as well as the chemical nature of surface capping ligands or coatings. Size, surface area and surface functionality clearly are major factors to take into account when considering any type of clinical application of nanoparticles. It seems from these initial investigations that through judicious design of corrosion protection coatings it should be possible to improve the cytocompatibility of nanoparticles. However, at this stage in our understanding of nanoparticle cytotoxicity it is wise to presume that no nanoparticle is safe until proven otherwise!

References

1. A.M. Derfus, W.C. Chan, S.N. Bhatia, Probing the Cytotoxicity of Semiconductor Quantum Dots, *Nano Lett.* 2004, **4**, 11.
2. A. Hoshino, K. Fujioka, T. Oku, M. Suga, Y.F. Sasaki, T. Ohta, M. Yashuhara, K. Suzuki, K. Yamamoto, Physicochemical Properties and Cellular Toxicity of Nanocrystal Quantum Dots Depend on their Surface Modification, *Nano Lett.* 2004, **4**, 2163.
3. C. Kirchner, T. Liedl, S. Kudera, T. Pellegrino, A.M. Javier, H.E. Gaub, S. Stolzle, N. Fertig, W.J. Parak, Cytotoxicity of Colloidal CdSe and CdSe/ZnS Nanoparticles, *Nano Lett.* 2005, **5**, 331.

Appendix C: Walking Macromolecules Through Colloidal Crystals

"My father considered a walk among the mountains
as the equivalent of churchgoing"
– Aldous Huxley (1894–1963)

Stationary phases used in chromatography usually have random pore structures, and their ability to separate molecules and macromolecules relies on selective permeability and/or diffusion. It has been predicted that by changing to an ordered porous media, the speed and resolution of a separation could be enhanced. As a first step in this direction, colloidal crystal capillary columns (C^4) and inverted versions thereof (IC^4), with high structural and optical quality, have recently been shown to function well as both a stationary phase and an optical Bragg detector, pointing the way to optical chromatography.[1] The C^4 and IC^4 materials consist of a three-dimensional periodic arrangement of solid spheres or air-spheres and constitute the stationary phase of these microcapillary chromatographic columns. These columns allow for the wavelength selective interaction of electromagnetic radiation with the stationary phase. The photonic stop-band responsible for the structural color of the column is monitored spectroscopically, and shifts of its wavelength generated by minute refractive index changes due to an analyte within the mobile phase can be immediately detected at any point along the column.

Structurally well-defined porous solids of this genre provide exciting opportunities for directly visualizing the motion of macromolecules in restricted geometries, knowledge of which has important implications when attempting to design an improved performance chromatography stationary phase.

An inverted poly-(acrylamide) hydrogel colloidal crystal monolayer turns out to be an ideally structured stationary phase for studying the diffusion of fluorescently labeled DNA, with different numbers of base pairs, from one cavity to another through adjoining necks as illustrated in the top of Figure C.1. To reduce this to practice it was necessary to tag DNA molecules in an aqueous buffer with a fluorescent dye, then coerce them into the cavities by the application of a small electric field orthogonal to the inverted poly-(acrylamide) colloidal crystal monolayer.

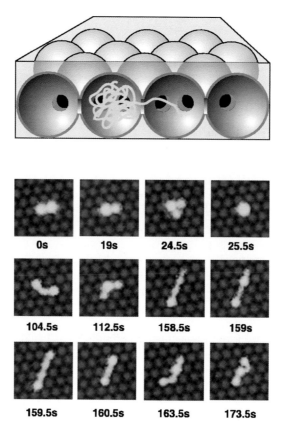

Figure C.1 *Schematic showing a DNA strand imbibed within a cavity of an inverted colloidal crystal monolayer, top. Bottom, configurations observed for a single 48.5 kbp DNA strand at different times*
(Reproduced with permission from Ref. 2)

A trajectory for a 7.25 kilobasepair (kbp) DNA strand moving from cavity to cavity within this kind of monolayer has been directly visualized by overlaying 500 consecutive images recorded every 0.5 s in a fluorescent optical microscope. Real time observations of the diffusion pathway seem to suggest that DNA motion through this ordered porous medium is occurring by short hops between cavities with long rest times within cavities. While a DNA rests in a cavity section it appears to continuously explore and sample the available adjacent spaces before it slides through a neck to a neighboring cavity.

DNA jump dynamics were studied in this way for samples with 2.69 to 48.5 kbp. Below 2.69 kbp it is difficult to discern the motion relative to the fluorescence background, while between 2.69 to 24 kbp the behavior is essentially the same as that described above. Above 24 kbp however, DNA chains start moving by a combination of diffusive hopping and reptation between cavities, more akin to bunching-stretching motion of an inchworm. Time resolved images of this larger DNA strand show that the more prevalent resting configurations involve two or more cavities where linear ones outnumber bent ones with just a few offshoots, as seen in the bottom of Figure C.1.

This study shows that a well-ordered stationary phase such as 2D inverted colloidal crystal can be used advantageously to determine the molecular weight dependence of diffusion and reptation mechanisms of DNA molecule movement through the reticulated cavities. It is worth noting that in a sieving process, smaller macromolecules diffuse more quickly than larger ones through cavity, channel and neck constrictions. Diffusion of a macromolecule by activated jumps through a neck is entropically driven as the configurational freedom is greater in a cavity than in a neck. As the molecular weight of the polymer increases, it becomes entropically disfavored for the macromolecule to reside in one cavity, and reptation starts to become the dominant mechanism of motion.

The ability to directly visualize, in real time, macromolecule diffusion in structurally well-ordered porous media has very important implications in the field of separation science and photonic crystal science. Understanding the control parameters for macromolecular diffusion in media with "crystalline porosity" will be especially valuable in the development of C^4 and IC^4 materials as improved performance chromatography stationary phases for simultaneously separating and detecting macromolecules.[1]

Information about macromolecular diffusion in such media is also vitally important in the quest for nanometer scale precision tuning of the photonic crystal properties of colloidal crystals using layer-by-layer electrostatic self-assembly of polyelectrolyte multi-layers within the spatially confined geometry of a colloidal crystal.[3] To achieve success in this endeavor it will be imperative to understand the diffusion mechanism of a polyelectrolyte within the colloidal crystal lattice as a function of the molecular weight and charge density of the polyelectrolyte, as well as the ionic strength of the polyelectrolyte solution and the surface potential of the colloidal crystal.

References

1. U. Kamp, V. Kitaev, G. von Freymann, G.A Ozin, S. Mabury, Colloidal Crystal Chromatography Columns – Towards Optical Chromatography, *Adv. Mater.* 2005, **17**, 438.
2. D. Nykypanchuk, H.H. Strey, D.A. Hoagland, Brownian Motion of DNA Confined Within a Two-Dimensional Array, *Science*, 2002, **297**, 987.
3. A.C. Arsenault, J. Halfyard, Z. Wang, V. Kitaev, G.A. Ozin, I. Manners, A. Mihi, H. Miguez, Tailoring photonic crystals with nanometer-scale precision using polyelectrolyte multilayers, *Langmuir*, 2005, **21**, 499.

Appendix D: Patterning Nanochannel Alumina Membranes With Single Channel Resolution

"One must be drenched in words, literally soaked in them, to have the right ones form themselves into the proper pattern at the right moment"
– Winston Churchill (1874–1965)

One of the most versatile and useful templates in the toolbox of nanochemistry used for making diverse kinds of functional nanowires, nanotubes and membranes with myriad uses is unquestionably nanochannel alumina (see Chapter 5). Films and membranes of this template, bearing periodic arrays of hexagonally close-packed narrow pore-size distribution nanochannels, are synthesized by lithographically directed anodic oxidation of aluminum.

An extremely well-ordered nanochannel array has recently been obtained by using a field ion-beam (FIB) for pre-patterning the aluminum to guide the growth of nanochannels in the alumina.[1] The sample can also be imaged in tandem, by collecting the secondary electrons generated during ion bombardment.

A remarkable advance that emerged from FIB processing of these templates was the discovery that the FIB process could also be used to pattern the template itself with custom designed arrays of nanochannels, Figure D.1. This was accomplished by ion-induced closing of the nanochannels in a process believed to involve either redistribution of ablated alumina, or lateral diffusion of alumina from the walls of the bombarded channels to their pore mouth. The ultimate spatial resolution achieved by FIB induced nanochannel closure was an impressive single line of pores and even a single open pore.

The demonstration that FIB can be utilized for guiding the growth, imaging the formation and custom designing patterns of open and closed nanochannels in alumina films and membranes, with single nanochannel spatial resolution, opens new vistas for innovation and invention in nanochemistry. Some envisioned uses for patterned nanochannel films include functional defects in 2D photonic crystals for localizing and guiding light, integrated nanoelectronic and nanophotonic

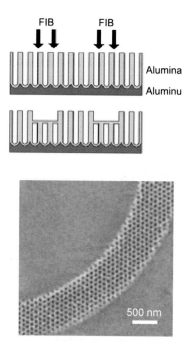

Figure D.1 *Process of selectively closing nanochannels using a focused ion beam, top.*
A section of a nanochannel annulus patterned by FIB, bottom.
(Reproduced with permission from Ref. 1)

devices and circuits made of nanowires and nanoclusters, spatially selective
filtration, high resolution lithographic masks and making use of encapsulated
cargos in the closed nanochannels for controlled release drug delivery chips.

Reference

1. N.-W. Liu, A. Datta, C.-Y. Liu, C.-Y. Peng, H.-H. Wang, Y.-L. Wang,
   Fabrication of Anodic-Alumina Films with Custom-Designed Arrays of
   Nanochannels, *Adv. Mater.* 2005, **17**, 222.

Appendix E: Muscle Powered Nanomachines

"I suppose that leadership at one time meant muscle but today it means getting along with people"
– Indira Gandhi (1917–1984)

One of Nature's greatest sources of biological machine power is muscle. It has enabled building of the great pyramids and body building, climbing Mount Everest and sub-10 s 100 meter sprinting, fighting battles and proliferation of humankind. But how about applying all this muscle power to accomplish machine tasks at the micro and nanoscale?

If muscle cells could be self-assembled together with synthetic micro- or nanostructures this might constitute a natural way to engineer tiny devices and machines powered by muscle. By this we mean the spontaneous organization and integration of muscle cells with small scale fabricated structures, rather than the manual process for achieving the same objective, the subject of earlier work. Because mature muscle is powered by massively parallel arrays of bimolecular motors, myosin sliding along acting filaments, they are ideal for performing machine tasks that demand very large amounts of power in the macroscopic world. This feat is not possible for individual biomolecular motors that working alone can only perform tiny jobs at the nanoscopic scale. Note that to power the motion of a device with a biological motor the available force of the motor must overcome all opposing forces, like drag and friction, which dominate in aqueous environments.

An important step towards the practical realization of a muscle powered nanomachine has been demonstrated by self-assembling living muscle cells with a free-standing micromechanical structure, used as a force transducer for measuring the mechanical properties of muscle as well as a device with movement powered by muscle.[1]

Three key steps underpin this accomplishment, first the ability to spatially direct the growth and maturation of muscle cells on specific locations of a microstructure, second control over the adhesion of muscle cells to the microstructure, and third enabling free movement of the hybrid muscle-microstructure while maintaining its integrity.

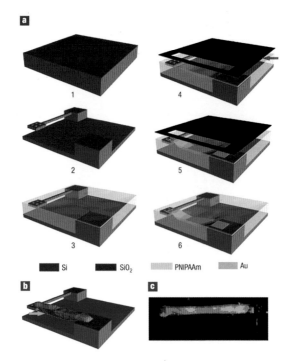

Figure E.1 *Process for making a hybrid muscle-micromechanical system.* **a)** (1–6) A Si (111) wafer with a surface layer of SiO$_2$ is used to fabricate cantilevers, coated next with a PNIPAA layer, using a shadow mask to etch a PNIPAA strip and deposit a nm thin Cr/Au layer therein (white arrow), removal of mask; **b)** A mature self-assembled muscle bundle (green arrow) traversing the space between cantilever and support by removing the PMIPAA layer. **c)** Fluorescence microscopy image of a muscle bundle stained with rhodamine– phalloidin
(Reproduced with permission from Ref. 1)

A scheme for achieving these objectives is outlined in Figure E.1. It utilizes a creative combination of silicon microfabrication and muscle self-assembly to coerce a bundle of mature muscle cells to spontaneously organize across the open space between a cantilever and a support. The three pivotal steps in this process involve making a silicon micromechanical structure, patterning of a sacrificial poly-N-isopropylacrylamide (PNIPAA) layer with a nanometer thin strip of Cr/Au and spatially directed culturing of muscle cells on the Au where they prefer to bind. Evidence that this could be achieved is seen in the fluorescence microscopy image of a dye-stained muscle bundle, which automatically assembled in the space between the mobile cantilever and the fixed support, as shown in Figure E.1c.

The self-assembly of muscle cell-microstructure hybrid constructs enabled creation of muscle based functional components, like a moveable foot on a cantilever without having to resort to earlier strategies of having to manually place muscle cells on the surfaces of microdevices. These kinds of biomechanical-actuators can be powered by physiological fluids containing glucose or they can be activated by an electrical stimulus, providing opportunities for the development of new classes of chemical and/or electrically driven devices and machines that integrate biological muscle with abiological micro and nanostructures.

Reference

1. J.Z. Xi, J.J. Schmidt, C.D. Montemagno, Self-Assembled Microdevices Driven by Muscle, *Nature Mater.*, 2005, **4**, 180.

Appendix F: Bacteria Power

"Bacteria keeps us from heaven and puts us there"
– Martin H. Fischer (1879–1962)

It is thought that bacteria inhabited the earth as far back as about 3.7 billion years ago. These ancient organisms can be both useful and harmful to animals and humans, plants and the environment. The role of bacteria is well known in disease and infection, nitrogen fixation to ammonia, degradation of organics, bioremediation of oil and toxic waste, fermentation of food and bioengineering of drugs. Bacterial shapes are quite varied, exemplified by rods, spheres and helices and aggregations thereof. They can be separated into two phyla by a staining technique named after the discoverer Hans Christian Gram, and are known as Gram-positive and Gram-negative, distinguished by their cell membrane composition. The former has a lipid membrane, while the latter has an additional peptidoglycan cell wall.

Bacteria reproduce asexually by cell division whereby two daughter cells are created with formation of a transverse cell membrane. The metabolism of bacteria is quite varied but can be divided into two main classes, those that depend on an organic source of carbon known as heterotrophs and those that are able to synthesize organic compounds from carbon dioxide and water called autotrophs. Autotrophs can be further divided into chemotrophs and phototrophs depending on whether they obtain their energy from chemistry (lithotrophs, inorganic or organotrophs, organic) or light or combinations thereof. For example, cyanobacteria are amongst the oldest organisms known from the fossil record. They are photolithoautotrophs and are believed to have played a central role in creating the Earth's oxygen atmosphere using the photosynthetic water splitting apparatus (the oxygen evolving complex). The growth behavior of bacteria in oxygen categorizes them into three groups, aerobes grow in oxygen, anaerobes grow in the absence of, and facultative anaerobes can grow in either.

Extremeophiles as their name implies are bacteria that can thrive in environments that are extreme for humankind, like hot springs (thermophiles), saline lakes (halophiles), acidic or alkaline environments (acidophilies and alkaliphiles) and even glaciers (psychrophiles). Bacteria motility is achieved in two main ways, the first *via* movement of a single, or cluster of, flagella arranged at one end of the cell or all over the cell. The second is achieved by gliding, tumbling, twisting or buoyancy motional modes of the cell. Finally, certain kinds of stimuli (taxes) attract or repel bacteria, exemplified by chemotaxis, phototaxis and magnetotaxis.

This is a brief synopsis of the "bacteria tool box" if one is to begin to imagine how bacteria power might be usefully employed in the emerging field of biomolecular nanotechnology.

Biological systems utilize the chemical energy stored in simple fuels, like glucose and ATP, to create mechanical and electrical energy. We have seen in Chapter 10 how biomolecular nanotechnology is beginning to make use of this stored energy for the creation of biomolecular devices and motors. So why not expand upon this theme to include bacterial power generation for performing useful nanomechanical tasks?

A step in this direction has recently been achieved through the anchoring of motile living bacteria to SAM patterned surfaces *via* designed interactions between their cell wall and flagella components and terminally functionalized alkanethiol SAMs on gold.

The key to forming patterned arrays of motile bacteria turns out to be the chemical linking strategy employed between bacteria and substrate. Proof-of-concept experiments were performed with *E. coli* bacteria that were directed to adhere to selected regions of a SAM by poly-L-lysine or *E. coli* antibody linkages, the bacteria being prevented from adhering to other regions of the support by PEG or alcohol terminated alkanethiol protection layers.

The SAM patterns were made by dip-pen nanolithography or microcontact printing, and the outcome of bacteria attachment to the SAMs was observed by AFM or LFM as well as optical and fluorescence microscopy. When contained in fresh liquid growth media, *E.coli* were seen to anchor by the aforementioned means to the patterned regions *via* cell wall and flagella. They were found to stay alive and motile for at least 4 hours, and they only adhered to the patterned areas of the substrate leaving the passivated regions untouched. Two-color fluorescence using two nucleic acid stains was a convenient way to differentiate live from dead

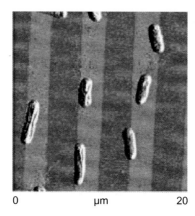

0   μm   20

Figure F.1   *LFM image of E. coli attached to a poly-L-lysine patterned gold surface, conforming to the size and shape of the pattern*
(Reproduced with permission from Ref. 1)

bacteria on SAM patterned arrays, thereby enabling the population of viable cells captured by the array to be elucidated. It was found that the majority ($>70\%$) of the anchored bacteria survived, showing that the biological activity of the cell is not adversely affected by the electrostatic, hydrogen bonding and van der Waals type interactions with the protein or antibody-antigen binding sites on the SAM. Single bacteria could be anchored on dots and organized into parallel lines (Figure F.1), and it was established by using a dot array with different dot sizes that there is a minimum feature size (1.3 μm) below which individual *E. coli* will not anchor.

This study demonstrates that the adhesion of motile living bacteria in specific numbers and orientation to functionalized patterned surfaces with features of pre-determined size and shape can be controlled. It is a first step towards the utilization of surface confined bacteria for performing specific machine tasks, like delivering a chemical or physical payload to a targeted destination or powering a MEMS device to perform a job. Bacteria are clearly muscling in on MEMS!

Reference

1. S. Rozhok, C.K.F. Shen, P.L.H. Littler, Z. Fan, C. Liu, C.A. Mirkin, R.C. Holz, Methods for Fabricating Microarrays of Motile Bacteria, *Small*, 2005, **4**, 445.

Appendix G: Chemically Driven Nanorod Motors

"The task ahead of us is never as great as the power behind us."
– Anonymous

One of the main hurdles in nanotechnology is the difficulty to interface nanoscale objects with power sources, which can endow them with a particular type of motion. The idea that anisotropic forces can be created on the body of a nano-object by an on-board chemical motor raises the exciting possibility of discovering whole new classes of entirely synthetic nanomachines, which can be "programmed" by synthesis and designed to perform specific tasks. Such "bottom-up" chemically powered nanomachines could provide the basis of a new genre of nanochem-omechanical systems (NCMS), in relation to "top-down" engineered silicon-based nanoelectromechanical systems known as NEMS.

It was recently demonstrated that the chemically powered linear motion of a nanorod could be achieved in solution, and that rotational motion could be realized when such a nanorod was tethered at one end to a surface. In these examples movement was provided by bubbles formed from the decomposition of hydrogen peroxide to oxygen and water, generated at a catalytic metal segment of a bimetal nanorod. Such self-propelled nanorods can also be subject to a rough "remote control", by incorporating a magnetic block into the structure and controlling its orientation with magnetic fields. However, it is still unclear whether the origin of the motion is gas propulsion, surface tension gradients, or something else entirely.[1-3]

It is important to note that while biologically powered movement of nano-objects has been demonstrated earlier, a classic case being an ATPase driven nickel nanopropeller,[4] the aforementioned examples of free-floating and substrate supported nanorod motion were the first to show that chemically powering a nano-object can be realized in a purely synthetic system!

While much needs to be clarified concerning the specific forces responsible for the mobility of these chemically powered nanorods, the possibilities for extending this concept are limitless. These include new nanorod compositions and catalytic reactions, where the nanorods may be freely suspended in solution, anchored to other nanorods or nano objects in pre-defined geometries, chemically tethered to

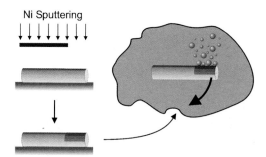

Figure G.1 *A metal such as nickel could be sputter deposited through a mask onto the end of a gold nanorod. After release from the substrate, addition of H_2O_2 could generate an asymmetric flow of oxygen from the nickel tip and cause axial rotation*

surfaces or contained within microfluidic channels. It is envisioned that exploring these cases will lead to chemically powered nanomachines designed to perform a range of tasks, some known yet others waiting to be discovered. This is fertile ground for innovation and invention and represents an exciting new direction in nanochemistry.

With a world of chemistry at our fingertips, there are countless types of chemical reactions and interactions we could use to power nanorods and other nanoparts. In addition, varying the geometry of the nanorods and considering nanorod-nanorod interactions can exponentially increase the achievable system complexity.

For instance, we could think about making a nanorotor which only spins axially. Ni metal can be deposited as a strip along one side of a Au nanorod by sputter deposition, or along the side of one end by sputtering through a mask (Figure G.1). When placed in aqueous H_2O_2 the evolved O_2 nanobubbles on the Ni side should propel the nanorods in an axial circular motion.

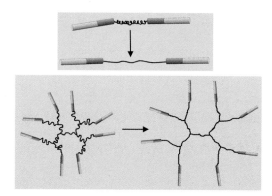

Figure G.2 *Chemically powered nanorods could be used to stretch out synthetic and biological macromolecules. The top image shows two nanorods stretching out a single coiled polymer chain, while the bottom image shows tip-affixed nanorods extending the size and changing the pores of a starburst dendrimer*

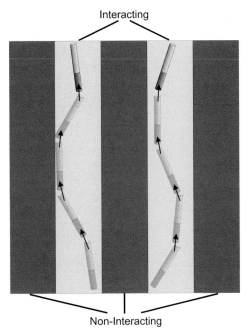

Figure G.3 *The movement of chemically powered nanorods on a substrate could be directed by surface interaction. By patterning lines with a surface functionality attractive to nanorods, they could move within these lines like cars on a road*

We could also think about using nanorods to unfold polymer strands, Figure G.2. Imagine two identical nanorods joined at their ends by an anchored polymer or biopolymer strand. Provide the nanorods with identical on-board chemical motors using one of the methods outlined above to cause the nanorods to move in opposite directions at one and the same speed. The adjoining polymer fiber will be brought under tension with equal forces acting at each end of the strand, creating a myriad of possibilities for studying polymer fiber mechanics and dynamics at the nanoscale. Prime examples include pulling apart a DNA double helix, changing the conformation of a protein, or forcing a cis-trans isomerization of a polyolefin or polyacetylene under tension. At a higher level of complexity imagine a collection of such nanorods each chemically anchored by one end to the terminal functional groups of a starburst dendrimer. It is conceivable that tugging on the periphery of the dendrimer will cause the internal pores to undergo changes in size and shape, which could conceivably release an imbibed guest. With this model it may even be possible to use strategically chemically tethered nanorod motors to unfold a globular protein.

Finally, chemically powered nanorods could be made to "slide" along micro-patterned "tracks", Figure G.3. Chemically driven nanorods could be made to slide along microchannels in PDMS, or on Si chips with appropriate surface functionalization of the rod and channel to keep them in the channels and to ensure they are mobile. This can be done by tuning the adhesion, friction, viscosity, etc...

Different catalytic segments, lengths and compositions will give these chemically driven nanorod sliders different channel speeds, which can be measured in a video optical microscope – the nanorod derby is about to begin!

These are only a few of the ideas which pop out when one starts to think about self-propelled, chemically powered nanoparts. It is likely that this area of research will yield incredibly diverse, creative discoveries, and perhaps even spawn a whole industry powered by tiny on-board chemical nanomotors!

References

1. W.F. Paxton, K.C. Kistler, C.C. Olmeda, A. Sen, S.K. St. Angelo, Y.Y. Cao, T.E. Mallouk, P.E. Lammert, V.H. Crespi, Catalytic nanomotors: Autonomous movement of striped nanorods, *J. Am. Chem. Soc.*, 2004, **126**, 13424.
2. S. Fournier-Bidoz, A.C. Arsenault, I. Manners, G.A. Ozin, Synthetic self-propelled nanorotors, *Chem. Commun.*, 2005, 441.
3. T.R. Kline, W.F. Paxton, T.E. Mallouk, A. Sen, Catalytic nanomotors: Remote-controlled autonomous movement of striped metallic nanorods, *Angew. Chem. Int. Ed.*, 2005, **44**, 744.
4. R.K. Soong, G.D. Bachand, H.P. Neves, A.G. Olkhovets, H.G. Craighead, C.D. Montemagno, Powering an inorganic nanodevice with a biomolecular motor, *Science*, 2000, **290**, 1555.

Subject Index